Geography of the Physical Environment

The *Geography of the Physical Environment* book series provides a platform for scientific contributions in the field of Physical Geography and its sub-disciplines. It publishes a broad portfolio of scientific books covering case studies, theoretical and applied approaches as well as novel developments and techniques in the field. The scope is not limited to a certain spatial scale and can cover local and regional to continental and global facets. Books with strong regional focus should be well illustrated including significant maps and meaningful figures to be potentially used as field guides and standard references for the respective area.

The series appeals to scientists and students in the field of geography as well as regional scientists, landscape planners, policy makers, and everyone interested in wide-ranging aspects of modern Physical Geography. Peer-reviewed research monographs, edited volumes, advance and under-graduate level textbooks, and conference proceedings covering the major topics in Physical Geography are included in the series. Submissions to the Book Series are also invited on the theme 'The Physical Geography of…', with a relevant subtitle of the author's/editor's choice.

More information about this series at http://www.springer.com/series/15117

Tobias Heckmann · David Morche
Editors

Geomorphology of Proglacial Systems

Landform and Sediment Dynamics in Recently Deglaciated Alpine Landscapes

Springer

Editors
Tobias Heckmann
Physical Geography
Catholic University of Eichstätt-Ingolstadt
Eichstätt, Germany

David Morche
Environmental Authority of Saalekreis District
Merseburg, Germany

ISSN 2366-8865 ISSN 2366-8873 (electronic)
Geography of the Physical Environment
ISBN 978-3-319-94182-0 ISBN 978-3-319-94184-4 (eBook)
https://doi.org/10.1007/978-3-319-94184-4

Library of Congress Control Number: 2018950204

Cover image by Sonja Weber, München

This Springer imprint is published by the registered company Springer Nature Switzerland AG
The registered company address is: Gewerbestrasse 11, 6330 Cham, Switzerland

Contents

Introduction

1

Tobias Heckmann, David Morche and Michael Becht

Abstract

Mountain regions are both sensitive to and disproportionally affected by recent climate change. Among the most important and most visible changes is glacier retreat. The latter entails the exposure of formerly glaciated terrain to subaerial conditions, with implications for hydrological, geomorphic and ecological processes. The geomorphic response to deglaciation has been conceptualised in paraglacial geomorphology, encompassing spatial and temporal changes in the activity of geomorphic processes, slope instability, and the build-up and depletion of sediment storage landforms. The transitional character of these adjustments to deglacial condition has been highlighted in recent research. In this chapter, we propose and discuss the definition of proglacial areas as the area that has been deglaciated since the glacial highstands at the end of the Little Ice Age. We then summarise the geomorphic response to deglaciation and recent geomorphological research in proglacial areas; based on this literature review, we identify avenues of future research. These include (i) investigations extending further into the past based on historical imagery; (ii) the assessment of the relative importance of glacial vs. non-glacial processes; (iii) the role of direct, local climate change impacts vs. the transient response to deglaciation; and (iv) the potential propagation of local geomorphic changes (with connectivity being an important system property moderating this propagation) with potential downstream effects on hydropower generation, freshwater ecosystems and natural hazards. Observing and understanding past- and present-day changes may provide templates for likely responses to future changes. The PROSA project conducted from 2012--2017 in the proglacial area of the Gepatsch glacier, Central Austrian Alps, forms the framework of several case studies presented in the present volume; therefore, we briefly outline the joint project, its study area, research problems and methods.

Keywords

Climate change · Glacier retreat · Little ice age · Paraglacial · PROSA project

T. Heckmann (✉) · M. Becht
Physical Geography, Catholic University of Eichstätt-Ingolstadt, Eichstätt, Germany
e-mail: tobias.heckmann@ku.de

M. Becht
e-mail: michael.becht@ku.de

D. Morche
University of Halle-Wittenberg, Halle, Germany

D. Morche
Environmental Authority of Saalekreis District, Merseburg, Germany

T. Heckmann and D. Morche (eds.), *Geomorphology of Proglacial Systems*,
Geography of the Physical Environment, https://doi.org/10.1007/978-3-319-94184-4_1

1.1 Motivation

Mountain regions worldwide host complex and fragile ecosystems, provide water and other resources and have always been important for mankind in terms of settlement, traffic and trade, culture and religion (Körner and Ohsawa 2005; FAO 2011). They have been shown to be disproportionally affected by recent climate change (Rangwala and Miller 2012; Beniston 2005). Together with the sensitivity of mountain ecosystems, their topographic and climatic conditions entailing intensive morphodynamics make mountains "sentinels of change" that allow us to observe the consequences of climatic and environmental changes more readily than elsewhere (Beniston and Stoffel 2014). Shrinking glaciers are one of the most emblematic manifestations of climate change today (Orlove et al. 2008; Smith and Joffe 2009), raising public attention for scientific work on glacier fluctuations. The latter have been observed on a global scale; the corresponding records of glacier length and mass balances are interpreted as consequences of measured climate change (Barry 2006; Zemp et al. 2008, 2009, 2015; Leclercq et al. 2014; Chap. 2). Dated evidence of past glacier extents and dynamics (trimlines, moraine landforms and sediments, varved lake sediments, soils and vegetation remains, etc.) has been used to infer climate changes of the past (Klok and Oerlemans 2004; Oerlemans 2005; Joerin et al. 2006). Hewitt (2002) emphasises the significance of mountain glaciers for water and sediment fluxes that is "out of all proportion to their share of the global ice cover. That applies to both landscape development and human affairs" (see also Milner et al. 2017 on the effects of glacier shrinkage on ecosystem services).

Many mountain ranges were shaped by Pleistocene glaciations. Their "glacial legacy" includes a characteristic topography (Sternai et al. 2011), including glacial cirques, U-shaped valleys with steep sidewalls and a wide valley bottom and large sediment stores (e.g. moraines, valley infill); it is, however, not limited to large-scale and long-term effects of multiple "Ice Ages" on the appearance of the present-day landscape. Deglaciation, both on the long and short temporal scale, exposes formerly glaciated terrain to subaerial conditions. In terms of landform and material properties, deglaciated terrain is not at equilibrium with non-glacial conditions and therefore prone to changes. Glacier retreat that started in the second half of the nineteenth century, after the end of a cold phase called the "Little Ice Age" (LIA; Mann 2002; Grove 2004; Matthews and Briffa 2005), has been accelerating (e.g. Zemp et al. 2008, 2009, 2015). Since only few decades ago, scientists have become increasingly aware of the consequences of global warming and deglaciation for "cold regions" at high latitudes and high elevation. These include changes of permafrost properties and distribution (e.g. Kneisel and Kääb 2007), river runoff (e.g. Moore et al. 2009), development of soil (e.g. Egli et al. 2006a, b) and vegetation (e.g. Moreau et al. 2008; Klaar et al. 2015) and the activity of a wide range of geomorphic processes (e.g. Ballantyne 2002b; Laute and Beylich 2014a; Beylich et al. 2016; Carrivick and Heckmann 2017).

Knight and Harrison (2014a) argue that observing and understanding past and present-day changes may provide templates for likely responses to future changes; although there are also differences between recent changes and those that occurred during the Holocene, they highlight the importance of the post-LIA transition for geomorphological research. As we outline in Sect. 1.2, the forefields of receding glaciers are hotspots with respect to the consequences of climate change since the end of the LIA. If these changes were to effectively propagate in the downstream direction, they would not remain restricted to the comparatively small proglacial areas but add to the geoecological, hydrological and geomorphic consequences of climate change on a more regional scale, affecting densely populated mountain ranges and entailing challenges for risk management in these regions (Keiler et al. 2010; Milner et al. 2017). Knight and Harrison (2014a) state that the geomorphic consequences of deglaciation "will become the most significant process controlling sediment supply and landscape change in the mid- to high latitudes over the next few hundred years".

1.2 Proglacial Areas and Paraglacial Dynamics

In alpine terrain, most glaciers are either cirque or valley glaciers, some of them emanating from larger plateaus. For a typical valley glacial landsystem, Benn et al. (2003) identify as most important controls (i) topography, (ii) debris supply to glacier surfaces and (iii) the efficiency of sediment transport from the glacier to the proglacial environment by the glacifluvial system. The elements of a (pro-)glacial landsystem can be categorised according to their geomorphic function as sediment sources, stores and sinks (c.f. Ballantyne 2002a). These functions are associated with geomorphic processes that transfer glacigenic (and other) sediments towards the outlet of the channel network.

Rockwalls, and glacigenic deposits on hillslopes and the valley floor, represent sediment sources. Deglaciated rockwalls are subject to different types of gravitational mass movements, ranging from small-scale rockfall to deep-seated gravitational deformations and rock avalanches (e.g. Cossart et al. 2008; Kellerer-Pirklbauer et al. 2010; McColl 2012; Chap. 8); evidence for destabilisation after deglaciation has been collected by studies that link measured rates of rockfall, rock mass strength and time since deglaciation (Vehling et al. 2016; Chap. 9). Mass movements also occur in surficial materials, both on massive moraine landforms (e.g. Hugenholtz et al. 2008; Klimeš et al. 2016) and comparatively shallow drift-mantled hillslopes (e.g. Holm et al. 2004). Especially steep lateral moraines are reworked by slope wash, fluvial erosion and debris flows (Palacios et al. 1999; Curry et al. 2006; Hürlimann et al. 2012; Haas et al. 2012; Chaps. 10 and 11); debris flows also initiate in steep proglacial gullies (Legg et al. 2014). "Dirty" snow avalanches transferring sediment to and within the proglacial area are another relevant process (Cossart 2008; Laute and Beylich 2014b). These processes supply sediment to storage landforms such as talus cones, debris flow or alluvial fans, and valley fluvial deposits. Sediment storage can be intermediate/short term, and the corresponding landform can function as sediment source for other processes—we may refer to these as sediment stores, while long-term storage landforms (e.g. a lake basin) are termed sinks. Ballantyne (2002a) further distinguishes primary and secondary sediment stores, with primary stores being derived directly from sediment sources and secondary stores being produced by the reworking of primary sediment stores.

The processes listed in the previous paragraph are non-glacial; i.e., they are gravitational, slope-aquatic, fluvial, periglacial, etc. In deglaciated terrain, however, the spatial domains and dynamics of many of them are directly conditioned by (de-)glaciation (Church and Ryder 1972; Ballantyne 2002a), that is, they "would operate at different rates (or not at all) had glaciation not occurred" (Ballantyne 2002a). Several authors have explained the response of geomorphic processes and sediment fluxes to deglaciation in a similar fashion as the reaction to a disturbance such as wildfires, developing a family of conceptual models of "paraglacial geomorphology" (Church and Ryder 1972; Church and Slaymaker 1989; Ballantyne 2002a, b; see also Church 2002). These models describe how morphodynamics and sediment transfer change over time: They highlight the transitional character of the response to deglaciation as topography adjusts, sediment stores are depleted, and/or sediment waves propagate from deglaciated areas through the downstream catchment. The response to deglaciation takes place on different temporal and spatial scales, reaching from decades (at the hillslope scale; Curry et al. 2006; Delaney et al. 2018) to millennia (at the catchment scale; Church and Slaymaker 1989; Buechi et al. 2014), and differs between deglaciated environments (Ballantyne 2002a, b).

Here, we adopt the clarifications made by Slaymaker (2009, 2011) in reaction to ongoing confusion in terminology (proglacial vs. paraglacial vs. periglacial): "Proglacial" refers to an area, "periglacial" is a processual term like fluvial or gravitational, defining geomorphic processes driven by frost. Finally, "paraglacial" addresses the specific morphodynamics (including their development over time) within a deglaciated landscape that were briefly summarised in the previous paragraphs. However, the area “close to the ice front” of a glacier, i.e. the proglacial area as defined by Penck and Brueckner (1909 *fide* Slaymaker 2011), is difficult to delineate in the field today. At present-day rates of glacier melting, the glacier snout in one year may be located tens of metres away from its position in the previous year (see Chap. 2). Moreover, under these conditions, the proglacial area is no longer a system at equilibrium between sediment delivery from the glacier and fluvial reworking (Slaymaker 2011). Where it does exist, this equilibrium is highly dynamic and prone to fundamental changes in system configuration, for example in the glacifluvial domain (c.f. Marren and Toomath 2014; Baewert and Morche 2014; Curran et al. 2017; Shugar et al. 2017). Proglacial systems have therefore been conceptualised as systems in transition from glacial to non-glacial conditions (Johnson 2002), or from an un- or metastable state (a state reached with deglaciation) towards a new equilibrium (periglacial or non-glacial conditions; Slaymaker 2011).

Regarding the delimitation of a present-day proglacial landsystem, we argue that the conspicuous termino-lateral moraines formed by many glaciers during the LIA serve as a reasonable distal boundary. Curry et al. (2006) observed that gullies on lateral moraines of a Swiss valley glacier reached their maximum level of incision approximately 50 years after deglaciation, stabilising within the following 30–90 years. Delaney et al. (2018) report a stabilisation of the Griesgletscher forefield (Switzerland) that has been ice-free since 1986 towards the end of their study period (2014), thus supporting a faster transition; however, the distal boundary of their study area is a reservoir lake, not the LIA maximum extent, which makes a direct comparison difficult. Using the results of Curry et al. (2006), the total of 80–140 years required for this transition is roughly consistent with the period after the end of the LIA ($\sim$1850; although other dates have been suggested for the LIA termination, see Matthews and Briffa 2005). Hence, today, the area that has become ice-free since the end of the LIA is characterised by paraglacial dynamics, the most distal parts approaching stability (Carrivick and Heckmann 2017). Under this assumption, the corresponding termino-lateral moraines may be taken as the boundary of a proglacial system, as implemented by Schiefer and Gilbert (2007), Heckmann et al. (2012a, b), Geilhausen et al. (2013), and Bosson et al. (2015). Zasadni (2007) cite Kinzl (1932) and Holzhauser (1982) to justify this definition of proglacial areas (“glacier forefields”) by stating that they are “clearly different from the surrounding area in respect to geomorphic setting, pedological characteristics, floristic succession and the degree of weathering”. In the definition adopted in this book, the ice-marginal criterion still applies, but the transient character of paraglacial adjustment in a deglaciating environment is given more importance (c.f. the “paraglacial landscape”; Slaymaker 2011). It has to be noted though that paraglacial dynamics are not strictly limited to the proglacial area; Church and Ryder (1972) state that paraglacial dynamics occur “around and within the margins of a former glacier”.

The use of LIA glacier extent as a frequently conspicuous boundary of proglacial landsystems facilitates their delineation (Fig. 1.1), quantitative analysis and comparison (Chap. 3). Glacier inventories containing the LIA extents of glaciers have been compiled on the basis of geomorphological evidence, for example in several parts of the Alps (Austria: Fischer et al. 2015; Groß and Patzelt 2015; Piedmont: Lucchesi et al. 2014; South Tyrol: Knoll et al. 2009; Trentino: Zanoner et al. 2017) and the Pyrenees (reviewed by Oliva et al. 2018).

Comparisons with present-day glacier extents show that the distance by which many glaciers have retreated since the end of the LIA is in the order of 10^2 to 10^3 m, and the area deglaciated is in the order of 10^3 to 10^6 m^2 (Chaps. 2 and 3). It should be noted, however, that moraine ridges may be ambiguous or obliterated, making their relation to former glacier dimensions difficult (see Table 1 of Kirkbride and Winkler 2012; Barr and Lovell 2014); moreover, the maximum extent may be of different age in different regions

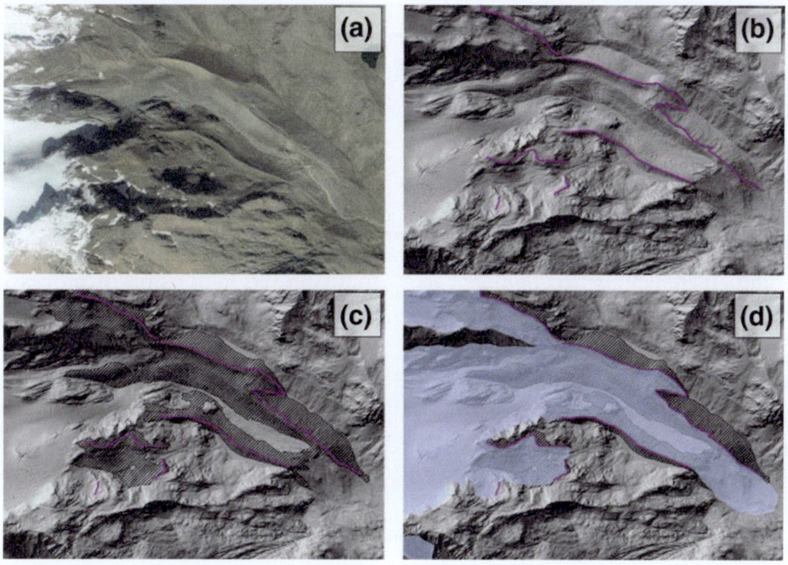

Fig. 1.1 Reconstruction of the LIA maximum glacier extent and LIA and post-LIA glacial deposits (La Mare glacier, Ortles-Cevedale group). **a** Analysis of archival data and visual interpretation of the 2006 orthophoto; **b** identification of the LIA moraine ridges (purple lines) and trimlines on the hillshade relief map derived from a LiDAR digital elevation model (DEM); **c** drawing of the complete area occupied by LIA and post-LIA glacial deposits (polygon areas with black fill symbol); and **d** final reconstruction of the LIA maximum extent of La Mare glacier (light-blue polygon area). *Source* Zanoner et al. (2017), with permission from Elsevier

of the world (Luckman 2000; Mann 2002; Matthews and Briffa 2005). Another problem of our delineation that is focused on post-LIA deglaciation and recent paraglacial dynamics is the neglect of the potentially very large area between the limits of last glacial maximum (LGM) glaciation and the LIA maxima that are characterised by multiple Holocene glacial advances and recessions (e.g. Zumbühl 1980; Nicolussi and Patzelt 2001). This area used to be close to the ice margin as well, experienced associated paraglacial dynamics during and after deglaciation, and may contain sediment sources that affect the present-day proglacial area. Glacier variations before and within the LIA (e.g. Le Roy et al. 2015, 2017) also may have induced multiple phases of paraglacial activity. However, we expect hillslopes adjacent to the LIA glacier forefields to be comparatively stable, given the long period of adjustment since the previous phase of deglaciation; in fact, historical paintings and photographs of nineteenth-century glaciers depict vegetated, even forested hillslopes in their immediate vicinity (e.g. pictures collected by Zumbühl 1980).

1.3 Recent Geomorphological Research in Proglacial Areas

Research in proglacial areas has long been focused on ecology (Matthews 1992). More recently, there has been an increase in geomorphological interest, reflected for example in a series of sessions at the EGU general assembly (2013–2016), a special issue of Earth Surface Processes and Landforms (Heckmann et al. 2016b), and last but not least in the publication of the present volume. In this section, we provide an overview of recent research in proglacial areas and outline avenues of future research.

Proglacial areas are natural laboratories where the initial development of soils and vegetation can be observed (Matthews 1992). This has mostly been based on the investigation of chronosequences (e.g. Egli et al. 2006a for soil;

Moreau et al. 2008 for vegetation). In this "ergodic reasoning" approach, space (=distance from present-day ice front) is substituting time since deglaciation, and sequences of soil and vegetation types and properties with increasing distance from the ice front are interpreted as representing temporal developments. More recent research has identified factors other than "just" time, and increasingly recognises the complex interplay of morphodynamics and soil formation (Temme and Lange 2014; Temme et al. 2015; Chap. 18), and the mutual interactions of vegetation development and geomorphic processes (Eichel et al. 2013, 2015; Chap. 17) that complicate chronosequence studies, particularly on smaller spatial scales (Burga et al. 2010). These disturbances to soil and vegetation development can be caused by glacial and glacifluvial dynamics, and non-glacial processes initiating both within and outside of the proglacial area. Where disturbances are not linked to glacial activity or deglaciation, they are consequently not controlled by terrain age (Matthews 1999 *fide* Eichel, Chap. 19). In geomorphology, chronosequences of landform morphological properties have been investigated to infer the development of morphodynamics (e.g. Curry et al. 2006). However, the interaction of different geomorphic processes, human impact and (dis-) connectivity of sediment transfer lead to site-specific and path-dependent behaviour that may strongly affect the applicability of the exhaustion model at the scale of specific locations or (sub-)catchments (Knight and Harrison 2014a). For example, repeated undercutting and sediment removal from secondary paraglacial storage landforms by the proglacial channel network may lead to sustained incision of lateral moraines because the accumulation of storage landforms is impeded and slope gradients are being kept high (c.f. Chap. 11). Cossart and Fort (2008, Fig. 1.2) use chronosequences and historical aerial photography to explain the geomorphic evolution of an alpine proglacial area; they illustrate how moraines that acted as a barrier to sediment flux at an earlier stage are breached, overspilled, or even act as sediment sources at a later stage. Thus, landforms can both store and release sediment (at different spatial and temporal scales), which adds complexity to the deglaciation response (Chap. 10). The influence of topographic factors on channel patterns in proglacial rivers through the modification of discharge and sediment supply is highlighted by Marren and Toomath (2014). Research projects should aim at monitoring these systems with a focus on the interaction of processes.

Changes in sediment fluxes occur on multiple temporal scales, from sub-daily to decadal (Hetherington et al. 2005; Milan et al. 2007; Mao et al. 2014; Bechet et al. 2016; Guillon et al. 2018; Geilhausen et al. 2012; Micheletti et al. 2015; Chap. 11). This has implications for the temporal scale, that is resolution and extent, of measurements. Investigations in highly dynamic proglacial areas have been facilitated by the advent and increased availability of high-resolution and high-accuracy surveying data. Terrestrial and airborne LiDAR and structure-from-motion (Westoby et al. 2012) enable the detection and quantification of surface changes and the computation of morphological sediment budgets (see recent review by Carrivick and Heckmann 2017; Chaps. 11 and 17). Terrestrial radar interferometry has proven capable of detecting subtle changes of surface elevation (Caduff et al. 2015 and references therein, Rouyet et al. 2016). With expected high rates of surface changes in proglacial areas, these measurement techniques allow for the detection of changes within comparatively short periods of time. A high temporal resolution supports the distinction of surface changes caused by different geomorphic processes, especially where they occur in response to events, or due to specific hydrometeorological conditions. Longer inter-survey periods are more likely to reflect the combined activity of multiple processes. The temporal extent of investigations has been expanded considerably by using historical imagery as a basis for multitemporal mapping (e.g. Ravanel and Deline 2011). Photogrammetric methods such as structure-from-motion are increasingly used to construct digital elevation models from these sources, providing new opportunities for quantitative appraisal of morphodynamics on decadal timescales

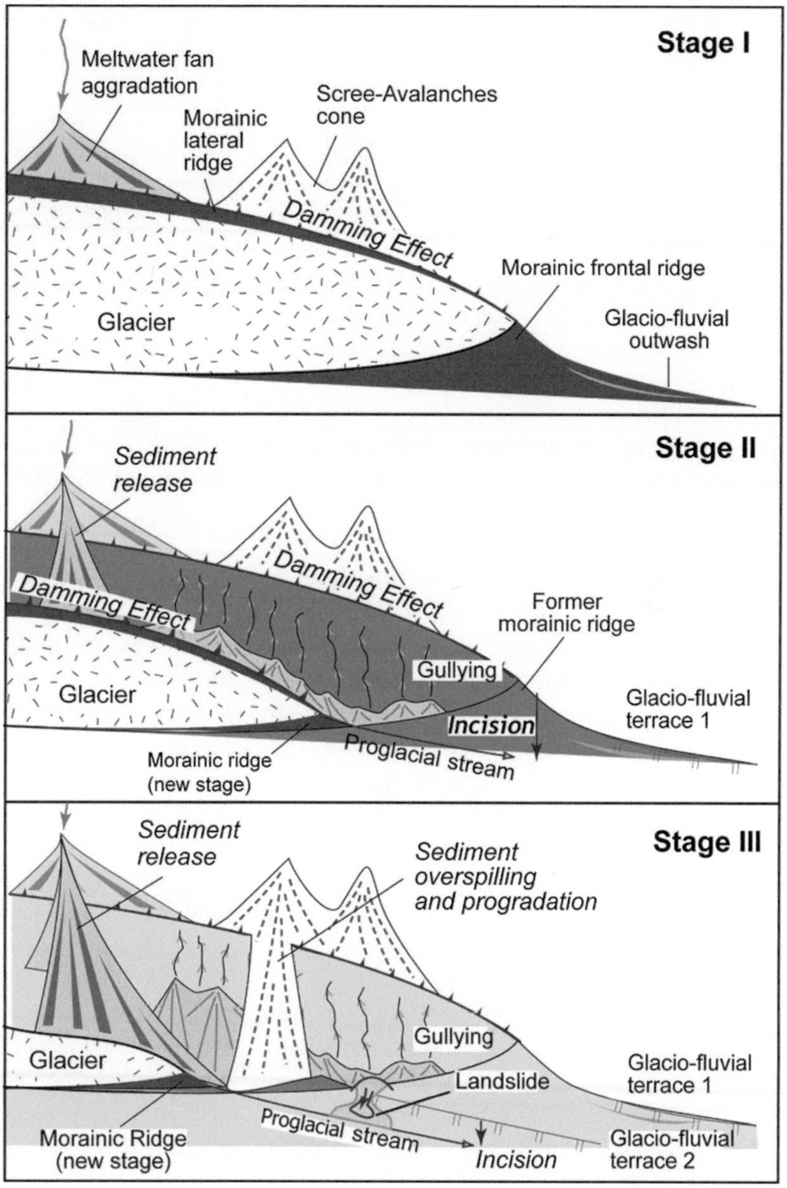

Fig. 1.2 Conceptual model of the geomorphic evolution of a recently deglaciated proglacial landsystem. The presence of moraines creates a damming effect, hence a fragmentation of the cascade sedimentary system, as shown by local aggradation of sediments usptream of both lateral and frontal moraines. The duration of such damming effects depends on (i) the number, volume and cohesion of moraines and (ii) the erosion processes at work on the moraine (note the difference between dammed meltwater cones and undammed scree cones). *Source* Cossart and Fort (2008: 128), with permission from Taylor and Francis

(Schiefer and Gilbert 2007; Micheletti et al. 2015) and model testing (Staines and Carrivick 2015). Future studies should use historical aerial photos and DEMs derived from them to provide present-day measurements with a context of longer-term evolution, e.g. in order to explore path-dependence. Present-day and historical observations and measurements of geomorphic processes and landforms can be complemented by absolute and relative dating of deglaciated surfaces or sediments. However, the short timescale and comparatively high temporal resolution needed to date post-LIA landforms and deposits determine the range of suitable dating techniques. Schimmelpfennig et al. (2014), for example, use cosmogenic nuclides exposure dating (see also Balco 2011) to distinguish pre-, within- and post-LIA surfaces. Sediments in proglacial lakes have been dated using annual varve counts [varves additionally yield sediment accumulation rates and palaeoenvironmental and palaeoclimatological proxies; e.g. Guyard et al. (2007), see also review by Zolitschka et al. (2015)] and radionuclides such as ^{137}Cs. Lichenometry has been frequently used to date moraines or the deposits of mass movements within proglacial areas (e.g. Thompson and Jones 1986; Winkler 2004; Loso et al. 2016). Dendrogeomorphological studies have been conducted to reconstruct glacier advances (by establishing the time when a tree was killed by the advancing glacier), but also, more indirectly, to date paraglacial responses further downstream (e.g. Hart et al. 2010).

Geomorphic change in proglacial areas is also related to the subsurface, specifically the changing distribution of ground ice (e.g. Bosson et al. 2015; Ribolini et al. 2010; Chaps. 6 and 7). This is typically investigated using geophysical methods such as electrical resistivity tomography (e.g. Kneisel and Kääb 2007) or ground-penetrating radar (e.g. Schwamborn et al. 2008). Ground ice dynamics have implications on measurements of changes in surface elevation as they cause changes that may compensate or add to changes that are due to erosion and deposition, providing big challenges for the interpretation of DEMs of difference and the quantification of erosion or deposition (e.g. Sailer et al. 2012; Micheletti and Lane 2016; Avian et al. 2018). The presence of snowcover in DEMs is another reason for the need to conduct such analyses with care (e.g. Carrivick et al. 2013).

The relative importance of different non-glacial processes occurring within proglacial areas needs further investigation, especially in order to compare them to glacial erosion (Harbor and Warburton 1993; O'Farrell et al. (2009) and references therein, Laute and Beylich 2014a, Chaps. 15 and 17). Delaney et al. (2018) conclude that denudation rates in the proglacial area of Griesgletscher, Switzerland, are almost 40 times the rates in the remaining catchment; they also highlight the importance of subglacial erosion for sediment export from the proglacial area. Using a cosmogenic nuclides and geomorphological mapping approach, Delunel et al. (2014) found that more than 75% of fluvial sediments in the glaciated Etages catchment, French Alps, were derived from glaciers (see also Guillon et al. (2015) with a 50–90% proportion derived from subglacial erosion). Other studies found a dominance of non-glacial processes (e.g. O'Farrell et al. 2009), highlighting persistent research needs regarding factors that control the relative importance (e.g. lithology, glacier dynamics, catchment area, glaciated area, relief, connectivity; see Guillon et al. 2018). Proglacial areas with glaciers remaining are judiciously the main focus of present-day research, because the glacial component of the runoff regime maintains high transport capacities and thus enhances the (potential) transfer of sediment beyond the proglacial area. In contrast, former proglacial areas whose glaciers have already disappeared (Chaps. 2 and 3), allow to study the assumed stabilisation of surface sediments and its implications for downstream fluvial systems where this component has already ceased to exist, which is arguably the fate of most catchments in the Alps that are still glaciated today (Haeberli et al. 2013). As the influence of glacial meltwater (see Chap. 3 for an assessment of spatial patterns) declines, the relative influence of snow melt and precipitation is bound to increase. Direct, local impacts of climatic change (e.g. changing intensity and/or frequency of forcing events such as heavy rain) may enhance or attenuate the effects of transient changes such as paraglacial dynamics—the onset and evolution of these does not require climate change except as the reason for deglaciation. Knight and Harrison (2014a) point out that, in case of deglaciation after the LIA, climate change and consequences of deglaciation are coeval rather than subsequent.

Proglacial areas are part of larger systems that have been described as cascading systems (Burt and Allison 2010; Chap. 15). Consequently, enhanced morphodynamics in proglacial areas may lead to increased sediment loads downstream (Knight and Harrison 2014a), representing a sedimentary disturbance that is being propagated through the system (Church 2002; Milner et al. 2017). These "off-site" effects include issues associated with reservoir sedimentation (Einsele and Hinderer 1997; Geilhausen et al. 2012) and natural hazards (O'Connor and Costa 1993; Richardson and Reynolds 2000; Moore et al. 2009). The propagation of geomorphic changes through mountain catchments, however, may be buffered by poor connectivity (Beylich et al. 2009; Lane et al. 2017; Chap. 16): Specific landforms and their topographic and material properties impede sediment transfer and consequently moderate the (transmission) sensitivity of catchments (Fryirs 2017) to climate and geomorphic change. Cossart et al. (2013) review the effects of landslides on sediment fluxes in deglaciated mountain slopes, stating that their importance as sediment source may be counteracted by disconnectivity: Landslides may not be coupled to other processes, sediments being stored in storage landforms; moreover, landslide deposits frequently construct barriers to sediment flux. At a large spatial and temporal scale, Bernhardt et al. (2017) investigated the propagation of sedimentary signals from the (deglaciating) Andes to marine sediment sinks on the Southern American continental shelf and in the deep ocean. While they inferred a maintained high connectivity in the north-central part of Chile that facilitated the propagation of decreasing sediment flux, the formation of piedmont lakes following deglaciation abruptly led to a decrease of sedimentation in the ocean. Connectivity, however, is not a time-invariant system property. Structural changes such as the re-direction of drainage, lake formation and drainage (Geilhausen et al. 2013; Bogen et al. 2015; Chap. 14), de- or recoupling of system components with the catchment outlet, and those affecting the interaction of processes are found more frequently within a limited area than in other geomorphic systems (c.f. Fig. 1.2). Such system-internal changes are likely to affect sediment yield as well, independent of, or in addition to, climate change. Shugar et al. (2017) describe how the retreat of Kaskawulsh glacier (Alaska) triggered a complete reorganization of the drainage network, which affects, among others, river discharge and sediment transport, water level and sediment influx of a lake. Lane et al. (2017) highlight that connectivity is not only influenced by topography and topographic change but also by the evolution of grain sizes and their implications for sediment transfer (see also Cossart 2008). They also recommend the investigation of the historical evolution of connectivity using historical data (e.g. DEMs reconstructed from historical aerial photos).

To conclude, more data from multiple proglacial areas across different mountain ranges and alpine regions are needed to explore the influence of local/regional (climate, lithology, etc.) versus contingent factors on the evolution of proglacial areas and off-site effects of proglacial morphodynamics. Datasets may include "documentary, geologic, sedimentary, radiocarbon and cosmogenic dating, dendrochronometric, instrumental climate and ecological data types" (Knight and Harrison 2014a: 1). Standardised sampling schemes and techniques should warrant comparability (Milner et al. 2017; see e.g. Beylich et al. 2007). Based on improving understanding of proglacial dynamics, knowledge gained from present-day and historical studies should be combined with the results of regional climate models in order to assess future trajectories of sediment fluxes (Micheletti and Lane 2016). In some study areas, long-standing records of glaciological, hydrological and geomorphological data exist that need to be organised and published (see Strasser et al. 2018 for an excellent example) in order to be leveraged for such endeavours.

1.4 The PROSA Project

The PROSA project (high-resolution measurements of morphodynamics in rapidly changing *PRO*glacial *S*ystems of the *A*lps) was designed to employ state-of-the-art techniques to quantify and analyse morphodynamics and sediment budgets in proglacial areas. It was run during the years 2012–2017 by five working groups at the universities of Eichstätt-Ingolstadt, Erlangen, Halle (Germany) and Vienna (Austria) and jointly funded by the German Research Foundation (DFG) and the Austrian Science Fund (FWF) (Fig. 1.3).

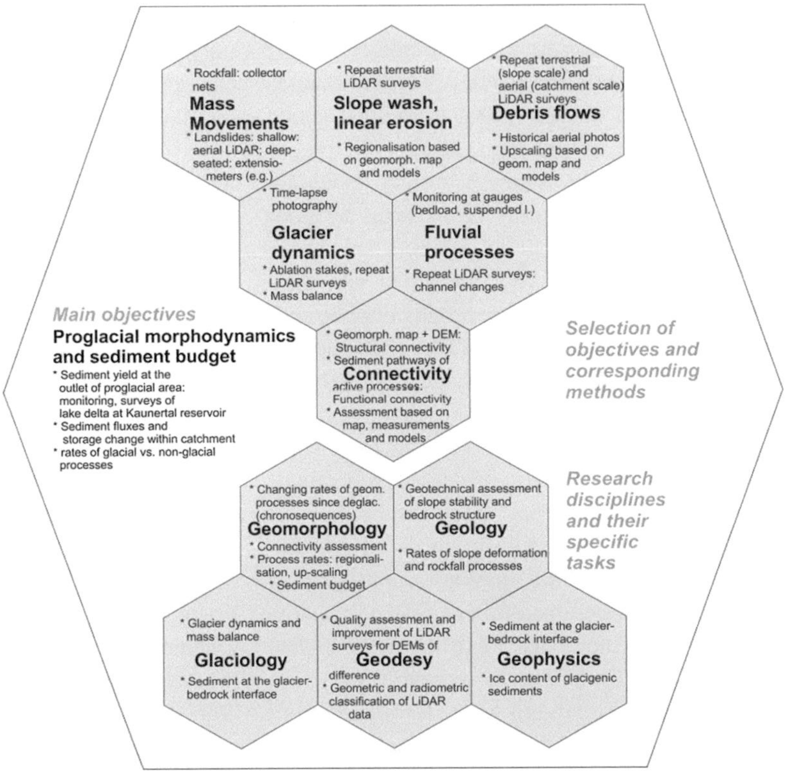

Fig. 1.3 Structure of the PROSA project. The main objectives are addressed from the perspectives of (i) partial objectives referring to processes in proglacial systems (top) and (ii) research disciplines taking part in the joint project, with their specific tasks (bottom)

The overarching research objective of the PROSA project was to set up a holistic sediment budget of the proglacial area of an Alpine glacier (see Heckmann et al. (2012a, b) for a project description). Sediment budgets are "quantitative statement[s] of the rates of production, transport, and discharge of detritus" (Dietrich et al. 1982: 6), and establishing a budget requires "(1) recognition and quantification of transport processes, (2) recognition and quantification of storage elements and (3) identification of linkages among transport processes and storage elements" (ibidem). In rough high-mountain terrain, the measurement of sediment fluxes is still a challenge and requests coordinated interdisciplinary research efforts. Hence, the PROSA team members worked in five research groups from different disciplines, including geomorphology, glaciology, geophysics, applied geology and geodesy. Although proglacial landsystems are regarded as highly dynamic, the short time scale of a publicly funded research project, and the low rate of surface change induced by the majority of geomorphic processes on a monthly scale require the reliable detection of subtle surface changes. The PROSA project aimed to take advantage of the advent of high-resolution, highly accurate and precise surveying techniques that have made such measurements possible since c. 15 years ago. The contribution of geodesy was to harness these techniques for geomorphological, geological and glaciological investigations, including accurate registration of both area-wide airborne and local terrestrial surveys and rigorous error assessment to distinguish significant changes from noise.

The upper Kaunertal (Kauner valley), defined as the catchment area of the river Fagge where it enters the Kaunertal reservoir (c. 62 km^2; Fig. 1.4), was selected as study area. The Kaunertal is a tributary valley of the river Inn, located in the Ötztal Alps, Tyrol, Austria, just north of the Italian border. It has experienced extensive glacier recession and features large, morphologically different proglacial areas (Fig. 1.5). Ample information is available

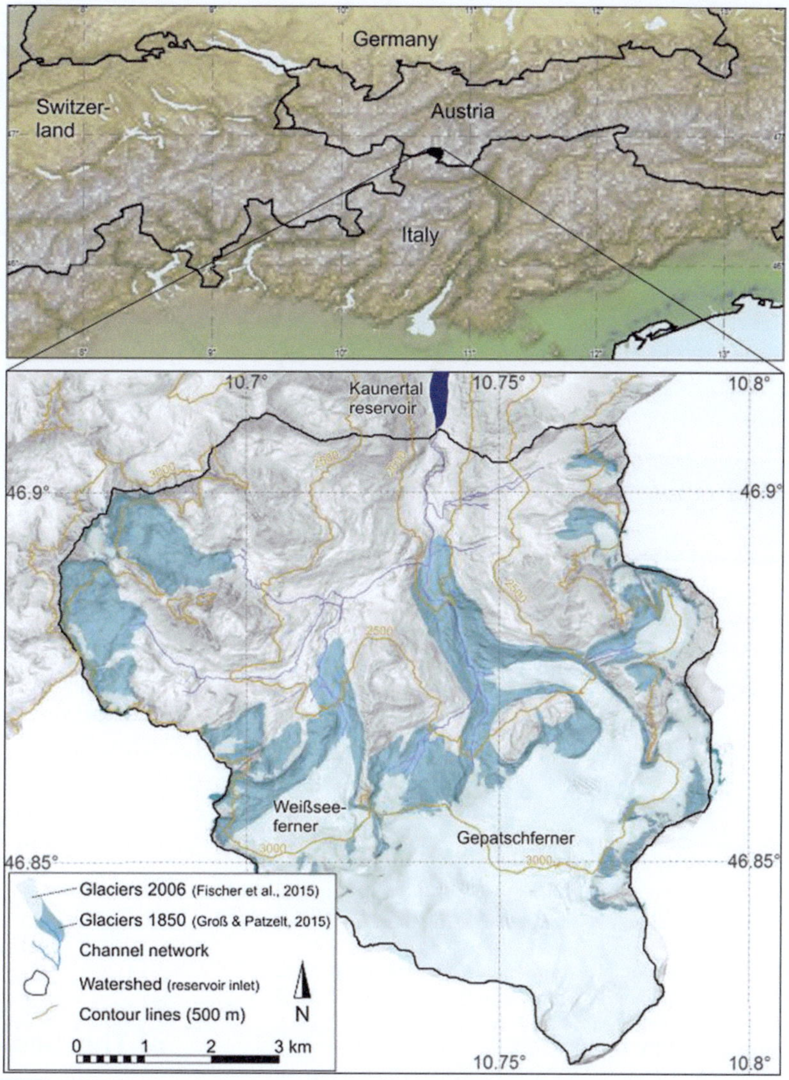

Fig. 1.4 Map of the study area showing the glacier extent at the end of the LIA (1850; Groß and Patzelt 2015) and 2006 (Fischer et al. 2015)

regarding historical glacier extent (e.g. Hartl 2010, and the Austrian Glacier Inventories, e.g. Groß and Patzelt 2015; Fischer et al. 2015). In the upper Kaunertal, glaciers covered 20.55 km^2 in 2006; the proglacial area, i.e. the area that has become ice-free after the end of the Little Ice Age, amounts to 13.3 km^2. The proportion of glaciated area decreased from c. 55% in 1850 to 34% in 2006. The tongue of the Gepatschferner retreated by c. 2.2 km, with only two short periods of stability or re-advance in the 1920s and 1980s (Nicolussi and Patzelt 2001). The large LIA lateral moraines deposited by the Gepatschferner reach over 200 m of height above the valley floor; they are characterised by intense geomorphodynamics that can be quantified using repeat surveys. The climate of Kaunertal is characteristic of the dry Central Alpine region; mean annual precipitation is between 920 and 1100 mm (at Weißsee and Gepatschalm gauges, respectively, data courtesy of TIWAG, Innsbruck), with a pronounced summer maximum and winter minimum. The mean annual temperature is between 2.8 °C (Gepatschalm, 1941 m elevation) and −0.9 °C (Weißsee, 2516 m).

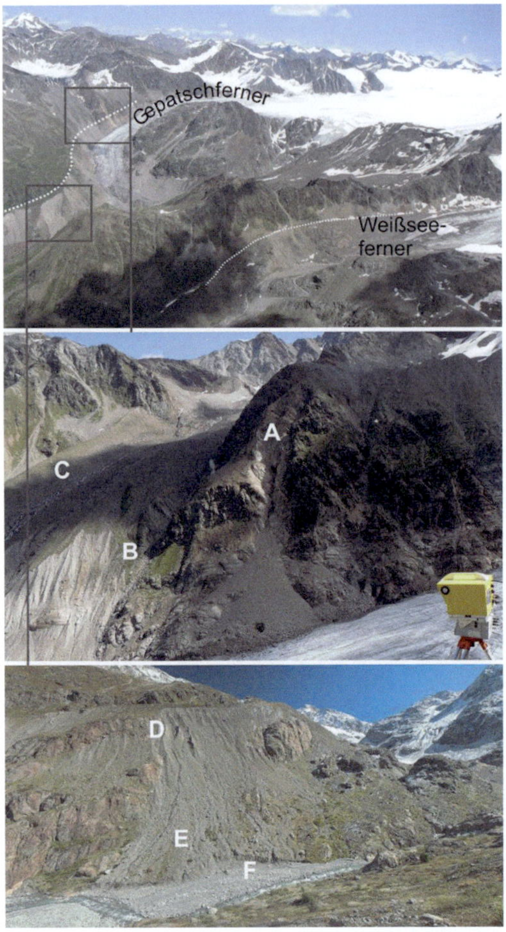

Fig. 1.5 Photographs from the Kaunertal study area. Top: Aerial photograph taken during the airborne LiDAR acquisition campaign on 18 July 2012. The dotted white lines show LIA lateral moraines of Gepatschferner and Weißseeferner glaciers. Contrary to the Gepatschferner, the proglacial area of Weißseeferner shows strong human impact (buildings, roads and construction works associated with ski runs). Photograph: TU Vienna. Middle: A terrestrial laserscanner surveys the unstable rock slope of "Schwarze Wand" on 12 August 2012 including the fresh scar (**a**) and deposits of a major rockfall (see Chap. 9). **b** and **c** denote LIA lateral moraines of Gepatschferner and a tributary glacier, respectively. Photograph: Morche. Bottom: **c** 200 m high section of the LIA right-hand lateral moraine of Gepatschferner with bedrock outcrops. The very steep slope is heavily dissected (**d**) by fluvial erosion and debris flows, the eroded sediment being partially deposited on alluvial/debris cones (**e**). The latter are being undercut by the proglacial river Fagge (**f**). Data from this section are also published in Chap. 10, Haas et al. (2012), Heckmann and Vericat (2018). Photo: Dusik

In addition to the case studies collated in this book, PROSA members have published papers in following fields of research:

- Quantification of rates of geomorphic processes, for example debris flows (Haas et al. 2012), rockfall (Vehling et al. 2016; Heckmann et al. 2016a; Vehling et al. 2017) and fluvial sediment dynamics (Morche et al. 2012; Baewert and Morche 2014; Morche et al. 2015).
- Rapid disintegration of glaciers (Stocker-Waldhuber et al. 2017).
- Genesis and dynamics of rock glaciers and their interaction with glaciers (Dusik et al. 2015).
- Radiometric calibration (Briese et al. 2012) and strip adjustment (Glira et al. 2015) of airborne LiDAR data. The modules "ICP" and "StripAdjust" of the airborne laserscanning processing software OPALS (http://geo.tuwien.ac.at/opals) were developed within the PROSA project.
- Use of DEMs of difference from airborne LiDAR to investigate snow cover on glaciers (Helfricht et al. 2014) and sediment delivery on hillslopes (Heckmann and Vericat 2018).
- Soil development and its interplay with geomorphodynamics (Temme et al. 2016).

1.5 Scope and Structure of This Book

This book focuses on the forefields of alpine glaciers as showcases of both climate-driven glacier retreat and the subsequent reaction of affected geomorphic and geoecological systems, at the small spatial and short temporal scale. The collection contains papers that deal with different aspects of geomorphic change and dynamics in high-mountain proglacial areas, specifically targeting areas that have become ice-free since the end of the LIA. It is intended as an "update" and complement of existing literature on proglacial areas (and paraglacial dynamics), namely Matthews (1992), Hewitt et al. (2002), Ballantyne

(2002b), Knight and Harrison (2009, 2014b). Furthermore, it contributes to the science of glaciers and environmental change (Knight 2006; Martini et al. 2011).

The book is structured by five parts, namely

1. Glaciers and ground ice in the proglacial zone (Chaps. 2–7).
2. Hillslope processes (Chaps. 8–11).
3. Proglacial rivers and lakes (Chaps. 12–14).
4. Proglacial sediment cascades and budgets (Chaps. 15–17), and
5. The role of time and morphodynamics in soil and vegetation development (Chaps. 18 and 19).

The parts and chapters are ordered along virtual sediment cascades (or more generally, cascades of processes), starting from glaciers and adjacent hillslopes (Chaps. 2–11), continuing along the glacifluvial system or ending on temporary or long-term sinks (Chaps. 12–14). The rates of geomorphic processes, and the way they interact spatially and functionally (that is, connectivity) determine the sediment budget of the proglacial area and sediment yield at its outlet (Chaps. 15–17). Soil formation and vegetation development (Chaps. 18 and 19) interact with morphodynamics and may exert considerable influence on these cascades.

Several chapters review the state of science in the respective field of research (Chaps. 2, 4, 6, 8, 10, 12, 14, 15, 16, 18 and 19), while others represent case studies (Chaps. 3, 5, 7, 9, 11, 13 and 17), most of which are derived from the PROSA joint research project (Sect. 1.4).

Acknowledgements The editors wish to thank all contributors who dedicated their time to make this book possible. Besides the chapter authors, we highly appreciate the efforts of the following colleagues (in alphabetical order) who reviewed the chapters and gave valuable advice to significantly improve the manuscripts: Jean-Baptiste Bosson, Jonathan Carrivick, Francesco Comiti, Michele D'Amico, Adam Emmer, Isabelle Gärtner-Roer, Wilfried Haeberli, Oliver Korup, Michael Krautblatter, Ingolf Kühn, Christophe Lambiel, Frédéric Liébault, Sven Lukas, Luca Mao, John Matthews, Sam McColl, Ronald Pöppl, Lothar Schrott, Jeff Warburton and Matt Westoby. The authors thank the German and Austrian Science Foundations (DFG and FWF) for funding the PROSA project (subproject numbers DFG 209753023, 209752370 and 209752836, see http://gepris.dfg.de).

References

Avian M, Kellerer-Pirklbauer A, Lieb GK (2018) Geomorphic consequences of rapid deglaciation at Pasterze Glacier, Hohe Tauern Range, Austria, between 2010 and 2013 based on repeated terrestrial laser scanning data. Geomorphology. https://doi.org/10.1016/j.geomorph.2018.02.003

Baewert H, Morche D (2014) Coarse sediment dynamics in a proglacial fluvial system (Fagge River, Tyrol). Geomorphology 218:88–97. https://doi.org/10.1016/j.geomorph.2013.10.021

Balco G (2011) Contributions and unrealized potential contributions of cosmogenic-nuclide exposure dating to glacier chronology, 1990–2010. Quatern Sci Rev 30:3–27. https://doi.org/10.1016/j.quascirev.2010.11.003

Ballantyne CK (2002a) Paraglacial geomorphology. Quatern Sci Rev 21:1935–2017. https://doi.org/10.1016/S0277-3791(02)00005-7

Ballantyne CK (2002b) A general model of paraglacial landscape response. The Holocene 12:371–376. https://doi.org/10.1191/0959683602hl553fa

Barr ID, Lovell H (2014) A review of topographic controls on moraine distribution. Geomorphology 226:44–64. https://doi.org/10.1016/j.geomorph.2014.07.030

Barry RG (2006) The status of research on glaciers and global glacier recession: a review. Prog Phys Geogr 30: 285–306. https://doi.org/10.1191/0309133306pp478ra

Bechet J, Duc J, Loye A, Jaboyedoff M, Mathys N, Malet J-P, Klotz S, Le Bouteiller C, Rudaz B, Travelletti J (2016) Detection of seasonal cycles of erosion processes in a black marl gully from a time series of high-resolution digital elevation models (DEMs). Earth Surf Dynam 4:781–798. https://doi.org/10.5194/esurf-4-781-2016

Beniston M (2005) Mountain climates and climatic change: an overview of processes focusing on the European Alps. Pure appl Geophys 162:1587–1606. https://doi.org/10.1007/s00024-005-2684-9

Beniston M, Stoffel M (2014) Assessing the impacts of climatic change on mountain water resources. Sci Total Environ 493:1129–1137. https://doi.org/10.1016/j.scitotenv.2013.11.122

Benn DI, Kirkbride MP, Owen LA, Brazier V (2003) Glaciated valley landsystems. In: Evans DJA (ed) Glacial landsystems. Arnold, London

Bernhardt A, Schwanghart W, Hebbeln D, Stuut J-BW, Strecker MR (2017) Immediate propagation of deglacial environmental change to deep-marine turbidite systems along the Chile convergent margin. Earth Planet Sci Lett 473:190–204. https://doi.org/10.1016/j.epsl.2017.05.017

Beylich AA, Warburton J (eds) (2007) SEDIFLUX Manual: analysis of source-to-sink fluxes and sediment budgets in changing high-latitude and high-altitude cold environments. Geological Survey of Norway Publication, vol 2007, 053

Beylich A, Laute K, Liermann S, Hansen L, Burki V, Vatne G, Fredin O, Gintz D, Berthling I (2009) Subrecent sediment dynamics and sediment budget of the braided sandur system at Sandane, Erdalen (Nordfjord, Western Norway). Norsk Geografisk Tidsskrift (Norwegian J Geogr) 63:123–131. https://doi.org/10.1080/00291950902907934

Beylich AA, Dixon JC, Zwoliński Z (eds) (2016) Source-to-sink-fluxes in undisturbed cold environments. Cambridge Univ Press, [Place of publication not identified]

Bogen J, Xu M, Kennie P (2015) The impact of pro-glacial lakes on downstream sediment delivery in Norway. Earth Surf Process Land 40:942–952. https://doi.org/10.1002/esp.3669

Bosson J-B, Deline P, Bodin X, Schoeneich P, Baron L, Gardent M, Lambiel C (2015) The influence of ground ice distribution on geomorphic dynamics since the Little Ice Age in proglacial areas of two cirque glacier systems. Earth Surf Process Land 40:666–680. https://doi.org/10.1002/esp.3666

Briese C, Pfennigbauer M, Lehner H, Ullrich A, Wagner W, Pfeifer N (2012) Radiometric calibration of muli-wavelength airborne laser scanning data. ISPRS Ann Photogramm Remote Sens Spatial Inf Sci 1:335–340

Buechi MW, Kober F, Ivy-Ochs S, Salcher B, Kubik PW, Christl M (2014) Denudation rates of small transient catchments controlled by former glaciation: the Hörnli nunatak in the northeastern Swiss Alpine Foreland. Quat Geochronol 19:135–147. https://doi.org/10.1016/j.quageo.2013.06.005

Burga CA, Krüsi B, Egli M, Wernli M, Elsener S, Ziefle M, Fischer T, Mavris C (2010) Plant succession and soil development on the foreland of the Morteratsch glacier (Pontresina, Switzerland): straight forward or chaotic? Flora—Morphol Distrib Funct Ecol Plants 205:561–576. https://doi.org/10.1016/j.flora.2009.10.001

Burt TP, Allison RJ (eds) (2010) Sediment cascades. Wiley, Chichester, UK

Caduff R, Schlunegger F, Kos A, Wiesmann A (2015) A review of terrestrial radar interferometry for measuring surface change in the geosciences. Earth Surf Process Land 40:208–228. https://doi.org/10.1002/esp.3656

Carrivick JL, Heckmann T (2017) Short-term geomorphological evolution of proglacial systems. Geomorphology 287:3–28. https://doi.org/10.1016/j.geomorph.2017.01.037

Carrivick JL, Geilhausen M, Warburton J, Dickson NE, Carver SJ, Evans AJ, Brown LE (2013) Contemporary geomorphological activity throughout the proglacial area of an alpine catchment. Geomorphology 188:83–95. https://doi.org/10.1016/j.geomorph.2012.03.029

Church M (2002) Fluvial sediment transfer in cold regions. In: Hewitt K, Byrne M-L, English M, Young G (eds) Landscapes of transition. Springer, Netherlands, Dordrecht, pp 93–117

Church M, Ryder JM (1972) Paraglacial sedimentation: a consideration of fluvial processes conditioned by glaciation. Geol Soc Am Bull 83:3059–3071

Church M, Slaymaker O (1989) Disequilibrium of Holocene sediment yield in glaciated British Columbia. Nature 337:452–454

Cossart E (2008) Landform connectivity and waves of negative feedbacks during the paraglacial period, a case study: the Tabuc subcatchment since the end of the Little Ice Age (massif des Écrins, France). Géomorphologie: relief, processus, environnement, 14(4):249–260

Cossart E, Fort M (2008) Sediment release and storage in early deglaciated areas: towards an application of the exhaustion model from the case of Massif des Ã‰crins (French Alps) since the Little Ice Age. Norsk Geografisk Tidsskrift—Norwegian Journal of Geography 62:115–131. doi: 10.1080/00291950802095145

Cossart E, Braucher R, Fort M, Bourlès DL, Carcaillet J (2008) Slope instability in relation to glacial debuttressing in alpine areas (Upper Durance catchment, southeastern France): evidence from field data and ^{10}Be cosmic ray exposure ages: paraglacial geomorphology: processes and paraglacial context. Geomorphology 95:3–26

Cossart É, Mercier D, Decaulne A, Feuillet T (2013) An overview of the consequences of paraglacial landsliding on deglaciated mountain slopes: typology, timing and contribution to cascading fluxes. Quaternaire 24 (1):13–24. https://doi.org/10.4000/quaternaire.6444

Curran JH, Loso MG, Williams HB (2017) Glacial conditioning of stream position and flooding in the braid plain of the Exit Glacier foreland, Alaska. Geomorphology 293:272–288. https://doi.org/10.1016/j.geomorph.2017.06.004

Curry AM, Cleasby V, Zukowskyj P (2006) Paraglacial response of steep, sediment-mantled slopes to post-Little Ice Age glacier recession in the central Swiss Alps. J Quat Sci 21:211–225. https://doi.org/10.1002/jqs.954

Delaney I, Bauder A, Huss M, Weidmann Y (2018) Proglacial erosion rates and processes in a Glacierized catchment in the Swiss Alps. Earth Surf Process Land 43:765–778. https://doi.org/10.1002/esp.4239

Delunel R, van der Beek PA, Bourlès DL, Carcaillet J, Schlunegger F (2014) Transient sediment supply in a high-altitude Alpine environment evidenced through a ^{10}Be budget of the Etages catchment (French Western Alps). Earth Surf Process Land 39:890–899. https://doi.org/10.1002/esp.3494

Dietrich WE, Dunne T, Humphrey NF, Reid LM (1982) Construction of sediment budgets for drainage basins. In: Swanson FJ, Janda RJ, Dunne T, Swanston DN (eds) Sediment budgets and routing in forested drainage basins: Gen. Tech. Rep. PNW-141. Pacific Northwest Forest and Range Experiment Station,

Forest Service, U.S. Department of Agriculture, Portland, pp 5–23
Dusik J-M, Leopold M, Heckmann T, Haas F, Hilger L, Morche D, Neugirg F, Becht M (2015) Influence of glacier advance on the development of the multipart Riffeltal rock glacier, Central Austrian Alps. Earth Surf Process Land 40:965–980. https://doi.org/10.1002/esp.3695
Egli M, Wernli M, Kneisel C, Haeberli W (2006a) Melting Glaciers and soil development in the Proglacial Area Morteratsch (Swiss Alps): I. soil type chronosequence. Arct Antarct Alp Res 38:499–509
Egli M, Wernli M, Kneisel C, Biegger S, Haeberli W (2006b) Melting Glaciers and soil development in the Proglacial Area Morteratsch (Swiss Alps): II. Modeling the present and future soil state. Arct Antarct Alp Res 38:510–521
Eichel J, Krautblatter M, Schmidtlein S, Dikau R (2013) Biogeomorphic interactions in the Turtmann glacier forefield, Switzerland. Geomorphology 201:98–110. https://doi.org/10.1016/j.geomorph.2013.06.012
Eichel J, Corenblit D, Dikau R (2015) Conditions for feedbacks between geomorphic and vegetation dynamics on lateral moraine slopes: a biogeomorphic feedback window. Earth Surf Process Land 41 (3):406–419. https://doi.org/10.1002/esp.3859
Einsele G, Hinderer M (1997) Terrestrial sediment yield and the lifetime of reservoirs, lakes and larger basins. Geol Rundsch 86:288–310
FAO (2011) Why invest in sustainable mountain development? Rome
Fischer A, Seiser B, Stocker Waldhuber M, Mitterer C, Abermann J (2015) Tracing glacier changes in Austria from the Little Ice Age to the present using a lidar-based high-resolution glacier inventory in Austria. The Cryosphere 9:753–766. https://doi.org/10.5194/tc-9-753-2015
Fryirs KA (2017) River sensitivity: a lost foundation concept in fluvial geomorphology. Earth Surf Process Land 42:55–70. https://doi.org/10.1002/esp.3940
Geilhausen M, Otto J-C, Morche D, Schrott L (2012) Decadal sediment yield from an Alpine proglacial zone inferred from reservoir sedimentation (Pasterze, Hohe Tauern, Austria). IAHS Publ 356:161–172
Geilhausen M, Morche D, Otto J-C, Schrott L (2013) Sediment discharge from the proglacial zone of a retreating Alpine glacier. Zeit fur Geo Supp 57:29–53. https://doi.org/10.1127/0372-8854/2012/S-00122
Glira P, Pfeifer N, Briese C, Ressl C (2015) A correspondence framework for ALS strip adjustments based on variants of the ICP algorithm Korrespondenzen für die ALS-Streifenausgleichung auf Basis von ICP. Photogrammetrie - Fernerkundung - Geoinformation 2015:275–289. https://doi.org/10.1127/pfg/2015/0270
Groß G, Patzelt G (2015) The Austrian glacier inventory for the Little Ice Age Maximum (GI LIA) in ArcGIS (shapefile) format
Grove JM (2004) Little ice ages: ancient and modern, 2nd edn. Routledge, London
Guillon H, Mugnier J-L, Buoncristiani J-F, Carcaillet J, Godon C, Prud'homme C, van der Beek P, Vassallo R (2015) Improved discrimination of subglacial and periglacial erosion using ^{10}Be concentration measurements in subglacial and supraglacial sediment load of the Bossons glacier (Mont Blanc massif, France). Earth Surf Process Land 40:1202–1215. https://doi.org/10.1002/esp.3713
Guillon H, Mugnier J-L, Buoncristiani J-F (2018) Proglacial sediment dynamics from daily to seasonal scales in a glaciated Alpine catchment (Bossons glacier, Mont Blanc massif, France). Earth Surf Process Land 287:126. https://doi.org/10.1002/esp.4333
Guyard H, Chapron E, St-Onge G, Anselmetti FS, Arnaud F, Magand O, Francus P, Mélières M-A (2007) High-altitude varve records of abrupt environmental changes and mining activity over the last 4000 years in the Western French Alps (Lake Bramant, Grandes Rousses Massif). Quatern Sci Rev 26:2644–2660. https://doi.org/10.1016/j.quascirev.2007.07.007
Haas F, Heckmann T, Hilger L, Becht M (2012) Quantification and modelling of debris flows in the Proglacial Area of the Gepatschferner/Austria using ground-based LIDAR. IAHS Publ 356:293–302
Haeberli W, Huggel C, Paul F, Zemp M (2013) 13.10 Glacial responses to climate change. In: Shroder JF, Switzer AD, Kennedy DM (eds) Treatise on Geomorphology. Elsevier, New York, pp 152–175
Harbor J, Warburton J (1993) Relative rates of glacial and nonglacial erosion in alpine environments. Arct Alp Res 25:1–7
Hart SJ, Clague JJ, Smith DJ (2010) Dendrogeomorphic reconstruction of Little Ice Age paraglacial activity in the vicinity of the Homathko Icefield, British Columbia Coast Mountains, Canada. Geomorphology 121:197–205. https://doi.org/10.1016/j.geomorph.2010.04.011
Hartl L (2010) The Gepatschferner from 1850–2006—changes in length, area and volume in relation to climate. Diploma thesis, University of Innsbruck, Innsbruck
Heckmann T, Vericat D (2018) Computing spatially distributed sediment delivery ratios: inferring functional sediment connectivity from repeat high-resolution digital elevation models. Earth Surf Process Land 218:88. https://doi.org/10.1002/esp.4334
Heckmann T, Haas F, Morche D, Schmidt K-H, Rohn J, Moser M, Leopold M, Kuhn M, Briese C, Pfeifer N, Becht M (2012a) Investigating an Alpine proglacial sediment budget using field measurements, airborne and terrestrial LiDAR data. IAHS Publication, pp 438–447
Heckmann T, Haas F, Morche D, Schmidt K-H, Rohn J, Moser M, Leopold M, Kuhn M, Briese C, Pfeifer N, Becht M (2012b) Quantifying proglacial morphodynamics and sediment budgets—the PROSA approach. In: EGU (ed) Geophysical Research Abstracts, vol 14

Heckmann T, Hilger L, Vehling L, Becht M (2016a) Integrating field measurements, a geomorphological map and stochastic modelling to estimate the spatially distributed rockfall sediment budget of the Upper Kaunertal, Austrian Central Alps. Geomorphology 260:16–31. https://doi.org/10.1016/j.geomorph.2015.07.003

Heckmann T, McColl S, Morche D (2016b) Retreating ice: research in pro-glacial areas matters. Earth Surf Process Land 41:271–276. https://doi.org/10.1002/esp.3858

Helfricht K, Kuhn M, Keuschnig M, Heilig A (2014) Lidar snow cover studies on glaciers in the Ötztal Alps (Austria): comparison with snow depths calculated from GPR measurements. The Cryosphere 8:41–57. https://doi.org/10.5194/tc-8-41-2014

Hetherington D, Heritage G, Milan DJ (2005) Daily fine sediment dynamics on an active Alpine glacier outwash plain. IAHS Publication 291:278–284

Hewitt K (2002) Introduction: landscape assemblages and transitions in cold regions. In: Hewitt K, Byrne M-L, English M, Young G (eds) Landscapes of transition. Springer, Netherlands, Dordrecht, pp 1–8

Hewitt K, Byrne M-L, English M, Young G (eds) (2002) Landscapes of transition. The GeoJournal library, vol 68. Springer, Netherlands, Dordrecht

Holm K, Bovis M, Jakob M (2004) The landslide response of alpine basins to post-Little Ice Age glacial thinning and retreat in southwestern British Columbia. Geomorphology 57:201–216

Holzhauser H (1982) Neuzeitliche Gletscherschwankungen. Geographica Helvetica 37:115–126. https://doi.org/10.5194/gh-37-115-1982

Hugenholtz C, Moorman B, Barlow J, Wainstein P (2008) Large-scale moraine deformation at the Athabasca Glacier, Jasper National Park, Alberta, Canada. Landslides 5:251–260. https://doi.org/10.1007/s10346-008-0116-5

Hürlimann M, Abancó C, Moya J (2012) Rockfalls detached from a lateral moraine during spring season. 2010 and 2011 events observed at the Rebaixader debris-flow monitoring site (Central Pyrenees, Spain). Landslides 9:385–393. https://doi.org/10.1007/s10346-011-0314-4

Joerin UE, Stocker TF, Schlüchter C (2006) Multicentury glacier fluctuations in the Swiss Alps during the Holocene. Holocene 16:697–704. https://doi.org/10.1191/0959683606hl964rp

Johnson PG (2002) Proglacial and paraglacial fluvial and lacustrine environments in transition. In: Hewitt K, Byrne M-L, English M, Young G (eds) Landscapes of transition. Springer, Netherlands, Dordrecht, pp 43–62

Keiler M, Knight J, Harrison S (2010) Climate change and geomorphological hazards in the eastern European Alps. Philos Trans Royal Soc A Math Phys Eng Sci 368:2461–2479. https://doi.org/10.1098/rsta.2010.0047

Kellerer-Pirklbauer A, Proske H, Strasser V (2010) Paraglacial slope adjustment since the end of the Last Glacial Maximum and its long-lasting effects on secondary mass wasting processes: Hauser Kaibling, Austria: Landslide geomorphology in a changing environment. Geomorphology 120:65–76. https://doi.org/10.1016/j.geomorph.2009.09.016

Kinzl H (1932) Die größten nacheiszeitlichen Gletschervorstöße in den Schweizer Alpen und in der Montblancgruppe: (the greatest post-ice age glacier advances in the Swiss Alps and the Montblanc mountain range; in German). Zeitschrift für Gletscherkunde 20

Kirkbride MP, Winkler S (2012) Correlation of late quaternary moraines: impact of climate variability, glacier response, and chronological resolution. Quatern Sci Rev 46:1–29. https://doi.org/10.1016/j.quascirev.2012.04.002

Klaar MJ, Kidd C, Malone E, Bartlett R, Pinay G, Chapin FS, Milner A (2015) Vegetation succession in deglaciated landscapes: implications for sediment and landscape stability. Earth Surf Process Land 40:1088–1100. https://doi.org/10.1002/esp.3691

Klimeš J, Novotný J, Novotná I, de Urries BJ, Vilímek V, Emmer A, Strozzi T, Kusák M, Rapre AC, Hartvich F, Frey H (2016) Landslides in moraines as triggers of glacial lake outburst floods: example from Palcacocha Lake (Cordillera Blanca, Peru). Landslides 13:1461–1477. https://doi.org/10.1007/s10346-016-0724-4

Klok EJ, Oerlemans J (2004) Climate reconstructions derived from global glacier length records. Arct Antarct Alp Res 36:575–583

Kneisel C, Kääb A (2007) Mountain permafrost dynamics within a recently exposed glacier forefield inferred by a combined geomorphological, geophysical and photogrammetrical approach. Earth Surf Process Land 32:1797–1810

Knight P (2006) Glacier science and environmental change. Blackwell Publications, Malden MA, Oxford

Knight J, Harrison S (eds) (2009) Periglacial and Paraglacial processes and environments, Special publications, vol 320. The Geological Society Publishing House, London

Knight J, Harrison S (2014a) Glacial and paraglacial environments. Geografiska Annaler: Ser A, Phys Geogr 96:241–244. https://doi.org/10.1111/geoa.12058

Knight J, Harrison S (2014b) Mountain Glacial and Paraglacial environments under global climate change: lessons from the past, future directions and policy implications. Geografiska Annaler: Ser A, Phys Geogr 96:245–264. https://doi.org/10.1111/geoa.12051

Knoll C, Kerschner H, Heller A, Rastner P (2009) A GIS-based reconstruction of Little Ice Age Glacier maximum extents for South Tyrol, Italy. Trans GIS 13:449–463. https://doi.org/10.1111/j.1467-9671.2009.01173.x

Körner C, Ohsawa M (2005) Mountain systems (Chap 24). In: Hassan RM, Scholes RJ, Ash N (eds) Ecosystems and human well-being: current state and trends: findings of the condition and trends working group of the millennium ecosystem assessment. Island Press, Washington, DC, pp 681–716

Lane SN, Bakker M, Gabbud C, Micheletti N, Saugy J-N (2017) Sediment export, transient landscape response

and catchment-scale connectivity following rapid climate warming and Alpine glacier recession. Geomorphology 277:210–227. https://doi.org/10.1016/j.geomorph.2016.02.015
Laute K, Beylich AA (2014a) Environmental controls, rates and mass transfers of contemporary hillslope processes in the headwaters of two glacier-connected drainage basins in western Norway. Geomorphology 216:93–113. https://doi.org/10.1016/j.geomorph.2014.03.021
Laute K, Beylich AA (2014b) Environmental controls and geomorphic importance of a high-magnitude/low-frequency snow avalanche event in Bodalen, Nordfjord. Western Norway Geogr Annaler A 96:465–484. https://doi.org/10.1111/geoa.12069
Le Roy M, Nicolussi K, Deline P, Astrade L, Edouard J-L, Miramont C, Arnaud F (2015) Calendar-dated glacier variations in the western European Alps during the Neoglacial: The Mer de Glace record, Mont Blanc massif. Quatern Sci Rev 108:1–22. https://doi.org/10.1016/j.quascirev.2014.10.033
Le Roy M, Deline P, Carcaillet J, Schimmelpfennig I, Ermini M (2017) ^{10}Be exposure dating of the timing of Neoglacial glacier advances in the Ecrins-Pelvoux massif, southern French Alps. Quatern Sci Rev 178:118–138. https://doi.org/10.1016/j.quascirev.2017.10.010
Leclercq PW, Oerlemans J, Basagic HJ, Bushueva I, Cook AJ, Le Bris R (2014) A data set of worldwide glacier length fluctuations. The Cryosphere 8:659–672. https://doi.org/10.5194/tc-8-659-2014
Legg NT, Meigs AJ, Grant GE, Kennard P (2014) Debris flow initiation in proglacial gullies on Mount Rainier, Washington. Geomorphology 226:249–260. https://doi.org/10.1016/j.geomorph.2014.08.003
Loso MG, Doak DF, Anderson RS (2016) Lichenometric dating of little ice age glacier moraines using explicit demographic models of lichen colonization, growth, and survival. Geografiska Annaler A 96:21–41. https://doi.org/10.1111/geoa.12022
Lucchesi S, Fioraso G, Bertotto S, Chiarle M (2014) Little Ice Age and contemporary glacier extent in the Western and South-Western Piedmont Alps (North-Western Italy). J Maps 10:409–423. https://doi.org/10.1080/17445647.2014.880226
Luckman BH (2000) The Little Ice Age in the Canadian rockies. Geomorphology 32:357–384. https://doi.org/10.1016/S0169-555X(99)00104-X
Mann ME (2002) Little Ice Age. In: MacCracken MC, Perry JS (eds) Encyclopedia of global environmental change: The Earth system: physical and chemical dimensions of global environmental change. Wiley, Chichester, New York, pp 504–509
Mao L, Dell'Agnese A, Huincache C, Penna D, Engel M, Niedrist G, Comiti F (2014) Bedload hysteresis in a glacier-fed mountain river. Earth Surf Process Land 39:964–976. https://doi.org/10.1002/esp.3563
Marren PM, Toomath SC (2014) Channel pattern of proglacial rivers: topographic forcing due to glacier retreat. Earth Surf Process Land 39:943–951. https://doi.org/10.1002/esp.3545
Martini IP, French H, Perez Alberti A (eds) (2011) Ice-marginal and periglacial processes and sediments. Geological society special publication, vol 354. Geological Society, London
Matthews JA (1992) The ecology of recently-deglaciated terrain: a geoecological approach to glacier forelands and primary succession. Cambridge studies in ecology. Cambridge University Press, Cambridge
Matthews JA, Briffa KR (2005) The "Little Ice Age": Re-evaluation of an evolving concept. Geografiska Annaler A 87:17–36. https://doi.org/10.1111/j.0435-3676.2005.00242.x
McColl ST (2012) Paraglacial rock-slope stability. Geomorphology 153–154:1–16. https://doi.org/10.1016/j.geomorph.2012.02.015
Micheletti N, Lane SN (2016) Water yield and sediment export in small, partially glaciated Alpine watersheds in a warming climate. Water Resour Res 52:4924–4943. https://doi.org/10.1002/2016WR018774
Micheletti N, Lambiel C, Lane SN (2015) Investigating decadal-scale geomorphic dynamics in an alpine mountain setting. J Geophys Res Earth Surf 120:2155–2175. https://doi.org/10.1002/2015JF003656
Milan DJ, Heritage G, Hetherington D (2007) Application of a 3D laser scanner in the assessment of erosion and deposition volumes and channel change in a proglacial river. Earth Surf Process Land 32:1657–1674. https://doi.org/10.1002/esp.1592
Milner AM, Khamis K, Battin TJ, Brittain JE, Barrand NE, Füreder L, Cauvy-Fraunié S, Gíslason GM, Jacobsen D, Hannah DM, Hodson AJ, Hood E, Lencioni V, Ólafsson JS, Robinson CT, Tranter M, Brown LE (2017) Glacier shrinkage driving global changes in downstream systems. Proc Natl Acad Sci U S A 114:9770–9778. https://doi.org/10.1073/pnas.1619807114
Moore RD, Fleming SW, Menounos B, Wheate R, Fountain ASK, Holm K, Jakob M (2009) Glacier change in western North America: influences on hydrology, geomorphic hazards and water quality. Hydrol Process 23:42–61
Morche D, Haas F, Baewert H, Heckmann T, Schmidt K-H, Becht M (2012) Sediment transport in the proglacial Fagge River (Kaunertal/Austria). IAHS Publ 356:72–80
Morche D, Schuchardt A, Dubberke K, Baewert H (2015) Channel morphodynamics on a small proglacial braid plain (Fagge River, Gepatschferner, Austria). Proc IAHS 367:109–116. https://doi.org/10.5194/piahs-367-109-2015
Moreau M, Mercier D, Laffly D, Roussel E (2008) Impacts of recent paraglacial dynamics on plant colonization: a case study on Midtre Lovénbreen foreland, Spitsbergen (79°N): Paraglacial Geomorphology: Processes and Paraglacial Context. Geomorphology 95:48–60
Nicolussi K, Patzelt G (2001) Untersuchungen zur holozänen Gletscherentwicklung von Pasterze und Gepatschferner (Ostalpen). Z Gletscherk Glazialgeol 36:1–87

O'Connor JE, Costa JE (1993) Geologic and hydrologic hazards in glacierized basins in North America resulting from 19th and 20th century global warming. Nat Hazards 8:121–140. https://doi.org/10.1007/BF00605437

Oerlemans J (2005) Extracting a climate signal from 169 Glacier records. Science 308:675–677. https://doi.org/10.1126/science.1107046

O'Farrell CR, Heimsath AM, Lawson DE, Jorgensen LM, Evenson EB, Larson G, Denner J (2009) Quantifying periglacial erosion: insights on a glacial sediment budget, Matanuska Glacier. Alaska Earth Surf Process Land 34:2008–2022. https://doi.org/10.1002/esp.1885

Oliva M, Ruiz-Fernández J, Barriendos M, Benito G, Cuadrat JM, Domínguez-Castro F, García-Ruiz JM, Giralt S, Gómez-Ortiz A, Hernández A, López-Costas O, López-Moreno JI, López-Sáez JA, Martínez-Cortizas A, Moreno A, Prohom M, Saz MA, Serrano E, Tejedor E, Trigo R, Valero-Garcés B, Vicente-Serrano SM (2018) The Little Ice Age in Iberian mountains. Earth Sci Rev 177:175–208. https://doi.org/10.1016/j.earscirev.2017.11.010

Orlove BS, Wiegandt E, Luckman BH (eds) (2008) Darkening peaks: Glacier retreat, science, and society. University of California Press, Berkeley

Palacios D, Parrilla G, Zamorano JJ (1999) Paraglacial and postglacial debris flows on a Little Ice Age terminal moraine: Jamapa Glacier, Pico de Orizaba (Mexico). Geomorphology 28:95–118

Rangwala I, Miller JR (2012) Climate change in mountains: a review of elevation-dependent warming and its possible causes. Clim Change 114:527–547. https://doi.org/10.1007/s10584-012-0419-3

Ravanel L, Deline P (2011) Climate influence on rockfalls in high-Alpine steep rockwalls: the north side of the Aiguilles de Chamonix (Mont Blanc massif) since the end of the 'Little Ice Age'. The Holocene 21:357–365. https://doi.org/10.1177/0959683610374887

Ribolini A, Guglielmin M, Fabre D, Bodin X, Marchisio M, Sartini S, Spagnolo M, Schoeneich P (2010) The internal structure of rock glaciers and recently deglaciated slopes as revealed by geoelectrical tomography: insights on permafrost and recent glacial evolution in the Central and Western Alps (Italy–France). Quatern Sci Rev 29:507–521. https://doi.org/10.1016/j.quascirev.2009.10.008

Richardson SD, Reynolds JM (2000) An overview of glacial hazards in the Himalayas. Quatern Int 65(66): 31–47

Rouyet L, Kristensen L, Derron M-H, Michoud C, Blikra LH, Jaboyedoff M, Lauknes TR (2016) Evidence of rock slope breathing using ground-based InSAR. Geomorphology. https://doi.org/10.1016/j.geomorph.2016.07.005

Sailer R, Bollmann E, Hoinkes S, Rieg L, Sproß M, Stötter J (2012) Quantification of geomorphodynamics in glaciated and recently deglaciated terrain based on airborne laserscanning data. Geografiska Annaler: Ser A, Phys Geogr 94:17–32. https://doi.org/10.1111/j.1468-0459.2012.00456.x

Schiefer E, Gilbert R (2007) Reconstructing morphometric change in a proglacial landscape using historical aerial photography and automated DEM generation. Geomorphology 88:167–178

Schimmelpfennig I, Schaefer JM, Akçar N, Koffman T, Ivy-Ochs S, Schwartz R, Finkel RC, Zimmerman S, Schlüchter C (2014) A chronology of Holocene and Little Ice Age glacier culminations of the Steingletscher, Central Alps, Switzerland, based on high-sensitivity beryllium-10 moraine dating. Earth Planet Sci Lett 393:220–230. https://doi.org/10.1016/j.epsl.2014.02.046

Schwamborn G, Heinzel J, Schirrmeister L (2008) Internal characteristics of ice-marginal sediments deduced from georadar profiling and sediment properties (Brøgger Peninsula, Svalbard). Geomorphology 95:74–83

Shugar DH, Clague JJ, Best JL, Schoof C, Willis MJ, Copland L, Roe GH (2017) River piracy and drainage basin reorganization led by climate-driven glacier retreat. Nat Geosci 10:95. https://doi.org/10.1038/NGEO2932

Slaymaker O (2009) Proglacial, periglacial or paraglacial? In: Knight J, Harrison S (eds) Periglacial and Paraglacial Processes and Environments. The Geological Society Publishing House, London, pp 71–84

Slaymaker O (2011) Criteria to distinguish between periglacial, proglacial and paraglacial environments. Quaest Geogr 30:85–94

Smith NW, Joffe H (2009) Climate change in the British press: the role of the visual. J Risk Res 12:647–663. https://doi.org/10.1080/13669870802586512

Staines KEH, Carrivick JL (2015) Geomorphological impact and morphodynamic effects on flow conveyance of the 1999 jökulhlaup at sólheimajökull, Iceland. Earth Surf Process Land 40:n/a. https://doi.org/10.1002/esp.3750

Sternai P, Herman F, Fox MR, Castelltort S (2011) Hypsometric analysis to identify spatially variable glacial erosion. J Geophys Res 116:2. https://doi.org/10.1029/2010JF001823

Stocker-Waldhuber M, Fischer A, Keller L, Morche D, Kuhn M (2017) Funnel-shaped surface depressions—indicator or accelerant of rapid glacier disintegration? a case study in the Tyrolean Alps. Geomorphology 287:58–72. https://doi.org/10.1016/j.geomorph.2016.11.006

Strasser U, Marke T, Braun L, Escher-Vetter H, Juen I, Kuhn M, Maussion F, Mayer C, Nicholson L, Niedertscheider K, Sailer R, Stötter J, Weber M, Kaser G (2018) The Rofental: A high Alpine research basin (1890–3770 m a.s.l.) in the Ötztal Alps (Austria) with over 150 years of hydrometeorological and glaciological observations. Earth Syst Sci Data 10: 151–171. https://doi.org/10.5194/essd-10-151-2018

Temme AJAM, Lange K (2014) Pro-glacial soil variability and geomorphic activity—the case of three Swiss valleys. Earth Surf Process Land 39:1492–1499. https://doi.org/10.1002/esp.3553

Temme AJAM, Lange K, Schwering MFA (2015) Time development of soils in mountain landscapes—

divergence and convergence of properties with age. J Soils Sediments 15:1373–1382. https://doi.org/10.1007/s11368-014-0947-8

Temme AJAM, Heckmann T, Harlaar P (2016) Silent play in a loud theatre—Dominantly time-dependent soil development in the geomorphically active proglacial area of the Gepatsch glacier, Austria. CATENA 147: 40–50. https://doi.org/10.1016/j.catena.2016.06.042

Thompson A, Jones A (1986) Rates and causes of proglacial river terrace formation in southeast Iceland: an application of lichenometric dating techniques. Boreas 15:231–246. https://doi.org/10.1111/j.1502-3885.1986.tb00928.x

Vehling L, Rohn J, Moser M (2016) Quantification of small magnitude rockfall processes at a proglacial high mountain site, Gepatsch glacier (Tyrol, Austria). Zeit fur Geo Supp 60:93–108. https://doi.org/10.1127/zfg_suppl/2015/S-00184

Vehling L, Baewert H, Glira P, Moser M, Rohn J, Morche D (2017) Quantification of sediment transport by rockfall and rockslide processes on a proglacial rock slope (Kaunertal, Austria). Geomorphology 287:46–57. https://doi.org/10.1016/j.geomorph.2016.10.032

Westoby MJ, Brasington J, Glasser NF, Hambrey MJ, Reynolds JM (2012) 'Structure-from-Motion' photogrammetry: a low-cost, effective tool for geoscience applications. Geomorphology 179:300–314. https://doi.org/10.1016/j.geomorph.2012.08.021

Winkler S (2004) Lichenometric dating of the 'Little Ice Age' maximum in Mt Cook National Park, Southern Alps, New Zealand. The Holocene 14:911–920

Zanoner T, Carton A, Seppi R, Carturan L, Baroni C, Salvatore MC, Zumiani M (2017) Little Ice Age mapping as a tool for identifying hazard in the paraglacial environment: the case study of Trentino (Eastern Italian Alps). Geomorphology 295:551–562. https://doi.org/10.1016/j.geomorph.2017.08.014

Zasadni J (2007) The Little Ice Age in the Alps: its record in glacial deposits and rock glacier formation. Studia Geomorphologica Carpatho-Balcanica 41:117–137

Zemp M, Paul F, Hoelzle M, Haeberli W (2008) Glacier fluctuations in the European Alps, 1850–2000: an overview and spatio-temporal analysis of available data. In: Orlove BS, Wiegandt E, Luckman BH (eds) Darkening peaks: Glacier retreat, science, and society. University of California Press, Berkeley

Zemp M, Hoelzle M, Haeberli W (2009) Six decades of glacier mass-balance observations: a review of the worldwide monitoring network. Ann Glac 50:101–111

Zemp M, Frey H, Gärtner-Roer I, Nussbaumer SU, Hoelzle M, Paul F, Haeberli W, Denzinger F, Ahlstrøm AP, Anderson B, Bajracharya S, Baroni C, Braun LN, Cáceres BE, Casassa G, Cobos G, Dávila LR, Delgado Granados H, Demuth MN, Espizua L, Fischer A, Fujita K, Gadek B, Ghazanfar A, Ove Hagen J, Holmlund P, Karimi N, Li Z, Pelto M, Pitte P, Popovnin VV, Portocarrero CA, Prinz R, Sangewar CV, Severskiy I, Sigurđsson O, Soruco A, Usubaliev R, Vincent C (2015) Historically unprecedented global glacier decline in the early 21st century. J Glaciol 61:745–762. https://doi.org/10.3189/2015JoG15J017

Zolitschka B, Francus P, Ojala AEK, Schimmelmann A (2015) Varves in lake sediments—a review. Quatern Sci Rev 117:1–41. https://doi.org/10.1016/j.quascirev.2015.03.019

Zumbühl HJ (1980) Die Schwankungen der Grindelwaldgletscher in den historischen Bild- und Schriftquellen des 12. bis 19. Jahrhunderts. Birkhäuser Basel, Basel

Part I

Proglacial Areas, Glaciers and Ground Ice

2 Glacier Changes Since the Little Ice Age

Frank Paul and Tobias Bolch

Abstract

The majority of glaciers are currently retreating globally but had been in an advanced position for several hundred years during the so-called Little Ice Age (LIA). During this period, the lateral accumulation of rock and debris created impressive moraine walls. Between these LIA moraines and the actual terminus position is the glacier forefield, which is growing as glaciers retreat. Whereas the forefields are constantly changing (e.g. due to the transport of sediment and rock, lake formation and growth, plant colonization), the outer boundary marked by the moraines changed little and has widely been used to reconstruct maximum LIA extents and volume for numerous glaciers around the world. Together with field and satellite measurements, a detailed time series of glacier fluctuations since the LIA has been obtained for hundreds of glaciers that indicate some regional and glacier-specific variability, but also robust global trends of shrinkage and volume loss. Overall, the kilometre-scale retreat and upward shift of glacier termini by several 100 m since the end of the LIA confirm a global temperature increase by about one degree. As most glaciers have not yet adjusted their geometry to current climatic conditions, they will further shrink while forefields will continue to grow.

Keywords

Climate change · Little Ice Age · Glacier fluctuations · Glacier mass balance
Glacier inventory

2.1 Introduction

In most regions of the world, glaciers reached a Holocene maximum extent at the end of the so-called Little Ice Age (LIA) and decreased in size more or less continuously afterwards (Grove 2004). This maximum extent was not reached everywhere at the same point in time. For example, in the Alps this was during the seventeenth to nineteenth century (often around 1820 or 1850), in Scandinavia in the mid-eighteenth century and in New Zealand or parts of Alaska in the early eighteenth century (Rabatel et al. 2008). The LIA is understood as a slightly cooler period in the Holocene, lasting from about 1300 to 1850 that is often explained by a reduced solar activity

F. Paul (✉) · T. Bolch
Department of Geography, University of Zurich, Zurich, Switzerland
e-mail: frank.paul@geo.uzh.ch

T. Bolch
e-mail: tobias.bolch@geo.uzh.ch

T. Heckmann and D. Morche (eds.), *Geomorphology of Proglacial Systems*,
Geography of the Physical Environment, https://doi.org/10.1007/978-3-319-94184-4_2

(e.g. the Maunder Minimum) along with increased volcanism and internal climatic variability over this period (Grove 2004; Wanner et al. 2008). However, the period was neither geographically nor temporarily homogenous at a global scale and details of the precise timing are still a matter of research (Matthews and Briffa 2005). Dating of moraines and reconstructions from a variety of sources such as pictorial and written documents revealed a detailed chronology of glacier fluctuations over the LIA period for selected glaciers in the Alps (e.g. Zumbühl and Holzhauser 1988; Nussbaumer et al. 2007), Norway (Nussbaumer et al. 2011a) or Patagonia and South America (Masiokas et al. 2009a, b). In general, terminus fluctuations during the period of maximum extent were within a few hundred metres. The exact timing and amplitude of these fluctuations are glacier-specific, e.g. a function of glacier size and slope as well as topographic conditions (e.g. hypsometry and shading) and mass balance sensitivity.

After 1850 however, glaciers globally decreased in size and volume and their fronts retreated by up to a few kilometres for the largest (land-terminating) glaciers (Oerlemans 2005; Vaughan et al. 2013) in response to a general increase in temperature. For smaller and more quickly adjusting glaciers, intermittent phases of re-advance were observed in several regions in the 1920s and 1980s (e.g. Alps, Alaska, Tropics) and 1990s (e.g. Norway, New Zealand, Caucasus), but the specific reasons for these fluctuations might have been different (Chinn et al. 2005; UNEP 2007; Zemp et al. 2015). In the Alps, some glaciers advanced several hundred metres during these periods, but none reached again the LIA maximum extent. Moreover, the 1970s advances did by far not reach the one from the 1920s advance so that glacier forefields (i.e. the ice-free terrain between the current glacier terminus and the 1850s maximum extent) show two further terminal moraine walls within the 1850s extent, a bigger one from the 1920s and a smaller one from the 1970s to 1980s (Fig. 2.1). Outside but very close to the extent from around 1850, glaciers in the Alps and elsewhere have further lateral or terminal moraines from the LIA period. They reveal larger glacier extents before 1850 as they were not buried by the latest advance. In the Alps, these are seldom well preserved and are only slightly larger than the 1850 extent (Fig. 2.2) so that the latter is often taken as a synonym for the maximum extent of the entire LIA. In other regions, however, this is wrong, as former extents (e.g. mid-eighteenth century in Scandinavia or mid-seventeenth century in Patagonia) were much larger than the mid-nineteenth-century extents (Nussbaumer et al. 2011a; Masiokas et al. 2009a). Hence, the '1850s maximum extent' mentioned above and in the following is only the *latest* LIA maximum extent, but often not the largest one in absolute terms. As the lateral LIA moraines from the latest advance are still comparably well preserved and often have vegetation-free inner sides, they can be identified on medium resolution (Landsat-type) optical satellite imagery and related extents can be mapped (Paul and Kääb 2005; Wolken 2006).

In response to a strong increase in global temperature around 1985 (Beniston 2006; Wild et al. 2007; Reid et al. 2016), glacier mass loss increased in many regions of the world (Zemp et al. 2009). As an immediate reaction, glaciers lost mass by surface lowering in the ablation area (Paul and Haeberli 2008). Additionally, several glaciers also started thinning in their upper parts, indicating a disequilibrium response where glaciers will ultimately melt away completely (Pelto 2010). For the time being, the accumulation area of glaciers is too small to sustain their current size and they will thus continue shrinking (Carturan et al. 2013b). This so-called committed area and volume loss will reduce the size of current glaciers by a further 30–60% within the next few decades, even without a further increase in temperature (Dyurgerov et al. 2009; Zemp et al. 2015).

With constantly shrinking glaciers, their forefields constantly grow (Heckmann et al. 2016). If not taken over by lakes that form between the glacier terminus and the LIA moraine or in local overdeepenings of the bedrock (e.g. Haeberli et al. 2016b; Loriaux and Casassa 2013; Chap. 14), the forefields provide new land where soil can develop (Egli et al. 2006; Chap. 18) and vegetation can grow (Chap. 19).

Fig. 2.1 Glacier forefields of Mutt (inset left) and Damma Glacier with moraines from the latest LIA maximum extent around 1850 (solid white arrows), the 1920s (light yellow) and the 1980s (black). Dotted black lines indicate current glacier extents. Short arrows in light blue for Damma Glacier (upper right corner) indicate a possible larger extent before 1850. The inset map in the lower right shows the location of Mutt (M), Damma (D) and Findelen Glacier (F), and the latter being displayed in Fig. 2.2. The image of Mutt Glacier is a satellite image (screenshot from Google Earth), and the Damma Glacier image is based on aerial photography (screenshot from map.geo.admin.ch)

Despite a large amount of studies that have reconstructed LIA glacier extents, little is known about the geomorphological and geomorphometric characteristics of the glacier forefields and their genesis over the past century, maybe apart from a few well-studied cases (see Chap. 3).

As a more glaciological background to glacier forefields, this study provides an overview on the response of glaciers to climate change with a focus on the centennial timescale (Sect. 2.2), observation of glacier fluctuations from the ground and from space (Sect. 2.3), the observed glacier changes since the LIA on a global scale and for the Alps (Sect. 2.4), and a discussion on potential future changes and related developments of glacier forefields (Sect. 2.5).

2.2 Glacier Response to Climate Change

2.2.1 Glacier Formation and Climate

As glaciers originate from compressed snow, they can be found where climatic conditions allow snow to survive summer melting and later accumulation over several decades. This results in three main factors required to build a glacier: temperatures must be sufficiently low so that precipitation falls in solid form and accumulates, precipitation must be sufficiently high that snow survives the summer melting (higher amounts can compensate higher temperatures), and there must

Fig. 2.2 Findelen Glacier in Switzerland has a series of further lateral moraines (arrows) outside the dominant moraine from its latest advance in the mid-nineteenth century. The image is a screenshot from Google Earth

be a place for the snow to accumulate. Such conditions are found in mountain and polar regions, whenever terrain is not too steep. Otherwise, snow will be transported downwards by avalanches and either contribute to glacier nourishment or melt. Hence, mountains must be high enough to receive solid precipitation that does not melt; i.e. towards warmer climatic zones, mountains must be higher to be glacier covered. Also, the amount of precipitation and several other factors play a role (e.g. von Humboldt 1845). When it is cold enough (e.g. in Arctic regions), also sparse precipitation can sustain glaciers. In Fig. 2.3, a map of global glacier distribution is shown with the colour-coding representing their mean elevation, as averaged from the global glacier inventory (Pfeffer et al. 2014) over the respective region. Lower mean elevations towards polar and more maritime regions can be clearly recognized.

Once glaciers are in place, their further existence depends on the balance between the mass gained by snowfall at their higher elevations (accumulation region) and the mass loss by melting and probably calving in their lower reaches (ablation zone). Thereby, the ice is constantly transported downwards by glacier flow. The mass balance is variable from year to year, and only long-term trends (decadal scale) in accumulation and/or ablation have finally an impact on glacier geometry. If over such a long period in nearly all years more ice is lost than replaced by glacier flow from the accumulation region, glaciers adjust their size to the new climatic conditions by shrinking (and vice versa for advance). This can take several decades, depending on the amplitude of the governing climate change and the size/slope of the glacier governing its response time (Oerlemans 2001). In effect, the resulting changes in glacier length can be seen as a delayed and smoothed, but also enhanced response to a longer-term (decadal to centennial scale) change of the climate, whereas the annual mass balance reflects the atmospheric

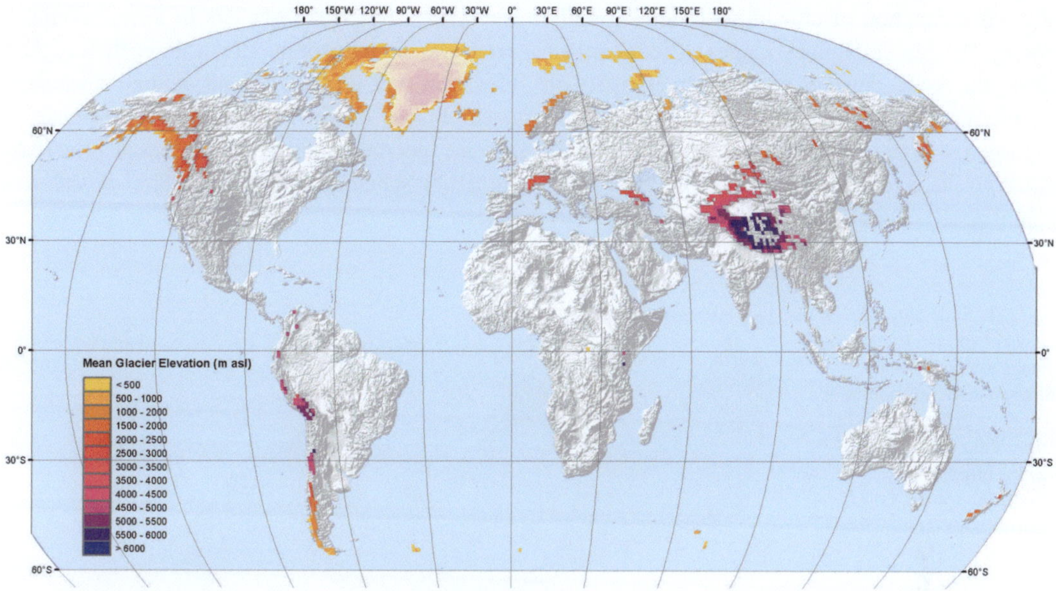

Fig. 2.3 Global Glacier distribution from the Randolph Glacier Inventory with colour-coding giving mean elevation (from Bolch and Christiansen 2014), reproduced with permission

conditions of the respective year and is thus the undelayed and direct response to the atmospheric forcing of that year (with possibly large variability from year to year). Hence, the response of a glacier to climate change (i.e. a trend over several decades) is becoming visible by its advance or retreat, which can be measured in the field and is widely recognized by the public (see Sect. 2.2.3).

2.2.2 Mass Balance Terms (ELA and AAR)

A key indicator for the annual conditions is the so-called equilibrium line (EL) that is dividing the accumulation region (above it) of a glacier from the ablation region (below it) and defined by a zero mass balance (i.e. mass gain equals mass loss). For mid-latitude glaciers, this line is roughly located at the boundary between the snow-covered and the bare ice region on a glacier. At the end of the summer (or the ablation period), the elevation of this boundary gives a good approximation of the annual mass balance as mass loss by melting linearly decreases with elevation for most glaciers (Oerlemans 2001). At a specific altitude of the EL, the mass balance averaged over the entire glacier (also called glacier-wide mass balance) is zero. This is the so-called balanced-budget ELA or ELA_0. If the ELA in a specific year is above this elevation, the mass balance will be negative (or positive if lower). The ELA_0 can be derived from mass balance measurements performed over several years and is approximately located at the mid-point elevation (maximum + minimum elevation / 2) of a glacier (Braithwaite and Raper 2009) where about 60% of the glacier area is above it and 40% below it (Bahr et al. 2009). For tropical and polar glaciers or glaciers mainly nourished by snow avalanches, these simple relations do not apply.

For mid-latitude glaciers (dominated by winter accumulation and summer ablation), the snow-covered area at the end of the ablation period can be used as a proxy for the accumulation area and the elevation of the snowline as a proxy for the EL. For glaciers with a known relationship between mass balance and the EL, snowlines (e.g. derived from satellite data) can be used to estimate glacier mass balance (e.g. Rabatel et al. 2005). Whereas the

ELA_0 is a function of climate and varies greatly among different mountain ranges (Pfeffer et al. 2014), the AAR_0 depends on the characteristics of the glacier and varies only between 0.4 and 0.7 (Dyurgerov et al. 2009). In general, a glacier is adjusting its geometry to climate change until annual AAR values over a longer period resemble the AAR_0. For small glaciers in lesser mountain ranges, an increase of the ELA_0 above their highest elevation will ultimately lead to their disappearance. Figure 2.4 is illustrating the response of glacier geometry to climate change (increase of ELA) schematically for glaciers of different sizes.

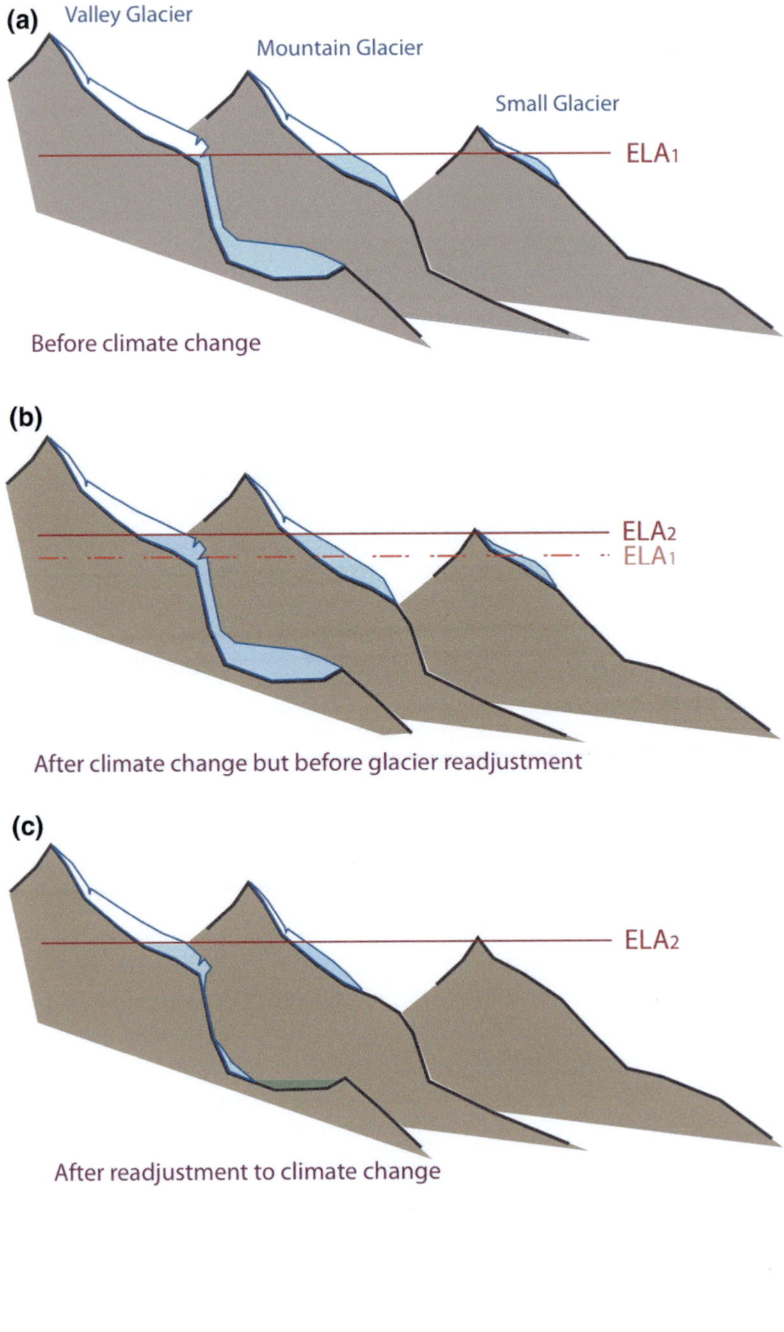

Fig. 2.4 Schematic sketch of glacier response to climate change for glaciers of different sizes in mountain ranges of different heights (from Vaughan et al. 2013). *Source* http://www.climatechange2013.org/report/reports-graphic/ch4-graphics

2.2.3 Glacier Response to Climate Change

The required adjustment to climate change can also be used quantitatively. For example, if we assume that temperature is increasing by one degree and precipitation is about constant, the ELA would approximately increase by about 150 m (given a mean 0.65 K per 100 m temperature lapse rate). To retain this as the mid-point elevation, the glacier terminus needs to shift upwards by about 300 m (twice as much for purely geometric reasons). Assuming that the glacier bed of a larger valley glacier is only weakly inclined (e.g. Linsbauer et al. 2012), for example 15°, the tongue would retreat by about 1.1 km for a one-degree temperature increase. When this retreat occurs over a period of about 100 years, the glacier would retreat 110 m per decade for a temperature increase of 0.1° over the same period. This strong change in extent for a small change in climate is one reason for naming glaciers sensitive climate indicators. Whereas a 0.1 °C increase in temperature over a 10-year period is difficult to measure (and to prove for a sceptic public), a 110 m retreat (11 m a^{-1}) can be accurately measured and is well recognizable by a large public. Indeed, only a few glaciers retreat constantly and the amplitude of the change is smaller for smaller glaciers, but the principle is the same and changes smaller than 11 m a^{-1} can be precisely determined as well.

For glaciers with a known AAR_0, its value can be used in a simple model to determine glacier-specific area changes for a given temperature change (Paul et al. 2007). This requires calculation of the area–elevation distribution (hypsometry) for each glacier and reducing the size of the respective accumulation area by the given shift of the ELA_0. For an assumed AAR_0 of 0.6, the new total area is the size of the resulting new accumulation area divided by 0.6. A simple summing up of the area in each elevation bin from top to bottom until this size is reached provides the new minimum elevation that can be used in a GIS to delete all glacier cells below this elevation. This allows visualizing the new glacier extent and at the same time the region that will get ice-free in future, i.e. the new glacier forefield (Paul et al. 2007). More complex methods have been developed in the past, which also consider an estimated bedrock topography, glacier flow dynamics and a transient climatic forcing (Huss and Hock 2015; Clarke et al. 2015; Radić and Hock 2011). Such models allow consideration of changes in precipitation and other climatic variables that might delay or enhance glacier shrinkage. So far unconsidered in these models is the reduction of melt and the related delay of the response for strongly debris-covered glaciers or the change of the radiation balance due to the geometric change.

The above approach is also inverted; i.e. changes in observed glacier extents are used as a proxy to derive the related changes in ELA and thus climate (e.g. Porter 1975; Seltzer 1994). Methods to determine the position of a former ELA_0 cover a wide range (Benn and Lehmkuhl 2000) and are in particular popular for reconstruction of temperature depressions during the Holocene, e.g. glacier advances in the Younger Dryas period (e.g. Ivy-Ochs et al. 2009; Kerschner and Ivy-Ochs 2008; Solomina et al. 2015). Thereby, the interplay with changes in precipitation (and radiation) is important (see Kuhn 1981 for details) but difficult to estimate for this time period. When applied to global glacier shrinkage from 1850 to the 1970s, the observed glacier retreat of a few kilometres for the largest glaciers and the increase in minimum elevation of a few hundred metres roughly corresponds to a 120–150 m increase of the ELA, thus confirming a related one-degree temperature increase over this period. This value varies from glacier to glacier and region to region, but the global picture is rather homogenous and consistent in this regard (WGMS 2008). From a dynamic point of view, one can note that most glaciers (except the largest ones) adjusted their extents to the warmer climatic conditions after 1850 until their re-advance phases occurred, e.g. in the Alps around the 1920s and 1970s to the 1980s. For the rapid warming after 1985, this adjustment is still ongoing.

A special characteristic of the LIA is its long duration. Despite some intermediate warming, glaciers seem to have been close to a maximum extent for more than 500 years (e.g. Zemp et al. 2011 and references therein). This long duration allowed creation of the huge and impressive moraines we see today in many regions of the world, confining glacier forefields and allowing us a reconstruction of their former extents. However, on a longer-term timescale (last 6000 years) the LIA can be seen as the culmination of a period with increasingly larger glaciers in response to a reduction in northern hemisphere solar insolation during summer (e.g. Wanner et al. 2008; Menounos et al. 2009; Nussbaumer et al. 2011b).

2.3 Glacier Observations from the Ground and Space

2.3.1 Ground-Based Information

Systematic and coordinated monitoring of glacier fluctuations (front variations or length changes) with annual temporal resolution were initiated back in 1893 at the 6th Geological Congress in Zurich with a sample of a few hundred glaciers globally. This sample increased to more than 900 for a few decades (1960 to 1990) and is now back to about 600 (Zemp et al. 2015). For a few dozen glaciers, information on their terminus fluctuations is also available since 1850, but often not with annual resolution (WGMS 2008; Vaughan et al. 2013). Since 1986, the World Glacier Monitoring Service (WGMS) is in charge of collecting and disseminating information on glacier fluctuations (front variations and mass balances) in a standardized format as provided by a global network of national correspondents and observers in the field (WGMS 2015 and earlier versions, wgms.ch/ggcb). Detailed information about the methods used to perform the measurements and the observed fluctuations can be found in WGMS (2008) or online at www.grid.unep.ch/glaciers. With each year being added, the time series becomes more and more valuable for numerous applications, including process understanding, climate change detection, and model calibration.

Detailed reconstructions of earlier glacier fluctuations are based on indirect evidence (such as paintings, dated moraines and dendrochronology) and have been compiled for several regions such as southern Scandinavia (Nussbaumer et al. 2011a), the Canadian Rocky Mountains (e.g. Luckman 2000; Koch et al. 2007), South America (e.g. Jomelli et al. 2009; Masiokas et al. 2009b; Rabatel et al. 2013) and the Alps (Maisch et al. 2000; Zumbühl et al. 2008). For the latter two regions, reconstructions go partly back into the sixteenth century (Zemp et al. 2011 and references therein). Röthlisberger (1986) gives an overview on glacier fluctuations of the past 10,000 years for selected glaciers globally. The focus in this comprehensive book is on dating past glacier extents with dendrochronology to obtain a time series of glacier fluctuations. For some of the glaciers, not only the frontal position (one-dimensional information) but also the lateral extent has been reconstructed, allowing a two-dimensional (2D) view at the former glacier area, at least for their lowermost parts (e.g. Zumbühl and Holzhauser 1988). To date, only a few of the reconstructed 2D LIA maximum extents (and their changes afterwards) are digitally available and most of the information is stored on printed maps (e.g. Grove 2004).

Somewhat more qualitative information about former terminus positions (1D information) has been derived over a period of several thousand years for a couple of glaciers in Norway (Nesje et al. 2008a) and the Swiss Alps using lake sediments and dendrochronology (Hormes et al. 2001; Holzhauser et al. 2005; Joerin et al. 2006). The studies for the Alps indicate that glaciers had a maximum extent similar to the LIA in the sixth century and were somewhat less extensive (the 1920s position) around 2600 years ago. In between these periods, the large valley glaciers had likely been smaller than today. How much smaller is difficult to say as minimum extents seldom provide quantitative evidence. During the middle Holocene (about 6000 to 7000 years ago), glaciers might have even completely disappeared as shown for several regions in Norway (Nesje et al. 2008b) or former glacier forefields

—currently still covered by glaciers—might have allowed trees to grow (Joerin et al. 2008). In contrast, there is evidence from the Alps or the Canadian Rocky Mountains and other places that ice cover at high elevation sites is now smaller than over the past 5500 years (e.g. Baroni and Orombelli 1996; Koch et al. 2014), indicating that the mid-Holocene warm period was followed by a continuously cooler period already mentioned above (Wanner et al. 2008). As pointed out by Koch et al. (2014), this is remarkable given that only a few hundred years before a Holocene maximum extent had been reached. Hence, at the timescale of centuries, glacier forefields are a very dynamic and constantly changing phenomenon.

2.3.2 Space-Based Information

Satellite images provide a different and largely complimentary access to the reconstruction of glacier fluctuations. Compared to ground-based information, they cover a shorter time period (e.g. Landsat Multi-Spectral Scanner (MSS) since 1972, Thematic Mapper (TM) since 1984) and they are not available in each year for a specific region (e.g. due to clouds or acquisition strategy). Furthermore, freely available satellite data with a long record such as Landsat have a limited spatial resolution (MSS: 79 m, TM: 30 m) that only allows following changes with reduced temporal resolution (e.g. every 5–10 years, depending on the rate of change). These disadvantages are compensated by the complete spatial coverage of a region; i.e. *all* glaciers in a region are captured rather than only a few selected ones with easy ground access.

Satellite images are mostly used for mapping of glacier extents (i.e. outlines) rather than front variations, but this has been done as well (see Sect. 2.4.1.1). As the information is available in regions or for time periods that are not covered by ground-based information, satellite data can complement existing data sets in this regard (e.g. Barandun et al. 2015). An important advantage of satellite images is that clean glacier ice can be classified automatically using a simple band ratio (e.g. dividing the red by the short-wave infrared band) and a threshold for segmentation into a binary image map. This map can henceforth be transformed to glacier outlines using raster–vector conversion (Paul 2002, 2015; Bolch and Kamp 2006). For the 30-year period with available Landsat data, such outlines have been generated for many regions in the world for multiple points in time (Bolch et al. 2010a; Paul and Mölg 2014; Narama et al. 2010), forming a base for our understanding of climate change impacts globally.

On the other hand, only a few studies have yet explored the large potential of satellite data to map LIA extents by digitizing moraines and trimlines (Baumann et al. 2009; Citterio et al. 2009; Davies and Glasser 2012; Loibl et al. 2014). This might be related to the time-consuming manual mapping, the limited spatial resolution, the unknown timing of the LIA extent, and because evidence of LIA extents might not be visible for all glaciers, among other factors. At least, the spatial resolution of (freely available) sensors is constantly increasing (now at 10 m with Sentinel 2) and commercial high-resolution images (along with aerial photography) have been used for this purpose as well (e.g. Fischer et al. 2015; Solomina et al. 2016). As an intermediate solution, images in Google Earth can be used to verify the interpretation of freely available but coarser resolution satellite images. As mentioned before, the timing of the maximum extent varies (Rabatel et al. 2008), but is rather homogenous for a larger mountain range as temperature trends (driving the long-term variability) are similar over a few hundred kilometres (Böhm et al. 2001). However, the small-scale variability in precipitation can impact on the magnitude and timing on a shorter (decadal) timescale. As only rates of relative area changes are comparable among different regions, the timing of the mapped extent has to be approximated as good as possible (e.g. from maybe available field measurements of length changes). Overall, the digital LIA extent data set is growing and its current limited availability will likely improve in the future.

2.4 Glacier Changes Since the Little Ice Age

2.4.1 A Global Overview

Numerous local to regional scale studies on glacier changes since the LIA have been performed, but only a few are providing a global overview. The most comprehensive study to date is the book 'Little Ice Age' by Jean M. Grove, published in 1988 (Grove 1988) and in a second edition with two volumes and a new title in 2004 (Grove 2004). As the level of detail provided in this book is far beyond what we can present here, we refer the interested reader to this book and focus here only on some aspects of the book along with the new findings reported in the more recent literature.

2.4.1.1 Length Changes

As described in Sect. 2.3, the most detailed knowledge we have on glacier fluctuations since the LIA is a result of direct measurements of length changes in the field with partly annual resolution for up to a few hundred glaciers globally (e.g. Zemp et al. 2015). Global overviews of the fluctuations can be found in (WGMS 2008, 2015) who include a regional analysis, in the study by Leclercq et al. (2014) who include a detailed statistical analysis of the fluctuations and the source data, and in IPCC AR5 (Vaughan et al. 2013) presenting a more selective and generalized picture. Regional scale analysis of the observed length changes since 1900 (or even earlier) is available for several regions such as for Norway by Andreassen et al. (2005), Italy (Citterio et al. 2007) or the Greater Himalaya (Bolch et al. 2012; Mayewski and Jeschke 1979). The measurements provide the following generalized insights:

- Glaciers retreated from their LIA maximum position in all regions of the world.
- Depending on their size, the terminus retreat is up to a few km for the largest glaciers.
- The general retreat was interrupted by intermediate phases of advance (at different times), in particular for more quickly responding medium-sized mountain glaciers.
- For larger glaciers, the amplitude of the change is also larger (but exceptions exist).
- Some regions have special glaciers that retreat or advance also for non-climatic reasons (e.g. dynamic thinning of marine/lacustrine terminating or rapid advance of surging glaciers).
- Such glaciers can retreat over much larger distances (tens of km) and advance much faster (in a few years) over distances of a few km, i.e. the typical range of LIA fluctuations.
- In case glaciers are covered by thick debris, their retreat can be much slower than for clean glaciers, at least when the terminus is not calving into water.
- Differences in regional trends can be largely explained with different climatic conditions (e.g. increased winter precipitation), whereas local differences can be explained by the specific characteristics of individual glaciers (e.g. size, response time, hypsometry).

For some regions of the world, such as Greenland (Leclercq et al. 2012), Gangotri Glacier in the Himalaya (Bhambri et al. 2012), or selected outlet glaciers of the Patagonian Icefields and the Cordillera Darwin (Lopez et al. 2010), satellite data were used to add further points in time to the length change record of several individual glaciers. The latter study partly revealed dramatic retreats (up to 10 km from about 1980 to 2011) for several larger glaciers (Nef, San Rafael, Uppsala, Jorge Montt, Marinelli), but they are all calving in pro-glacial lakes that likely enhanced their retreat. On the other hand, very high rates of glacier retreat (4 km from 1986 to 2011) and massive shrinkage of larger glaciers have also been documented for several land-terminating glaciers in Chile (Rivera et al. 2012) and northern Patagonia (Paul and Mölg 2014) using satellite data. In this region, the lakes that grow as glaciers retreat often cover the resulting glacier forefields more or less completely.

2.4.1.2 Area Changes

A large number of studies have reconstructed time series of changes in glacier extents since the LIA (cf. Grove 2004). Whereas these studies generally confirm what is known about glacier fluctuations from length changes, their special advantage of also providing two-dimensional information of past glacier extents has not yet been fully exploited, basically as they lack digital availability. Hence, our knowledge is limited to regions where LIA extents have also been derived from topographic maps. Detailed studies of glacier area changes since the LIA have been performed for individual glaciers in the Pyrenees (Cía et al. 2005), the French Alps (Marti et al. 2015), the Italian Alps (Carturan et al. 2014), the Drangajøkull ice cap in Iceland (Brynjólfsson et al. 2015) or the Caucasus (Solomina et al. 2016) and small samples of glaciers around Mt. Kenya (Hastenrath 2005), Jotunheimen in Norway (Baumann et al. 2009), the Pyrenees (González Trueba et al. 2008), the Bavarian Alps (Hagg et al. 2012) or the Wind River Range (DeVisser and Fountain 2015) using field surveys, historic topographic maps, aerial photography and also satellite images.

Samples over entire mountain ranges are available for Austria (Fischer et al. 2015), the Swiss Alps (Maisch et al. 2000) and western Italy (Lucchesi et al. 2014) (see Sect. 2.4.2.1 for results). Satellite data have been used to map LIA extents from trimlines on Baffin Island (Svoboda and Paul 2009), for local glaciers in parts of western Greenland (Citterio et al. 2009), for large parts of Patagonia (Davies and Glasser 2012) or for south-east Tibet (Loibl et al. 2014). The obtained values of area change are not directly comparable as time periods, climatic regimes, glacier characteristics and sample sizes differ, but most of them indicate an area loss of about 30–60% since their mapped 1850 maximum extent.

A most comprehensive and regionalized overview on glacier changes can be found in the GLIMS book (Kargel et al. 2014) along with detailed descriptions of the techniques applied in each region. Vaughan et al. (2013) are giving a more generalized overview of the derived area change rates (in per cent per year) from about 1940 to 2010. This overview study is sorted for the RGI regions and revealed the following main findings:

- In the mean over an entire study region, glaciers decreased in size everywhere.
- Several hundred glaciers melted completely.
- Relative area change rates are similar in most study regions, but there is also variability of the rates within a single RGI region.
- Most of the relative change rates varied between −0.05 and −0.5% per year (1940–2010).
- Several regions also showed higher rates (−0.5 to −1% per year), but for a more recent and shorter time period (1970s–2010).
- The largest shrinkage rates (−1 to −2% per year) are found in the European Alps, the Low Latitudes and western Canada, and the latter over a comparably short period of 15 years.
- The majority of the studies with multi-temporal assessments show an increase in shrinkage rates for the more recent period (but several also have no trends or a slight decrease).

These averaged values confirm the results obtained from the more detailed investigations at individual glaciers, but their changes can be much higher or lower. Considering that relative area changes depend on glacier area (e.g. Paul et al. 2004; Bolch et al. 2010b; Gardent et al. 2014), area changes should always only be compared across entire mountain ranges or for glaciers of the same size class.

2.4.2 Changes in the Alps

2.4.2.1 Length

As mentioned above, the record of directly observed glacier length changes in the Alps is one of the longest on record and most comprehensive worldwide. More than 300 glaciers have been observed over a certain time period, and several glaciers provide more or less (allowing gaps of up to 5 years) continuous time series since 1895 (Zemp et al. 2015). Zemp et al. (2008) provided a generalized Alpine-wide overview of length changes from 1850 to 2000 (along with area changes and mass changes from 1950 to 2000), and Citterio et al. (2007) presented a detailed analysis of length changes in the Italian Alps over the 1908 to 2000 period. These studies indicate stationary or advancing glaciers in the Alps around 1890 (only Zemp et al. 2008), in the 1920s and from the 1970s to 1980s as intermittent phases superimposed on a general trend of retreat (see Fig. 2.5). In general, mostly medium-sized and steep (mountain) glaciers with comparably short response times took part in the advance phases, whereas larger and flat (valley) glaciers with long response times continued their general retreat (Fig. 2.5). For the 1970s advance phase also differences in hypsometry played a role to explain the partly very different responses (strong advance vs. continuous retreat) of neighbouring glaciers (Kuhn et al. 1985).

Cumulative retreats reached about 1–2.5 km for the largest glaciers (e.g. Great Aletsch, Gorner, Pasterze), and advances reached several hundred metres for some strongly reacting mountain glaciers such as Trient in the Mont Blanc region and Kesselwand in the Öetztaler Alps. Whereas the specific reasons for the advances of individual glaciers are complex and depending on many factors, atmospheric conditions acting on a mountain range scale play a role as well. Lüthi and Bauder (2010) used the dense time series of length change measurements in Switzerland for a reconstruction of the corresponding variability in the ELA with a macroscopic glacier flow model. They found that a single time series of ELA fluctuations could explain the observed terminus fluctuations after 1850 but not the variability during the LIA. In general, the advances of the 1920s and 1970s to the 1980s seem to follow several years of balanced or positive mass budgets (see Sect. 4.2.3) with a decade or so, compliant with the typical response times of steep mountain glaciers (Haeberli and Hoelzle 1995).

2.4.2.2 Area

Reconstruction of LIA maximum glacier extents in the Swiss Alps is largely based on comparably accurate topographic maps that have been derived from land surveys in the period 1845 to 1865 and are known as the Dufourmap (swisstopo.ch). Together with field surveys and high-resolution aerial photography (for confirmation of moraines), Maisch et al. (2000) used this information for a precise reconstruction of maximum LIA extents for all Swiss glaciers. This data set was at first only prepared in analogue form and thus analysed using cartographic and planimetric techniques (Maisch 1992). However, later glacier extents were also digitized

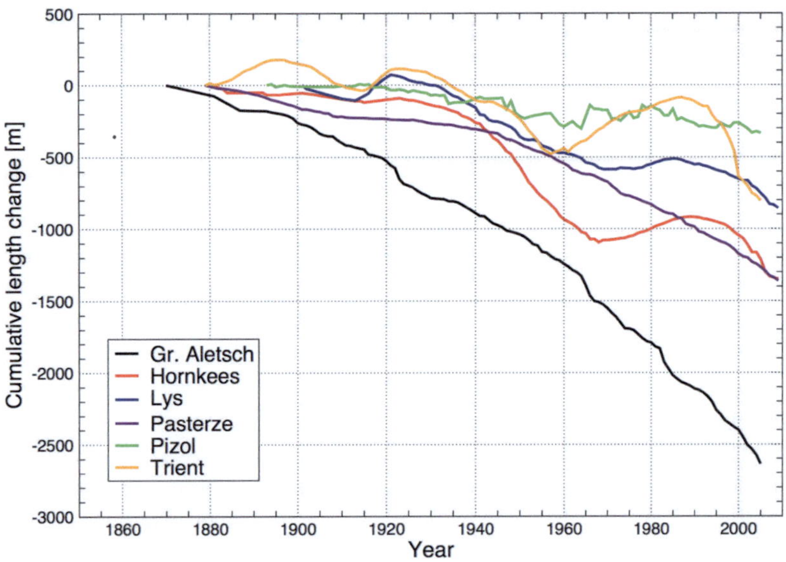

Fig. 2.5 Glacier length changes since the end of the LIA for selected glaciers in the Alps (source data: WGMS). For a global overview, see Vaughan et al. (2013) that is also available online at http://www.climatechange2013.org/report/reports-graphic/ch4-graphics

together with extents from the 1973 Swiss Glacier Inventory (Paul 2004) and analysed digitally in regard to glacier area changes (Paul et al. 2004). More recently, LIA extents were also digitized and analysed for Austria (Fischer et al. 2015) and western Italy (Lucchesi et al. 2014).

The above studies reveal an area loss of about 30–40% (or 0.3%/a) from 1850 to the 1970s, a further 15–20% (or 0.6%/a) from 1973 until 2000 (that mostly occurred after 1985), a further 10–15% from 2000 to 2010 (or 1.2%/a) and a total of 50–60% for the entire period (Zemp et al. 2008; Fischer et al. 2015; Maisch et al. 2000). The extrapolation by Zemp et al. (2008) for the entire Alps gives a total area change from 4400 km^2 in 1850 to 2900 km^2 in the 1970s and a further reduction to 2200 km^2 by 2000. Hence, an area of about 2200 km^2 has transformed from being glacier covered to glacier forefield in 150 years. Gardent et al. (2014) compiled from the body of the literature relative area change rates for various regions in the Alps and found rates of about −1.5% per year for the 1990–2010 period and −2 and −3% for the most recent decade. Hence, area loss rates have seemingly accelerated recently, indicating that glaciers are not yet in balance with current climatic conditions and that non-climatic feedback effects due to surface lowering, and disintegration might play a role (e.g. Carturan et al. 2013b).

2.4.2.3 Mass Balance

For the Alps, we have also a rather good overview on the temporal development of mass balances, as several long-term time series are available (e.g. WGMS 2015). Huss et al. (2010) extrapolated annual mass balances of 30 Swiss glaciers back to 1900 using climate data, direct measurements and the available geodetic surveys for calibration. This reconstruction shows a period of mass gain from 1910 to 1920 and 1970 to 1980, rather balanced conditions between 1950 and 1970 and periods of strong mass loss from 1940 to 1950, and since 1985. However, strong interannual fluctuations were observed in each of the periods after 1950 (Zemp et al. 2015), indicating that some longer-term averaging (5–10 years) is required for trend detection. When analysing 10-year means, mean annual mass budgets in the Alps are getting increasingly negative since 1980, about −0.33 m w.e. for 1980–1989, −0.81 m w.e. for 1990–1999 and −1.25 m w.e. for 2000–2009. This increasingly negative trend reveals that glacier extents cannot adjust quickly enough and ablation areas get increasingly larger (seasonal mass balance measurements indicate that there is no trend in winter precipitation) combined with continued surface lowering. Under current climatic conditions, this process will continue for decades, in particular for the larger glaciers with long response times. With current AARs of about 30%, total areas have to be diminished by a further 50% to get again an AAR_0 of 0.6 and stable glacier extents (Zemp et al. 2015; Dyurgerov et al. 2009; Carturan et al. 2013a).

2.5 Discussion and Future Challenges

2.5.1 Length and Area Changes

Although glacier area and length are correlated, changes in the respective parameters provide complimentary information. This also applies to their methods of determination (often field-based for length vs. remote sensing for area), their spatial sampling (individual vs. all glaciers in a mountain range), period of available observations (centuries vs. decades) and temporal resolution (annual vs. decadal). When combined, a most complete picture of climate change impacts on glaciers and our environment can be derived (e.g. Vaughan et al. 2013). However, it has to be noted that area changes are per se not really a good climate indicator, as the relation to climate change is complex and depending on several other factors (e.g. ice thickness distribution, hypsometry, debris cover). This is different for length changes that can be well reconstructed from climate data using glacier flow models and vice versa (e.g. Jouvet et al. 2009).

But there is also some climatic information in area changes: they can be globally compared, show the potential variability among and within regions (Vaughan et al. 2013) or help in interpreting current trends (see Sect. 2.4.2.1) and recognize potential accelerations (e.g. Paul et al. 2004). Area change rates are useful to extrapolate glacier inventories to other time periods, for example to close observational gaps (Zemp et al. 2008) or initialization of climate models for a given time (Kotlarski et al. 2010). Once outlines from different points in time (e.g. LIA and today) are digitally available, numerous additional calculations can be performed, for example the related volume changes and sea-level contribution (Glasser et al. 2011), the impacts of geometric changes on measured and modelled mass budgets (Paul 2010; Huss et al. 2012), length changes from digital intersection of extents with central flow lines, changes in mean elevation as a proxy for ELA and thus former climatic conditions (Wolken et al. 2008), and calibration of models simulating past and future glacier extents (Marzeion et al. 2012; Huss and Hock 2015). Hence, despite their limited use as a direct indicator of climate change, multi-temporal glacier extents are very useful for numerous calculations related to climate change and a globally more complete data set of digitally available extents (back to the LIA) is highly desirable (Marzeion et al. 2017).

2.5.2 Glacier Forefields

Owing to the committed future area loss and the very likely further increasing global temperatures, glacier forefields will further grow. The new terrain will consist not only of unconsolidated debris but also of bare rock and lakes. On the positive side, it will in some cases (at elevations below the tree line) only take a few decades before plants stabilize the debris (Burga 1999) and new lakes might develop and get touristic attention and value. However, shrinking glaciers and expanding forefields also have some downsides. Apart from the increasing potential for debris flows due to the reduced retention of rain by glaciers and the increasing area of steep lateral moraines (see Chaps. 10 and 11) and bare rocks (for rock slope instability see Chaps. 8 and 9), there are several further likely negative impacts. For example, access to mountain huts might get difficult when glaciers shrink and the terrain follows or when the glacier is replaced by a lake or a steep moraine wall (Ritter et al. 2012).

Pro-glacial lakes (Chap. 14) might also be a growing hazard as the dams holding them back can burst creating so-called glacier lake outburst floods (GLOFs) (Richardson and Reynolds 2000; Bolch et al. 2011). Such bursts can be triggered by mass movements (rockfall), which might occur as a result of reduced buttressing by the now missing glacier (Oppikofer et al. 2008; Kos et al. 2016), from steep moraine walls and thawing permafrost above such a lake (Frey et al. 2010; Haeberli et al. 2017). The GLOFs can travel downstream for dozens of kilometres, thus reaching and potentially damaging human infrastructure or settlements (e.g. Clague and Evans 2000; Huggel et al. 2003). On a positive note, these lakes can be automatically mapped from satellite images and their extents are thus digitally available (e.g. Bolch et al. 2008; Gardelle et al. 2011; ICIMOD 2011). In the case they are classified as dangerous, they are sometimes also monitored from the ground so that early warnings can be released (e.g. Carey et al. 2012). A special problem is that in many regions glacier shrinkage is so fast that new lakes are constantly formed, and a frequent reassessment is required. Such lakes might also have a potential to be used for hydropower purposes (Haeberli et al. 2016a), or they can build an important sediment trap (at least for some years) that keeps the sediment away from hydropower dams downstream (Haeberli et al. 2012). In short, growing glacier forefields can influence the sediment balance in both ways, strongly increasing it (no lake) or reducing it.

2.5.3 Future Challenges

Current rates of glacier area loss are in some regions (Alps, Andes) an order of magnitude higher than in the decades following the LIA maximum extent. Current 2D/3D glacier flow models start to correctly model also changes in glacier area on a regional scale (Huss and Hock 2015; Clarke et al. 2015), but this is highly challenged due to missing information about the bedrock or ice thickness distribution as well as missing multi-temporal validation data. However, for some well-studied glaciers such as Rhone or Great Aletsch in Switzerland modelling results (for past extents) are in good agreement with observations (e.g. Jouvet et al. 2011). Not yet included are effects on the radiation balance (replacing bright glacier ice with dark gravel or rock) that might cause additional heating in case glaciers shrink (Carturan et al. 2013a). This would enhance glacier melt, might change snowfall to rain and increase thawing of permafrost higher up (Gobiet et al. 2014). The resulting overall destabilization of the environment towards a state of imbalance will challenge computation of related effects and cause more severe and unforeseen impacts of extreme events due to missing historic knowledge and nonlinear effects in cascading response systems.

2.6 Summary

We described basic principles of glacier response to climate change, glacier changes as measured in the field and from space, and presented an overview of glacier fluctuations around the world with a particular focus on the Alps. As not all details could be included in this overview, we focused on key references under each topic and the more general relations and results. With the ongoing glacier retreat, glacier forefields are further growing. They are continuously changing on timescales of minutes (flooding) to millennia and record past glacier fluctuations. Detailed investigations about this very dynamic environment exist only locally, and a quantification of the new land that has been exposed after glacier retreat is only regionally available, as the digital mapping of LIA extents started only recently. Nevertheless, simplified schemes allow reconstruction of the regional climatic history using ELA depressions derived from the topographic analysis of former glacier extents. For the time after the LIA, the observed glacier retreat of a few kilometres for the largest glaciers is in good agreement with the required increase in temperature of about one degree since the LIA. As all glaciers are different, such numbers show substantial glacier-specific variability, but consideration of such differences provides consistent results. The pro-glacial lakes that are currently growing in many mountain regions of the world as glaciers retreat are both a source of hazards (GLOFs) and a landscape element with touristic and economic (hydropower) potential. It remains to be seen which of these elements will prevail.

Acknowledgements This study has been funded by the ESA project Glaciers_cci (4000109873/14/I-NB). We would like to thank an anonymous reviewer and S. Nussbaumer for their helpful comments to improve this contribution.

References

Andreassen LM, Elvehøy H, Kjøllmoen B, Engeset RV, Haakensen N (2005) Glacier mass balance and length variations in Norway. Ann Glaciol 42:317–325

Bahr DB, Dyurgerov M, Meier MF (2009) Sea-level rise from glaciers and ice caps: A lower bound. Geophys Res Lett 36(3):L03501. https://doi.org/10.1029/2008GL036309

Barandun M, Huss M, Sold L, Farinotti D, Asisov E, Salzmann N, Usubaliev R, Merkushkin A, Hoelzle M (2015) Re-analysis of seasonal mass balance at Abramov glacier 1968–2014. J Glaciol 61 (230):1103–1117

Baroni C, Orombelli G (1996) The Alpine "iceman" and holocene climatic change. Quat Res 46(1):78–83

Baumann S, Winkler S, Andreassen LM (2009) Mapping glaciers in Jotunheimen, South-Norway, during the "Little Ice Age" maximum. Cryosphere 3(2):231–243

Beniston M (2006) Mountain weather and climate: a general overview and a focus on climatic change in the Alps. Hydrobiologia 562:3–16

Benn DI, Lehmkuhl F (2000) Mass balance and equilibrium-line altitudes of glaciers in high-mountain environments. Quatern Int 65/66:15–29

Bhambri R, Bolch T, Chaujar RK (2012) Frontal recession of Gangotri Glacier, Garhwal Himalayas, from 1965 to 2006, measured through high resolution remote sensing data. Curr Sci India 102(3):489–494

Böhm A, Auer I, Brunetti M, Maugueri M, Nanni T, Schöner W (2001) Regional temperature variability in the European Alps: 1760–1998 from homogenized instrumental time series. Int J Climatol 21(14):1779–1801

Bolch T, Chrsitiansen H (2014) Large scale settings: the physiography of the cryosphere. In: Haeberli W, Whiteman C (eds) Snow and ice-related hazards, risks and disasters. Elsevier, New York, pp 201–217
Bolch T, Kamp U (2006) Glacier mapping in high mountains using DEMs, landsat and ASTER data. Grazer Schriften der Geographie und Raumforschung 41. In: Proceedings of the 8th Int. Symp. on High Mountain Remote Sensing Cartography, 20–27 March 2005, La Paz, Bolivia, pp. 13–24
Bolch T, Buchroithner MF, Peters J, Baessler M, Bajracharya SR (2008) Identification of glacier motion and potentially dangerous glacier lakes at Mt. Everest area/Nepal using spaceborne imagery. Nat Hazards Earth. Syst Sci 8(6):1329–1340
Bolch T, Menounos B, Wheate RD (2010a) Landsat-based inventory of glaciers in western Canada, 1985–2005. Remote Sens Environ 114(1):127–137
Bolch T, Yao T, Kang S, Buchroithner MF, Scherer D, Maussion F, Huintjes E, Schneider C (2010b) A glacier inventory for the western Nyainqentanglha Range and Nam Co Basin, Tibet, and glacier changes 1976–2009. Cryosphere 4:419–433
Bolch T, Peters J, Pradhan B, Yegorov AB, Buchroithner MF, Blagoveshchenskiy VP (2011) Identification of potentially dangerous glacial lakes in the northern Tien Shan. Nat Hazards 59(3):1691–1714
Bolch T, Kulkarni A, Kääb A, Huggel C, Paul F, Cogley JG, Frey H, Kargel JS, Fujita K, Scheel M, Bajracharya S, Stoffel M (2012) The state and fate of Himalayan glaciers. Science 336(6079):310–314
Braithwaite RJ, Raper S (2009) Estimating equilibrium-line altitude (ELA) from glacier inventory data. Ann Glaciol 50(53):127–132
Brynjólfsson S, Schomacker A, Guðmundsdóttir ER, Ingólfsson Ó (2015) A 300-year surge history of the Drangajökull ice cap, northwest Iceland, and its maximum during the 'Little Ice Age'. Holocene 25 (7):1076–1092
Burga C (1999) Vegetation development on the glacier forefield Morteratsch (Switzerland). Appl Veg Sci 2:17–24
Carey M, Huggel C, Bury J, Portocarrero C, Haeberli W (2012) An integrated socio-environmental framework for glacier hazard management and climate change adaptation: lessons from Lake 513, Cordillera Blanca, Peru. Clim Change 112(3):733–767
Carturan L, Baroni C, Becker M, Bellin A, Cainelli O, Carton A, Casarotto C, Dalla Fontana G, Godio A, Martinelli T, Salvatore MC, Seppi R (2013a) Decay of a long-term monitored glacier: Careser Glacier (Ortles-Cevedale, European Alps). Cryosphere 7 (6):1819–1838
Carturan L, Filippi R, Seppi R, Gabrielli P, Notarnicola C, Bertoldi L, Paul F, Rastner P, Cazorzi F, Dinale R, Dalla Fontana G (2013b) Area and volume loss of the glaciers in the Ortles-Cevedale group (Eastern Italian Alps): controls and imbalance of the remaining glaciers. Cryosphere 7(5):1339–1359
Carturan L, Baroni C, Carton A, Cazorzi F, Fontana GD, Delpero C, Salvatore MC, Seppi R, Zanoner T (2014) Reconstructing fluctuations of La Mare Glacier (Eastern Italian Alps) in the Late Holocene: new evidence for a Little Ice Age maximum around 1600 AD. Geogr Ann A 96(3):287–306
Chinn T, Winkler S, Salinger MJ, Haakensen N (2005) Recent glacier advances in Norway and New Zealand: a comparison of their glaciological and meteorological causes. Geogr Ann A 87(1):141–157
Cía JC, Andrés AJ, Sánchez MS, Novau JC, Moreno JL (2005) Responses to climatic changes since the Little Ice Age on Maladeta Glacier (Central Pyrenees). Geomorphology 68(3–4):167–182
Citterio M, Diolaiuti G, Smiraglia C, D'Agata C, Carnielli T, Stella G, Siletto GB (2007) The fluctuations of Italian glaciers during the last century: a contribution to the knowledge about alpine glacier changes. Geogr Ann A 89(3):167–184
Citterio M, Paul F, Ahlstrøm AP, Jepsen HF, Weidick A (2009) Remote sensing of glacier change in West Greenland: accounting for the occurrence of surge-type glaciers. Ann Glaciol 50(53):70–80
Clague JJ, Evans SG (2000) A review of catastrophic drainage of moraine-dammed lakes in British Columbia. Quat Sci Rev 19:1763–1783
Clarke GKC, Jarosch AH, Anslow FS, Radic V, Menounos B (2015) Projected deglaciation of western Canada in the twenty-first century. Nature Geosci 8 (5):372–377
Davies BJ, Glasser NF (2012) Accelerating shrinkage of Patagonian glaciers from the Little Ice Age (AD 1870) to 2011. J Glaciol 58(212):1063–1084
DeVisser MH, Fountain AG (2015) A century of glacier change in the Wind River Range, WY. Geomorphology 232:103–116
Dyurgerov M, Meier MF, Bahr DB (2009) A new index of glacier area change: a tool for glacier monitoring. J Glaciol 55(192):710–716
Egli M, Wernli M, Kneisel C, Biegger S, Haeberli W (2006) Melting glaciers and soil development in the proglacial area Morteratsch (Swiss Alps): II. Modeling the present and future soil state. Arct Antarct Alp Res 38(4):510–521
Fischer A, Seiser B, Stocker Waldhuber M, Mitterer C, Abermann J (2015) Tracing glacier changes in Austria from the Little Ice Age to the present using a lidar-based high-resolution glacier inventory in Austria. Cryosphere 9:753–766
Frey H, Haeberli W, Linsbauer A, Huggel C, Paul F (2010) A multi-level strategy for anticipating future glacier lake formation and associated hazard potentials. Nat Hazards Earth Syst Sci 10:339–352
Gardelle J, Arnaud Y, Berthier E (2011) Contrasted evolution of glacial lakes along the Hindu Kush Himalaya mountain range between 1990 and 2009. Global Planet Change 75:47–55
Gardent M, Rabatel A, Dedieu J, Deline P (2014) Multitemporal glacier inventory of the French Alps

from the late 1960s to the late 2000s. Global Planet Change 120:24–37
Glasser NF, Harrison S, Jansson KN, Anderson K, Cowley A (2011) Global sea-level contribution from the Patagonian Icefields since the Little Ice Age maximum. Nat Geosci 4(11):303–307
Gobiet A, Kotlarski S, Beniston M, Heinrich G, Rajczak J, Stoffel M (2014) 21st century climate change in the European Alps—a review. Sci Total Environ 493:1138–1151.
González Trueba JJ, Moreno RM, Martínez de Pisón E, Serrano E (2008) 'Little Ice Age' glaciation and current glaciers in the Iberian Peninsula. Holocene 18 (4):551–568
Grove JM (1988) The Little Ice Age. Methuen, London and New York
Grove JM (2004) Little Ice Ages: ancient and modern, 2nd edn. Routledge, London
Haeberli W, Hoelzle M (1995) Application of inventory data for estimating characteristics of and regional climate-change effects on mountain glaciers: a pilot study with the European Alps. Ann Glaciol 21:206–212
Haeberli W, Schleiss A, Linsbauer A, Künzler M, Bütler M (2012) Gletscherschwund und neue Seen in den Schweizer Alpen. Wasser Energie Luft 104(2):93–102
Haeberli W, Buetler M, Huggel C, Friedli TL, Schaub Y, Schleiss AJ (2016a) New lakes in deglaciating high-mountain regions – opportunities and risks. Clim Change 139(2):201–214
Haeberli W, Linsbauer A, Cochachin A, Salazar C, Fischer UH (2016b) On the morphological characteristics of overdeepenings in high-mountain glacier beds. Earth Surf Process Landforms 41(13):1980–1990
Haeberli W, Schaub Y, Huggel C (2017) Increasing risks related to landslides from degrading permafrost into new lakes in de-glaciating mountain ranges. Geomorphology 293:405–417
Hagg W, Mayer C, Mayr E, Heilig A (2012) Climate and glacier fluctuations in the Bavarian Alps in the past 120 years. Erdkunde 66(2):121–142
Hastenrath S (2005) The glaciers of Mount Kenya 1899–2004. Erdkunde 59(2):120–125
Heckmann T, McColl S, Morche D (2016) Retreating ice: research in pro-glacial areas matters. Earth Surf Process Land 41(2):271–276
Holzhauser H, Magny M, Zumbuühl HJ (2005) Glacier and lake-level variations in west-central Europe over the last 3500 years. Holocene 15(6):789–801
Hormes A, Müller BU, Schlüchter C (2001) The Alps with little ice: evidence for eight Holocene phases of reduced glacier extent in the Central Swiss Alps. Holocene 11(3):255–265
Huggel C, Kääb A, Haeberli W, Krummenacher B (2003) Regional-scale GIS-models for assessment of hazards from glacier lake outbursts: evaluation and application in the Swiss Alps. Nat Hazards Earth Syst Sci 3:647–662
Huss M, Hock R (2015) A new model for global glacier change and sea-level rise. Front Earth Sci 3:54. https://doi.org/10.3389/feart.2015.00054
Huss M, Hock R, Bauder A, Funk M (2010) 100-year mass changes in the Swiss Alps linked to the Atlantic Multidecadal Oscillation. Geophys Res Lett 37(10): L10501. https://doi.org/10.1029/2010GL042616
Huss M, Hock R, Bauder A, Funk M (2012) Conventional versus reference-surface mass balance. J Glaciol 58 (208):278–286
ICIMOD (2011) Glacial lakes and glacial lake outburst floods in Nepal. Kathmandu
Ivy-Ochs S, Kerschner H, Maisch M, Christl M, Kubik PW, Schlüchter C (2009) Latest Pleistocene and Holocene glacier variations in the European Alps. Holocene and latest pleistocene Alpine Glacier fluctuations: a global perspective. Quat Sci Rev 28(21–22):2137–2149
Joerin UE, Stocker TF, Schlüchter C (2006) Multicentury glacier fluctuations in the Swiss Alps during the Holocene. Holocene 16(5):697–704
Joerin UE, Nicolussi K, Fischer A, Stocker TF, Schlüchter C (2008) Holocene optimum events inferred from subglacial sediments at Tschierva Glacier, Eastern Swiss Alps. Quat Sci Rev 27(3–4):337–350
Jomelli V, Favier V, Rabatel A, Brunstein D, Hoffmann G, Francou B (2009) Fluctuations of glaciers in the tropical Andes over the last millennium and palaeoclimatic implications: a review. Palaeogeogr Palaeocl 281(3–4):269–282
Jouvet G, Huss M, Blatter H, Picasso M, Rappaz J (2009) Numerical simulation of Rhonegletscher from 1874 to 2100. J Comput Phys 228(17):6426–6439
Jouvet G, Huss M, Funk M, Blatter H (2011) Modelling the retreat of Grosser Aletschgletscher, Switzerland, in a changing climate. J Glaciol 57(206):1033–1045
Kargel JS, Leonard GJ, Bishop MP, Kääb A, Raup BH (eds) (2014) Global land ice measurements from space. Springer, Berlin Heidelberg
Kerschner H, Ivy-Ochs S (2008) Palaeoclimate from glaciers: examples from the Eastern Alps during the Alpine Lateglacial and early Holocene. Historical and Holocene glacier – climate variations. Global Planet Change 60(1–2):58–71
Koch J, Clague JJ, Osborn G (2007) Glacier fluctuations during the last millennium in Garibaldi Provincial Park, southern Coast Mountains, British Columbia. Can J Earth Sci 44:1215–1233
Koch J, Clague JJ, Osborn G (2014) Alpine glaciers and permanent ice and snow patches in Western Canada approach their smallest sizes since the mid-Holocene, consistent with global trends. Holocene 24(12):1639–1648
Kos A, Amann F, Strozzi T, Delaloye R, von Ruette J, Springman S (2016) Contemporary glacier retreat triggers a rapid landslide response, Great Aletsch Glacier, Switzerland. Geophys Res Lett 43:12466–12474. https://doi.org/10.1002/2016GL071708

Kotlarski S, Jacob D, Podzun R, Paul F (2010) Representing glaciers in a Regional Climate Model. Clim Dyn 34(1):27–46
Kuhn M (1981) Climate and glaciers. IAHS Publ 131:3–20
Kuhn M, Markl G, Kaser G, Nickus U, Obleitner F, Schneider H (1985) Fluctuations of climate and mass balance:different responses of two adjacent glaciers. Z Gletscherkd Glazialgeol 21:409–416
Leclercq PW, Weidick A, Paul F, Bolch T, Citterio M, Oerlemans J (2012) Historical glacier length changes in West Greenland. Cryosphere 6:1339–1343
Leclercq PW, Oerlemans J, Basagic HJ, Bushueva I, Cook AJ, Le Bris R (2014) A data set of worldwide glacier length fluctuations. Cryosphere 8(2):659–672
Linsbauer A, Paul F, Haeberli W (2012) Modeling glacier thickness distribution and bed topography over entire mountain ranges with GlabTop: application of a fast and robust approach. J Geophys Res 117:F03007. https://doi.org/10.1029/2011JF002313
Loibl D, Lehmkuhl F, Grießinger J (2014) Reconstructing glacier retreat since the Little Ice Age in SE Tibet by glacier mapping and equilibrium line altitude calculation. Geomorphology 214:22–39
Lopez P, Chevallier P, Favier V, Pouyaud B, Ordenes F, Oerlemans J (2010) A regional view of fluctuations in glacier length in southern South America. Global Planet Change 71(1–2):85–108
Loriaux T, Casassa G (2013) Evolution of glacial lakes from the Northern Patagonian Icefield and terrestrial water storage in a sea-level rise context. Global Planet Change 102:33–40
Lucchesi S, Fioraso G, Bertotto S, Chiarle M (2014) Little Ice Age and contemporary glacier extent in the Western and South-Western Piedmont Alps (North-Western Italy). J Maps 10(3):409–423
Luckman BH (2000) The Little Ice Age in the Canadian rockies. Geomorphology 32(3):357–384
Lüthi M, Bauder A (2010) Analysis of alpine glacier length change records with a macroscopic glacier model. Geogr Helv 2:92–102
Maisch M (1992) Die Gletscher Graubündens. Rekonstruktion und Auswertung der Gletscher und deren Veränderung seit dem Hochstand von 1850 im Gebiet der östlichen Schweizer Alpen (Bündnerland und angrenzende Regionen). Teil A: Grundlagen, Analysen, Ergebnisse; Teil B: Verzeichnisse, Datenkataloge, Gletscherkarten. Physische Geographie Zurich
Maisch M, Wipf A, Denneler B, Battaglia J, Benz C (2000) Die Gletscher der Schweizer Alpen - Gletscherhochstand 1850 - Aktuelle Vergletscherung - Gletscherschwundszenarien, Schlussbericht NFP 31, 2nd edn. Hochschulverlag ETH Zürich
Marti R, Gascoin S, Houet T, Ribière O, Laffly D, Condom T, Monnier S, Schmutz M, Camerlynck C, Tihay JP, Soubeyroux JM, René P (2015) Evolution of Ossoue glacier (French Pyrenees) since the end of the Little Ice Age. Cryosphere 9(5):1773–1795
Marzeion B, Jarosch AH, Hofer M (2012) Past and future sea-level change from the surface mass balance of glaciers. Cryosphere 6(6):1295–1322
Marzeion B, Champollion N, Haeberli W, Langley K, Leclercq P, Paul F (2017) Observation of glacier mass changes on the global scale and its contribution to sea level change. Surv Geophys 38(1):105–130
Masiokas MH, Luckman BH, Villalba R, Delgado S, Skvarca P, Ripalta A (2009a) Little Ice Age fluctuations of small glaciers in the Monte Fitz Roy and Lago del Desierto areas, south Patagonian Andes. Argentina. Palaeogeogr Palaeocl 281(3–4):351–362
Masiokas MH, Rivera A, Espizua LE, Villalba R, Delgado S, Aravena JC (2009b) Glacier fluctuations in extratropical South America during the past 1000 years. Palaeogeogr Palaeocl 281(3–4):242–268
Matthews JA, Briffa KR (2005) The 'Little Ice Age': re-evaluation of an evolving concept. Geogr Ann A 87 (1):17–36
Mayewski PA, Jeschke PA (1979) Himalayan and Trans-Himalayan Glacier Fluctuations Since AD 1812. Arctic Alpine Res 11(3):267–287
Menounos B, Osborn G, Clague JJ, Luckman BH (2009) Latest Pleistocene and Holocene glacier fluctuations in western Canada. Quat Sci Rev 28(21–22):2049–2074
Narama C, Kääb A, Duishonakunov M, Abdrakhmatov K (2010) Spatial variability of recent glacier area changes in the Tien Shan Mountains, Central Asia, using Corona (1970), Landsat (2000), and ALOS (2007) satellite data. Global Planet Change 71(1–2):42–54
Nesje A, Bakke J, Dahl SO, Lie Ø, Matthews JA (2008a) Norwegian mountain glaciers in the past, present and future. Historical and Holocene glacier – climate variations. Global Planet Change 60(1–2):10–27
Nesje A, Dahl SO, Thun T, Nordli Ø (2008b) The 'Little Ice Age' glacial expansion in western Scandinavia: summer temperature or winter precipitation? Clim Dyn 30(7):789–801
Nussbaumer SU, Zumbühl HJ, Steiner D (2007) Fluctuations of the Mer de Glace (Mont Blanc area, France) AD 1500–2050. Part I: The history of the Mer de Glace AD 1570–2003 according to pictorial and written documents. Z Gletscherkunde Glazialgeol 40:5–140
Nussbaumer SU, Nesje A, Zumbühl HJ (2011a) Historical glacier fluctuations of Jostedalsbreen and Folgefonna (Southern Norway) reassessed by new pictorial and written evidence. Holocene 21(3):455–471
Nussbaumer SU, Steinhilber F, Trachsel M, Breitenmoser P, Beer J, Blass A, Grosjean M, Hafner A, Holzhauser H, Wanner H, Zumbühl HJ (2011b) Alpine climate during the Holocene: a comparison between records of glaciers, lake sediments and solar activity. J Quat Sci 26(7):703–713
Oerlemans J (2001) Glaciers and climate change. Balkema Lisse
Oerlemans J (2005) Extracting a climate signal from 169 glacier records. Science 308:675–677

Oppikofer T, Jaboyedoff M, Keusen H (2008) Collapse at the eastern Eiger flank in the Swiss Alps. Nat Geosci 1 (8):531–535

Paul F (2002) Changes in glacier area in Tyrol, Austria, between 1969 and 1992 derived from Landsat 5 TM and Austrian Glacier Inventory data. Int J Remote Sens 23(4):787–799

Paul F (2004) The new Swiss glacier inventory 2000 - application of remote sensing and GIS. Dissertation, Physical Geography, University of Zurich, Zurich

Paul F (2010) The influence of changes in glacier extent and surface elevation on modeled mass balance. Cryosphere 4(4):569–581

Paul F (2015) Kartierung von Gletschern mit Satellitendaten und das globale Gletscherinventar. In: Lozán JL, Grassl H, Kasang D, Notz D Escher-Vetter H (Hrsg.): Warnsignal Klima: Das Eis der Erde (Kap. 4.1):103–110

Paul F, Haeberli W (2008) Spatial variability of glacier elevation changes in the Swiss Alps obtained from two digital elevation models. Geophys Res Lett 35: L21502. https://doi.org/10.1029/2008GL034718

Paul F, Kääb A (2005) Perspectives on the production of a glacier inventory from multispectral satellite data in Arctic Canada: Cumberland Peninsula, Baffin Island. Ann Glaciol 42:59–66

Paul F, Mölg N (2014) Hasty retreat of glaciers in northern Patagonia from 1985 to 2011. J Glaciol 60 (224):1033–1043

Paul F, Kääb A, Maisch M, Kellenberger T, Haeberli W (2004) Rapid disintegration of Alpine glaciers observed with satellite data. Geophys Res Lett 31 (21):L21402. https://doi.org/10.1029/2004GL020816

Paul F, Maisch M, Rothenbüler C, Hoelzle M, Haeberli W (2007) Calculation and visualisation of future glacier extent in the Swiss Alps by means of hypsographic modelling. Global Planet Change 55(4):343–357

Pelto MS (2010) Forecasting temperate alpine glacier survival from accumulation zone observations. Cryosphere 4(1):67–75

Pfeffer W, Arendt AA, Bliss A, Bolch T, Cogley JG, Gardner AS, Hagen J, Hock R, Kaser G, Kienholz C, Miles ES, Moholdt G, Mölg N, Paul F, Radić V, Rastner P, Raup BH, Rich J, Sharp MJ, The Randolph Consortium (2014) The Randolph Glacier Inventory: a globally complete inventory of glaciers. J Glaciol 60 (221):537–552

Porter SC (1975) Equilibrium-line altitudes of late Quaternary glaciers in the Southern Alps, New Zealand. Quat Res 5(1):27–47

Rabatel A, Dedieu J, Vincent C (2005) Using remote-sensing data to determine equilibrium-line altitude and mass-balance time series: validation on three French glaciers. J Glaciol 51(175):539–546

Rabatel A, Francou B, Jomelli V, Naveau P, Grancher D (2008) A chronology of the Little Ice Age in the tropical Andes of Bolivia (16°S) and its implications for climate reconstruction. Quat Res 70(2):198–212

Rabatel A, Francou B, Soruco A, Gomez J, Cáceres B, Ceballos JL, Basantes R, Vuille M, Sicart J, Huggel C, Scheel M, Lejeune Y, Arnaud Y, Collet M, Condom T, Consoli G, Favier V, Jomelli V, Galarraga R, Ginot P, Maisincho L, Mendoza J, Ménégoz M, Ramirez E, Ribstein P, Suarez W, Villacis M, Wagnon P (2013) Current state of glaciers in the tropical Andes: a multi-century perspective on glacier evolution and climate change. Cryosphere 7(1):81–102

Radić V, Hock R (2011) Regionally differentiated contribution of mountain glaciers and ice caps to future sea-level rise. Nat Geosci 4(2):91–94

Reid PC, Hari RE, Beaugrand G, Livingstone DM, Marty C, Straile D, Barichivich J, Goberville E, Adrian R, Aono Y, Brown R, Foster J, Groisman P, Hélaouët P, Hsu H, Kirby R, Knight J, Kraberg A, Li J, Lo T, Myneni RB, North RP, Pounds JA, Sparks T, Stübi R, Tian Y, Wiltshire KH, Xiao D, Zhu Z (2016) Global impacts of the 1980s regime shift. Glob Change Biol 22(2):682–703

Richardson S, Reynolds J (2000) An overview of glacial hazards in the Himalayas. Quat Int 65/66(1):31–47

Ritter F, Fiebig M, Muhar A (2012) Impacts of global warming on mountaineering: a classification of phenomena affecting the Alpine Trail Network. Mt Res Dev 32(1):4–15

Rivera A, Bown F, Carrión D, Zenteno P (2012) Glacier responses to recent volcanic activity in Southern Chile. Environ Res Lett 7(1):14036. https://doi.org/10.1088/1748-9326/7/1/014036

Röthlisberger F (1986) 10000 Jahre Gletschergeschichte der Erde. Sauerländer, Aarau

Seltzer GO (1994) Climatic interpretation of alpine snowline variations on millennial time scales. Quat Res 41(2):154–159

Solomina ON, Bradley RS, Hodgson DA, Ivy-Ochs S, Jomelli V, Mackintosh AN, Nesje A, Owen LA, Wanner H, Wiles GC, Young NE (2015) Holocene glacier fluctuations. Quat Sci Rev 111:9–34

Solomina O, Bushueva I, Dolgova E, Jomelli V, Alexandrin M, Mikhalenko V, Matskovsky V (2016) Glacier variations in the Northern Caucasus compared to climatic reconstructions over the past millennium. Global Planet Change 140:28–58

Svoboda F, Paul F (2009) A new glacier inventory on Southern Baffin Island, Canada, from ASTER data: I. Applied methods, challenges and solutions. Ann Glaciol 50(53):11–21

UNEP (2007) Global outlook for ice and snow. UNEP/GRID-Arendal Norway

Vaughan DG, Comiso JC, Allison I, Carrasco J, Kaser G, Kwok R, Mote P, Murray T, Paul F, Ren J, Rignot E, Solomina O, Steffen K, Zhang T (2013) Observations: cryosphere. In: Stocker T, Qin D, Plattner G, Tignor M,

Allen S, Boschung J, Nauels A, Xia Y, Bex V, Midgley P (eds) Climate change 2013: the physical science basis. Contribution of working group I to the fifth assessment report of the intergovernmental panel on climate change. Cambridge University Press, Cambridge, United Kingdom and New York, NY, USA, pp 317–382

von Humboldt A (1845) Kosmos. Entwurf einer physischen Weltbeschreibung. Bd. 1. Cotta Stuttgart u. Tübingen

Wanner H, Beer J, Bütikofer J, Crowley TJ, Cubasch U, Flückiger J, Goosse H, Grosjean M, Joos F, Kaplan JO, Küttel M, Müller SA, Prentice IC, Solomina O, Stocker TF, Tarasov P, Wagner M, Widmann M (2008) Mid- to Late Holocene climate change: an overview. Quat Sci Rev 27(19–20):1791–1828

WGMS (2008) Global glacier changes: facts and figures. UNEP, World Glacier Monitoring Service, Zurich, Switzerland

WGMS (2015) Global Glacier Change Bulletin No. 1 (2012–2013). World Glacier Monitoring Service, Zurich, Switzerland

Wild M, Ohmura A, Makowski K (2007) Impact of global dimming and brightening on global warming. Geophys Res Lett 34(4):L04702. https://doi.org/10.1029/2006GL028031

Wolken GJ (2006) High-resolution multispectral techniques for mapping former Little Ice Age terrestrial ice cover in the Canadian High Arctic. Remote Sens Environ 101(1):104–114

Wolken GJ, England JH, Dyke AS (2008) Changes in late-Neoglacial perennial snow/ice extent and equilibrium-line altitudes in the Queen Elizabeth Islands, Arctic Canada. Holocene 18(4):615–627

Zemp M, Paul F, Hoelzle M, Haeberli W (2008) Alpine glacier fluctuations 1850–2000: an overview and spatio-temporal analysis of available data and its representativity. In: Orlove B, Luckman B, Wiegandt E (eds) Darkening peaks: glacier retreat, science, and society. University of California Press, Berkeley and Los Angeles, pp 152–167

Zemp M, Hoelzle M, Haeberli W (2009) Six decades of glacier mass balance observations - a review of the worldwide monitoring network. Ann Glaciol 50:101–111

Zemp M, Zumbühl H, Nussbaumer SI, Masiokas MEL, Pitte P (2011) Extending glacier monitoring into the Little Ice Age and beyond. PAGES news 19(2):67–69

Zemp M, Frey H, Gärtner-Roer I, Nussbaumer SU, Hoelzle M, Paul F, Haeberli W, Denzinger F, Ahlstrøm AP, Anderson B, Bajracharya S, Baroni C, Braun LN, Cáceres BE, Casassa G, Cobos G, Dávila LR, Delgado Granados H, Demuth MN, Espizua L, Fischer A, Fujita K, Gadek B, Ghazanfar A, Hagen JO, Holmlund P, Karimi N, Li Z, Pelto M, Pitte P, Popovnin VV, Portocarrero CA, Prinz R, Sangewar CV, Severskiy I, Sigurðsson O, Soruco A, Usubaliev R, Vincent C (2015) Historically unprecedented global glacier decline in the early 21st century. J Glaciol 61(228):745–762

Zumbühl HJ, Holzhauser H (1988) Alpengletscher in der kleinen Eiszeit: Sonderheft zum 125jährigen Jubiläum des SAC. Die Alpen 64(3):129–322

Zumbühl HJ, Steiner D, Nussbaumer SU (2008) 19th century glacier representations and fluctuations in the central and western European Alps: an interdisciplinary approach. Global Planet Change 60:42–57

An Inventory of Proglacial Systems in Austria, Switzerland and Across Patagonia

3

Jonathan Carrivick, Tobias Heckmann, Mauro Fischer and Bethan Davies

Abstract

Deglaciation since the Little Ice Age has exposed only a small areal proportion of alpine catchments, but these proglacial systems are disproportionately important as sediment sources. Indeed sediment yields from proglacial rivers are amongst the highest measured anywhere in the World. Motivated by a desire to understand where exactly within catchments this sediment is coming from and how it might evolve, this chapter presents the first digital inventories of proglacial systems and the first comparative inter- and intra-catchment comparison of their geometry, topography and geomorphology. Whilst focussing on the description of these inventories and on descriptive statistics, it highlights the potential of these data, with examples from Austria, Switzerland and Patagonia, for interpreting landscape evolution status, predominant earth surface processes and glacial meltwater inundation patterns. Switzerland has by far the highest proportion of very small (<0.5 km^2) proglacial systems. Patagonia has the most even distribution of proglacial systems in terms of areal size and elevation above sea level. We found no east–west or north–south spatial pattern in geometric or topographical metrics, such as hypsometric index, thereby refuting a straightforward control of climate-driven precipitation or air temperature. However, we note that geology, particularly rock hardness, could be a major factor in proglacial system character. Likely sediment sinks can occupy up to ~30 and 20% of proglacial systems in Austria and Switzerland, respectively, but up to 90% of those in Patagonia where many systems terminate on a coastline. Meltwater influence maps and landform maps, derived from contributing area and slope data, respectively, will not only be useful for many environmental science disciplines but also for water resources, landscape managers and natural hazards authorities. Overall, this chapter presents an objective and easily implemented method for making proglacial systems inventories and for characterising inter- and intra-catchment geomorphological form and function.

J. Carrivick (✉)
School of Geography, University of Leeds, Woodhouse Lane, Leeds, West Yorkshire LS2 9JT, UK
e-mail: j.l.carrivick@leeds.ac.uk

T. Heckmann
Physical Geography, Catholic University of Eichstätt-Ingolstadt, Eichstätt, Germany

M. Fischer
Unit of Geography, University of Fribourg, Chemin Du Musée 4, 1700 Fribourg, Switzerland

B. Davies
Centre for Quaternary Research, Department of Geography, Royal Holloway, University of London, Egham, Surrey TW20 0EX, UK

T. Heckmann and D. Morche (eds.), *Geomorphology of Proglacial Systems*,
Geography of the Physical Environment, https://doi.org/10.1007/978-3-319-94184-4_3

Keywords
Little Ice Age · Deglaciation · Proglacial areas
Geomorphometry · Glacier inventory

3.1 Introduction

Proglacial systems are amongst the most rapidly adjusting landscapes in the World. They are geomorphologically dynamic and are platforms for rapid ecological succession and biological extremes. Whilst individual proglacial systems occupy a small areal proportion of the major alpine catchments, they are disproportionately important as sediment sources (Carrivick and Heckmann 2017). Their form and function are consequences of global climate change as manifest in ongoing deglaciation, glacier mass loss and terminus retreat, and in consequent meltwater production, permafrost degradation and hillslope activity. These processes are of concern for natural landscape management, natural hazard risk assessment, and for understanding water and sediment fluxes that concern water resource management.

Given these characteristics and importance of proglacial systems (Heckmann et al. 2016; Carrivick and Heckmann 2017), it is surprising that prevailing approaches to studying them are generally lacking regional analysis or synthesis. Indeed, the literature on proglacial systems, whether it is within the domain of glacial geology, geomorphology, soil science, biology or ecology, is profoundly dominated by single-site studies. International initiatives such as the IAG-AIG working group sediment budgets in cold climates 'SEDIBUD' (Beylich et al. 2010) have commendable aims to standardise monitoring and to gain representative global coverage, but they necessarily trade spatial for temporal density of data collection and analysis.

If multiple proglacial systems are studied simultaneously, using a common framework of analysis, then spatial patterns in proglacial system geomorphology can be identified and used to infer regional controls on geomorphic processes and trends in landscape evolution. Thus, multiple hypotheses of landscape evolution in response to deglaciation, i.e. paraglacial theory (e.g. Church and Ryder 1972; Ballantyne 2002), can be quantitatively tested. However, for such analyses to be performed, firstly a definition of proglacial systems must be made that spatially discriminates them from adjacent parts of a landscape. Secondly, an inventory of proglacial systems must be constructed (in much the same manner as global glacier inventories) to include spatial extent and spatially distributed topographic data.

Then, with this proglacial system spatial and topographic information, the exposure of land surfaces and the modification of landforms in proglacial systems with time since deglaciation can be analysed. Specifically, proglacial systems can be treated as natural laboratories to aid understanding of short-term land stability and associated sediment fluxes. Furthermore, with the premise that form is a proxy of process type or of process rates, then proglacial systems inventories will be both indicative of the recent geomorphological activity, and indicative of the conditioning of geomorphological processes in the near future.

The aim of this chapter is to present the first inventories of proglacial systems and the first comparative examination (inter- and intra-catchment) of their geometry and topography. We view this chapter, this inventory, as a first step towards a comparative analysis of the geomorphological composition and predominant geomorphological processes within these systems. Furthermore, the spatial datasets and quantitative descriptions of these proglacial systems will have (i) interest and utility in other related environmental sciences and (ii) opportunities for local site-specific measurements to be regionalised or upscaled.

3.2 Methods and Study Sites

3.2.1 Proglacial System Spatial Definition

In this study, proglacial systems were defined automatically by subtracting modern glacier outlines from LIA glacier outlines. On the one hand, this is a simple spatial definition motivated to facilitate automatic analysis of multiple spatial entities across entire regions. On the other hand, the definition of proglacial areas as the area between the present day and the LIA glacier extent is quite old and has been implemented in a number of previous studies (Chap. 1; Schiefer and Gilbert 2007; Bosson et al. 2015).

To minimise extraneous features and parts of proglacial systems where (i) some glaciers have reduced in width at relatively high elevations, (ii) some glaciers have reduced in ice extent on plateaux or on cols as a result of fragmentation, and (iii) portions of the landscape presently in transition between glacially dominated and non-glacial process regimes, and (iv) with regard to the resolution of the imagery from which the outlines were mapped (see below), we specified a semi-arbitrary 100 m buffer around the modern outlines and excluded this area from our analyses. Nonetheless, a few proglacial systems as automatically defined in this manner were composed of multiple parts. In those cases, each part was analysed as a separate spatial entity. In this manner, a proglacial system is a recently (i.e. since the end of the LIA) deglaciated area.

Published mapping of LIA glacier outlines tends to be restricted to relatively small sub-regions; examples include the eastern Qilian Shan in north–west China (Tibet) (Shiyin et al. 2003), Jotunheimen in southern Norway (Baumann et al. 2009) and parts of the Italian Alps (Knoll et al. 2009; Lucchesi et al. 2014; Zanoner et al. 2017). For other large-scale inventories of proglacial systems to be compiled, much more inventory-style mapping of LIA glacier outlines is required. We note that very little regional information on the LIA extents of glaciers is available for high-altitude and polar regions where climate change is (and will be) most pronounced (Solomon 2007: IPCC).

3.2.2 Study Sites

In this study, proglacial systems in Austria, Switzerland and across Patagonia are analysed, due to these regions having availability of digital glacier inventories of both Little Ice Age (LIA) glacier outlines and of recent glacier outlines. Across all these three study areas, the two glacier inventories cover the entire (de)glaciated area; i.e. in this study, we are dealing with a population, not a sample of proglacial systems. Comparison between the LIA and recent inventories has been used to quantify glacier retreat and glacier area change (e.g. for Switzerland: Fischer et al. 2015c), but not to investigate the deglaciated terrain itself; one exception is Maisch et al. (1999) who studied bedrock, sedimentary and mixed glacier forefields.

Swiss glacier outlines for both the LIA and the year 2010 were obtained by Fischer et al. (2014, 2015a) by manual digitization from 25-cm aerial orthophotographs acquired between 2008 and 2011. Our topographic data for Swiss proglacial systems came from a 10-m cell-size gridded Digital Elevation Model (DEM) derived from Airborne Laser Scanning (ALS), as published by Fischer et al. (2014, 2015a).

Glacier outlines in Austria were obtained from Groß and Patzelt (2015; LIA) and Fischer et al. (2015b, c; year 2006); Daten Österreichs (2016) provide a 10-m DEM of Austria that was downsampled from 2-m ALS data.

For Patagonia, we used glacier outlines during the LIA and from the years 2010–2011 mapped by Davies and Glasser (2012) from orthorectified Landsat Enhanced Thematic Mapper Plus (ETM+) colour images. They have a spatial resolution of 30 m and a geopositional accuracy of ±50 m (Davies and Glasser 2012). We obtained an Advanced Spaceborne Thermal Emission and Reflection Radiometer Global Digital Elevation Model Version 2 (ASTER

GDEM V2) mosaic (http://asterweb.jpl.nasa.gov/gdem.asp) to cover the whole of Patagonia.

3.2.3 Metrics

As a study producing an inventory of proglacial systems, we focus herein on calculating zonal descriptive statistics (central tendency and dispersion) of terrain attributes of the proglacial zones. The proglacial system outline attributes included were kept to a minimum being mindful of this chapter size and the thematic focus of this book. We extracted latitude and longitude, the size of the proglacial area (Table 3.1), and for each proglacial zone elevation (Figs. 3.1, 3.2; Table 3.2) and slope (Figs. 3.4 and 3.5) derived from the DEMs (Fig. 3.3).

We calculated the hypsometric index (HI; Fig. 3.3; Table 3.3) of each zone as a consideration of the geomorphological energy within each proglacial system, with the concept that systems with predominant area at higher elevation have proportionally more potential energy per unit area than systems with elevations predominantly at lower elevation. HI was calculated following Jiskoot et al.'s (2009) approach for glaciers.

Additionally, and to demonstrate the utility of proglacial system inventories, we present a first-order attempt at classifying proglacial systems based on prevailing geomorphological processes. We calculated the contributing area per grid cell within each zone, as an estimate of hydrological connectivity and relative likely 'wetness'. Contributing area was determined per grid cell via the D-infinity flow direction and flow accumulation algorithm (Tarboton 1997). To separate parts of proglacial zones that receive glacier meltwater from unglaciated patches, we differenced two maps of flow accumulation; one was calculated using the full DEM, the other one using the DEM with the present-day glacier extent masked. This approach discriminated grid cells that cannot receive runoff from present-day glaciers, as coloured greens and reds in Fig. 3.6, versus grid cells that could, according to flow routing, receive runoff from glaciers, as coloured shades of blue in Fig. 3.6. We visually checked these 'potential meltwater maps' against the meltwater inundation during mid-summer at several field sites, including the Ödenwinkelkees (e.g. Carrivick et al. 2013, 2015a) and the Kaunertal (Chap. 1), both in Austria, and found excellent visual agreement.

Another first-order geomorphological classification was implemented following Loye et al. (2009) by decomposing the empirical distribution of slope gradient of a large sample of proglacial DEM cells. The intersections of four slope distributions associated with (i) rockwalls, (ii) steep slopes, talus and morainic landforms, (iii) alluvial and debris cones and (iv) fluvial landforms (braidplain, terraces) and lakes were selected as thresholds for a simple slope-based classification; these thresholds were set to 42°, 22° and 8°, respectively. This data-based, nevertheless, coarse classification was validated by visual comparison with the geomorphological map of Val d'Herens, Switzerland (Lambiel et al. 2016; Fig. 3.7).

3.3 Results

In total, we identified 3398 areas deglaciated since the end of the LIA which hereon we refer to as proglacial systems (Table 3.1). To emphasise, this is a complete population, *not* a sample of discrete areas. Of these, 19% was in Austria,

Table 3.1 Descriptive statistics of the area of proglacial systems in Austria, Switzerland and across Patagonia

Proglacial area (km^2)	*N*	Mean	St. Dev.	Sum	Minimum	Median	Maximum
Austria	638	0.53	0.68	341.26	0.0001	0.28	5.23
Switzerland	2174	0.27	0.56	591.58	0.0001	0.10	11.25
Patagonia	586	6.25	14.75	3663.42	0.001	2.27	207.97

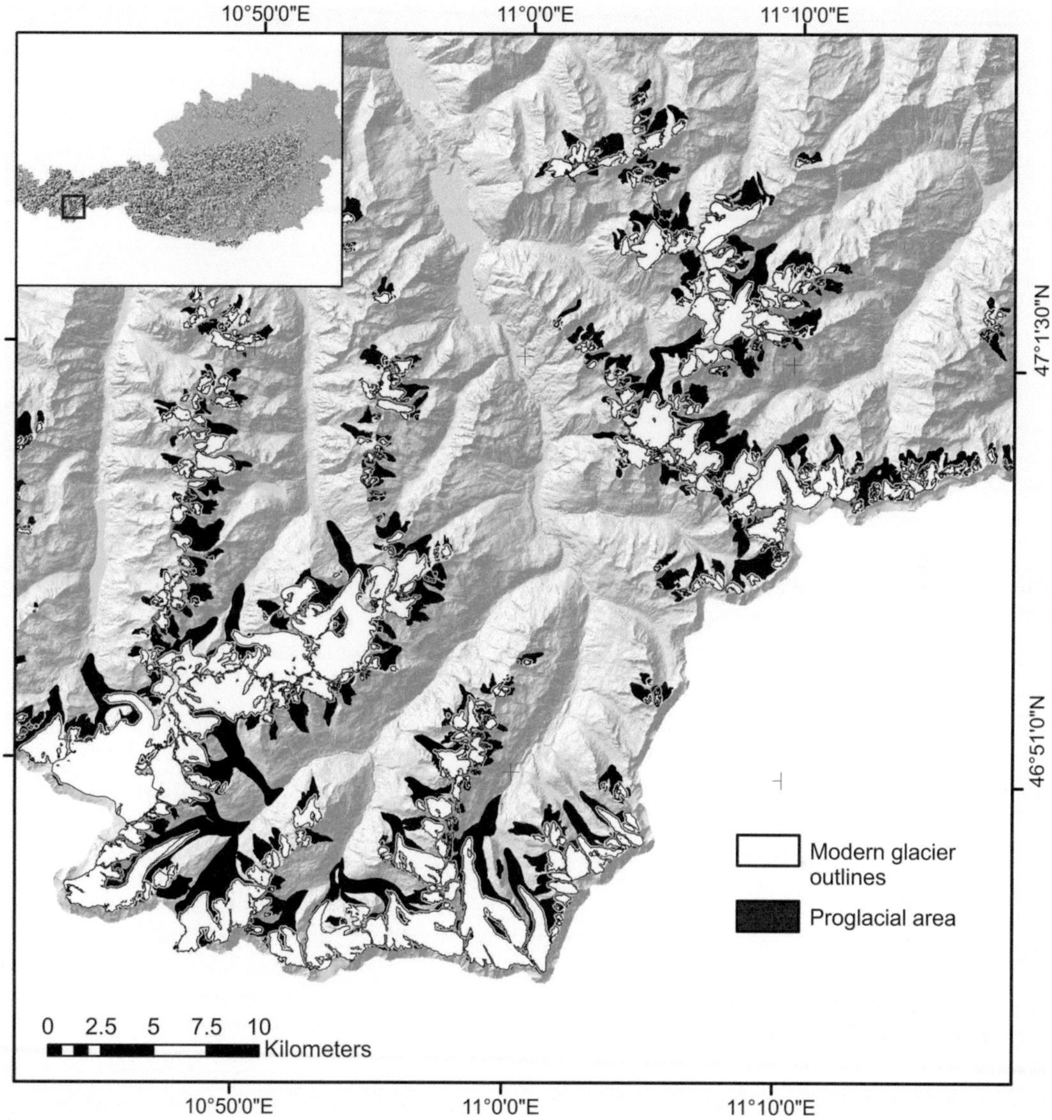

Fig. 3.1 Extent of proglacial systems in the Vent-Obergurgl region of Austria. *Note* this area illustrated is only part of the full inventory for Austria

64% was in Switzerland, and 17% was across Patagonia. These proglacial systems have a combined area of 341.3 km^2 in Austria, 591.6 km^2 in Switzerland and 3663.4 km^2 across Patagonia (Table 3.1). They have a small contribution to the area of major mountain catchments; for example, they occupy just 11% of the Mattertal (containing Zermatt) catchment in Switzerland, or just 2.2% of the Swiss Alps 26,830 km^2 area. About 95% of proglacial systems in Austria and Switzerland are >1 km^2, whereas across Patagonia 95% of the proglacial systems are >5 km^2 (Fig. 3.2). Our discrimination of proglacial systems highlights that they are not just *in front* of a glacier, but are also *beside* a glacier (as a result of glacier trunk thinning and narrowing). In Switzerland, whilst 60% of the proglacial systems are still connected to a present-day glacier, 40% have reached their theoretical maximum size where the

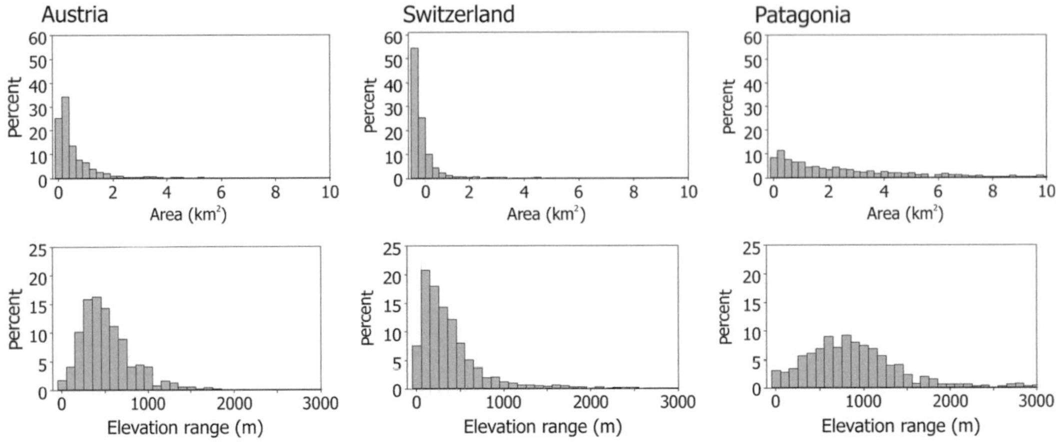

Fig. 3.2 Size and elevation range histograms of proglacial systems in the Austria Alps, Swiss Alps and across Patagonia

Table 3.2 Descriptive statistics of the elevation distribution of proglacial systems in Austria, Switzerland and across Patagonia

Elevation (m.asl)	Regional mean	Regional St. Dev.	Regional median	Regional maximum
Austria				
Zonal min	2442.8	273.3	2458	3271
Zonal range	514.4	297.5	463	1814
Zonal mean	2641.3	228.7	2643.3	3304.9
Zonal St. Dev.	104.9	66.3	90.6	587.2
Zonal median	2635	230.5	2638.5	3306
Switzerland				
Zonal min	2597.7	416.8	2587.5	4152
Zonal range	370.2	367.8	270	2891
Zonal mean	2740.5	377.6	2720.7	4201
Zonal St. Dev.	77.0	73.9	59	738.2
Zonal median	2732.9	383.4	2717	4210
Patagonia				
Zonal min	399.1	344	330	2101
Zonal range	989.7	779.6	839.5	4846
Zonal mean	704.3	386.5	695.1	2244
Zonal St. Dev.	204.1	148.1	178.6	1083.9
Zonal median	666.3	397.5	670.0	2101.0

corresponding glaciers have completely disappeared. In Austria, just 5% of proglacial systems deglaciated since the LIA are no longer connected to an active glacier.

Proglacial systems occur at elevations above sea level lower and within a narrower range in Austria than in Switzerland, and at lower elevations but across a wider range of elevations in

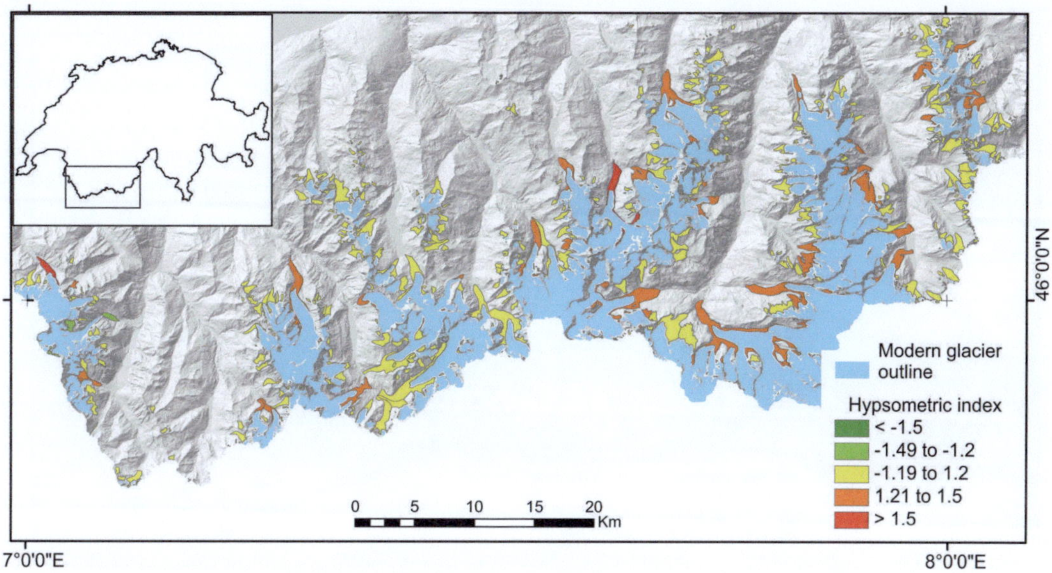

Fig. 3.3 Spatial pattern of the hypsometric index (HI) of proglacial systems in the Valais region of Switzerland. *Note* this area illustrated is only part of the full inventory for Switzerland

Table 3.3 Proportion of proglacial systems per country within each hypsometric index (HI) class for Austria, Switzerland and Patagonia. Very top heavy is where most area is situated at higher elevations, and very bottom heavy is where most area is situated at lower elevations

Statistic	Austria	Switzerland	Patagonia
Very top heavy (%)	0	<1	0
Top heavy (%)	0	<1	0
Equidimensional (%)	31	95	22
Bottom heavy (%)	14	4	12
Very bottom heavy (%)	55	<1	66

Patagonia than in either of the European countries (Fig. 3.2; Table 3.2). In general, larger proglacial systems have lower elevations (Fig. 3.1).

Analysis of the distribution of elevation per proglacial system, i.e. its hypsometry, shows that those in Austria and Patagonia have elevation distributions dominated by relatively low values; i.e. they are bottom heavy (Table 3.3) and thus with an overall elevation distribution that is concave up. In contrast, proglacial systems in Switzerland are virtually all equidimensional, i.e. with a uniform distribution of area with altitude (Table 3.3). A Pearson's correlation (not illustrated) of $r^2 = 0.05$ ($p = 0.17$) of latitude or longitude versus HI suggests that there is no spatial pattern to the HI of proglacial systems (Fig. 3.3). We did find a significant correlation ($r^2 = 0.76$, $p = 0.01$) of the HI with proglacial system size such that larger proglacial systems have a trend towards a more concave up distribution.

Parts of proglacial systems that might be considered to be sediment sinks were mapped for all our study sites using local (grid cell) slopes below 8°. This slope value was selected manually and subjectively to include lakes, gravel braidplains, river floodplains, and also the footslopes of alluvial and debris cones. It is in accordance with the 3° to 9° slope value used by

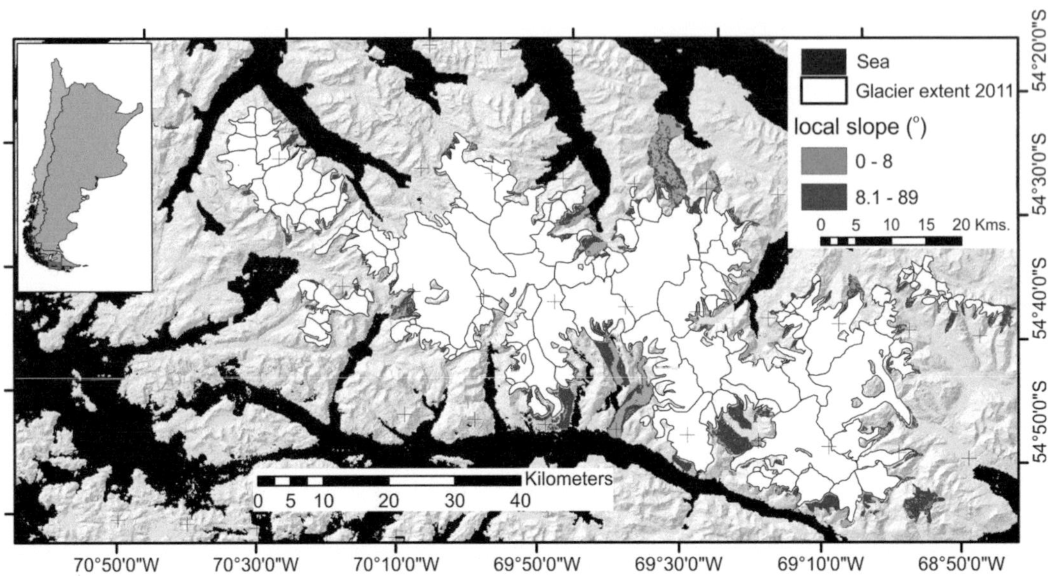

Fig. 3.4 Example from the Cordillera Darwin region of Patagonia of parts of proglacial systems that might be considered to be sediment sinks, based on a simple local slope criteria. *Note* this is only part of the full inventory across Patagonia and is a part particularly close to the sea fjords

Clubb et al. (2017), and with an analysis of slope distributions after Loye et al. (2009; see also Fig. 3.7). An example for a part of Patagonia is depicted in Fig. 3.4, because proglacial areas in Patagonia occupy a total area of over 2000 km^2, which is a far larger area (55% of all Patagonian proglacial systems) than for those of Austria and Switzerland combined. The areal size distribution of likely sediment sinks in Patagonia is far greater than for those in Austria or in Switzerland (Fig. 3.5a, b, c).

From the contributing area information (Fig. 3.6a), a simple ratio of glacial meltwater-influenced versus non-glacial meltwater-influenced parts of each proglacial system can be calculated. The comparison of Switzerland and Austria (Fig. 3.6b) reflects the larger number of completely deglaciated (and hence, with a zero meltwater influence) proglacial areas in Switzerland. Leaving the disappeared glaciers aside, the distributions suggest that, in Austria, there is a near symmetric distribution with a mode of ca. 50% of the area influenced by meltwater; Swiss proglacial areas show a right-skewed distribution (mode ca. 15%) with a higher percentage of less meltwater-dominated areas. In Patagonia, there are not only many much larger catchments but there is also much more dispersion in the histogram of this measure.

3.4 Discussion

3.4.1 Inventories of Proglacial Systems

Digital spatial inventories of proglacial systems are needed because of their intense geomorphodynamics (Carrivick and Heckmann 2017), according to the paraglacial concept (Church and Ryder 1972; Ballantyne 2002). Hence, proglacial systems are very likely important sources of sediment with impacts on riverine geomorphology, river ecology and on water resources management potentially reaching far beyond the source area (Knight and Harrison 2014).

Digital spatial inventories of proglacial systems should include areal size (Table 3.1) because that metric is a proxy for the size of potential sediment sources; i.e. sediment yield. The areal size histograms for Austria, Switzerland and Patagonia (Fig. 3.2) could therefore be

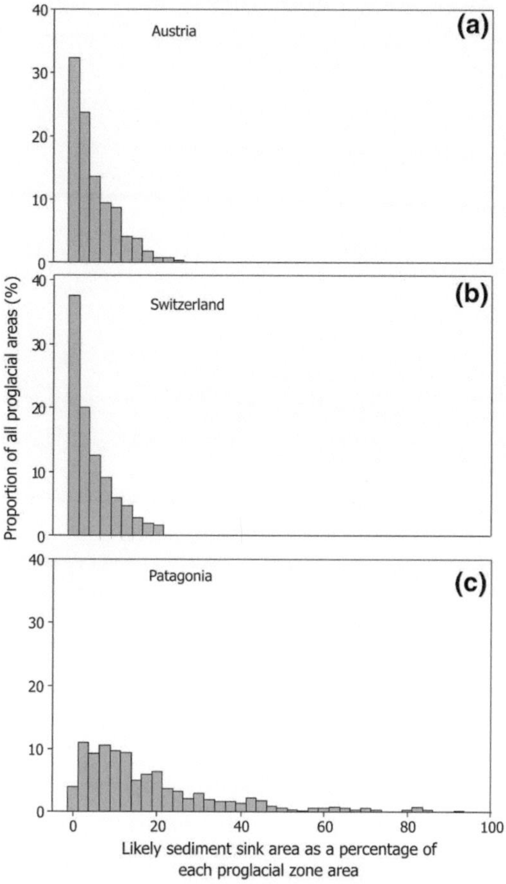

Fig. 3.5 Histograms of likely sediment-sink areas in each proglacial system, based on a threshold of 8° local slope

suggestive of the potential sediment yield from each of those proglacial systems, and also the degree to which there is variability within a country. Proglacial systems inventories should include lithology (sediment vs. bedrock) because that metric could inform on relative sediment availability (Hallet et al. 1996; Maisch et al. 1999; Jaeger and Koppes 2016). Elevation metrics (max., min., range, distribution: Table 3.2; Fig. 3.3) will be important for assessing the susceptibility of individual proglacial systems to changes in air temperature and precipitation. Landforms can be used to infer the type of predominant earth surface process and process dynamics/rates; while geomorphological mapping can produce this kind of information on a catchment scale (Lambiel et al. 2016; Chap. 17), the application to regional scales requires automated approaches based on digital geomorphometry (Loye et al. 2009; Clubb et al. 2017; Carrivick and Heckmann 2017).

The shape of a proglacial area and the configuration of sediment sources, sinks, and the channel network influence the transfer of mobilised glacial sediment through the system (sediment connectivity; Chap. 16). In topographically confined settings, there is no accommodation space, storage landforms are regularly undercut, gradients are being kept high, and there is increased slope-channel coupling (Chaps. 10, 11 and 16). Contrastingly, in topographically unconfined areas, there is accommodation space for growing storage landforms and there is stabilisation of primary sediment sources.

Thus overall, it is the topographical characteristics of proglacial systems that can aid definition of system energy and of gravity-driven processes, and it is the geomorphological characteristics of proglacial systems that control which type of processes predominate, process rates, and how much sediment is available, both on hillslopes and on valley floors.

Proglacial inventories therefore have great utility for the assessment of potential geomorphological reactions to continued deglaciation or to the propagation of earth surface process changes aided by the assessment of connectivity. They should be used for making assessments of how representative local site-specific case studies are. They will aid regionalisation of local measurements of process rates from particular study sites to other similar proglacial systems.

3.4.2 Inter- and Intra-catchment Form and Function

Proglacial system position, areal size and hypsometric distribution are largely conditioned by processes of (LIA) glaciation. Whilst we found no simple spatial (east–west, north–south) trend in these metrics, we suggest that inclusion of climatic (mean annual air temperature, total annual rainfall) and geological parameters

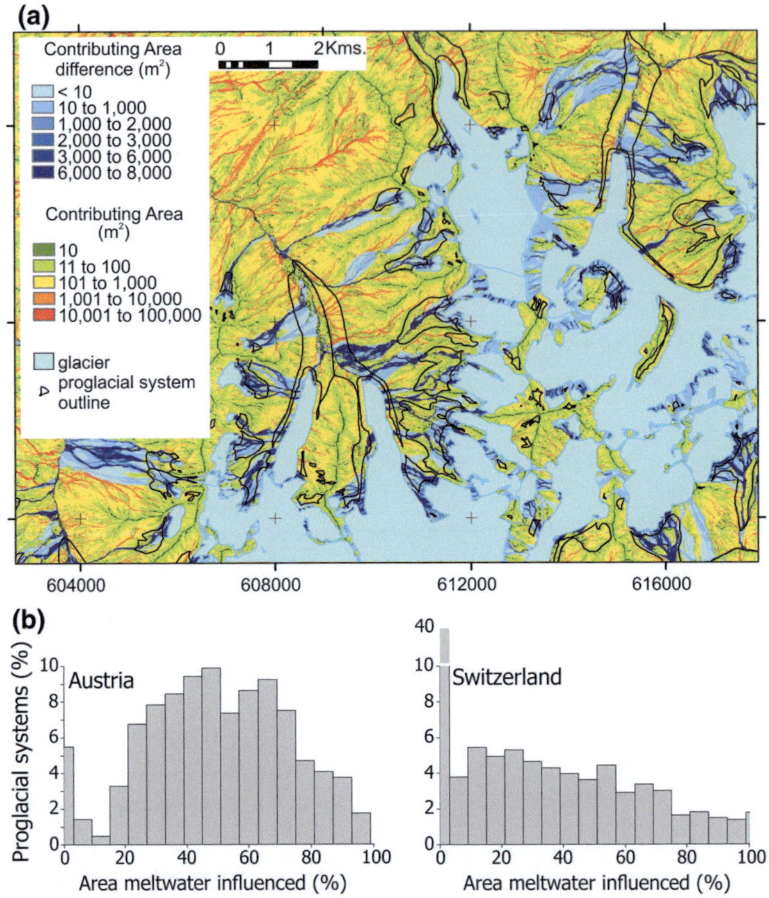

Fig. 3.6 Contributing (hydrological) area per grid cell calculated both with and also without glacier surfaces in the DEM. A cell-by-cell difference between the two calculations produces a map of grid cells that cannot receive glacial meltwater versus those that probably do (**a**). This example in panel A is for the Glacier du Mont Miné and Glacier de Ferpècle area in Switzerland. Note this is only part of the full inventory across Switzerland. Panel B depicts the histograms of the areal proportion of proglacial systems likely to be influenced by glacial meltwater

(hardness) may reveal significant spatial clustering due to the effects of those on sediment availability (Maisch et al. 1999), respectively. Derivation of the climatic and geological parameters for each proglacial system will also be important for understanding the form and function of these catchment zones.

Inventories of proglacial systems and of their geomorphology offer potential to determine landscape evolution since the end of the LIA. As one example, our finding that larger proglacial systems have a trend towards a more concave up distribution can be explained by larger proglacial systems being developed where long glacier tongues have retreated by up to 2–3 km. These proglacial systems are bottom heavy. Jiskoot et al. (2009) found that glaciers with top-heavy hypsometry were more sensitive to retreat, as an increase in equilibrium line altitude causes higher losses in accumulation area; similarly, proglacial areas with a top-heavy hypsometry could be associated with (previously) top-heavy glaciers. Many of these proglacial systems are characterised by a "T" shape, indicating either former areas of confluence of multiple glaciers, or former zones convergence of very wide glaciers.

Caution must be applied when making these interpretations of proglacial area shape and HI though, because similar HI values can be produced by complex interactions of climate, tectonism, sedimentation and rock resistance (Bishop et al. 2002) and in the case of proglacial systems also glacial history. Carrivick and Rushmer (2009), for example, noted that despite similar catchment size, shape, hypsometry and glacier cover, Fox Glacier and Franz Josef Glacier proglacial systems exhibited quantifiably different geomorphological compositions and functioning. Nonetheless, the HI can highlight anomalous proglacial systems within a mountain range.

As a second example, Glasser et al. (2009) manually mapped the geomorphological components and inferred the evolution of proglacial systems across Patagonia. However, with digital inventories of proglacial systems, such as those in this study, landforms that have evolved since the end of the LIA could be identified automatically. For glacial landforms, Smith et al. (2006) provide a subjective methodology and testing. Objective slope-based classifications can be efficient for application across large areas at high spatial resolution (Loye et al. 2009; Fig. 3.7). With the slope thresholds derived from this analysis, we identified parts of proglacial systems probably dominated by lakes and floodplains (slope < 8°), alluvial and debris cones (8–22°), talus (22–42°) and rockwalls (>42°). In contrast, there are few (manually mapped) geomorphological maps of (areas containing) proglacial systems but some examples include Kjær et al. (2008), Geilhausen et al. (2012), Lambiel et al. (2016) and Chap. 17 of this book.

Our consideration of slope (Fig. 3.4) and of contributing area (Fig. 3.6) herein is preliminary only to demonstrate potential utility of such proglacial systems inventories, for example for making objective automatic landform classifications (Fig. 3.7) and for making automatic glacial meltwater routing estimates (Fig. 3.6), respectively. The former is a step towards estimating landscape stability and sediment production as a function of landform type (see Carrivick and Heckmann 2017: their Fig. 10). The latter is also useful for landscape stability estimates, but also for considering the conveyance of sediment through and out of a proglacial system. Both landform and meltwater maps will be useful for many other environmental sciences related to geomorphology and to landscape, water and natural hazard managers too.

The paraglacial reworking of unconsolidated glacigenic sediments in proglacial systems means that proglacial rivers have some of the highest sediment yields reported (e.g. Hallet et al. 1996; Jaeger and Koppes 2016), thereby affecting the river channel network and including downstream effects such as reservoir infilling. Sediment connectivity is an important property moderating the propagation of local, e.g. hillslope, geomorphological changes through proglacial systems towards and along the channel network (Micheletti and Lane 2016; Lane et al. 2017). Proglacial systems exhibit different morphologies, leading to differences in sediment connectivity (Chap. 16). Connectivity will also change with time, as landforms evolve, and thus must be monitored for the potential downstream consequences, such as detrimental reservoir infilling, and natural hazards. Therefore, we suggest future analysis of the proglacial inventories presented herein should include the application of DEM-based connectivity indices, such as that by Cavalli et al. (2013). Additionally, a characterisation of proglacial zones in terms of lateral confinement (e.g. Fryirs et al. 2016) could be used to estimate their buffering capacity with respect to the propagation of geomorphological changes that occur within their boundaries.

3.4.3 Further Study

Historic (since the end of the LIA) evolution of proglacial systems, as represented by the inventories of system areal size, topography and geomorphological composition, shows us that we can expect that the next few decades will see profound changes to proglacial systems worldwide as glaciers continue to diminish in extent and volume, as the consequent meltwater regime adjusts, and as permafrost degradation intensifies and perhaps extends spatially. Detecting and understanding these changes are important for the purposes of natural landscape management and conservation, and natural hazards assessment, as well as for water and sediment effluxes that concern water resource management-related applications.

Re-surveys of proglacial systems, either airborne or ground-based, of glaciers (e.g. Fischer et al. 2014, 2015a, b, c) and of proglacial topography (e.g. Carrivick et al. 2013) will permit comparisons with the inventories of this

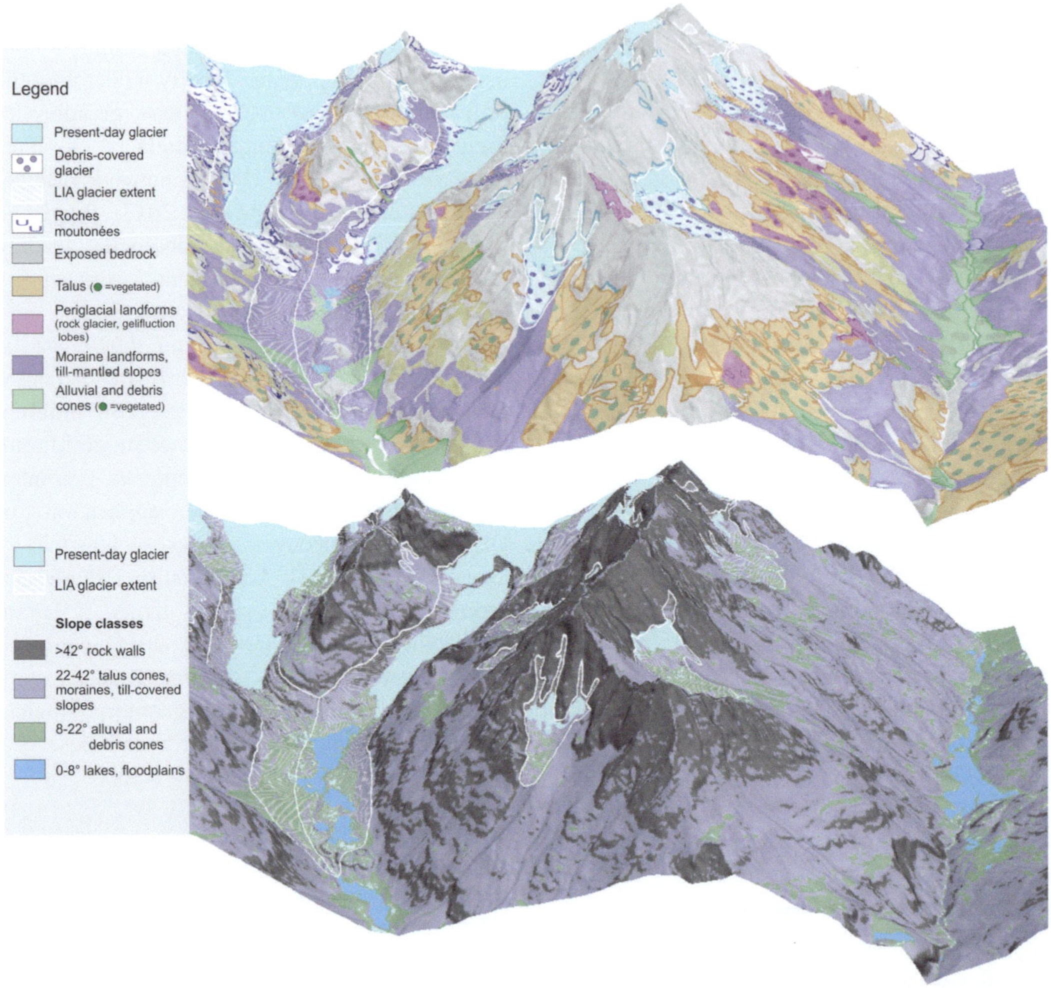

Fig. 3.7 Three-dimensional perspective visualisation of the Val d'Herens, Switzerland. The upper part shows a generalised version of a geomorphological map published by Lambiel et al. (2016). The lower panel presents a simple slope-based classification; the thresholds separating the slope categories were derived from the distribution of slope of the Swiss DEM10 (except present-day glaciers) following Loye et al. (2009), leading to a first-order classification of proglacial systems morphology that is, amongst others, consistent with the 8° threshold for sediment sinks used in Figs. 3.4 and 3.5

study to detect changes in area, elevation and thus (landform) volume. New techniques being applied in the geosciences such as structure from motion offer the potential for obtaining topography with a spatial coverage, spatial density, and point accuracy comparable to that of terrestrial laser scanners (Carrivick et al. 2015b; Smith et al. 2016). Therefore, they could consider geomorphological changes (via high-resolution point-to-point comparisons or via gridded digital elevation models) that are beyond the level of detection within photogrammetry-derived data and satellite-based elevation products.

The proglacial system outlines presented herein, and others that we hope will be calculated for other parts of the World, can be used to spatially delimit future topographic-based spatial enquiries. For example, aerial photographs can be used to map multi-temporal snapshots (Schiefer and Gilbert 2007; Micheletti et al. 2015). If these are with respect to a known time and position of glacier extent, then a chronosequence can be mapped, which is informative for soil development (e.g. Egli et al. 2006), vegetation succession (e.g. D'Amico et al. 2014) and ecology (e.g. Carlson et al. 2010). Multi-spectral imagery can yield information on proglacial system land cover, which could quite easily be obtained via an automatic clustering algorithm applied to LandSat imagery (Bands 2–7) although admittedly at a spatial resolution somewhat coarser than the DEMs for Austria and Switzerland used in this study.

An especially important land cover type to identify and monitor is that of lakes. Proglacial lakes are presently increasing in number and size around the world (Carrivick and Tweed 2013) and since they act as a buffer to water and sediment are a control on geomorphological activity and aquatic ecology both within and beyond a proglacial system (Chap. 14). Some also are a potential hazard to society as sources of sudden glacier outburst floods (Carrivick and Tweed 2016).

Upscaling of field-derived results, and other regional modelling, should be enabled with this inventory-style identification and compilation of metrics of proglacial systems. For example, (i) sediment yield is correlated with catchment size and with maximum elevation above sea level (e.g. Hallet et al. 1996; Jaeger and Koppes 2016), (ii) rates of proglacial volume change can be classed according to the corresponding landform (Carrivick and Heckmann 2017), and (iii) the representativity of case studies can be assessed by viewing the respective study areas and their morphometric properties in context of the inventory.

3.5 Conclusions

This chapter presents an objective and easily implemented method for spatially discriminating proglacial systems based on glacier outlines from the LIA and those from the present day. Hopefully, it will serve as an inspiration to compile or if necessary to create digital LIA outlines for other regions of the World, which can then be compared to contemporary glacier outlines to generate proglacial system areas for those regions.

This chapter also presents a preliminary analysis of digital elevation models of these proglacial systems to provide the platform for how future understanding of inter- and intra-catchment topography and geomorphology can be facilitated. Thus, a capability to quantify the spatial distribution of past, present and future proglacial geomorphological functioning has been developed. Tests of paraglacial theory should have recourse to these inventories, as will upscaling or regionalisation of local plot-scale measurements of geomorphological activity. Otherwise, isolated case studies can be placed in context to similar topography and landform configuration within the inventory, in order to assess their representativeness and to extrapolate results to larger regions. Related disciplines including vegetation, ecology, soil science and the freshwater sciences will all have a use for these inventories because proglacial systems are natural laboratories for rapid succession and community adaptation. We anticipate that these and future proglacial systems inventories will have immediate usage within applied water resource and natural hazard studies.

Acknowledgements Andrea Fischer is thanked for pointing us towards the sources of Austria glacier outline and DEM data. Oliver Korup provided a critical yet constructive review, and Editor David Morche is thanked for his assistance.

References

Ballantyne CK (2002) Paraglacial geomorphology. Quat Sci Rev 21:1935–2017. https://doi.org/10.1016/S0277-3791(02)00005-7

Baumann S, Winkler S, Andreassen LM (2009) Mapping glaciers in Jotunheimen, South-Norway, during the "Little Ice Age" maximum. Cryosphere 3:231–243. https://doi.org/10.5194/tc-3-231-2009

Beylich AA, Lamoureux SF, Decaulne A, Dixon JC, Orwin JF, Otto JC, Overeem I, Saemundsson T, Warburton J, Zwolinski Z (2010) Sedimentary fluxes and budgets in changing cold environments: the global IAG/AIG sediment budgets in cold environments (SEDIBUD) programme. Geografiska Annaler:

Ser A, Phys Geogr 92:151–153. https://doi.org/10.1111/j.1468-0459.2010.00386.x

Bishop M, Shroder JF, Bonk R, Olsenholler J (2002) Geomorphic change in high mountains: a western Himalayan perspective. Global Planet Change 32:311–329. https://doi.org/10.1016/S0921-8181(02)00073-5

Bosson J-B, Deline P, Bodin X, Schoeneich P, Baron L, Gardent M, Lambiel C (2015) The influence of ground ice distribution on geomorphic dynamics since the Little Ice Age in proglacial areas of two cirque glacier systems. Earth Surf Process Land 40:666–680. https://doi.org/10.1002/esp.3666

Carlson ML, Flagstad LA, Gillet F, Mitchell EAD (2010) Community development along a proglacial chronosequence: are above-ground and below-ground community structure controlled more by biotic than abiotic factors? Above- and below-ground communities in succession. J Ecol 98:1084–1095. https://doi.org/10.1111/j.1365-2745.2010.01699.x

Carrivick JL, Heckmann T (2017) Short-term geomorphological evolution of proglacial systems. Geomorphology 287:3–28. https://doi.org/10.1016/j.geomorph.2017.01.037

Carrivick JL, Rushmer EL (2009) Inter- and intra-catchment variations in proglacial geomorphology: an example from Franz Josef Glacier and Fox Glacier, New Zealand. Arct Antarct Alp Res 41:18–36. https://doi.org/10.1657/1938-4246(07-099)%5b CARRIVICK%5d2.0.CO;2

Carrivick JL, Tweed FS (2013) Proglacial lakes: character, behaviour and geological importance. Quat Sci Rev 78:34–52. https://doi.org/10.1016/j.quascirev.2013.07.028

Carrivick JL, Tweed FS (2016) A global assessment of the societal impacts of glacier outburst floods. Global Planet Change 144:1–16. https://doi.org/10.1016/j.gloplacha.2016.07.001

Carrivick JL, Geilhausen M, Warburton J, Dickson NE, Carver SJ, Evans AJ, Brown LE (2013) Contemporary geomorphological activity throughout the proglacial area of an alpine catchment. Geomorphology 188:83–95. https://doi.org/10.1016/j.geomorph.2012.03.029

Carrivick JL, Berry K, Geilhausen M et al (2015a) Decadal-Scale Changes of the Ödenwinkelkees, Central Austria, Suggest Increasing Control of Topography and Evolution Towards Steady State. Geografiska Annaler: Ser A, Phys Geogr 97:543–562. https://doi.org/10.1111/geoa.12100

Carrivick JL, Smith MW, Carrivick DM (2015b) Terrestrial laser scanning to deliver high-resolution topography of the upper Tarfala valley, arctic Sweden. GFF 137:383–396. https://doi.org/10.1080/11035897.2015.1037569

Cavalli M, Trevisani S, Comiti F, Marchi L (2013) Geomorphometric assessment of spatial sediment connectivity in small Alpine catchments. Geomorphology 188:31–41. https://doi.org/10.1016/j.geomorph.2012.05.007

Church M, Ryder JM (1972) Paraglacial sedimentation: a consideration of fluvial processes conditioned by glaciation. Geol Soc Am Bull 83:3059. https://doi.org/10.1130/0016-7606(1972)83%5b3059:PSACOF%5d2.0.CO;2

Clubb FJ, Mudd SM, Milodowski DT, Valters DA, Slater LJ, Hurst MD, Limaye AB (2017) Geomorphometric delineation of floodplains and terraces from objectively defined topographic thresholds. Earth Surf Dyn 5:369–385. https://doi.org/10.5194/esurf-5-369-2017

D'Amico ME, Freppaz M, Filippa G, Zanini E (2014) Vegetation influence on soil formation rate in a proglacial chronosequence (Lys Glacier, NW Italian Alps). CATENA 113:122–137. https://doi.org/10.1016/j.catena.2013.10.001

Daten Österreichs (2016) Digitales Geländemodell (DGM) Österreich. https://www.data.gv.at/katalog/dataset/b5de6975-417b-4320-afdb-eb2a9e2a1dbf. Last visited Mar, 2016

Davies BJ, Glasser NF (2012) Accelerating shrinkage of Patagonian glaciers from the Little Ice Age (~AD 1870) to 2011. J Glaciol 58:1063–1084. https://doi.org/10.3189/2012JoG12J026

Egli M, Wernli M, Kneisel C, Haeberli W (2006) Melting glaciers and soil development in the proglacial area morteratsch (Swiss Alps): I. soil type chronosequence. Arct Antarct Alp Res 38:499–509. https://doi.org/10.1657/1523-0430(2006)38%5b499:MGASDI%5d2.0.CO;2

Fischer M, Huss M, Barboux C, Hoelzle M (2014) The New Swiss Glacier Inventory SGI2010: relevance of using high-resolution source data in areas dominated by very small glaciers. Arct Antarct Alp Res 46:933–945. https://doi.org/10.1657/1938-4246-46.4.933

Fischer A, Seiser B, Stocker Waldhuber M, Mitterer C, Abermann J (2015a) Tracing glacier changes in Austria from the Little Ice Age to the present using a lidar-based high-resolution glacier inventory in Austria. Cryosphere 9:753–766. https://doi.org/10.5194/tc-9-753-2015

Fischer M, Huss M, Hoelzle M (2015b) Surface elevation and mass changes of all Swiss glaciers 1980–2010. Cryosphere 9:525–540. https://doi.org/10.5194/tc-9-525-2015

Fischer A, Seiser B, Stocker-Waldhuber M, Abermann J (2015c) The Austrian Glacier Inventory GI 3, 2006. In: ArcGIS (shapefile) format. https://doi.org/10.1594/pangaea.844985

Fryirs KA, Wheaton JM, Brierley GJ (2016) An approach for measuring confinement and assessing the influence of valley setting on river forms and processes: measuring confinement along fluvial corridors. Earth Surf Proc Land 41:701–710. https://doi.org/10.1002/esp.3893

Geilhausen M, Otto J-C, Schrott L (2012) Spatial distribution of sediment storage types in two glacier landsystems (Pasterze & Obersulzbachkees, Hohe Tauern, Austria). J Maps 8:242–259. https://doi.org/10.1080/17445647.2012.708540

Glasser NF, Harrison S, Jansson KN (2009) Topographic controls on glacier sediment–landform associations around the temperate North Patagonian Icefield. Quat Sci Rev 28:2817–2832. https://doi.org/10.1016/j.quascirev.2009.07.011

Groß G, Patzelt G (2015) The Austrian Glacier inventory for the Little Ice Age maximum (GI LIA) in ArcGIS (shapefile) format. https://doi.org/10.1594/pangaea.844987

Hallet B, Hunter L, Bogen J (1996) Rates of erosion and sediment evacuation by glaciers: a review of field data and their implications. Global Planet Change 12:213–235. https://doi.org/10.1016/0921-8181(95)00021-6

Heckmann T, McColl S, Morche D (2016) Retreating ice: research in pro-glacial areas matters: research in pro-glacial areas. Earth Surf Proc Land 41:271–276. https://doi.org/10.1002/esp.3858

Jaeger JM, Koppes MN (2016) The role of the cryosphere in source-to-sink systems. Earth Sci Rev 153:43–76. https://doi.org/10.1016/j.earscirev.2015.09.011

Jiskoot H, Curran CJ, Tessler DL, Shenton LR (2009) Changes in Clemenceau Icefield and Chaba Group glaciers, Canada, related to hypsometry, tributary detachment, length–slope and area–aspect relations. Ann Glaciol 50:133–143. https://doi.org/10.3189/172756410790595796

Kjær KH, Korsgaard NJ, Schomacker A (2008) Impact of multiple glacier surges—a geomorphological map from Brúarjökull, East Iceland. J Maps 4:5–20. https://doi.org/10.4113/jom.2008.91

Knoll C, Kerschner H, Heller A, Rastner P (2009) A GIS-based Reconstruction of Little Ice Age Glacier maximum extents for South Tyrol, Italy: reconstruction of Little Ice Age Glacier maximum extents. Trans GIS 13:449–463. https://doi.org/10.1111/j.1467-9671.2009.01173.x

Lambiel C, Maillard B, Kummert M, Reynard E (2016) Geomorphology of the Hérens valley (Swiss Alps). J Maps 12:160–172. https://doi.org/10.1080/17445647.2014.999135

Lane SN, Bakker M, Gabbud C, Micheletti N, Saugy JN (2017) Sediment export, transient landscape response and catchment-scale connectivity following rapid climate warming and Alpine glacier recession. Geomorphology 277:210–227. https://doi.org/10.1016/j.geomorph.2016.02.015

Loye A, Jaboyedoff M, Pedrazzini A (2009) Identification of potential rockfall source areas at a regional scale using a DEM-based geomorphometric analysis. Nat Hazards Earth Syst Sci 9:1643–1653. https://doi.org/10.5194/nhess-9-1643-2009

Lucchesi S, Fioraso G, Bertotto S, Chiarle M (2014) Little Ice Age and contemporary glacier extent in the Western and South-Western Piedmont Alps (North-Western Italy). J Maps 10:409–423. https://doi.org/10.1080/17445647.2014.880226

Maisch M, Haeberli W, Hoelzle M, Wenzel J (1999) Occurrence of rocky and sedimentary glacier beds in the Swiss Alps as estimated from glacier-inventory data. Ann Glaciol 28:231–235. https://doi.org/10.3189/172756499781821779

Micheletti N, Lane SN (2016) Water yield and sediment export in small, partially glaciated Alpine watersheds in a warming climate. Water Resour Res 52:4924–4943. https://doi.org/10.1002/2016WR018774

Micheletti N, Lambiel C, Lane SN (2015) Investigating decadal-scale geomorphic dynamics in an alpine mountain setting. J Geophys Res Earth Surf 120:2155–2175. https://doi.org/10.1002/2015JF00 3656

Schiefer E, Gilbert R (2007) Reconstructing morphometric change in a proglacial landscape using historical aerial photography and automated DEM generation. Geomorphology 88:167–178. https://doi.org/10.1016/j.geomorph.2006.11.003

Shiyin L, Wenxin S, Yongping S, Gang L (2003) Glacier changes since the Little Ice Age maximum in the western Qilian Shan, Northwest China, and consequences of glacier runoff for water supply. J Glaciol 49:117–124. https://doi.org/10.3189/172756503781830926

Smith MJ, Rose J, Booth S (2006) Geomorphological mapping of glacial landforms from remotely sensed data: an evaluation of the principal data sources and an assessment of their quality. Geomorphology 76:148–165. https://doi.org/10.1016/j.geomorph.2005.11.001

Smith MW, Carrivick JL, Quincey DJ (2016) Structure from motion photogrammetry in physical geography. Prog Phys Geogr 40:247–275. https://doi.org/10.1177/0309133315615805

Solomon S (ed) (2007) Climate change 2007-the physical science basis: Working group I contribution to the fourth assessment report of the IPCC. Cambridge University Press, Cambridge

Zanoner T, Carton A, Seppi R et al (2017) Little Ice Age mapping as a tool for identifying hazard in the paraglacial environment: the case study of Trentino (Eastern Italian Alps). Geomorphology 295:551–562. https://doi.org/10.1016/j.geomorph.2017.08.014

Debris-Covered Glaciers

4

Elisabeth Mayr and Wilfried Hagg

Abstract

Debris-covered glaciers are characterised by a mantle of rock material, the supraglacial debris, spread over part of the ablation zone. The debris material origins from the catchment above and the bedrock below the glacier and appears at the glacier surface in the ablation zone. It often forms medial moraines in the upper areas which increase down glacier until the glacier is covered across its full width. Debris-covered glaciers are located in all high relief areas around the world, but quantitative information on global scale does not exist. A debris cover affects the ablation of the subjacent glacier ice by increasing it below thin covers and protecting the ice below thicker ones. The critical debris thickness is usually a few centimetres. The surface morphology of debris-covered glaciers can be hilly and ruptured by supraglacial lakes and ice cliffs, which play an important role in glacier mass balance as they are regions of significantly enhanced ice melt. The glacier snouts of debris-covered glaciers are often very slow or stagnant, and the terminus positions are stable over long periods. In this case, only small amounts of debris are transported from the glacial into the proglacial system. It is still a matter of discussion, if debris covers protect glaciers from mass loss due to climate change. They are important sediment sources and are deposited as very large lateral moraines, as lateral–frontal moraines or, after glacier wastage, as supraglacial melt-out. Additionally, a thick debris cover can serve as habitat for animate beings.

Keywords

Debris-covered glaciers · Debris supply
Debris transport · Morphology · Vegetation

4.1 Definition, Development and Global Distribution of Debris-Covered Glaciers

Debris-covered glaciers are characterised by the rock material, the so-called supraglacial debris, which covers part of the ablation zone across its full width (Kirkbride 2011). The dynamics of bare ice glaciers in alpine areas were already

E. Mayr (✉) · W. Hagg
Department of Geography,
Ludwig-Maximilians-University, Munich, Germany
e-mail: e.mayr.geographie@mailbox.org

T. Heckmann and D. Morche (eds.), *Geomorphology of Proglacial Systems*,
Geography of the Physical Environment, https://doi.org/10.1007/978-3-319-94184-4_4

discussed in Chap. 2. A debris cover strongly influences and sometimes even changes these dynamics, like mass balance and ice flow. In the following chapter, we will describe the development of debris-covered glaciers, show their global distribution and explain their special characteristics.

4.1.1 Debris Origin and Development of Debris-Covered Glaciers

The debris covering the glacier snout originates from the erosion of the relief and sediment accumulations in the glacier catchment and bed. The majority of debris from above origins in the steep walls of the glacier catchment and is transported gravitationally to the glacier surface (Anderson 2000). Responsible processes are rockfalls, rock avalanches, debris flows as well as debris-laden ice and snow avalanches (Hambrey et al. 2008). Accordingly, the weathering rates of the surrounding rocks, controlled by climate and lithology, and the topography of the surrounding terrain are the controlling factors of debris supply (see Fig. 4.1). Another debris source are lateral moraines of the glacier itself (for more detail about lateral moraines, see below), which become unstable when the glacier surface lowers (Hambrey et al. 2008). In special cases, debris can also be deposited by aeolian dust, tephra of volcanoes, salt, micro-organisms, meteorites and anthropogenic pollution (black carbon) (Benn and Evans 1998). The deposited debris is transported at the ice surface or inside the glacier, but relatively close to its surface, in the so-called high-level transport zone (Benn and Evans 2010).

The bedrock beneath the glacier plays a minor role for debris supply. Here, rock particles are collected by glacial erosion (abrasion and plucking) or the entrainment of basal debris (Benn and Evans 1998). Controlling factors are substrate erodibility and the ice sliding velocity (Herman et al. 2015). This material usually remains in the so-called basal transport zone (Benn and Evans 1998) but in some cases can also be elevated to the high-level transport zone close to the glacier terminus. There, basal debris is moved along shear planes between active and stagnant ice to the glacier surface (Benn and Evans 1998) or at glacier termini that are obstructed, for example, by moraine dams (Kirkbride and Deline 2013). Also in zones of ice confluence or in the lee of bedrock obstacles,

Fig. 4.1 Debris-covered terminus of Pasterze Glacier, Austria. More debris is provided by the high slope to the left than by the low slope to the right. The debris cover protects the glacier ice against melting

upward debris transport is possible: here, flow lines from the basal transport zone can be elevated to the high-level transport zone. The material transported by this process is then called 'elevated debris' (Benn and Evans 1998). Debris-covered glaciers are often associated with such elevated glacier beds (Kirkbride 2000).

In the upper ablation zone, debris material starts appearing as debris stripes parallel to the flow line, which grow downglacier until the whole glacier width is covered. These so-called medial moraines can be distinguished by debris supply and morphological development into three main types. Material that is deposited in the accumulation area of a glacier is covered by snow and disappears from the surface due to submergence. Below the equilibrium line, in the ablation zone, this debris returns to the surface due to emergence and downwasting (Anderson 2000). This type is called ablation-dominant moraine (Eyles and Rogerson 1978). Ice–stream interaction moraines form below ice confluences where two supraglacial lateral moraines merge into a medial moraine. Exceptional rockfall events form the so-called avalanche-type moraines (Eyles and Rogerson 1978). They can be recognised as megablocks on the medial moraines, for example on Unteraargletscher in Switzerland (Deline 2005) or as limited bands of thicker debris cover caused by rock slide episodes, like on Miage Glacier in Italy (Mihalcea et al. 2008). In general, debris can give hints on its area of origin and type of transport. In catchments with a strongly varying geology, different areas of origin are visible as distinct debris bands for long distances along the glacier until they are totally redistributed.

Moraine bands usually grow in width downglacier, while their depth remains nearly constant. This is caused by the isolating effect of the debris (see Sect. 4.2.1). The ice below the moraine bands melts less than the adjacent bare ice. On the resulting slopes between debris-covered and bare ice, the debris slides towards the bare ice areas and the width of the moraine increases. As soon as adjacent moraine bands connect, debris thickness starts to increase downglacier (Anderson 2000). The appearance of supraglacial debris ranges from thin dust layers to a cover of several metres thickness, e.g. 4 m in the Himalaya (Shroder et al. 2000).

As mentioned above, the majority of supraglacial debris origins in the walls above the glacier and is transported onto the glacier surface by gravitational processes. For this reason, the composition of debris is dominated by angular clasts (Bolch 2011), which may become subangular or even subrounded by redistribution due to ablation. Debris transport by meltwater also leads to some fluvially rounded gravel and sand (Hambrey et al. 2008). Subglacial debris in the lower transport zone experiences more stress. If the material is elevated to the upper transport zone, it shows signs of abrasion, fraction and comminution (Benn et al. 2003).

The size of the debris material ranges from fine silt and clay up to blocks with a several metres in diameter. The material is often sorted vertically, with a plaster of bigger clasts at the surface and more matrix-rich layers below (Kirkbride 2011). Due to melting at the ice surface, the lower part of the debris cover often is wet.

In summary, if a glacier develops a significant debris cover depends on the weathering rates of the surrounding rocks, the height and steepness of the surrounding terrain and the effectivity of the other debris sources. But additionally, if the provided material is sufficient to form an extended debris cover strongly depends on ice velocity. Glaciers with high bare ice ablation rates but slow ice flow velocities are unable to evacuate the accumulating debris cover (Kirkbride 2011). This relationship is explained by the model for debris cover expansion by Kirkbride (2000). In periods with positive mass balance, debris covers form further downglacier due to faster ice flow and lower bare ice ablation. If the mass balance is negative, debris covers spread upglacier due to reduced ice flow and increased ablation. Additionally, more debris is provided due to deglaciation of the surrounding relief and possible permafrost degradation. For this reason, debris covers appear on former debris-free glaciers or expand in area on already debris-covered glaciers due to global warming (Deline 2005).

As debris covers of sufficient thickness decrease ice melt and therefore mass loss, this leads to a negative feedback to climate change (Bosson and Lambiel 2016).

4.1.2 Differentiation Between Debris-Covered Glaciers and Rock Glaciers

Debris-covered glaciers are often associated with another debris-dominated landform, the so-called *rock glaciers* (see Chaps. 6 and 7). They are lobate, tongue-shaped debris masses which slowly move downslope. The surface is characterised by ridges and furrows, and the front of active rock glaciers is steep enough for debris sliding and falling down (Benn and Evans 2010). Even if there are some similarities between the morphology of rock glaciers and debris-covered glaciers, they usually differ considerably from debris-covered glaciers (Berthling 2011).

One possible differentiation is based on the source of the containing ice into *periglacial rock glaciers* (ice from permafrost) and *glacial rock glaciers/debris-covered glaciers* (ice from glaciers). As it is difficult to distinguish 'glacier ice' from 'permafrost ice' (Haeberli et al. 2006), other authors (e.g. Berthling 2011) suggest that not the origin of the ice but the existence of permafrost ('cryo-conditioning') should be the primary condition for the formation of rock glaciers.

The prerequisites for a rock glacier to form are discontinuous permafrost, availability of sufficient ice-supersaturated debris and a relief steep enough to allow the permafrost body to creep (Barsch 1996). The debris origins are quite similar to those of debris-covered glaciers. The contained ice can be formed by several processes: at the surface by snow metamorphism and surface icing and at the subsurface by freezing of liquid water from groundwater, rain or various kinds of meltwater (Haeberli et al. 2006; Barsch 1996). Additionally, rock glaciers can evolve from glaciers and vice versa (Benn and Evans 2010). Inactive rock glaciers still contain ice, but not enough to allow creeping, while fossil forms are completely melted-out debris bodies (Wahrhaftig and Cox 1959).

4.1.3 Global Distribution of Debris-Covered Glaciers

Mountain regions with debris-covered glaciers are spread all over the globe. As described above, steep slopes and high denudation rates are prerequisites for the formation of debris-covered glaciers, as they are able to supply the necessary quantities of debris (Kirkbride 2011). Such situations are especially found at geologically young, uplifting orogenic belts (Foster et al. 2012) in recently tectonically active regions (Benn et al. 2003). Extensive data about the global distribution of debris-covered glaciers and the glacier area covered by debris are not available. This is mainly due to the difficulties in the delineation of debris-covered areas by remote sensing (for more details, see below). Nevertheless, numerous studies about individual debris-covered glaciers or their occurrence in special regions have already been published.

In the Himalaya, the glaciers are characteristically covered with debris (Shroder et al. 2000). For example, in the Khumbu Himal Mountain range in Nepal, 95% of the glaciers with a total area >2 km^2 are debris-covered (Moribayashi and Higuchi 1977). The same is true for the adjacent mountain ranges Karakoram and Pamir (Bolch et al. 2012; Gardelle et al. 2013; Shroder et al. 2000) as well as for the Tian Shan mountain range north of the Taklamakan desert (Juen et al. 2014; Mayr et al. 2014; Hagg et al. 2008). Bolch et al. (2012) estimated for the Himalaya and Karakoram region a total debris cover of ~10%. Other regions with debris-covered areas are the Southern Alps in New Zealand and Southern Alaska (Kirkbride 1993; Brook et al. 2012). In the Central Caucasus, the individual glaciers have a debris cover of <5% up to >25% of the total glacier area (Stokes et al. 2007). The Andean glaciers are only slightly (2.5% of glacier area) debris-covered (Janke et al. 2015). At the Cordillera Blanca in Peru, only 3% of the glacierised area is debris-covered (Racoviteanu et al. 2008). Glaciers in the European Alps are not typically debris-covered, but their number and the share of debris-covered areas increased in recent years (Caccianiga et al. 2011; Kirkbride and Deline 2013). Such a debris-covered area expansion has been documented in many glaciated areas with retreating glaciers due to climate warming (Foster et al. 2012; Mayer et al. 2011). In Central Asia, the general glacier degradation leads to increasing debris covers on many glaciers (Ageta et al. 2000). In the Central Caucasus, the debris-covered area increased by 3–6% between 1985 and 2000 (Stokes et al. 2007).

4.2 Characteristics and Dynamics of Debris-Covered Glaciers

4.2.1 Subdebris Ablation

Supraglacial debris strongly affects the ablation of the subjacent glacier ice. Debris covers usually have a lower albedo than clean ice and absorb more incoming radiation. In contrast to ice, their surface temperature can reach values far above 0 °C, resulting in a conduction of heat into lower layers. If the debris mantle is thin, this flux causes an energy gain at the debris/ice interface and increases melt (Fig. 4.2). Additionally, a debris cover decreases the wind speed at the ice surface. This reduces the energy loss due to evaporation, which is now available for melting (Evatt et al. 2015). Above a critical debris thickness of few centimetres, the isolating effect of the debris predominates this surplus of incoming melt energy and ice melt decreases significantly with increasing debris thickness. This dependency of debris thickness and melt rates was first described by Østrem (1959) and was since then confirmed by many authors (e.g. Mattson 2000; Nicholson and Benn 2013). The protecting effect of thick debris covers is shown in Fig. 4.1, where the debris-covered glacier area shows less ice loss than the bare ice parts.

The critical debris thicknesses and the magnitude of the isolating effect are governed by the thermal conductivity of the debris material. Thermal conductivity depends on albedo, grain size, density of the material as well as air humidity. For this reason, the effect of a certain debris thickness on the melt rates can vary significantly. For debris covers of 10 cm, Lambrecht et al. (2011) discovered thermal resistances (the reciprocal of thermal conductivity, in $°C\ m^2\ W^{-1}$) between 0.04 and 0.12 on Asian glaciers, proving a variation in the energy flux by a factor of three. As described above, both debris thickness and thermal conductivity vary considerably along the glacier surface. This so-called *selective ablation* has strong influence of the surface morphology of debris-covered glacier areas (see below) (Benn and Evans 2010).

The influence of debris cover on melt rates impacts on the ablation gradient of debris-covered glaciers. At clean ice glaciers, highest melt rates are measured in the lowest glacier areas where temperatures are highest. At debris-covered glaciers, debris thickness increases towards glacier terminus and its isolating effect exceeds the melt increase due to higher temperatures. This leads to a reverse of the normal ablation gradient (Kirkbride 1993). The areas with highest melt rates are measured in areas with thin debris covers (Benn et al. 2012) which are usually located below but close to the equilibrium line.

4.2.2 Surface Morphology and Melt Hot Spots

Debris-covered glaciers usually show a complicated surface morphology. A debris-covered glacier tongue is not flat but hilly, and its surface is ruptured by supraglacial lakes (c.f. Chap. 14 on proglacial lakes) and ice cliffs. Supraglacial lakes are hollow moulds filled with water. Ice cliffs are ice surfaces steep enough, so debris slides down, revealing bare ice only covered with a very thin dust layer. These typical phenomena on debris-covered glaciers are called *glacial karst* (Mavlyudov 2006). They are formed by the above-described selective ablation and by the collapse of englacial run-off channels.

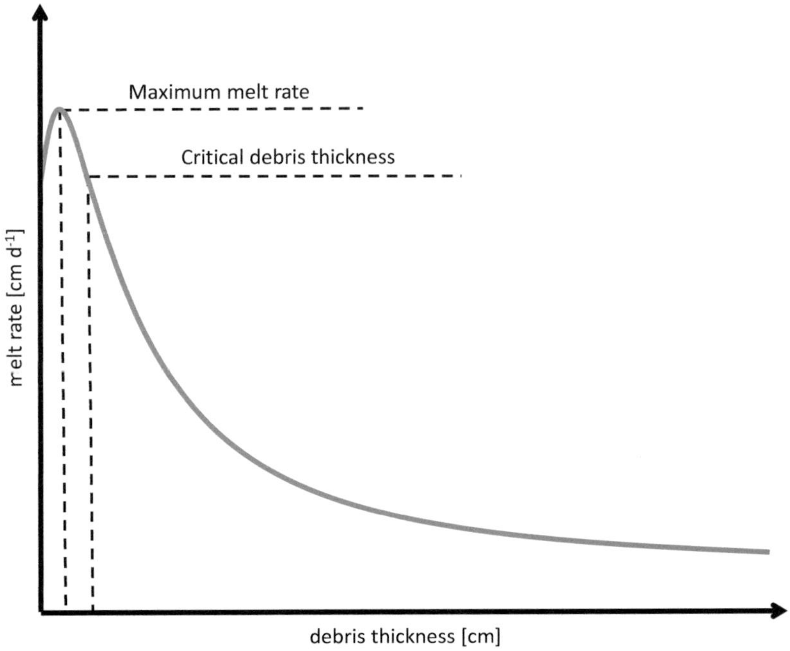

Fig. 4.2 Relationship between debris thickness and melt rate. The maximum melt rate is reached below a thin debris cover. As soon as a certain critical debris thickness of few centimetres is reached, melt below debris is decreased compared to bare ice (e.g. Østrem 1959; Nakawo and Young 1982)

The latter happens particularly in the lower ablation zone, where the channel roof is thinned due to ablation (Kirkbride 1993).

The distribution of supraglacial lakes depends on the inclination of the glacier surface. On slopes between 6 and 10°, only small and isolated lakes appear. More small lakes appear in areas of 2–6° inclination (Benn et al. 2012). This type of glacier lakes is called perched lakes. They form in hollows on impermeable ice without connection to the en- or subglacial channel system and are filled by the precipitation and meltwater provided by their own small supraglacial catchments. They are emptied as soon as they are connected to the glacier run-off system. For this reason, these lakes are more common in uncrevassed flat glacier areas with slow ice flow. Also, they cannot grow unlimited, because growing sooner or later cuts a crevasse or channel (Benn et al. 2001, 2012).

Small lakes tend to connect and form bigger lakes as soon as the hydrological base level of the glacier is intersected by the downwasting glacier surface (Benn et al. 2012). These lakes are called *base-level lakes* and have a connection to the englacial run-off system (Roehl 2008). They form in flat glacier regions with inclination of 0–2° and have a larger lifespan than perched lakes (Benn et al. 2012). Their existence plays an important role in fast terminus disintegration, predominantly due to calving retreat (Roehl 2008). Base-level lakes are often associated with closed terminal moraine dams—sometimes ice-cored—in front of the glacier snout where the glacial run-off is collected (Benn et al. 2012; Kirkbride 1993), then called *moraine-dammed lakes* (Kirkbride 2011).

The melt rates at ice cliffs and their evolution are strongly dependent on their exposition (Juen et al. 2014; Sakai et al. 2002). Compared to bare ice, melt is enhanced at south-, east- and west-exposed cliffs, but reduced at north-exposed ones. South-exposed cliffs are melting mainly due to shortwave radiation which is smaller at the lower cliff areas due to shadowing effects. That is why these cliffs loose inclination and get buried by debris rather quickly. In contrary, north-exposed cliffs receive their melt energy from longwave radiation of the surrounding debris which is more effective in the lower areas and in general smaller than at south-exposed cliffs. For this reason, north-exposed cliffs

Fig. 4.3 Supraglacial lake with ice cliff at Koxcar Glacier, Tian Shan, China

remain steeper, last longer and are more numerous (Sakai et al. 2002; Reid and Brock 2014).

Ice cliffs are not necessarily associated with supraglacial lakes, but many cliffs emerge directly out of them (see Fig. 4.3). Due to its low albedo, the lake water warms at the surface, which is an important energy source for the melting of the ice cliffs at the waterline, where the so-called *thermo-erosional notches* form. These notches are a prerequisite for calving processes from the ice cliff into the lake (Sakai et al. 2009). Wind, which transports the warm surface water to the lake margin, plays an important role within this energy transfer (Sakai et al. 2009; Benn et al. 2001). The lake floor is predominantly covered by fine sediment that is nearly impermeable and inhibits heat convection. Melt rates at the lake bottoms are thus dependent on water temperature and debris thickness and are supposed to be rather small (Sakai et al. 2000).

While debris cover generally leads to decreasing overall melt rates on a glacier, supraglacial lakes as well as ice cliffs are melt hot spots which can explain downwasting even on heavily debris-covered glaciers. On Koxcar Glacier in the Chinese Tian Shan, ice cliffs cover only 1.7% of the debris-covered area, but yield 12% of total meltwater (Juen et al. 2014).

4.2.3 Mass Balance of Debris-Covered Glaciers

Melt below debris as well as at melt hot spots clearly influence the glacier mass balance of debris-covered glaciers. Due to the shielding effect of thick debris covers, glacier termini of debris-covered glaciers descend to lower elevations than bare ice glaciers in the same region (Deline 2005). In the Khumbu Himal for example, the difference in terminus elevation is 300–400 m (Moribayashi and Higuchi 1977). Due to the decreased ablation below debris, a larger ablation area and thus a lower glacier terminus are required to reach net balances comparable to those of neighbouring bare ice glaciers. For the same reason, a debris-covered glacier needs to advance under constant climate to maintain equilibrium (Benn et al. 2003).

On debris-covered glacier snouts, the ice velocity is often very slow or the ice is completely stagnant (Bolch et al. 2012). This is caused by the low or even inversed surface gradients and the general reduction of ice thickness during the past decades (Benn et al. 2012). Generally, debris-covered glaciers show a smaller amplitude of frontal fluctuations than bare ice glaciers (Deline 2005) and their terminus is usually stable over long periods (Benn et al. 2003; Schmidt and Nüsser 2009). While mass fluctuations on bare ice glaciers are expressed in advance or retreat of the terminus, debris-covered glaciers reflect these changes only partly (Mayer et al. 2006). Instead, debris-covered glacier snouts tend to react by thickening or thinning (Kirkbride 2011). Additionally, they often show a propagation of mass gain or loss on decadal timescales (e.g. Capt et al. 2016). Exceptions are ice contact lakes, which can cause very rapid terminus degradation by calving (Kirkbride 1993).

It is still unclear if debris-covered glaciers have less negative mass balances than bare ice glaciers. While on heavily debris-covered areas the melt is clearly decreased, thin debris covers and especially melt hot spots locally increase melt. For this reason, the climatic sensitivity of areas with thin debris and hot spots is increased, while a thick debris cover strongly reduces this sensitivity. Scherler et al. (2011) observed that retreat rates at Himalayan glaciers can vary from high at debris-free glaciers to zero at debris-covered glaciers. In contrast and in contrary to expectations, Kääb et al. (2012) found, for the Hindu Kush–Karakoram–Himalaya region, that regional averaged mass losses at debris-covered areas were similar to those of clean ice, which they explained with the enhanced melt processes at ice cliffs and glacial lakes. More research is necessary to examine the effect of climate change to debris-covered glaciers, especially on possible back coupling effects between enhanced melt due to climate warming and increasing debris thickness due to enhanced melt-out.

4.2.4 Debris Transport, Deposition and Input into the Proglacial System

At debris-covered glaciers, debris is transported glacially and glaciofluvially. How much debris is transported by glacier ice depends on the ice velocity and the debris load (Benn et al. 2003). At the lateral glacier margin below the equilibrium line, this material forms lateral moraines by debris dumping from the glacier margin (Winkler 2009). Their large vertical size can be explained by the relatively stable location of lateral glacier margins and is further caused by the fact that these moraines are often formed during several advances of similar magnitudes (Winkler 2009). They are usually much bigger than those of bare ice glaciers, because of the high debris load and the long-term stability of the ice contact zone due to low terminus velocity at debris-covered glaciers (Benn et al. 2003). Near the glacier terminus, these lateral moraines can reach an elevation of tens to hundreds of metres (Kirkbride 2011). Sometimes, they are developed asymmetrically due to differences in valley geometry and geology or due to different patterns of ice flow (Winkler 2009). They are major sediment sinks at debris-covered glaciers (Benn et al. 2003) and can be reactivated as a source of supraglacial debris if the glacier surface lowers (Hambrey et al. 2008). In the lower glacier areas, terminal moraines often are separated from the adjacent walls. In the space in between the moraine and the wall, debris from the valley side can be trapped (Benn et al. 2003) and even lakes can develop. The material of lateral moraines consists of many angular components, because most of it has a supraglacial origin. Due to the passive dumping process, it shows characteristically subparallel alignment to the external moraine slope and is then called *layered lateral moraine* (Humlum 1978).

Glacially transported material, which is not deposited in lateral moraines, is transported by the moving ice down glacier to the glacier snout. There it forms a lateral–frontal moraine as long as the glacier run-off is not strong enough to remove it (Benn et al. 2003). However, as discussed above, many debris-covered glaciers experience a slowing down of ice movement at the glacier tongue or are even stagnant (Benn et al. 2012). For this reason, the majority of supraglacial debris remains on the glacier surface where it is redistributed and reworked during ice ablation (Benn et al. 2003). If it is finally sedimented, it is called *supraglacial melt-out till.* Compared to subglacial deposited till, it is less compacted, contains less rounded components and shows no alignment and striation because of the missing subglacial transport and ice pressure during sedimentation. Additionally, it is characterised by bigger grain size, because fine material was washed out by meltwater on the glacier surface (Winkler 2009). The landform built by melt-out till is called *hummocky moraine* and is characterised by an irregular surface, crossed by moulds and ridges (Benn et al. 2003). Also other ice disintegration and negative relief features like pits and depressions are common in this landscape, which is therefore also referred to as *kame and kettle topography* (Sugden and John 1976).

As the glacial transport to the proglacial system is limited, as mentioned above, the input of sediment into the proglacial system is dominated by glaciofluvial transport in englacial and supraglacial pathways. This coupling is strong during periods of glacier advance or standstill, but small during periods of glacier retreat (Benn et al. 2003). In the latter case, the material cannot be evacuated downslope because of the small slope angle of the glacier terminus and a possible surrounding moraine dam, forming sediment sinks and transport-limited systems (Bosson et al. 2015). Usually only small amounts of debris are directly transported from the glacial into the proglacial system (Kirkbride 2011). An important exception is sediment trapped in glacial lakes, which can be mobilised quickly during glacier lake outburst floods (GLOFs).

4.2.5 Vegetation on Debris-Covered Glaciers and Their Role as Plant Refugia

Supraglacial debris is not only an important sediment source. On many heavily debris-covered glacier tongues, it is also a—more or less mobile—habitat for animate beings such as micro-organisms, animals, yeasts and plants (Fickert et al. 2007; Vezzola et al. 2016). In particular, the vegetation is often very rich in both, species and individuals (Caccianiga et al. 2011). For example, Miage Glacier in Italy reveals substantial supraglacial vegetation; in total, 45 vascular plant species were found, among them even several shrub species and well-developed trees (Vezzola et al. 2016; Caccianiga et al. 2011). In south-central Alaska, more than a dozen debris-covered glaciers with mature forests and trees of more than 50 cm in diameter on their surface exist (Fickert et al. 2007). At Koxkar Glacier in the Chinese Tian Shan, even grazing sheep of the local nomads can be spotted on the glacier.

If a debris cover is populated by vegetation or not depends on the different parameters. Most important are the elevation of the ablation zone and the associated climatic conditions and substrate stability, which is controlled by glacier velocity as well as slope angle and thinning rate (Vezzola et al. 2016; Caccianiga et al. 2011). Debris thickness is important as it influences the root frost occurrence in summer. At the ice/debris interface, temperatures are always at the melting point; therefore, the roots in thick debris covers are in warmer substrate than in thin covers (Vezzola et al. 2016). At Miage Glacier, Vezzola et al. (2016) investigated the following threshold values for tree growth: a maximum ice velocity of 7 m/a, a minimum debris thickness of 19 cm and a slope angle maximum of 10°.

It is discussed if, in the past, debris-covered glaciers have served as plant refugia during cold or warm periods. During glacial periods in the Pleistocene, former mountain plants lost their living environment due to growth of extended ice sheets. Debris-covered glaciers may have provided refugia for those plants (Fickert et al. 2007). However, due to high ice velocities of advancing glaciers and shrinking debris source areas due to increasing ice covers, the existence of extensive debris covers during the glacier maximum is questionable (Caccianiga et al. 2011). Plant refugia for high-altitude species during warm periods in the Holocene are more probable (Caccianiga et al. 2011). Today, alpine vegetation grows in lower elevations on debris-covered glaciers than on neighbouring ice-free slope, which can be explained by the cooling effect of the glacier ice (Fickert et al. 2007). During interglacial periods, these areas might have acted as refugia for species occurring near the glacier edge during the glacier maximum.

These interactions should be considered in scenarios aiming at the response of alpine ecosystems to climate change (Caccianiga et al. 2011). Additionally, those plant covers may have other consequences: a more or less closed forest probably protects the adjacent ice from melting and could reduce the speed of glacier downwasting and thereby attenuates the effect of climate warming. If the glacier tongue is finally melted down completely, the already existing plant cover may skip several steps of vegetation succession compared to other recently deglacierised areas (see Chap. 19 for a review of vegetation development in proglacial areas).

4.2.6 Difficulties in Observation and Modelling

As mentioned above, the delineation of debris-covered areas by remote sensing causes some difficulties. Bare ice and snow can be mapped by means of multispectral classification. This approach does not work in debris-covered areas, because they have the same spectral properties as debris outside the glacier margin such as moraines or fluvioglacial deposits (Paul et al. 2004). Thin debris covers can be detected automatically using the near-infrared bands of, e.g. ASTER satellite data because debris covers above ice are colder than pure debris. Unfortunately, this method is limited to areas with debris thicknesses below 0.5 m (Ranzi et al. 2004), because above this value, the debris surface temperature becomes independent from debris thickness (Foster et al. 2012). Including surface shape parameters like slope angle derived from DEMs helps significantly in those areas (Bolch et al. 2007). Another promising approach is the use of differential synthetic aperture radar interferometry (DInSAR) techniques (Fischer et al. 2014).

For ablation modelling, debris thickness distribution is always an indispensable prerequisite. This information can be retrieved from an empirical approach, where debris thickness measured in the field is related to remotely sense surface temperatures derived by near-infrared bands (Mihalcea et al. 2008). If empirical data are missing, it is also possible to solve the energy balance at the debris surface using the same remote sensing data (Foster et al. 2012). Both methods underlie the debris thickness limit of 0.5 m.

For melt- and run-off simulation scenarios, the future development of the glacier area is of crucial importance. On clean ice glaciers, this area change can be calculated for example based on the relationship between glacier size and the steady-state accumulation area ratio. Climate warming leads to an uplift of the equilibrium line and to a reduction of the accumulation area, which can be used to calculate the new glacier geometry (Paul et al. 2007). As debris-covered glaciers react in a different way to mass balance changes, as described above, such a simple method cannot be applied. Several studies aim at modelling mass balance and glacier dynamics of debris-covered glaciers (e.g. Huss and Fischer 2016; Rowan et al. 2015; Jouvet et al. 2011), but it remains a difficult task to model the development of debris-covered glacier area reliably.

Acknowledgements Wilfried Hagg is supported by the DFG Heisenberg programme (project HA 5061/3-1). The valuable comments by Jean-Baptiste Bosson are highly acknowledged.

References

Ageta Y, Iwata S, Yabuki H, Naito N, Sakai A, Narama C, Karma T (2000) Expansion of glacier lakes in recent decades in the Bhutan Himalayas. Debris-Covered Glaciers IAHS Publ 264:165–175

Anderson RS (2000) A model of ablation-dominated medial moraines and the generation of debris-mantled glacier snouts. J Glaciol 46(154):459–469

Barsch D (1996) Rockglaciers: indicators for the present and former geoecology in high mountain environments, vol 16. Springer, Berlin

Benn D, Evans D (1998) Glaciers and glaciation. United Kingdom, London

Benn D, Evans D (2010) Glaciers and glaciation. Hodder Education, London

Benn DI, Wiseman S, Hands KA (2001) Growth and drainage of supraglacial lakes on debris-mantled Ngozumpa Glacier, Khumbu Himal, Nepal. J Glaciol 47(159):626–638

Benn DI, Kirkbride MP, Owen LA, Brazier V (2003) Glaciated valley landsystems. Glacial Landsystems, pp. 372–406. Arnold, London

Benn DI, Bolch T, Hands K, Gulley J, Luckman A, Nicholson LI, Quincey D, Thompson S, Toumi R, Wiseman S (2012) Response of debris-covered glaciers in the Mount Everest region to recent warming, and implications for outburst flood hazards. Earth-Sci Rev 114(1–2):156–174

Berthling I (2011) Beyond confusion: Rock glaciers as cryo-conditioned landforms. Geomorphology 131 (3):98–106

Bolch T (2011) Debris. In: Singh VP, Singh P, Haritashya UK (eds) Encyclopedia of snow, ice and glaciers. Springer, Berlin, pp 186–188

Bolch T, Buchroithner MF, Kunert A, Kamp U (2007) Automated delineation of debris-covered glaciers based on ASTER data. In: Geoinformation in Europe, Proceedings of the 27th EARSeL Symposium, pp 4–6

Bolch T, Kulkarni A, Kaab A, Huggel C, Paul F, Cogley JG, Frey H, Kargel JS, Fujita K, Scheel M, Bajracharya S, Stoffel M (2012) The state and fate of Himalayan Glaciers. Science 336(6079):310–314. https://doi.org/10.1126/science.1215828

Bosson J-B, Lambiel C (2016) Internal structure and current evolution of very small debris-covered glacier systems located in alpine permafrost environments. Front Earth Sci 4:39

Bosson JB, Deline P, Bodin X, Schoeneich P, Baron L, Gardent M, Lambiel C (2015) The influence of ground ice distribution on geomorphic dynamics since the Little Ice Age in proglacial areas of two cirque glacier systems. Earth Surf Proc Land 40(5):666–680

Brook MS, Hagg W, Winkler S (2012) Debris cover and surface melt at a temperate alpine glacier: Franz Josef Glacier, New Zealand. New Zealand J Geol Geophys, https://doi.org/10.1080/00288306.2012.736391

Caccianiga M, Andreis C, Diolaiuti G, D'Agata C, Mihalcea C, Smiraglia C (2011) Alpine debris-covered glaciers as a habitat for plant life. Holocene 21(6):1011–1020

Capt M, Bosson J-B, Fischer M, Micheletti N, Lambiel C (2016) Decadal evolution of a very small heavily debris-covered glacier in an Alpine permafrost environment. J Glaciol 62(233):535–551

Deline P (2005) Change in surface debris cover on Mont Blanc massif glaciers after the 'Little Ice Age' termination. Holocene 15(2):302–309

Evatt GW, Abrahams ID, Heil M, Mayer C, Kingslake J, Mitchell SL, Fowler AC, Clark CD (2015) Glacial melt under a porous debris layer. J Glaciol 61 (229):825–836

Eyles N, Rogerson RJ (1978) A framework for the investigation of medial moraine formation: Austerdalsbreen, Norway, and Berendon Glacier, British Columbia. Canada. J Glaciol 20(82):99–113

Fickert T, Friend D, Grüninger F, Molnia B, Richter M (2007) Did debris-covered glaciers serve as Pleistocene refugia for plants? a new hypothesis derived from observations of recent plant growth on glacier surfaces. Arct Antarct Alp Res 39(2):245–257

Fischer M, Huss M, Barboux C, Hoelzle M (2014) The new Swiss Glacier Inventory SGI2010: relevance of using high-resolution source data in areas dominated by very small glaciers. Arct Antarct Alp Res 46 (4):933–945

Foster LA, Brock BW, Cutler MEJ, Diotri F (2012) A physically based method for estimating supraglacial debris thickness from thermal band remote-sensing data. J Glaciol 58(210):677–691

Gardelle J, Berthier E, Arnaud Y, Kaab A (2013) Region-wide glacier mass balances over the Pamir-Karakoram-Himalaya during 1999–2011. Cryosphere 7(4):1263–1286

Haeberli W, Hallet B, Arenson L, Elconin R, Humlum O, Kääb A, Kaufmann V, Ladanyi B, Matsuoka N, Springman S (2006) Permafrost creep and rock glacier dynamics. Permafrost Periglac Process 17(3):189–214

Hagg W, Mayer C, Lambrecht A, Helm A, Michailjlow W (2008) Glaciological results of the 2005 expedition to Inylchek Glacier, Central Tian Shan. Geog Environ Sustain 1:38–45

Hambrey MJ, Quincey DJ, Glasser NF, Reynolds JM, Richardson SJ, Clemmens S (2008) Sedimentological, geomorphological and dynamic context of debris-mantled glaciers, Mount Everest (Sagarmatha) region, Nepal. Quat Sci Rev 27(25–26):2361–2389

Herman F, Beyssac O, Brughelli M, Lane SN, Leprince S, Adatte T, Lin JY, Avouac J-P, Cox SC (2015) Erosion by an Alpine glacier. Science 350(6257):193–195

Humlum O (1978) Genesis of layered lateral moraines: implications for palaeoclimatology and lichenometry. Geografisk Tidsskrift-Danish J Geog 77(1):65–72

Huss M, Fischer M (2016) Sensitivity of very small glaciers in the Swiss Alps to future climate change. Front Earth Sci 4:34

Janke JR, Bellisario AC, Ferrando FA (2015) Classification of debris-covered glaciers and rock glaciers in the Andes of central Chile. Geomorphology 241:98–121

Jouvet G, Huss M, Funk M, Blatter H (2011) Modelling the retreat of Grosser Aletschgletscher, Switzerland, in a changing climate. J Glaciol 57(206):1033–1045

Juen M, Mayer C, Lambrecht A, Han H, Liu S (2014) Impact of varying debris cover thickness on ablation: a case study for Koxkar Glacier in the Tien Shan. Cryosphere 8(2):377–386. https://doi.org/10.5194/tc-8-377-2014

Kääb A, Berthier E, Nuth C, Gardelle J, Arnaud Y (2012) Contrasting patterns of early twenty-first-century glacier mass change in the Himalayas. Nature 488 (7412):495–498

Kirkbride MP (1993) The temporal significance of transitions from melting to calving termini at glaciers in the central Southern Alps of New Zealand. The Holocene 3:232–240

Kirkbride MP (2000) Ice-marginal geomorphology and holocene expansion of debris-covered Tasman Glacier, New Zealand. Debris-Covered Glaciers 264:211–217

Kirkbride MP (2011) Debris-covered glaciers. In: Singh VP, Singh P, Haritashya UK (eds) Encyclopedia of snow, ice and glaciers. Springer, Berlin, pp 180–182

Kirkbride MP, Deline P (2013) The formation of supraglacial debris covers by primary dispersal from transverse englacial debris bands. Earth Surf Proc Land 38(15):1779–1792

Lambrecht A, Mayer C, Hagg W, Popovnin V, Rezepkin A, Lomidze N, Svanadze D (2011) A comparison of glacier melt on debris-covered glaciers in the Northern and Southern Caucasus. Cryosphere 5(3):525–538

Mattson LE (2000) The influence of a debris cover on the midsummer discharge of Dome Glacier, Canadian Rocky Mountains. Debris-Covered Glaciers IAHS Publ 264:25–33

Mavlyudov B (2006) Glacial karst, why it important to research. Acta Carsologica 35(1):55–67

Mayer C, Lambrecht A, Belo M, Smiraglia C, Diolaiuti G (2006) Glaciological characteristics of the ablation zone of Baltoro glacier, Karakoram, Pakistan. Ann Glaciol 43(43):123–131

Mayer C, Lambrecht A, Hagg W, Narozhny Y (2011) Glacial debris cover and melt water production for glaciers in the Altay, Russia. Cryosphere Discuss 5 (1):401–430

Mayr E, Juen M, Mayer C, Usubaliev R, Hagg W (2014) Modeling runoff from the Inylchek Glaciers and filling of Ice-Dammed Lake Merzbacher, Central Tian Shan. Geogr Ann A 96(4):609–625. https://doi.org/10.1111/Geoa.12061

Mihalcea C, Brock BW, Diolaiuti G, D'Agata C, Citterio M, Kirkbride MP, Cutler MEJ, Smiraglia C (2008) Using ASTER satellite and ground-based surface temperature measurements to derive supraglacial debris cover and thickness patterns on Miage Glacier (Mont Blanc Massif, Italy). Cold Reg Sci Technol 52(3):341–354

Moribayashi S, Higuchi K (1977) Characteristics of glaciers in the Khumbu region and their recent variations. 雪氷 39 (Special):3–6

Nakawo M, Young GJ (1982) Estimate of Glacier ablation under a debris layer from surface-temperature and meteorological variables. J Glaciol 28(98):29–34

Nicholson L, Benn DI (2013) Properties of natural supraglacial debris in relation to modelling sub-debris ice ablation. Earth Surf Proc Land 38 (5):490–501

Østrem G (1959) Ice melting under a thin layer of moraine, and the existence of ice cores in moraine ridges. Geogr Ann 41:228–230

Paul F, Huggel C, Kaab A (2004) Combining satellite multispectral image data and a digital elevation model for mapping debris-covered glaciers. Remote Sens Environ 89(4):510–518

Paul F, Maisch M, Rothenbuhler C, Hoelzle M, Haeberli W (2007) Calculation and visualisation of future glacier extent in the Swiss Alps by means of hypsographic modelling. Global Planet Change 55 (4):343–357

Racoviteanu AE, Arnaud Y, Williams MW, Ordonez J (2008) Decadal changes in glacier parameters in the Cordillera Blanca, Peru, derived from remote sensing. J Glaciol 54(186):499–510

Ranzi R, Grossi G, Iacovelli L, Taschner S (2004) Use of multispectral ASTER images for mapping debris-covered glaciers within the GLIMS Project. In: Geoscience and Remote Sensing Symposium, 2004. IGARSS'04. Proceedings. 2004 IEEE International. IEEE, pp 1144–1147

Reid TD, Brock BW (2014) Assessing ice-cliff backwasting and its contribution to total ablation of debris-covered Miage glacier, Mont Blanc massif, Italy. J Glaciol 60(219):3–13. https://doi.org/10.3189/2014jog13j045

Roehl K (2008) Characteristics and evolution of supraglacial ponds on debris-covered Tasman Glacier, New Zealand. J Glaciol 54(188):867–880

Rowan AV, Egholm DL, Quincey DJ, Glasser NF (2015) Modelling the feedbacks between mass balance, ice flow and debris transport to predict the response to climate change of debris-covered glaciers in the Himalaya. Earth Planet Sci Lett 430:427–438

Sakai A, Takeuchi N, Fujita K, Nakawo M (2000) Role of supraglacial ponds in the ablation process of a debris-covered glacier in the Nepal Himalayas. Debris-Covered Glaciers IAHS Publ 264:119–130

Sakai A, Nakawo M, Fujita K (2002) Distribution characteristics and energy balance of ice cliffs on debris-covered glaciers, Nepal Himalaya. Arct Antarct Alp Res 34(1):12–19

Sakai A, Nishimura K, Kadota T, Takeuchi N (2009) Onset of calving at supraglacial lakes on

debris-covered glaciers of the Nepal Himalaya. J Glaciol 55(193):909–917
Scherler D, Bookhagen B, Strecker MR (2011) Spatially variable response of Himalayan glaciers to climate change affected by debris cover. Nat Geosci 4(3):156–159
Schmidt S, Nüsser M (2009) Fluctuations of Raikot Glacier during the past 70 years: a case study from the Nanga Parbat massif, northern Pakistan. J Glaciol 55 (194):949–959
Shroder JF, Bishop MP, Copland L, Sloan VF (2000) Debris-covered glaciers and rock glaciers in the Nanga Parbat Himalaya. Pakistan. Geogr Ann A 82a(1):17–31
Stokes CR, Popovnin V, Aleynikov A, Gurney SD, Shahgedanova M (2007) Recent glacier retreat in the Caucasus Mountains, Russia, and associated increase in supraglacial debris cover and supra-/proglacial lake development. Ann Glaciol 46:195–203
Sugden DE, John BS (1976) Glaciers and landscape. E. Arnold, London
Vezzola LC, Guglielmina AD, D'Agata C, Smiraglia C, Pelfini M (2016) Assessing glacier features supporting supraglacial trees: a case study of the Miage debris-covered Glacier (Italian Alps). The Holocene OnlineFirst
Wahrhaftig C, Cox A (1959) Rock glaciers in the Alaska Range. Geol Soc Am Bull 70(4):383–436
Winkler S (2009) Gletscher und ihre Landschaften. Wissenschaftliche Buchgesellschaft, Darmstadt

5 Closing the Balances of Ice, Water and Sediment Fluxes Through the Terminus of Gepatschferner

Martin Stocker-Waldhuber and Michael Kuhn

Abstract

The terminus of Gepatschferner (46°52′ N, 10° 46′ E) was subject to detailed glaciological investigations in the joint project PROSA. Both direct and geodetic methods were applied. Specifically, ice surface lowering measured at ablation stakes determined mass loss at the glacier surface, ice surface velocity was measured directly at the same stakes with differential GPS, and geodetic radar and vibroseismic soundings came into operation to investigate ice thickness and thickness of subglacial sediments. Multiple high-resolution airborne laser scanning (ALS) surveys document total volume changes. In contrast to a differentiated examination of the balances of ice water and sediments, the combination of these balances was an appropriate approach for investigating glacier mass exchanges and identifying dominant processes within the glacial system. The calculation or estimate of the fluxes of ice, water and sediments entering the narrow terminus at an elevation of 2875 m and leaving it at the glacier snout at 2200 m was based on glacier motion, surface and basal melt rates and on the lateral mass transport to the glacier from rock face and moraine bedrock erosion recorded from repeated terrestrial laser scans. In the course of the investigations on Gepatschferner, multiple rockfall events and the rapid evacuation of subglacial sediments were observed. The highest mass fluxes within the glacial system of Gepatschferner were associated with these extreme or episodic events, which exceeded the normal annual processes by multiple orders of magnitude. The relevant geophysical processes in this study period were thus not representative of long-term averages, if these ever existed. They did, however, display an interesting spectrum of naturally occurring situations. In that period, the mean velocity through the cross section at 2875 m was 22.5 m per year. Below that profile, the ice loss at the terminus corresponds to a mean surface lowering of 3.61 m per year between 2012 and 2015.

Keywords

PROSA project · Glacier mass balance
Surface velocity · Subglacial sediment
Sediment flux

M. Stocker-Waldhuber (✉)
Institute for Interdisciplinary Mountain Research, Austrian Academy of Sciences, Innsbruck, Austria
e-mail: martin.stocker-waldhuber@oeaw.ac.at

M. Stocker-Waldhuber
Department of Geography, Physical Geography, Catholic University of Eichstätt-Ingolstadt, Eichstätt-Ingolstadt, Germany

M. Kuhn
Institute of Atmospheric and Cryospheric Sciences, University of Innsbruck, Innsbruck, Austria

T. Heckmann and D. Morche (eds.), *Geomorphology of Proglacial Systems*,
Geography of the Physical Environment, https://doi.org/10.1007/978-3-319-94184-4_5

List of variables and indices

Variables	
a	Ablation
A	Area
B	Glacier width
C	Cross section
E	Evaporation
h	Elevation
H	Thickness
M	Mass flux
P	Precipitation
Q	Volume flux
R	Run-off
S	Storage
t	Time
u	Velocity
V	Volume
ρ	Density

Indices	
b	Basal
bed	Bedload
d	Deformation
i	Ice
lat	Lateral
r	Bedrock
s	Surface
sed	Sediment
sus	Suspension
top	From the area above 2875 m to the terminus
w	Water

5.1 Introduction

Changes in glacier mass can be quantified with geodetic, hydrological and direct glaciological measurements (Hoinkes 1970; Cuffey and Paterson 2010). A combination of such measurements and investigations (Hubbard and Glasser 2005) is the basis for an approach to close the respective balances of ice, water and sediments. Estimating the magnitudes of each term of the balancing approach helps to separate dominant processes from less important ones and their contribution to the ongoing change of proglacial systems under conditions of rapid deglaciation (Heckmann et al. 2012). This chapter is based on the idea of conceptualizing a connection between the budgets of ice, water and sediment and their seasonal variation within a glacial system on the basis of state-of-the-art measurements.

Surface and total mass balance measurements at several glaciers in the Alps and their production of meltwater have been investigated for a long time (e.g. Hoinkes 1970; Thibert et al. 2008; WGMS 2012; Fischer et al. 2013) as well as the fluvial sediment transport and the transport of suspended solids in the glacial stream (e.g. Gurnell and Clark 1987; Hofer 1987; Warburton 1990; Chaps. 12 and 13). Likewise, measurements of glacier surface velocity have a long history (Span et al. 1997; Span and Kuhn 2003) and provide important information about glacier dynamics (Van der Veen 2013). Geophysical methods offer a wide range of possibilities for investigating glaciological issues such as physical properties of ice or the internal structures of glaciers, ice thickness distribution and bedrock topography or subglacial sediments (Hauck and Kneisel 2008). The potential of Airborne Laser Scanning (ALS) data for glaciological questions has been demonstrated in several publications (e.g. Abermann et al. 2010; Bollmann et al. 2011; Helfricht et al. 2012) and can be taken as one of the most important databases for direct and indirect investigations. Carrivick et al. (2015) show that combining multiple methods yields insights into glacier evolution.

Gepatschferner, located in the Kaunertal in the Ötztal Alps (46°52′ N, 10°46′ E), is the second biggest glacier of the Eastern Alps (Patzelt 1980; Groß 1987; Lambrecht and Kuhn 2007; Kuhn et al. 2012) and extended over 16.4 km^2 in 2006, 15.6 km^2 in 2012 (Fischer et al. 2015) and 15.3 km^2 in 2015. It was subject to a wealth of measurements in past decades. Length changes of the glacier terminus have been measured since

1856 (e.g. Fritzsch 1898; Fischer 2016). Seismic investigations were carried out between 1953 and 1958 (Förtsch et al. 1955) and were continued in 1960 and 1961 (Giese 1963); ice thickness measurements with ground penetrating radar were conducted in 1996 (Span et al. 2005). Changes in length, area and volume between 1850 and 2006 were calculated by Hartl (2010). Since 2012, the glacier is part of a catchment where detailed investigations of geomorphodynamics have been carried out (Heckmann et al. 2012; Chap. 1). The glaciological measurements in this study are mainly confined to the terminus, defined as the glacier below 2875 m, except calculations of area, elevation and volume changes from ALS data, which are calculated for all glacier areas. Detailed investigations on the terminus of the glacier focus on surface melt, surface velocities, ice thickness and the thickness of subglacial sediments.

With a synopsis of these measurements at the terminus of Gepatschferner, an approach to close the balances of ice, water and sediments is given in the first part of this chapter. That means, we need to know how much ice is flowing into the glacier terminus at 2875 m surface elevation, what is the balance of accumulation and of ice melt at the terminus below, how much water enters the terminus from above and what is the run-off from the glacier, how much sediment reaches the terminus from above and from the lateral slopes and how much leaves the glacier as bedload and as suspension in the river. In the second part, we attempt an estimate of the magnitudes of the turnover volumes.

5.2 Methods

High-resolution airborne laser scans (Abermann et al. 2010) are fundamental databases to arrive at distributed surface elevation changes and to georeference measurements of ablation, glacier motion, ice thickness and thickness of subglacial sediments. In this study, ALS data from 04.07.2012 and 22.08.2015 are used; they set the time frame of the balances. The digital elevation models (DEMs) were aggregated as mean values form the ALS point clouds on a raster size of 1 × 1 m. Gaps were closed by a stepwise resampling using a bilinear interpolation. The DEMs, which are the basis of the calculations, aim at a vertical accuracy of ±0.2 m for each DEM and ±0.4 m for the DEM of difference (DoD) (Bollmann et al. 2011), which is in the same order as the basal change rates. An additional DEM is available from orthophotographs from 18.07.2014. Differential Global Positioning System (DGPS) measurements and direct stake readings were carried out on the glacier terminus at 15 stake positions (Fig. 5.1). Five of the stakes are located at a cross section in the root zone of the terminus at an elevation of 2875 m. Additionally, vibroseismic measurements (Eisen et al. 2010) were carried out at the terminus of Gepatschferner in 2012, 2013 and 2014 to investigate the thickness of subglacial sediment layers. A DEM of the glacier bed is provided by Span et al. (2005). It was produced by interpolating between ground penetrating radar (GPR) measurements.

The mean hydrological balance was calculated with a semi-distributed, conceptual model (Kuhn and Batlogg 1998, 1999; Kuhn 2000, 2003; Kuhn et al. 2016) for the period 2006/07–2014/15, based on the geodetic total glacier mass balance and measured values of temperature and precipitation. The model provides monthly values of basin precipitation, snow cover, glacier mass balance and run-off, separated into meltwater and rain for 100 m elevation bands partly verified by the basin run-off measured at Gepatschalm gauging station, situated at 1900 m.

Bedload transport and the transport of suspended solids are measured directly in the proglacial river at the glacier snout (Fig. 5.2; see details in Chap. 13). The lateral mass flux from adjacent hillslopes to the glacier surface was calculated from ALS and terrestrial laser scanning (TLS) data (Fig. 5.2; see details in Chaps. 9, 11).

5.2.1 Ice and Water Fluxes

Several factors contribute to elevation changes of the glacier surface over time (dh_s/dt) at a specific

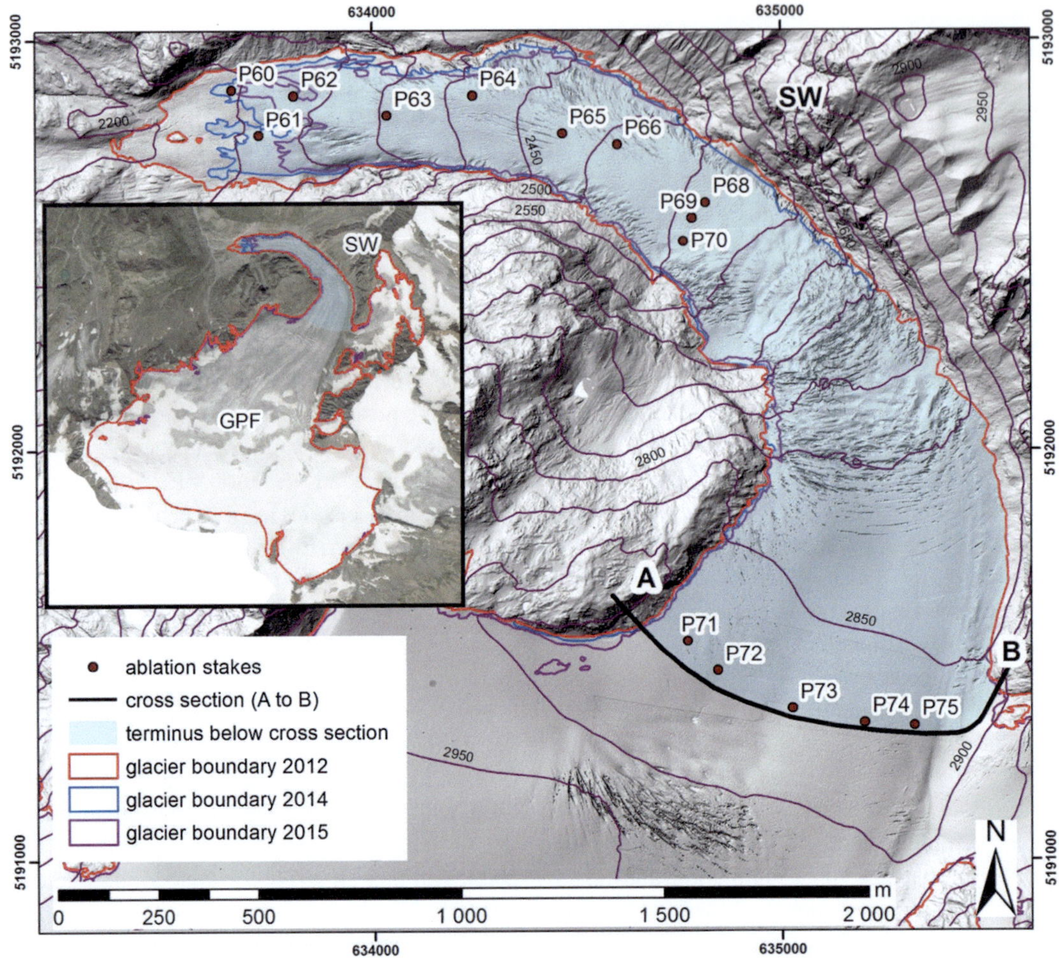

Fig. 5.1 Position of ablation stakes at the terminus of Gepatschferner 2014 and glacier boundaries 2012, 2014 and 2015 as well as the cross section at the root zone of the terminus between point A and B. Hill shade dated July 2012. Inset: the entire Gepatschferner (GPF) on orthophotograph (*data source* Land Tirol—data.tirol.gv.at). SW: Schwarze Wand

point, including the ablation on the surface ($a_s < 0$), basal melt ($a_b < 0$), thickness changes of the subglacial sediment layer (dH_{sed}/dt) and bedrock (dh_r/dt) erosion, advection of ice thickness ($u * dH_i/dx$), convergence or divergence of ice ($H_i * du/dx$). Englacial and subglacial water is seasonally stored, mostly in the matrix of sediments and ice, and has negligible changes from year to year (Jansson et al. 2003).

At any point (x, y) on the glacier surface

$$\frac{\partial H_i}{\partial t} = (a_s + a_b) - \frac{\partial Q_i}{B \partial x} \tag{5.1}$$

$$\frac{\partial Q_i}{B \partial x} = \frac{\partial}{\partial x}(\bar{u} * H_i) = H_i \frac{\partial \bar{u}}{\partial x} + \bar{u} \frac{\partial H_i}{\partial x} \tag{5.2}$$

The flux of ice to the terminus (Q_i) through a cross section (C) at a mean glacier width (B) and

Fig. 5.2 Photograph of the terminus of Gepatschferner. A: transport of ice and sediments through the cross section to the terminus (Eqs. 5.3, 5.4 and 5.11); B: lateral mass transport on to the terminus from rock slopes and moraines (Eq. 5.11); C: bedload transport and transport of suspended solids within the glacial stream (Eq. 5.11); SW: Schwarze Wand

the mean ice thickness (H_i) at the root zone of the terminus depends on the velocity of ice ($\bar{u}$), which is the sum of the mean deformation velocity ($\bar{u}_d$) and the mean velocity of basal sliding ($\bar{u}_b$).

$$Q_i = B * H_i(\bar{u}_d + \bar{u}_b) \tag{5.3}$$

$$Q_i = C * \bar{u} \tag{5.4}$$

The volume of ice (V_i) stored in the terminus in relation to the area (A) each year is

$$V_i = Q_i + (a_s + a_b) * A \tag{5.5}$$

from the glaciological method, equivalent to

$$V_i = \frac{\partial H_i}{\partial t} * A \tag{5.6}$$

from the geodetic method. This means that the amount of subglacial melt (a_b) can be inferred in metres of ice per year as

$$a_b = \frac{\partial H_i}{\partial t} - \frac{Q_i}{A} - a_s \tag{5.7}$$

where H_i, a_s and a_b are means over the entire terminus, and

$$\frac{\partial H_i}{\partial t} = \frac{\partial h_s}{\partial t} - \frac{\partial H_{sed}}{\partial t} \tag{5.8}$$

where h_s is determined from ALS and H_{sed} from seismic soundings.

The densification of snow or firn can be neglected at the terminus, except for a minor volume change due to closing crevasses, and also because the cross section is located within the ablation zone; thus, the density of ice (ρ_i) is kept constant, giving a straightforward way of converting volume to mass fluxes. The conventional glaciological mass balance measurements refer indeed to the weighed mass of snow where accumulation is concerned. All other measurements, i.e. ablation, changes of surface elevation, ice flow velocity and sediment thickness, are measured in terms of metres. This is why we prefer to express the balance terms of thickness ($H[m]$), volume ($V[m^3]$), volume fluxes per year ($Q[m^3/a]$) and volume flux densities per year ($[m^3/m^2a]$, which is equivalent to mean velocities horizontally or to accumulation or ablation vertically, in metres per year). Only in those cases where several components (ice, water, sediments) are involved do we refer to mass fluxes per year ($[kg/a]$ or water equivalent (w.e.) per year $[m\ w.e./a]$).

Mass losses at the terminus can be calculated from the absolute values of surface and subglacial ablation ($a_s + a_b$) over the terminus area (A). The 'remaining' ice of the specified area corresponds to the storage of ice ($S[m\ w.e.]$).

$$S = Q_i \frac{\rho_i}{\rho_w} + (a_s + a_b)\frac{\rho_i}{\rho_w} * A \tag{5.9}$$

The storage of water (S) in the form of glacier ice (Jansson et al. 2003) is the connection to the hydrological balance of precipitation (P), run-off (R) and evaporation (E).

$$S = P - R - E \tag{5.10}$$

5.2.2 Sediment Fluxes

The mass flux of the bedload transport (M_{bed}) and the transport of suspended solids within the glacial stream to the snout of the glacier (M_{sus}) equals the thickness changes of the subglacial sediments ($(\partial H_{sed}/\partial t)\rho_{sed}$) of the terminus area ($A$), the flux of sediments from the firn area to the terminus (M_{top}) and the lateral input to the glacier from landslides and rockfalls of the surrounding slopes (M_{lat}). Neglecting bedrock erosion, the thickness change is a result of the divergence in the sediment transport. Assuming a stationary case, the terms of mass fluxes are balanced by thickness changes of the subglacial sediments (s. Table 5.1).

$$M_{bed} + M_{sus} = M_{top} + M_{lat} - \frac{\partial H_{sed}}{\partial t}\rho_{sed} * A \tag{5.11}$$

5.2.3 Balancing Approach

The absolute elevation of the glacier surface above sea-level (h_s) depends on the elevation of the bedrock topography (h_r), the thickness of ice (H_i), the thickness of subglacial sediments (H_{sed}) and of liquid water (H_w) and changes in time. In this case, bedrock erosion is neglected, so the bedrock topography does not change. Likewise, annual changes of englacial and subglacial water are neglected.

$$h_s = h_r + H_{sed} + H_i + H_w \tag{5.12}$$

$$\frac{\partial h_s}{\partial t} = \frac{\partial H_{sed}}{\partial t} + \frac{\partial H_i}{\partial t} \tag{5.13}$$

Of the terms in Eqs. 5.12 and 5.13, h_s is determined from ALS data, h_r from GPR data, H_{sed} from seismic data, H_i follows from Eq. 5.12 if H_w is neglected, because englacial water does not change the surface elevation and subglacial water is mostly contained within the sediment matrix.

5.3 Estimating Magnitudes of the Fluxes

At a cross section at the root zone of the terminus, at an elevation of 2875 m, the mean ice thickness is about 100 m and the maximum ice thickness 150 m. The cross-sectional area (C) between the bed topography and the ice surface represented in the DEM 2014 is 125,547 m^2 (Fig. 5.3). Between 2012 and 2015, the cross-sectional area varied by only 2–3% to the sectional area of 2014. Maximum ice surface velocities were found at the central stakes 'P73' and 'P74' (Fig. 5.1) of the cross-profile at the centre flow line of the glacier, with 35 m per year as a mean value between 2012 and 2015. The mass flux of ice to the terminus (Eq. 5.3) is thus calculated from the cross-sectional area of 2014 and the maximum velocity at the central stake. The mean velocity through the section is taken as approximately 64% of the maximum ice surface velocity (Span and Kuhn 2003). So the mean annual velocity through the section is 22.5 m per year, or 67.6 m for the period (04.07.2012–22.08.2015), which leads to a total volume gain of the terminus of 8,489,060 m^3, or 6.20 m/m^2 during the period, calculated for the 1.369 km^2 area of the terminus in 2014. For the investigated

Table 5.1 Measured, calculated and estimated quantities within the PROSA catchment with respect to the balancing approach to the terminus of Gepatschferner

Measured quantities		
Glacier area of Gepatschferner 2015	15.281 ± 0.23	km^2
Terminus area 2014 (<2875 m)	1.369 ± 0.03	km^2
Cross-sectional area (2875 m), bedrock to surface 2012	0.128 ± 0.01	km^2
Cross-sectional area (2875 m), bedrock to surface 2015	0.122 ± 0.01	km^2
Mean ice thickness cross section	100 ± 10	m
Max. ice thickness at cross section	150 ± 15	m
Max. velocity on cross-sectional surface	35 ± 0.2	m/a
Max. thickness of subglacial sediment layers (Stocker-Waldhuber et al. 2017)	20 ± 5	m
Min. surface melt rate at the terminus	−1.14 ± 0.10	m/a
Max. surface melt rate at the terminus	−8.72 ± 0.10	m/a
Calculated and estimated quantities (04.07.2012–22.08.2016)		
Mean velocity through the cross section per year	22.5	m/a
Ice volume flux to the terminus	8.489	10^6 m^3
Mean specific ice mass gain of the terminus	6.20	m
Mean elevation change at the terminus (DoD)	–10.84	m
Mean specific surface mass balance at the terminus	–15.59	m
Subglacial changes at the terminus (basal melt and bedrock erosion)	–1.45	m
Hydrological balance based on Kuhn et al. (2016) (2006/07–2014/15)		
Mean specific annual glacier mass balance	–868	mm w.e.
Mean annual storage of the catchment; negative part of the glaciers	–314	mm w.e.
Mean annual catchment run-off	1,740	mm
Contribution of the glaciers to the total run-off	18	%
Contribution of Gepatschferner terminus to total run-off	6.1	%
Lateral contribution of Schwarze Wand	66,000	t/a
Estimated transport duration at central flow line	10–30	a
Estimated sediment production of Gepatschferner	10^4–10^6	m^3/a
Mean erosion Gepatschferner 2012–2015 (assuming: 1 mm, ρ = 2500 kg/m^3)	38,618	t/a

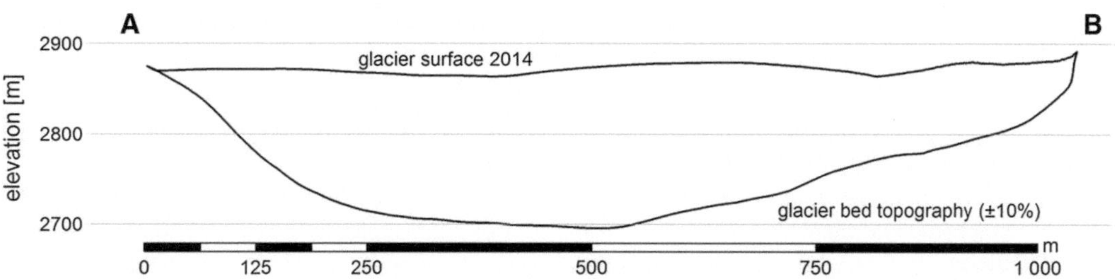

Fig. 5.3 Vertical cross section from point A to point B (Fig. 5.1) between the glacier surface 2014 and the interpolated glacier bed topography calculated from ice thicknesses values determined by ground penetrating radar with an estimated uncertainty of ±10% by Span et al. (2005)

three-year period, the geodetic determination of the change in surface elevation of the terminus out of the DoD is −10.84 m of ice. The direct investigations at the glacier surface by stake readings returned a specific surface balance of the terminus of −15.59 m of ice within three years (04.07.2012–22.08.2015). From Eq. 5.1, the subglacial changes are −1.45 m for this period or about −48 cm per year as the remainder of the DoD, the mass gain due to the specific mass flux of ice to the terminus and the specific surface mass balance. The basal rate of change of −0.48 m per year consists of the basal melt rate and minor contributions of the bedrock erosion.

A comparison of the uncertainty of each term of the ice mass balance of the terminus reveals different orders of magnitude. The uncertainty of the surface mass balance measured with the direct glaciological method (Hoinkes 1970) is assumed to be smaller than 0.1 m of ice per year (Kuhn et al. 1999). The surface velocities measured with DGPS aim at a horizontal precision of a few centimetres (Monteiro et al. 2005) and are in the range of ±0.2 m per year at the cross section on Gepatschferner. The variation in the maximum surface velocity per year at the cross section during the three years of stake measurements was ±5% and affects the basal change rate by about ±0.11 m. The uncertainty of GPR measurements and the bedrock topography at the cross section is an order of magnitude higher and can only be estimated. However, we assume that about 10% of the given ice thickness should be a good approximation. In the case of Gepatschferner, a calculation with ±10% of the mean ice thickness within the cross section at the root zone of the terminus leads to a calculated range of the basal change rate at the terminus between −0.28 and −0.69 m per year.

Water storage within an Alpine catchment can be separated into liquid water and ice components. Within a highly glaciated catchment, the dominant storage of liquid water is within snow or firn (Kuhn et al. 2016), but those components are not as significant for total run-off as the volume of ice. For the whole catchment, the storage term of the hydrological balance (Eq. 5.10) equates the ice mass loss of the glacier. This negative part of the specific glacier mass balance between 2006/07 and 2014/15 amounts to −314 mm w.e. (water equivalent) per m^2 and year or 18% of the total headwater run-off. By calculating the specific contribution of the terminus relative to the whole catchment area, the negative balance of the terminus of Gepatschferner amounts to −107 mm w.e. per year. So the terminus below 2875 m makes a specific contribution of about 6.1% to the total run-off. A model run with higher temperatures (+2 °C) and constant glacier area shows a significant increase of 36% in the meltwater production (Kuhn et al. 2007). An assumed reduction of the ice-covered area until 2050 due to higher temperatures still leads to an increase of 27% of the total run-off.

The bedload transport and the transport of suspended solids within the proglacial stream (see details in Chap. 13) represent the mass loss or export from a catchment and thus lead to a mean surface lowering in the order of magnitude of millimetres to centimetres per year (Carrivick et al. 2013). The sediment mass originates from active rock slopes, moraines and from subglacial bedrock erosion. However, only a small part of the surrounding steep slopes and moraines provide a lateral mass transport directly on to the surface of the terminus of Gepatschferner. Most of the material from active rock slopes (>40°) is deposited on moraines, and often, the material from moraines is transported underneath the terminus at the peripheral zones (see details in Chap. 11).

On Gepatschferner, the very active rock slope Schwarze Wand is responsible for most of the deposited material on the glacier surface, with approximately 66,000 tons a year between 2012 and 2015 (see details in Chap. 9). Such active rockfall zones or comparable ‘extreme’ events exceed the average transport processes many times over. On Alpine glaciers, a ‘final’ deposition of solid rocks in the glacier forefield, which were previously deposited on the glacier surface, is delayed by the glacier motion and/or retreat by up to several hundred years (Span et al. 1997). In the case of Gepatschferner, solid rocks from the Schwarze Wand, deposited at the central flow line of the glacier, will arrive in the forefield of the glacier in approximately 30 years under present conditions of glacier motion and the present extent of the terminus. Given retreat rates of the terminus of about 100 m per year between 2012 and 2015 (Fischer 2016), the transport time to the forefield can be assumed as approx. 10 years. Material from rock faces or moraines deposited on dead ice bodies has not been taken into account, because these regions are dynamically almost decoupled from the terminus itself.

The distribution and thickness of subglacial sediment layers at specific profiles was investigated with vibroseismic soundings on the terminus of Gepatschferner. In one profile at the lowermost part of the glacier, a sediment trap measured maximum thicknesses of 20 ± 5 m. During a heavy precipitation event (Baewert and Morche 2014), the subglacial sediment layers were flushed out rapidly, resulting in an increased deepening process of a surface depression (Stocker-Waldhuber et al. 2017). In contrast to this extreme event, the specific erosion of the glacier should be in the order of magnitude of millimetres up to a few decimetres per year (Hallet et al. 1996). The volume of sediments evacuated during the extreme event mentioned above would be in the same order of magnitude as a specific erosion rate of a few centimetres per year in relation to the area of the terminus. The glacier area further up the terminus (Gepatschferner >2875 m) was not taken into account because the sediment transport from the firn area is unknown. However, the estimated magnitude of the sediment production from bedrock erosion of the whole Gepatschferner is assumed to be in the range of 10^4–10^6 m^3 per year, with a calculated range of specific erosion rates of about one millimetre to a five centimetres per year (c.f. Chap. 17).

5.4 Conclusions

The balancing approach presented here aims at combining measured quantities and magnitudes only estimated on an annual basis. A regionalized analysis of each parameter should be taken into account and is as important as the separation of periodic/annual from episodic processes. On Gepatschferner, the glaciological measurements are mainly confined to the terminus or even more specific locations, such as the vibroseismic investigations at the surface depression. On the terminus, we found from geodetic investigations that the mean surface lowering from 04.07.2012 to 22.08.2015 was −10.84 m. In contrast, the directly measured surface ablation of −15.59 m within the same period is largely compensated by the ice mass flux to the terminus, calculated out of surface velocities and ice thickness data, which leads to a gain of 6.20 m. The remainder of this balance suggests a basal change rate of −0.48 m per year. These basal changes consist of subglacial ice melt and the erosion of sediments and the bedrock. However, the three-year time window is a snapshot that may be short compared to the recurrence of extreme events. Thus, an extrapolation of the measured and calculated changes and processes to the whole glacier area remains problematic, as indicated by examples of 'extreme' or episodic events, which are multiple orders of magnitude higher than changes within 'normal' daily or annual cycles, like the rockfalls on Schwarze Wand or the rapid evacuation of subglacial sediment layers during a heavy precipitation event.

Acknowledgements The glaciological investigations on Gepatschferner are part of the joint project PROSA, funded by the German Research Foundation (DFG) (SCHM 472/17-1, SCHM 472/17-2 and BE 1118/33-3) and the Austrian Science Fund (FWF) (I 894-N24 and I 1646-N19). Numerous people were involved in the fieldwork to whom we are deeply grateful for their valuable contribution. Thanks to J. Carrivick for his review on the manuscript which helped us to improve this chapter. We thank B. Scott for editing the English.

References

Abermann J, Fischer A, Lambrecht A, Geist T (2010) On the potential of very high-resolution repeat DEMs in glacial and periglacial environments. Cryosphere 4:53

Baewert H, Morche D (2014) Coarse sediment dynamics in a proglacial fluvial system (Fagge River, Tyrol). Geomorphology 218:88–97. https://doi.org/10.1016/j.geomorph.2013.10.021

Bollmann E, Sailer R, Briese C, Stötter H (2011) Potential of airborne laser scanning for geomorphologic feature and process detection and quantifications in high alpine mountains. Z Geomorphol, Supplementary Issues 55:83–104

Carrivick JL, Berry K, Geilhausen M, James WHM, Williams C, Brown LE, Rippin DM, Carver SJ (2015) Decadal-scale changes of the Ödenwinkelkees, Central Austria, suggest increasing control of topography and evolution towards steady state. Geogr Ann: Ser A, Phys Geogr 97:543–562. https://doi.org/10.1111/geoa.12100

Carrivick JL, Geilhausen M, Warburton J, Dickson NE, Carver SJ, Evans AJ, Brown LE (2013) Contemporary geomorphological activity throughout the proglacial area of an alpine catchment. Geomorphology 188:83–95. https://doi.org/10.1016/j.geomorph.2012.03.029

Cuffey K, Paterson WSB (2010) The physics of glaciers, 4th edn. Butterworth-Heinemann/Elsevier, Burlington, MA

Eisen O, Hofstede C, Miller H, Kristoffersen Y, Blenkner R, Lambrecht A, Mayer C (2010) A new approach for exploring ice sheets and sub-ice geology. Eos, Trans Am Geophys Union 91:429–430. https://doi.org/10.1029/2010EO460001

Fischer A (2016) Gletscherbericht 2014/2015-Sammelbericht über die Gletschermessungen des Österreichischen Alpenvereins im Jahre 2015. Bergauf 02(2016):6–13

Fischer A, Markl G, Kuhn M (2013) Glacier mass balances and elevation zones of Hintereisferner, Ötztal Alps, Austria, 1952/1953 to 2010/2011

Fischer A, Seiser B, Stocker-Waldhuber M, Mitterer C, Abermann J (2015) Tracing glacier changes in Austria from the Little Ice Age to the present using a lidar-based high-resolution glacier inventory in Austria. Cryosphere 9:753–766. https://doi.org/10.5194/tc-9-753-2015

Förtsch O, Schneider H, Vidal H (1955) Seismische Messungen auf dem Gepatsch-und Kesselwand-Ferner in den Ötztaler Alpen. Gerlands Beitr Geophys 64:233–261

Fritzsch M (1898) Verzeichnis der bis zum Sommer 1896 in den Ostalpen gesetzten Gletschermarken. Verlag des Dtsch Österr Alpenvereins, Wien, p 131

Giese P (1963) Some results of seismic refraction work at the Gepatsch glacier in the Oetztal Alps. IAHS Publ 61:154–161

Groß G (1987) Der Flächenverlust der Gletscher in Österreich 1850–1920–1969. Z Gletscherk Glazialgeol 23(2):131–141

Gurnell AM, Clark MJ (1987) Glacio-fluvial sediment transfer—an Alpine perspective. Wiley, West Sussex

Hallet B, Hunter L, Bogen J (1996) Rates of erosion and sediment evacuation by glaciers: a review of field data and their implications. Global Planet Change 12:213–235. https://doi.org/10.1016/0921-8181(95)00021-6

Hartl L (2010) The Gepatschferner from 1850 to 2006-changes in length, area and volume in relation to climate. Unpubl. Diploma thesis, Innsbruck University, p 82

Hauck C, Kneisel C (2008) Applied geophysics in periglacial environments. Cambridge University Press Cambridge

Heckmann T, Haas F, Morche D, Schmidt KH, Rohn J, Moser M, Leopold M, Kuhn M, Briese C, Pfeiffer N, Becht M (2012) Investigating an Alpine proglacial sediment budget using field measurements, airborne and terrestrial LiDAR data. IAHS-AISH publication, pp 438–447

Helfricht K, Schöber J, Seiser B, Fischer A, Stötter J, Kuhn M (2012) Snow accumulation of a high alpine catchment derived from LiDAR measurements. Adv Geosci 32:31

Hofer B (1987) Der Feststofftransport von Hochgebirgsbächen am Beispiel des Pitzbaches. Österr Wasserwirtsch 39:30–38

Hoinkes H (1970) Methoden und Möglichkeiten von Massenhaushaltsstudien auf Gletschern: Ergebnisse der Messreihe Hintereisferner (Ötztaler Alpen) 1953–1968. Z Gletscherk Glazialgeol 6:37–90

Hubbard B, Glasser NF (2005) Field techniques in glaciology and glacial geomorphology. Wiley, Chichester, West Sussex, England ; Hoboken, NJ

Jansson P, Hock R, Schneider T (2003) The concept of glacier storage: a review. J Hydrol 282:116–129

Kuhn M (2000) Verification of a hydrometeorological model of glacierized basins. Ann Glaciol 31:15–18

Kuhn M (2003) Redistribution of snow and glacier mass balance from a hydrometeorological model. J Hydrol 282:95–103

Kuhn M, Batlogg N (1998) Glacier runoff in Alpine headwaters in a changing climate. International Association of Hydrological Sciences, Publication, pp 79–88

Kuhn M, Batlogg N (1999) Modellierung der Auswirkung von Klimaänderung auf verschiedene Einzugsgebiete in Österreich. Schriftenreihe Forschung im Verbund, Wien; p 98

Kuhn M, Dreiseitl E, Hofinger S, Markl G, Span N, Kaser G (1999) Measurements and models of the mass balance of Hintereisferner. Geogr Ann: Ser A, Phys Geography 81:659–670

Kuhn M, Helfricht K, Ortner M, Landmann J (2016) Liquid water storage in snow and ice in 86 Eastern Alpine basins and its changes from 1970–97 to 1998–2006. Ann Glaciol 57:11–18

Kuhn M, Lambrecht A, Abermann J, Patzelt G, Groß G (2012) The Austrian Glaciers 1998 and 1969, area and volume changes. Z Gletscherk Glazialgeol 43(44):3–107

Kuhn M, Olefs M, Fischer A (2007) Auswirkungen von Klimaänderungen auf das Abflussverhalten von vergletscherten Einzugsgebieten im Hinblick auf die Speicherkraftwerke. Project report: StartClim2007.E, p 49

Lambrecht A, Kuhn M (2007) Glacier changes in the Austrian Alps during the last three decades, derived from the new Austrian glacier inventory. Ann Glaciol 46:177–184

Monteiro LS, Moore T, Hill C (2005) What is the accuracy of DGPS? J Navig 58:207–225

Patzelt G (1980) The Austrian glacier inventory: status and first results. IAHS Publication 126, (Riederalp Workshop 1978—World Glacier Inventory), pp 181–183

Span N, Fischer A, Kuhn M, Massimo M, Butschek M (2005) Radarmessungen der Eisdicke österreichischer Gletscher [1]: Messungen 1995 bis 1998. Österreichische Beiträge zur Meteorologie und Geophysik 33, p 145

Span N, Kuhn M (2003) Simulating annual glacier flow with a linear reservoir model. J Geophys Res: Atmos 108

Span N, Kuhn M, Schneider H (1997) 100 years of ice dynamics of Hintereisferner, Central Alps, Austria, 1894–1994. Ann Glaciol 24:297–302

Stocker-Waldhuber M, Fischer A, Keller L et al (2017) Funnel-shaped surface depressions—indicator or accelerant of rapid glacier disintegration? A case study in the Tyrolean Alps. Geomorphology 287:58–72. https://doi.org/10.1016/j.geomorph.2016.11.006

Thibert E, Blanc R, Vincent C, Eckert N (2008) Instruments and methods glaciological and volumetric mass-balance measurements: error analysis over 51 years for Glacier de Sarennes, French Alps. J Glaciol 54:522–532

Van der Veen CJ (2013) Fundamentals of glacier dynamics. CRC Press

Warburton J (1990) Comparison of bed load yield estimates for a glacial meltwater stream. Proc Int Conf Water Resour Mountainous Reg, IAHS Publ 193:315–322

WGMS (World Glacier Monitoring Service) (2012) Fluctuations of glaciers 2005–2010, vol 10. World Glacier Monitoring Service, University of Zürich, http://www.wgms.ch/

(Ground) Ice in the Proglacial Zone

6

Isabelle Gärtner-Roer and Alexander Bast

Abstract

In mid-latitude mountains, most of the valley glaciers currently experience distinct and enhanced volume and area loss. In parallel with the glacier retreat, the related proglacial areas enlarge, leaving unconsolidated sediments and ground ice of different origins and thus forming a transitional landscape, as developing from a glacial to a non-glacial environment. The erosion, transport and accumulation of sediment in these proglacial areas are characterized by high spatio-temporal dynamics, which are typically highest in the direct glacier forefield and become more inactive with increasing distance to the glacier front. Glacial, periglacial, fluvial and gravitational processes occur and highly interact in space and time. The glacial history of recently deglaciated zones influences the complex thermal regime of the subsurface and determines the current ground ice occurrence. Besides the glacio-fluvial processes, low-temperature conditions, as well as the occurrence of ground ice, are the most effective drivers for geomorphic dynamics and related landform evolution in these proglacial areas. A deeper knowledge of ongoing processes as well as of the amounts of sediment and ground ice is decisive to assess the availability of unconsolidated sediment for potential hazardous processes (e.g. debris flows) and the availability of water from ground ice bodies. There is an increasing need for high-resolution data (e.g. repeated topographic data) of proglacial areas as well as the systematic monitoring of these environments.

Keywords

Ground ice · Dead ice · Permafrost
Rockglaciers · Geophysical measurements

6.1 Introduction

In mid-latitude mountains, most of the valley glaciers have experienced distinct volume and area losses since the end of the so-called Little Ice Age (LIA) (Grove 2004) and an enhanced loss since the 1990s (WGMS 2015; Chap. 2). Glaciers in the European Alps lost about 35% of their total area between 1850 and 1975 and about 50% by the year 2000 (Zemp et al. 2006).

I. Gärtner-Roer (✉)
Department of Geography, University of Zurich, Zurich, Switzerland
e-mail: isabelle.roer@geo.uzh.ch

A. Bast
Swiss Federal Research Institute WSL, Birmensdorf, Switzerland

T. Heckmann and D. Morche (eds.), *Geomorphology of Proglacial Systems*,
Geography of the Physical Environment, https://doi.org/10.1007/978-3-319-94184-4_6

In parallel with the glacier retreat, the related proglacial areas have enlarged (Chap. 3), leaving unconsolidated sediments and ground ice of different origins and thus forming a transitional landscape, as developing from a glacial to a non-glacial environment (Ballantyne 2002). The sediments are in an unstable or metastable state and are susceptible to release, reworking and redeposition at rates exceeding typical denudation rates (Curry et al. 2006; Lane et al. 2017). Proglacial areas are characterized as geomorphologically active areas downslope of the respective glacier, including the ice-marginal settings around the glacier such as lateral moraines (Curry et al. 2006). Glacial, periglacial, fluvial and gravitational processes occur and highly interact in space and time. This landscape as well as the time of adjustment from a glacial to a non-glacial condition (to a new equilibrium) is defined as "paraglacial" (Ballantyne 2002) and characterized by "non-glacial processes that are directly conditioned by glaciation" (Church and Ryder 1972). Depending on the size of the glacier, its catchment and the local situation (in a valley bottom or on a steeper slope), the importance and activity of single processes vary. The transitional character is represented by landform changes, sediment transfer and ice degradation indicating an enhanced activity (Vaughan et al. 2013). Low-temperature conditions, as well as the occurrence of ground ice, are the most effective drivers for geomorphic dynamics and related landform evolution in proglacial areas (Swift et al. 2014).

6.1.1 Occurrence and Origin of Ground Ice

The occurrence of ice in the proglacial zone encompasses all possible origins from glacier ice remnants (sedimentary ice), segregation ice, to polygenetic ice (a mixture of both) in debris and bedrock. The latter includes ice-filled fissures and joints that are highly susceptible to temperature changes and control bedrock stability (Deline et al. 2014). The physical, thermal and mechanical properties of ground ice and its coalescence with the surrounding material strongly influence the dynamics of associated processes, as weathering, backwasting or periglacial creep (Schomacker 2008; Arenson et al. 2014). The persistence, aggradation or degradation of ice in a recently deglaciated area strongly depends on its thermal state, initially controlled by the glaciers location, aspect and elevation and its respective type (cold-based, temperate or polythermal glacier) (Benn and Evans 2010; Cuffey and Paterson 2010). While cold-based and polythermal glacier enable the occurrence of ground ice underneath, temperate glaciers prevent the ground from cooling. Nevertheless, permafrost conditions may lead to the aggradation of ground ice after the glaciers retreat (Kneisel 2010). Beside the glacier characteristics, the persistence of ground ice is mainly driven by thermal and hydrological conditions, including positive and negative feedbacks. In this context, the topography, the surface substrate and the snow cover are the most important influencing factors (French 2013; Schomacker 2008).

Massive ice, ice-cemented and ice-free debris typically coexist in the proglacial zone and are highly variable in space and time (Reynard et al. 2003; Haeberli 2005; Kneisel and Kääb 2007). The occurrence of ice in proglacial areas is evidenced by indicative landforms (e.g. ice-cored moraines or thrust moraines) or visible degrading processes (thermal or hydrological erosion). Depending on the origin of the ice, the landforms are either ice-marginal directly formed from glacier ice or periglacial as formed by ice aggradation, or a mixture thereof. While the occurrence of massive ice bodies in glacier forefields is often related to buried glacier ice, ice-cemented areas reflect the genesis of polygenetic ground ice and hence the occurrence of permafrost conditions (Kneisel 1999; Reynard et al. 2003). This can be related to active ice aggradation or to relict permafrost features, as deduced from deep permafrost bodies covered by unfrozen deposits from the LIA advance (Reynard et al. 2003). The interaction of glacial and periglacial landforms determines the proglacial zone in many mountain environments as continuous transition from glaciers to rockglaciers occur in some places (Ackert 1998; Berthling 2011; Lilleøren et al. 2013; Bosson et al. 2015).

6.1.2 Characteristic Landforms and Controls

Within the proglacial zones of valley glaciers as well as cirque glaciers, characteristic landforms are described in relation to the existence of ground ice. The occurrence and persistence of these landforms are controlled by the thermal properties of the ground (as influenced by the thermal properties of the former glacier as well as the climatic conditions), ground ice conditions, characteristics of the ground material/sediment and the hydrological processes.

Ice-cored moraines are very common landforms, especially in forefields of (large) valley glaciers. They are described as bodies of ice within sediment accumulations at or near the glacier front (Østrem 1959) and disconnected from the active ice margin (Lukas 2011). They are characterized as transitional landforms, since the ice melts out before the final landform, a stable moraine, will establish (Lukas 2011). The development of *dead-ice bodies* is most commonly related to debris-covered glacier tongues and their gradual disconnection from the active glacier body (Benn and Evans 2010). The properties of the ice are very similar to the ones of ice-cored moraines, but less influenced by the ice movement and the related accumulation material building a moraine. Backwasting processes are largely related to climate and topographic parameters (Schomacker 2008). For both landforms, the preservation potential strongly depends on the ice–debris ratio as well as the superficial debris thickness. Very prominent landforms are *rockglaciers* deriving from debris-covered glaciers (Chap. 4), as a seamless transition from one landform to another and indicating the occurrence of permafrost conditions (Lilleøren et al. 2013; Bosson and Lambiel 2016). *Thrust moraines* are the result of frozen sediments which were pushed and thrust upwards by the advancing glacier (Baumhauer and Winkler 2014) and are reflected by the tectonic-like, bulged ridges at the surface. All the features described so far represent the active and intact landforms. As soon as the internal ice is melting, the surface is lowering and typical *settlement features* (in permafrost environment: *thermokarst*), such as ponds or small lakes, appear.

Other dominant landforms are related to gravitational processes (not compulsory related to the occurrence of ground ice), such as landslides (Cossart et al. 2013) or debris flows (Chiarle et al. 2007) shaping steep slopes in the deglaciated area (Chaps. 10 and 11). The flatter areas in front of the active glacier tongue are predominantly influenced by glacio-fluvial processes (Chaps. 12 and 13). Ground ice in this zone often persists only for short time periods, due to thermal erosion.

6.1.3 Characteristics of Proglacial Areas

The impact of ground ice on geomorphological processes and related landforms within the proglacial zone strongly depends on the environmental characteristics such as topography, climate, glacier type, as well as hydrology and weathering processes. For the different landforms and processes, related relevant spatio-temporal scales exist, which characterize the complex landscape response to deglaciation. While glacio-fluvial processes may shape the landscape within short time spans or even single events, whereas lateral moraines often persist over decades and centuries. In a very simplified manner, some major characteristics occur in a sequence as the landscape responds to deglaciation:

- Glaciers are retreating (e.g. Zemp et al. 2006).
  - Paraglacial adjustment starts (Church and Ryder 1972; Ballantyne 2002).
- Size of proglacial area is increasing.
  - More sediment is in transfer (erosion and accumulation) and is also becoming more variable in space and time (Lane et al. 2017).

- Run-off is changing and seems to increase at least in the medium term (e.g. Huss et al. 2014; Pellicciotti et al. 2014).
- Amount of ice in the proglacial area is decreasing (melt, change in hydrology).
  - The amount of ice in the proglacial area can also increase due to aggradation (permafrost conditions) (e.g. Kneisel and Kääb 2007).
- Depending on the occurrence or degradation of ground ice, the sediment can be bonded or released (Schomacker 2008; Lane et al. 2017).
- Ice-related movements of sediment may increase or decrease, depending on local temperature conditions (Kääb and Kneisel 2006; Bosson and Lambiel 2016).

So far, only a few attempts have been made to characterize and analyse the occurrence, distribution and dynamics of ground ice in recently deglaciated areas (Kneisel 1999; Reynard et al. 2003; Haeberli 2005; Schomacker 2008; Kneisel 2010; Bosson et al. 2015). Since proglacial areas are rapidly enlarging, there is an increasing interest in ongoing processes, both from geomorphologists and hydrologists.

6.2 Ensembles of Proglacial Landforms

We use the term "ensemble" in a geomorphological context as the occurrence of landforms coexisting in close vicinity in a specific environment. The ensemble of proglacial landforms results from site-specific conditions and interactions of the glacial history, topography, climate, sediment characteristics and amounts as well as run-off characteristics and amounts and characterizes the landscape in transition. Different sites in the Swiss Alps were selected to give qualitative and quantitative impressions on diverse proglacial areas with typical sets of landforms, their characteristics and dynamics. Certain patterns can be deduced for valley glaciers and cirque glaciers/glacier remnants from the European Alps.

6.2.1 Alpine Valley Glaciers

Alpine valleys are experiencing rapidly changing geomorphological and hydrological processes as glacier masses diminish and consequently landforms on related hillslopes and in proglacial areas adjust (Carrivick et al. 2015). While the glacier retreat is often very obvious and relatively easy to quantify, changes in proglacial areas are various and occur on different spatio-temporal scales and a great variety of landforms indicate the transition of the landscape (Ballatyne 2002). We selected two example areas to introduce different landforms and their environmental settings.

The Turtmann glacier (Valais, Switzerland) is a temperate valley glacier situated in the head of the Turtmann Valley, Southern Switzerland. The glacier extends from the Bishorn at 4140 m a.s.l. over a length of about 5.9 km (in 2010) down to 2260 m a.s.l. and covers an area of 5.2 km^2 (in 2010) with an average inclination of 18° (VAW 2015). The position of the glacier front is monitored since 1880. Since then, the glacier lost about 1.4 km in length and since the year 2000 it has retreated by about 200 m (WGMS 2015). In the 1980s and 1990s, the glacier experienced some small advances forming a couple of small moraines clearly visible in the glacier forefield (Fig. 6.1). The lithology of the area mainly consists of Palaeozoic gneisses and schists of the Penninic Siviez-Mischabel nappe as well as Mesozoic dolomites, marbles and limestones (Labhart 1998; Bearth 1980). Due to the inner alpine location, the climate of the Turtmann Valley shows a continental character. The glacier forefield is characterized by a variety of glacial, glacio-fluvial, periglacial and geomorphological processes (Otto and Dikau 2004) and is gradually recaptured by vegetation (Eichel et al. 2013).

Beside the front variation measurements (A. Brigger and colleagues, WGMS 2015), the observations of the Turtmann Glacier are of qualitative character. The changes in the forefield are investigated by mapping of repeated terrestrial and aerial photographs. Large LIA moraines are framing the proglacial zone. The one on the orographic right side of the glacier is well

Fig. 6.1 Proglacial zone of the Turtmann Glacier, Valais, Swiss Alps (photographs taken in August 2003)

preserved, only breached by the outflow of the adjacent Brunegg Glacier (a former tributary of Turtmann Glacier). On the other side, the trimline of the LIA moraine is still visible on the slope even if large parts are eroded by denudation processes, such as debris flows, or slope wash. During the first decade of the twenty-first century, the glacier retreated strongly and exposed lateral ice-cored moraines and dead-ice bodies in front of the active glacier tongue (Fig. 6.1). The observed ice-cored moraines occur within the margins of the 1970s–1990s extend. Hence, the connection to the active glacier body seems to be interrupted since about 15–30 years. With the ongoing degradation, the ice becomes partially and only temporarily visible, due to the erosion of the surface material. The erosion typically includes gravitational processes such as small debris flows or landslides and leaves unconsolidated moraine material behind.

The dead-ice bodies observed at Turtmann Glacier in the first decade of the twenty-first century are concentrated on the orographic right side of the proglacial area. This part is shadowed to the east by the nearby flank of the Adlerflüe summit and experienced some larger rockfall events in former times, as indicated by large rocks at the surface. Until today, most of the ice has gone, most probably resulting from the reiterating shift of the glacier outlet as running water is enhancing melt processes. Geomorphologic indicators of the former dead-ice bodies are hardly recognizable since the material simply subsided or was eroded by glacio-fluvial processes. Hence, these landforms characterize the transitional character of the proglacial zone.

The Findelen glacier (Fig. 6.2) is a temperate valley glacier located close to the village of Zermatt, only a few kilometres south of the Turtmann Glacier. With an area of more than

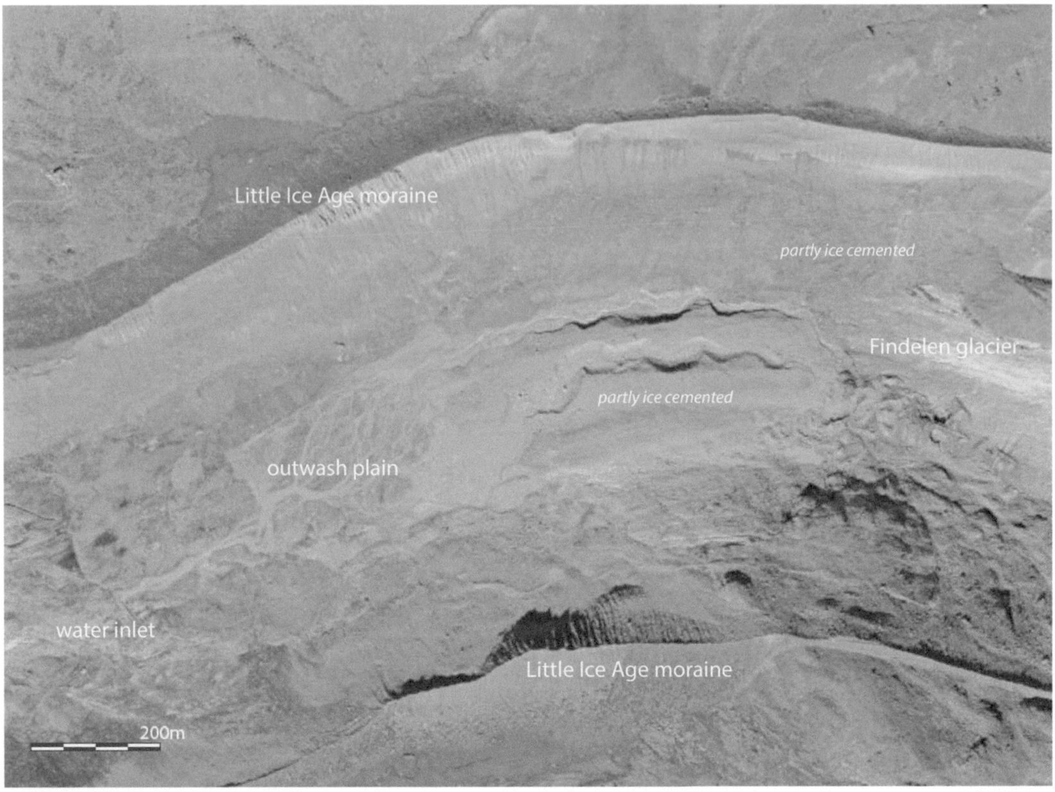

Fig. 6.2 Proglacial zone of the Findelen Glacier, Valais, Swiss Alps. Underlying aerial of 2017, reproduced by permission of swisstopo (BA17137)

13 km^2 and a length of 6.9 km (in 2010), it is one of the largest valley-type glaciers in the Alps (Maisch et al. 2000). It extends from 4190 m a.s.l. down to 2500 m a.s.l. (in 2013). Front variations are documented since 1885 and indicate a retreat of 1.9 km up to 2010 (VAW 2015). Regarding its area, the glacier lost about 30% of its LIA extent (Maisch et al. 2000). Between 2005 and 2010, the retreat of the front amounts to 200 m (Joerg et al. 2012). The mass balance of Findelen Glacier is measured by direct glaciological methods since 2005 (Huss and colleagues, WGMS 2015). In the context of glacier monitoring, repeated airborne light detection and ranging (LiDAR) surveys were conducted in 2005, 2009 and 2010 providing high-resolution (1 m) surface information of the entire catchment (Joerg et al. 2012). The uncertainty of the elevation data sets amounts to 0.2 m as given by the standard deviation after co-registration (Rastner et al. 2016). Beside the investigation of glacier changes, the data allow also for the analyses of geomorphological changes in the deglaciated area surrounding the glacier. For a recent analysis of the proglacial zone of Findelen Glacier, the data series was extended by an additional survey in 2014 using an unmanned aerial vehicle (UAV) and delivered an orthophotograph as well as a digital elevation model (DEM) of the area (Ruff 2015). The multi-temporal geodetic data allow for the quantification of vertical change as well as the mapping and characterization of ongoing geomorphic and in particular fluvial processes.

The proglacial zone is characterized by the prominent lateral moraines standing up to 140 m above the valley floor. These moraines are very stable over time, especially on the external side. Between 2005 and 2014, changes mainly occurred in the connection zone with the active glacier where vertical changes are quantified to

1–4 ma^{-1}. The partly debris-covered glacier is directly associated with ice-cored moraines. The glacier mass loss is destabilizing the adjacent slope and causes redistribution of sediment and probably also melt of ground ice. Thus, the quantified changes represent not only the ice loss, but also the transfer of sediment from the upper slope downwards. In addition, less pronounced changes occur in the steepest part of the moraines, where ongoing active gullying is related to precipitation and appendant linear erosion (Fig. 6.3; Ruff 2015). In the first investigation period (2005–2009), the vertical thickening at the foot of the orographic right moraine is striking (Fig. 6.3). This mass gain is clearly related to the sediment transport from the moraine into the proglacial zone. Since this part becomes more and more inactive with the vanishing ice and has been very close to the former glacier front position, the sediment transfer might be related to the pressure release of the retreating glacier.

The vertical changes in the valley bottom represent glacio-fluvial processes including erosion, redistribution and accumulation of sediment, mainly of smaller particle sizes. This area is in constant change and characterized by shifting outlet streams, with building up and draining of small lakes, ponds and outwash plains (Figs. 6.2 and 6.3). Beside the constant changes also single formative events occur. The deep channel appearing in the data set of 2014 (Fig. 6.3b) was built in a single rainstorm event, as discovered by the analysis of daily webcam information from that side (Huss and colleagues, University of Fribourg). In this case, the eroded material can be linked directly to the accumulated material in the outwash plain (Fig. 6.3).

Directly outside the orographic right moraine, and thus initially not coupled to the proglacial zone, another prominent landform occurs indicating strong dynamics. During the nine-year observation period, this landform, looking like a landslide or rock slide, moved down the slope and pushed the moraine onto the glacier. In the field, the top crest of the moraine is still identifiable due to a striking difference in sediment characteristics. The moving mass is represented in the pattern of positive and negative vertical changes as the mass is pushed downslope. Strongest movements occurred in the period 2009/2010.

6.2.2 An Alpine Cirque Glacier

The glacier Vadret Muragl (Grisons, Switzerland) is located in the Val Muragl, Upper Engadine, Eastern Swiss Alps. Here, continental conditions with daily and seasonal high-temperature amplitudes and rather low precipitation characterize the climate (Ott et al. 1997). Except for ridges above 2600 m a.s.l., the Val Muragl was fully glacier-covered during the Late Glacial Maximum (Maisch et al. 2003), and subsequent warming led to lateglacial retreat stages. Today, the Vadret Muragl is a small remnant of the former valley glacier and occupies only a very small area within the upper part of the cirque beneath the Piz Muragl, ranging from 2920 to 3080 m a.s.l. (Fig. 6.4a).

The most prominent landforms in the upper part of the Val Muragl are talus cones, lateglacial moraines, a glacier forefield and a rockglacier (Fig. 6.4a). The rockglacier was intensively investigated in several studies (e.g. Arenson et al. 2002; Maurer and Hauck 2007; Springman et al. 2012; PERMOS 2013). It was developed over millennia and is the result of periglacial debris cones that started to creep due to permafrost conditions after glacier retreat (Maisch et al. 2003).

The LIA glacier extent and the following rapid ice-decay since the 1850s resulted in a well-delimited and several meter elevated sediment bed (Fig. 6.4a), ranging in elevation between 2650 m and 2920 m a.s.l. (Fig. 6.4a). Estimated mean annual air temperatures of approximately −2.5 to −5 °C (Bast 2009) expose the glacier forefield to climate conditions, which promote permafrost occurrence and led, in combination with the glaciers' characteristics, to an area of complex glacier/permafrost interactions (e.g. Kneisel 2010). These interactions are in particular reflected by a thrust moraine complex on the orographic right side and glacial

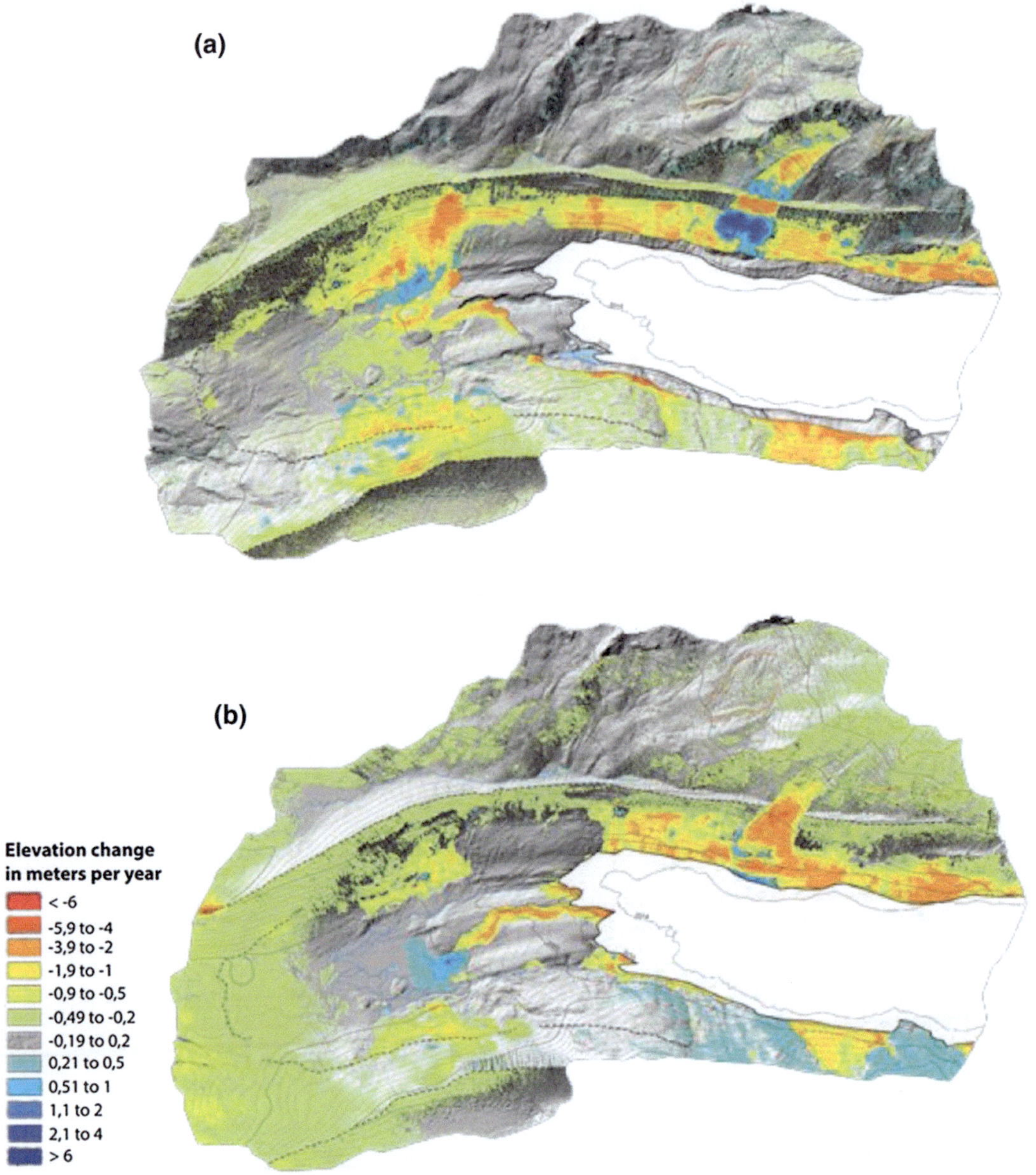

Fig. 6.3 Vertical changes (ma^{-1}) between 2005 and 2009 (**a**) and 2010–2014 (**b**), as derived from multi-temporal DTMs (Ruff 2015). The given glacier outline is of 2010

flutes in the central part of the proglacial zone (Haeberli 1983) (Fig. 6.4a). The occurrence of both landforms supplies the geomorphic evidence that the former glacier was polythermal with a temperate accumulation zone and partially cold ablation areas (Haeberli 1983). Close to the active glacier tongue, debris-covered ice is observable (Fig. 6.4c). Related to the ice in the subsurface, small debris flows and/or shallow landslides (active layer slides) are frequently occurring and lead to the exposure and erosion of the ground ice. Debris-covered ground ice appears also in close proximity of the small glacieret within a niche on the orographic left site (Fig. 6.4a). Melt water from snow, ground and glacier ice resulted in the formation of stream

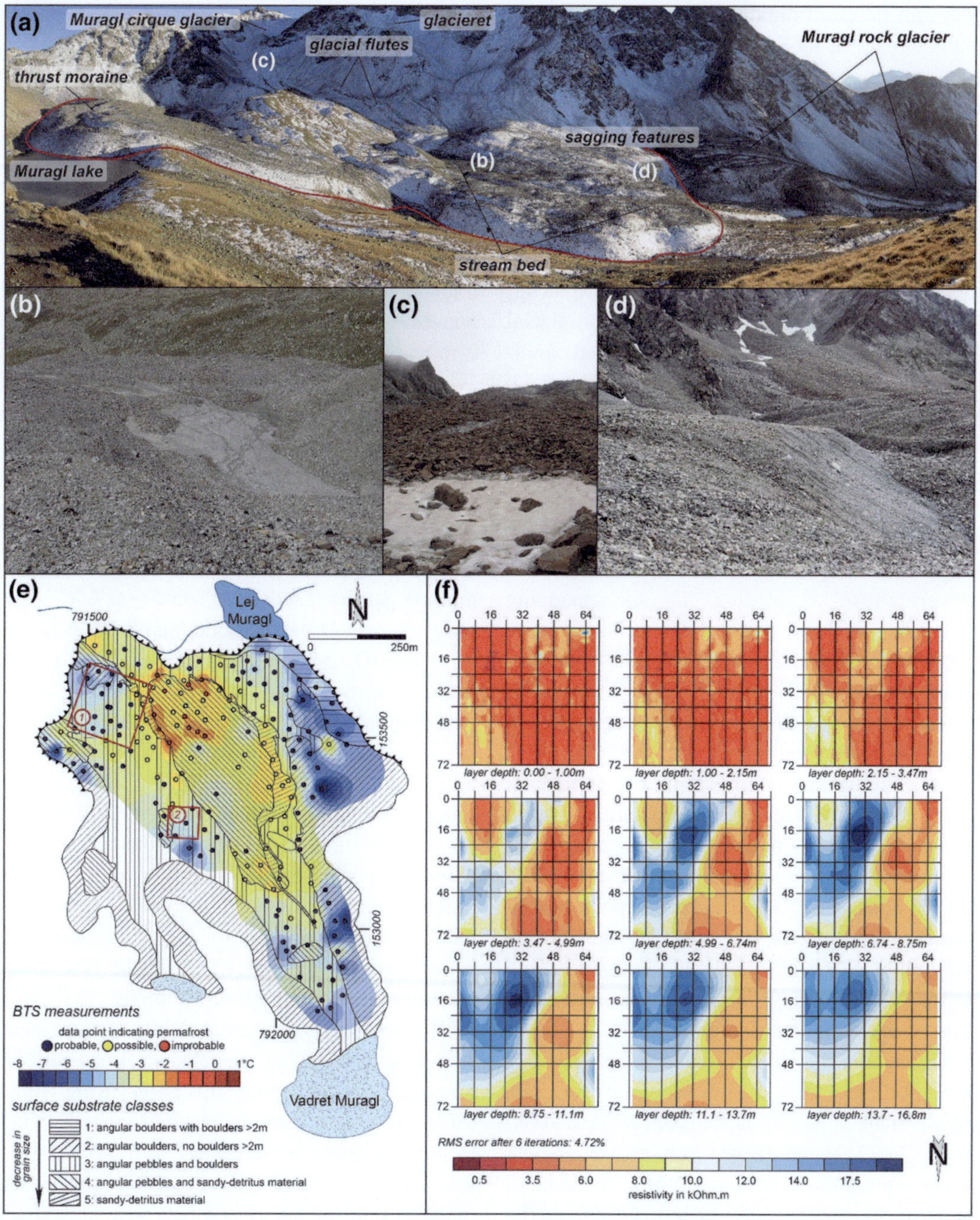

Fig. 6.4 **a–d** Photograph of the upper Val Muragl with the most prominent landforms (**a**). The red solid line marks the approximate LIA extent of the Vadret Muragl (glacier forefield). Red letters refer to the proglacial landforms given in the photographs below. The three images show the braided outlet stream (**b**), the debris-covered glacier (**c**) and the transitional landform of a talus- and moraine-derived rockglacier (**d**). **e** BTS measurements (n = 192, gathered in March 2008) and surface substrate distribution (mapped in July 2008) within the Muragl Glacier forefield. Single BTS data points were mapped and dots indicate the occurrence of ground ice following Haeberli (1973), whereas inverse distance weighting (IDW) allowed the multivariate interpolation of the BTS values (colour bar ranging from blue to red). Surface substrate was mapped in five substrate classes, which are plotted as line pattern. The red rectangles delineate the location of the two described quasi-3D images. **f** Horizontal model slices of the quasi-3D image performed on a shaded mountain ridge on the orographic left site (rectangle 2 in Fig. 6.4e). The model is based on 17 ERT soundings (black lines). Threshold value between ground ice and non-frozen conditions is approximately 9 kOhm m. (data of rectangle 1 are not shown. For further information, see Bast (2009), Bast and Kneisel 2011). *Note* Figure 6.4e, f were modified after Bast (2009)

beds and hence in the development of a braided outlet stream in the lower part of the proglacial zone at the orographic right site (Fig. 6.4a, b). At the orographic left side, a relative steep slope characterizes the lower end of the glacier forefield. This area is creeping with surface velocities of up to 0.55 ma^{-1} due to permafrost conditions (Kääb and Kneisel 2006) (Fig. 6.4d). Kneisel (2010) described this area as a transitional landform of a talus- and moraine-derived rockglacier. In the central and lower part of the proglacial zone, settlement features (funnel-shaped) indicate recent thawing of ground ice (Fig. 6.4a; Kneisel and Kääb 2007).

Basal Temperature of Snow (BTS) measurements (Haeberli 1973) gathered in March 2008 (Bast 2009) allowed the delineation of the small-scale pattern of ground surface temperatures and the potential occurrence of ground ice/permafrost. According to the mapping (Fig. 6.4e), in large parts of the central glacier forefield permafrost is improbable (BTS > −2 °C) or possible (BTS −2 to −3 °C), whereas the upper parts close to the cirque, the orographic left side, the transitional landform and the thrust moraine complex are underlain by permafrost (BTS < −3 °C). Coldest temperatures were recorded within the thrust moraine complex and the cirque area. Following Kneisel and Kääb (2007), Kneisel (2010), these areas possess sedimentary ice (glacier ice) and polygenetic ice (mixture of sedimentary ice and segregation ice). Warmest temperatures were measured within the zone of the settlement features.

Throughout the entire proglacial area, the surface substrate shows a high variability in grain size distribution within short distances (Bast 2009) (Fig. 6.4e). Substrate ranges from coarse, angular blocks with grain sizes above 2 m at the zone of the thrust moraine to fine-grained, detritus-sandy/loamy substrate within the braided river streambed (Fig. 6.4e). Surface substrate affects both, the ground surface temperature and the snow cover evolution and duration. Hence, it acts as a buffer system in the energy balance between atmosphere and lithosphere and, along with climate, topography as well as potential incoming solar radiation (PISR); it determines and influences the patchy occurrence of ground ice (e.g. Harris and Pedersen 1998; Juliussen and Humlum 2008; Dusik et al. 2015). The importance of the surface substrate on the ground thermal regime becomes more apparent by comparing Spearman's rank correlation coefficients of BTS and surface substrate ($r_{BTS/substrate} = 0.69$) with BTS and PISR ($r_{BTS/PISR} = 0.15$) (Bast 2009). Year-round ground surface temperature measurements of miniature temperature data loggers, placed between coarse-grained angular blocks of up to 2 m and fine-grained sandy surface, point out an average temperature offset of 2.3 °C in the period 10/2006–10/2007 within a distance of 120 m (Bast 2009; Bast and Kneisel 2011). Borehole temperature measurements, recorded between 2006 and 2011, emphasize the different temperature regimes of the surface material and show minimum temperatures of −1.0 °C (coarse-grained surface substrate), and −0.7 °C (fine grained), respectively (Rödder and Kneisel 2012).

Quasi-3D resistivity imaging enables the link between surface substrate and subsurface structure and delineates the synergy between surface substrate and ground ice distribution. In summer 2008, Bast (2009) conducted 22 Electric Resistivity Tomography (ERT) profiles consisting of 10 parallel lines and 12 perpendicular tie lines (Wenner array, 5 m electrode spacing, 4356 data points) to build-up a high-resolution subsurface model covering an area of 175 × 175 m in the lower part of the glacier forefield (Fig. 6.4e, rectangle 1). For the quasi-3D image, the single ERT soundings were collated into a 3D dataset within a routine in Res2Dinv (Loke 2015). Smoothness-constrained inversion using finite difference forward modelling and quasi-Newton techniques were implemented with the software Res3Dinv to create the quasi-3D subsurface model (Loke and Barker 1995, 1996). The model reveals that higher resistivities below the coarse blocky material indicate higher ice contents and/or colder ice temperatures (Bast 2009; Bast and Kneisel 2011; data not shown). With decreasing grain size, resistivities decreases, suggesting lower ice contents and/or lower temperatures or permafrost-free areas, respectively. Simultaneously, the thickness of the active layer increases from 2 to 7 m (Bast 2009; Bast and Kneisel 2011; data not shown).

In September 2009, further 17 ERT soundings (10 parallel lines and 7 perpendicular tie lines, Wenner Schlumberger array, 2 m spacing, 4896 data points) were performed on a shaded mountain ridge to establish another quasi-3D image (Bast 2009; Kneisel et al. 2009) (Fig. 6.4e, rectangle 2). Within the covered area of 72 × 72 m, resistivities range between 1 and 22 kOhm m (Fig. 6.4f). A threshold, inferred from borehole temperature measurements, of approximately 9 kOhm m mark the transition between ice-free conditions (red to dark yellow colours) and ground ice occurrence (light yellow to blue colours) (Bast 2009; Kneisel et al. 2009). Following such a threshold, two separate ice bodies are recognizable; a larger one in the centre and eastern part of the model as well as a smaller one at the western end. Resistivities between 9 and 12 kOhm m (light yellow and light blue colours) indicate a high content of unfrozen water along the fringe area of the ground ice lenses, whereas higher resistivities (>12 kOhm m) suggest higher ice contents in the central part of the lenses (dark blue colours). In some parts, the active layer thickness exceeds 5 m. In addition, the small ground ice body in the west tends to degrade (Bast 2009; Kneisel et al. 2009). Again, fine-grained material plays an important role since it favours the formation of segregation ice (e.g. French 2013). Following Kneisel (2010), segregation ice has developed after the 1850s glacier retreat due to the warm-basal ice conditions within that area of the proglacial zone (Kneisel 2010). In contrast to the site described before, the coarse blocky material is of secondary importance, whereas topographical aspects (mountain ridge, shadow/reduced PISR, snow cover/snow drift) become significant (Bast 2009; Kneisel et al. 2009).

6.3 Implications and Conclusions

The glacial history of recently deglaciated, proglacial zones influences the complex thermal regime of the subsurface and determines the current ground ice occurrence. Polythermal glacier regimes enable the coexistence of ground ice of different ages. Under a cold ablation zone, ground ice was formed already before LIA conditions, whereas under warm-basal glacier conditions, one has to assume that the ice has developed recently or is debris-covered glacier ice (sedimentary ice). Ground ice distribution and its properties are very heterogeneous in between small distances, due to the complex surface characteristics like substrate, PISR, topography or snow cover (Bast 2009). In addition, it can form or degrade within relatively short timescales. There is some evidence that permafrost, which was formed after the glacial retreat (permafrost aggradation), is nowadays in equilibrium with local climate conditions and coexists side by side with degrading permafrost (Kneisel and Kääb 2007). Thus, warm ground ice with lower ice content and slightly cold ice with higher ice content may appear in close proximity. The knowledge on the occurrence of ground ice in proglacial areas is currently of great importance, since the areas are rapidly enlarging with the shrinking glaciers. The increase in area is accompanied by an increasing amount of unconsolidated material and partly also of ground ice (Lane et al. 2017). The erosion, transport and accumulation of sediment in the proglacial zone of mid-latitude mountain ranges show high spatio-temporal dynamics and emphasizes the transient character of the system. The dynamics are typically highest in the direct glacier forefield and become more inactive with increasing distance to the glacier front. This might be related to the occurrence of debris-covered ice, which impede the circulation of water and lead to shallow landslides and glacio-fluvial erosion. Depending on the topography, the glacier-derived sediment can actively influence the magnitude or frequency of downslope geomorphological processes such as debris flows or rockfalls. In combination with ground ice, this can lead to either the bonding of sediment in frozen bodies, such as rockglaciers or ice-cored moraines and hence the stabilization of the ground. Or, it can influence the release/destabilization of sediment with the thawing of formerly frozen debris (Haeberli et al. 2016). In addition, degrading ground ice can also significantly influence the amount of meltwater (predominantly in summer) leading to consequences for hydropower production and drinking water.

A deeper knowledge of ongoing processes as well as the amounts of sediment and ground ice is decisive to assess the availability of unconsolidated sediment for potential hazardous processes (e.g. debris flows; Chiarle et al. 2007) and the availability of water from ground ice bodies. The latter is probably less important on the catchment scale, but rather on the scale of entire mountain ranges and probably only for a limited time period (depending on the climate system development). Nevertheless, it might be an essential livelihood for humans living in or close to glaciated mountains. Thus, there is an increasing need for high-resolution data (e.g. repeated topographic data) of proglacial areas (Carrivick et al. 2015) as well as the systematic monitoring of these environments.

References

Ackert RPJ (1998) A rock glacier/debris-covered glacier system at Galena Creek, Absaroka Mountains, Wyoming. Geogr Ann A 80:267–276

Arenson L, Hoelzle M, Springman S (2002) Borehole deformation measurements and internal structure of some rock glaciers in Switzerland. Permafrost Periglac Process 13:117–135

Arenson LU, Colgan W, Marshall HP (2014) Physical, thermal, and mechanical properties of snow, ice and permafrost. In: Haeberli W, Whiteman C (eds) Snow and ice-related hazards, risks, and disasters. Elsevier, Amsterdam, pp 35–75

Ballantyne CK (2002) Paraglacial Geomorphology. Quatern Sci Rev 21:1935–2017

Bast A (2009) Kleinräumige Permafrostverbreitung in einem alpinen Gletschervorfeld, Val Muragl/ Oberengadin, Schweiz. Diplomarbeit Thesis, Universität Würzburg, 271 pp

Bast A, Kneisel C (2011) The surface in the subsurface? —Towards small-scale permafrost distribution and quasi-3D resistivity imaging. Geophys Res Abstr 13

Baumhauer R, Winkler S (2014): Glazialgeomorphologie: Formung der Landoberfläche durch Gletscher. Borntraeger

Bearth P (1980) Erläuterungen zu Atlasblatt 71 (1308 St. Niklaus) des „Geologischen Atlas der Schweiz 1:25.000". Schweizerische Geologische Kommission. Zürich

Benn DI, Evans DJA (2010) Glaciers and glaciation. Hodder Arnold Publication, 802 pp

Berthling IT (2011) Beyond confusion: rock glaciers as cryo-conditioned landforms. Geomorphology 131(3–4)

Bosson JB, Lambiel C (2016) Internal structure and current evolution of very small debris-covered glacier systems located in alpine permafrost environments. Frontiers Earth Sci 4(39)

Bosson JB, Deline P, Bodin X, Schoeneich P, Baron L, Gardent M, Lambiel C (2015) The influence of ground ice distribution on geomorphic dynamics since the Little Ice Age in proglacial areas of two cirque glacier systems. Earth Surf Proc Land 40:666–680

Carrivick JL, Smith MW, Carrivick DM (2015) Terrestrial laser scanning to deliver high-resolution topography of the upper Tarfala valley, Arctic Sweden. GFF 137:4. https://doi.org/10.1080/11035897.2015.1037569

Chiarle M, Iannotti S, Mortara G, Deline P (2007) Recent debris flow occurrences associated with glaciers in the Alps. Glob Planet Change 56:123–136

Church M, Ryder JM (1972) Paraglacial sedimentation: a consideration of fluvial processes conditioned by glaciation. GSA Bull 83(10):3059–3072

Cossart E, Mercier D, Decaulne A, Feuillet T (2013) An overview of the consequences of paraglacial landsliding on deglaciated mountain slopes: typology, timing and contribution to cascading fluxes. Quaternaire 24(1):13–24

Cuffey K, Paterson WSB (2010) The physics of glaciers. Academic Press, 704 pp

Curry AM, Cleasby V, Zukowskyj P (2006) Paraglacial response of steep, sediment-mantled slopes to post-'Little Ice Age' glacier recession in the central Swiss Alps. J Quat Sci 21(3):211–225

Deline P, Gruber S, Delaloye R, Fischer L, Geertsema M, Giardino M, Hasler A, Kirkbride M, Krautblatter M, Magnin F, McColl S, Ravanel L, Schoeneich P (2014) Ice loss and slope stability in high-mountain regions. In: Haeberli W, Whiteman C (eds) Snow and ice-related hazards, risks, and disasters. Elsevier, Amsterdam, pp 35–75

Dusik J-M, Leopold M, Heckmann T, Haas F, Hilger L, Morche D, Neugrig F, Brecht B (2015) Influence of glacier advance on the development of the multipart Riffeltal rock glacier, Central Austrian Alps. Earth Surf Proc Land 40(7):965–980

Eichel J, Krautblatter M, Schmidtlein S, Dikau R (2013) Biogeomorphic interactions in the Turtmann glacier forefield, Switzerland. Geomorphology 201:98–110

French HM (2013) The periglacial environment. Wiley

Grove JM (2004) Little ice ages: ancient and modern. Vol. I + II, 2nd edn. Routledge, London and New York

Haeberli W (1973) Die Basis-Temperatur der winterlichen Schneedecke als möglicher Indikator für die Verbreitung von Permafrost. Z Gletscherk Glazialgeol 9:221–227

Haeberli W (1983) Permafrost–glacier relationships in the Swiss Alps—today and in the past. In: Proceedings of the fourth international conference on permafrost. National Academy Press, Washington, DC, Fairbanks, pp 415–420

Haeberli W (2005) Investigating glacier-permafrost relationships in high-mountain areas: historical background, selected examples and research needs. Glaciers and Permafrost, In Cryospheric Systems

Haeberli W, Schaub Y, Huggel C (2016) Increasing risks related to landslides from degrading permafrost into new lakes in de-glaciated mountain ranges

Harris SA, Pedersen DE (1998) Thermal regimes beneath coarse blocky materials. Permafrost Periglac Process 9:107–120

Huss M, Zemp M, Joerg PC, Salzmann N (2014) High uncertainty in 21st century runoff projections from glacierized basins. J Hydrol 510:35–48

Joerg PC, Morsdorf F, Zemp M (2012) Uncertainty assessment of multi-temporal airborne laser scanning data: A case study on an Alpine glacier. Remote Sens Environ 127:118–129

Juliussen H, Humlum O (2008) Thermal regime of openwork block fields on the mountains Elgåhogna and Sølen, Central-Eastern Norway. Permafrost Periglac Process 19:1–18

Kääb A, Kneisel C (2006) Permafrost creep within a recently deglaciated glacier forefield: Muragl, Swiss Alps. Permafrost Periglac Process 17:79–85

Kneisel C (1999) Permafrost in Gletschervorfeldern. Eine vergleichende Untersuchung in den Ostschweizer Alpen und Nordschweden. Universität Trier: Trierer Geographische Studien 22

Kneisel C (2010) The nature and dynamics of frozen ground in alpine and subarctic periglacial environments. Holocene 20:423–445

Kneisel C, Kääb A (2007) Mountain permafrost dynamics within a recently exposed glacier forefield inferred by a combined geomorphological, geophysical and photogrammetrical approach. Earth Surf Proc Land 32:1797–1810

Kneisel C, Bast A, Schwindt D (2009) Quasi-3-D resistivity imaging—mapping of heterogeneous frozen ground conditions using electrical resistivity tomography. Cryosphere Discuss 3:895–918

Labhart TP (1998) Geologie der Schweiz. Thun, 211 pp

Lane SN, Bakker M, Gabbud C, Micheletti N, Saugy J-N (2017) Sediment export, transient landscape response and catchment-scale connectivity following rapid climate warming and Alpine glacier recession. Geomorphology 277:210–227

Lilleøren KS, Etzelmüller B, Gärtner-Roer I, Kääb A, Westermann S, Gudmundsson A (2013) The distribution, thermal characteristics and dynamics of permafrost in Tröllaskagi, Northern Iceland, as inferred from the distribution of rock glaciers and ice-cored moraines. Permafrost Periglac Process 24:322–335

Loke MH (2015) Tutorial: 2-D and 3-D electrical imaging surveys. www.geotomosoft.com/coursenotes.zip

Loke MH, Barker RD (1995) Least-squares deconvolution of apparent resistivity pseudosections. Geophysics 60:1682–1690

Loke MH, Barker RD (1996) Practical techniques for 3D resistivity surveys and data inversion. Geophys Prospect 44:499–523

Lukas S (2011) Ice-cored moraines. In: Singh VP, Singh P, Haritashya UK (eds) Encyclopedia of snow, ice and glaciers. Encyclopedia of earth science series, part 3, pp 616–619. https://doi.org/10.1007/978-90-481-2642-2_78

Maisch M, Wipf A, Denneler B, Battaglia J, Benz C (2000) Die Gletscher der Schweizer Alpen. Gletscherhochstand 1850, Aktuelle Vergletscherung, Gletscherschwundszenarien (2 edn). Zürich: vdf Hochschulverlag AG an der ETH Zürich

Maisch M, Haeberli W, Frauenfelder R, Kääb A (2003) Lateglacial and holocene evolution of glaciers and permafrost in the Val Muragl, Upper Engadin, Swiss Alps. In: Phillips M, Springman SM, Arenson L (eds) Proceedings of the eighth international conference on permafrost. Balkema, Zurich, pp 717–722

Maurer H, Hauck C (2007) Geophysical imaging of alpine rock glaciers. J Glaciol 53:110–120

Østrem G (1959) Ice melting under a thin layer of moraine, and the existence of ice cores in moraine ridges. Geogr Ann 51:228–230

Ott E, Frehner M, Frey H-U, Lüscher P (1997) Gebirgsnadelwäler. Paul Haupt Verlag, Bern, Stuttgart, Wien, Ein praxisorientierter Leitfaden für eine standortgerechte Waldbehandlung

Otto JC, Dikau R (2004) Geomorphologic system analysis of a high mountain valley in the Swiss Alps. Zeitschrift für Gemorphologie, N.F. 48(3):323–341

Pellicciotti F, Carenzo M, Bordoy R, Stoffel M (2014) Changes in glaciers in the Swiss Alps and impact on basin hydrology: current state of the art and future research. Sci Total Environ 493:1152–1170

PERMOS (2013) Permafrost in Switzerland 2008/2009 and 2009/2010

Rastner P, Joerg P-C, Huss M, Zemp M (2016) Historical analysis and visualization of the retreat of Findelengletscher, Switzerland, 1859–2010. Global Planet Change 145:67–77

Reynard E, Lambiel C, Delaloye R, Devaud G, Baron L, Chapellier D, Marescot L, Monnet R (2003) Glacier/permafrost relationships in forefields of small glaciers (Swiss Alps). In: Phillips M, Springman SM, Arenson LU (eds) 8th international conference on Permafrost, vol 1, Zurich. A.A. Balkema, Lisse, pp 947–952

Rödder T, Kneisel C (2012) Influence of snow cover and grain size on the ground thermal regime in the discontinuous permafrost zone, Swiss Alps. Geomorphology 175–176:176–189

Ruff A (2015) Temporal and spatial quantification of geomorphological processes in the recently deglaciated area surrounding the Findelengletscher. Master thesis, Department of Geography, University of Zurich, 78 pp

Schomacker A (2008) What controls dead-ice melting under different climate conditions) a discussion. Earth Sci Rev 90:103–113

Springman SM, Arenson L, Yamamoto Y, Maurer H, Kos A, Buchli T, Derungs G (2012) Multidisciplinary investigations on three rock glaciers in the Swiss Alps: legacies and future perspecitves. Geogr Ann Ser A Phys Geogr 94:215–243

Swift DA, Cook S, Heckmann T, Moore J, Gärtner-Roer I, Korup O (2014) Ice and snow as land-forming agents. In: Haeberli W, Whiteman C (eds) Snow and ice-related hazards, risks, and disasters. Elsevier, Amsterdam, pp 167–199

Vaughan DG et al (2013) Observations: cryosphere. In: Stocker TF et al (eds) Climate change 2013: the physical science basis. Cambridge University Press, Cambridge, United Kingdom and New York, NY, USA, 2013, Chapter 4. http://www.climatechange2013.org/report/full-report/

VAW (2015) The Swiss Glaciers 2009/2010 and 2010/2011. In: Bauder A (ed) Glaciological Report No. 131/132. Publication of the Cryospheric Commission (EKK) of the Swiss Academy of Sciences (SCNAT), 113 pp

WGMS (2015) Global glacier change bulletin no. 1 (2012–2013). In: Zemp M, Gärtner-Roer I, Nussbaumer SU, Hüsler F, Machguth H, Mölg N, Paul F, Hoelzle M (eds) ICSU(WDS)/IUGG(IACS)/UNEP/UNESCO/WMO, World Glacier Monitoring Service, Zurich, Switzerland, 230 pp

Zemp M, Haeberli W, Hoelzle M, Paul F (2006) Alpine glaciers to disappear within decades? Geophys Res Lett 33:L13504

7 Periglacial Morphodynamics in the Upper Kaunertal

Jana-Marie Dusik, Matthias Leopold and Florian Haas

Abstract

Proglacial areas represent transitional zones between glacial and periglacial landscapes. Non-glacial processes in the proglacial zone might evolve under periglacial conditions, and vice versa periglacial processes often occur in proglacial zones. Hence, periglacial morphodynamics constitute an inherent part of proglacial studies and are therefore presented here in the context of the research project PROSA. To get an overview of periglacial morphodynamics in the Upper Kaunertal, we (i) established a permafrost model based on BTS measurements, (ii) analysed meteorological data in order to verify ground indications for thermal-contraction cracking and (iii) analysed both geophysical and remote sensing data for the identification of permafrost creep and thermokarst phenomena. In addition, this chapter includes a case study on the morphodynamics of a complex rock glacier in the Riffltal, a tributary valley to the Upper Kaunertal. Results show a diverse genesis and development of different rock glacier parts and lobes, which are mainly related to glacial advances during the Egesen and LIA stadials. More recent developments are largely dependent on (i) sediment sources and pathways constituting diverse states of activity, (i) dynamic interactions between the tongues and (iii) the topographic surrounding, that is already causing dynamic inactivity of parts of the rock glacier.

Keywords

PROSA project · Geophysical measurements Permafrost model · Rock glaciers · Surface changes

7.1 Introduction

Periglacial morphodynamics in general are defined as landscape evolution by seasonal and perennial frost action, snow and azonal processes (French and Thorn 2006; French 2007). In mountainous landscapes, which are characterized by high-relief energy and steep temperature gradients—periglacial processes are mainly expressed in slope mass wasting by frost creep and frost-induced heaving and subsidence as well

J.-M. Dusik (✉)
Bavarian State Agency for Environment (LfU), Geological Survey, Hof/Saale, Germany
e-mail: jana.dusik@gmx.de

F. Haas
Physical Geography, Catholic University of Eichstätt-Ingolstadt, Eichstätt, Germany

M. Leopold
School of Agriculture and Environment, The University of Western Australia, Crawley, Australia

T. Heckmann and D. Morche (eds.), *Geomorphology of Proglacial Systems*,
Geography of the Physical Environment, https://doi.org/10.1007/978-3-319-94184-4_7

as sorting. Physical weathering and cryoclastic processes are the main drivers for scree production and guarantee the availability of moveable sediments. A common slope process in the absence of permafrost would be solifluction. Its counterpart for surfaces underlain by permafrost is gelifluction (Washburn 1967). The corresponding landforms are rather shallow lobate structures of often less than 1 m thickness and transport rates of only a few cm yr^{-1}. On a larger scale, mass wasting processes arise in form of creeping ice-rich talus/moraines and rock glaciers, or transitional landforms in between the named ones (see Chap. 6). Rock glaciers may reach km in length. Some of the most active examples in the European Alps show movement rates in the range of a few m yr^{-1} (Haeberli 1985; Hausmann et al. 2007a). The ice-rich landforms are temperature sensitive and hence react to climate change with accelerated creep velocities (Kääb et al. 2007). Retreating glaciers expose large areas with ground moraine sediments, which in consequence are prone to seasonal and diurnal frost processes. Supplied with moisture from melt and precipitation water, frost heave and sorting processes are a common phenomenon of the alpine periglacial zone.

Compared to the polar sciences, where permafrost and periglacial processes have been studied since 1906 (Anderson 1906), high mountain cryospheric science depicts a rather young discipline. The term *periglacial* was first introduced by Lozinski (1909) (French and Thorn 2006). Periglacial processes have been subject to alpine geomorphology studies since the 1960s, starting with Jäckli (1957), followed by Barsch's work on rock glaciers in the lower Engadin, Swiss Alps (Barsch 1969). Since then, the local permafrost distribution and periglacial processes have become the focus of a wide range of studies in high mountain areas. Especially, the numerous studies of Haeberli and associated researchers on the principles of high mountain periglacial environments, geomorphodynamics and glaciology enhanced these fields of research from nearly the beginning to the present (i.a. Haeberli 1975, 1985, 2000, 2013; Haeberli et al. 1998, 2006). Following Haeberli, the increasing number of studies about permafrost in high mountain regions must be seen as the consequence of the clearly visible changes in high mountain landscapes (e.g. melting glaciers, destabilisation of rock faces), following the global warming since the second half of the last century. Amongst them are numerous studies with different foci, including

- geophysical investigations of mountain permafrost and periglacial processes (Hauck et al. 2004; Hauck and Kneisel 2008; Hausmann et al. 2007a, c, 2012; Hilbich et al. 2008; Kääb and Kneisel 2006; Kneisel 2004; Leopold et al. 2008a, b, 2011),
- mountain permafrost distribution and characteristics (Boeckli et al. 2012; Etzelmüller et al. 2006, 2007; Gruber and Hoelzle 2001; Guodong and Dramis 1992; Hoelzle et al. 1999, 2001; Ishikawa and Hirakawa 2000; King et al. 1992; Verleysdonk et al. 2011; Lerjen et al. 2003; Lewkowicz and Ednie 2004; Lewkowicz et al. 2011; Leopold et al. 2014; Noetzli et al. 2007; Otto et al. 2012; Ruiz and Trombotto 2012)
- the ground thermal regime of high mountain areas (Farbrot et al. 2013; Gorbunov et al. 2004; Harris and Pedersen 1998; Ishikawa 2003; Nötzli 2011; Rödder and Kneisel 2012; Vonder Mühll and Haeberli 1990). Solifluction, that belongs to the processes driven by seasonal frost mechanisms in periglacial environments that are not necessarily underlain by permafrost or ground ice, is subject to studies by Kinnard and Lewkowicz (2006), Lewkowicz (1992), Matsuoka (2001), Matsuoka et al. (2005) or Walsh et al. (2003).

As air temperature is predicted to rise by 2.3–3.3 °C by 2055 (Nogués-Bravo et al. 2007; Marcott et al. 2013), mountain permafrost—as "a central element" (French and Thorn 2006) of cold regions where periglacial processes occur—recently has not only become of special interest in the context of climate change, but especially for sediment flux studies (Beylich and Warburton 2007; Chap. 15). Cold-frozen sediments are immobile, and unlike warm permafrost bodies, show only little surface deformation (Beylich and Warburton 2007; Kääb et al. 2007). Increasing air temperatures lead to a warming of permafrost, causing according slope instabilities: (1) steep bedrock walls are destabilised and induce increased rockfall hazards (Nötzli et al. 2003; Krautblatter 2008, 2009), (2) creep velocities of rock glaciers are accelerated as reported by Kaufmann and Ladstädter (2003), Kääb et al. (2007) and Roer et al. (2008). Roer et al. (2008) consider higher ground temperatures to be the reason for "[…]changes in ice content or ice characteristics; changes in the shear horizons (e.g. number, position, frictional behaviour, occurrence of unfrozen water); or changes in the internal structure of the permafrost body." The higher the deformational changes within frozen ground are in a short period, the higher the probability for material failure gets (Arenson et al. 2014). Freeze-thaw processes undergo changes in magnitude and frequency caused by steeper day–night temperature gradients at high altitudes (French 2007; Arenson and Jakob 2015). As the state and the future changes of permafrost must be seen as one main influencing factor for the mobilization of sediment on slopes, it is of major importance to analyse the sensitivity of permafrost to increasing air temperatures. Relating to this topic, research programs like PACE (Harris et al. 2001) were set up to monitor changes of the active layer depth, snow cover duration, extent and thickness, the thermal regime of permafrost and other important parameters for permafrost temperature interactions and the associated feedbacks (Hoelzle et al. 2001; French 2007; Ebohon and Schrott 2008; Verleysdonk et al. 2011). Glacier–permafrost interactions are becoming more and more interesting for geomorphological purpose with decreasing size of glaciers that are cold-based or the acceleration of glaciers, as this interplay causes processes that transform and transport sediments in ice-marginal and proglacial areas, including glacial thrusting, resulting in the formation of push moraines or extrusion moraines (Harris and Murton 2005).

Investigations of periglacial morphodynamics in the Upper Kaunertal, that depicts the research area of the project PROSA (Chap. 1), are summarized in the following. The Kaunertal is located in the Western Austrian Alps and like all the large Austrian tributary valleys of the river Inn south–north striking. Formed by Pleistocene glaciers, the trough valley drains to the Inn in the North at an altitude of 864 m a.s.l. To the South, the Kaunertal is bordered by mountains of the alpine main ridge with the Weißseespitze (3526 m a.s.l.) as highest peak of the large catchment, including the Gepatschferner as second longest glacier of the Eastern European Alps and several smaller glaciers (World Glacier Monitoring Service 2015). The high alpine valley is part of the central Alpine dry climate zone, with a mean annual precipitation of 1095 mm.[1] Mean annual air temperatures (MAAT) decrease from 4.4 °C at 1770 m a.s.l., over 2.7 °C at 1900 m a.s.l. to −0.6 °C at 2600 m a.s.l., indicating the cold conditions of the upper catchment with a predominant, periglacial morphodynamic (French 2007). Rock glaciers are one of the dominating landforms with 3.9% (123) of all Tyrolean rock glaciers located in this catchment and have therefore intensely been studied throughout the Ötztal Alps (Krainer and Ribis 2012; Krainer and Mostler 2006; Hausmann et al. 2007b). In addition to these large and partly slow moving sediment bodies, many small-scaled periglacial forms can be found in the proglacial zone of the Kaunertal, e.g. stone polygons, thermal-contraction cracks and soli-/gelifluction lobes.

7.2 Distribution of Periglacial Phenomena in the Proglacial Zone

7.2.1 Permafrost Modelling

Permafrost modelling is an important tool for many purposes in the Alps. Tourism requires construction measures for the accessibility and practice of leisure opportunities. Other construction measures are necessary to protect

[1] 1980 to 2014, at 1900 m a.s.l.

infrastructure and livelihood from natural hazards provoked by warming permafrost (Phillips et al. 2009; Etzelmüller 2013). Modelling the distribution of permafrost serves as a basis to find locations where protection measures would be useful. With respect to sediment budgets, the ice content of frozen sediments has to be examined to get the true sediment volume of ice supersaturated sediment bodies, like rock glaciers or ice-rich talus, without the proportion of interstitial ice. And of course, permafrost models give an estimate about where periglacial processes occur. Models on permafrost distribution are manifold. Mainly, there are two kinds of models to be distinguished: (1) empirical-statistical models that rest upon permafrost observations in the field related to topographic and climatic factors and (2) process-oriented models that rely on physical interactions of permafrost with the atmosphere and, hence take energy fluxes into account (Hoelzle et al. 2001). Hoelzle et al. (2001) give a summary about the development of permafrost distribution models from the first study by Haeberli (1975) to the present.

To reveal the spatial distribution of permanently frozen ground in the Upper Kaunertal, an empirical-statistical model structure was chosen: it is based on a multivariable linear regression of the topoclimatic parameters elevation, total insolation during snow-free conditions, surface roughness and profile curvature, all derived from an airborne LiDAR DEM (Digital Elevation Model) from 2012 with 1 m resolution. The model is informed by 432 measurements of the base temperature of the winter snow cover (BTS), which isolates the ground from atmospheric influence when it is settled and at least 80 cm deep. Therefore, it gives the opportunity to measure near-surface ground temperatures (Haeberli 1973). The predictors were selected according to their known influence on permafrost. Altitudinal values serve as a proxy for the dry-adiabatic temperature gradient and therefore an air temperature decrease with increasing elevation. Total insolation is a measure for energy fluxes and was calculated depending on aspect and several adjusted atmospheric parameters. Profile curvature was chosen because it reflects slope foot positions where the snow cover persistence may persist until or even survive summertime and thus preserves cool ground temperatures. Surface roughness helps to separate coarse from fine sediments and therefore reflects influences on permafrost occurrence like bulk ventilation and pore space (Keller 1992; Hoelzle 1992; Keller and Hoelzle 1996). The model (Fig. 7.1, left) gives a reasonable estimate of the permafrost distribution with an adjusted R^2 of 60% (p-Value $< -2e - 16$). Predicted near surface temperatures range from −10 to 1 °C. The model is validated by 10-fold cross-validation, which yields an RMSE of 2.3 °C. After the classification of ground temperature values by Haeberli (1975), 37.2% of the distribution are within the permafrost probable class (class 1: BTS < -3 °C), 25.7% belong to terrain with possible permafrost occurrence (class 2: BTS from −3 to −2 °C) and 37.1% predict no probability for permafrost (class 3: BTS $>$ −2 °C). Landforms and locations of the first class include mountain ridges, north exposed slopes, rock glaciers, blocky terrain/sedimentary deposits of mainly coarse grain sizes. Permafrost is predicted as possible in all aspects, in flat as well as in steep terrain. Areas where permafrost occurrence is rather improbable are mostly south-facing slopes and flat terrain in low elevations with smooth surface structures.

7.2.2 Thermal-Contraction Cracks

Many of the recently deglaciated areas—especially those of small glaciers—are located within the actual permafrost belt, starting from 2600 m a.s.l. and upwards in the Upper Kaunertal. Proglacial areas of glaciers with pronounced subglacial moraines are subject to seasonal frost processes. Where moisture is available and a sediment body of sufficient size is present, freeze–thaw cycles lead to characteristic phenomena, visible at the surface in form of a furrow–ridge structure and known as thermal-contraction cracks.

French (2007) summarizes that some important controls on thermal-contraction cracking are

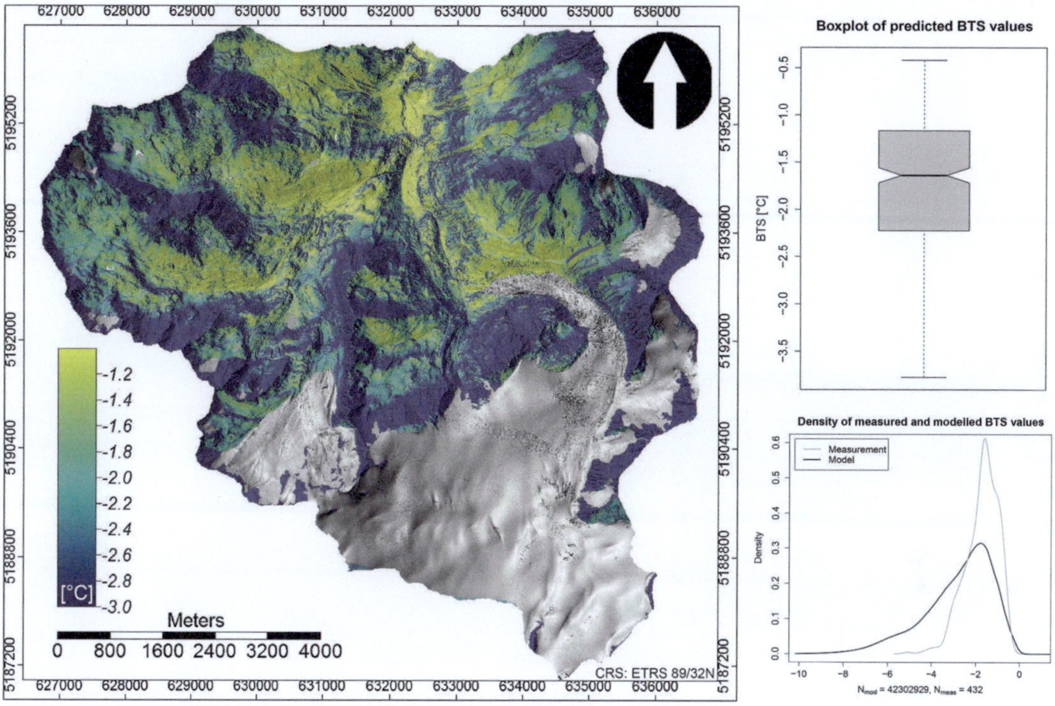

Fig. 7.1 Left panel: permafrost model of the Upper Kaunertal; top right: boxplot of modelled BTS values; bottom right: comparison of densities (via kernel density estimation) of modelled and measured BTS values

dropping air temperatures on at least 4 following days with rates of 1.8 °C/day (Fig. 7.2), a thin snow cover, changing micro-relief and the absence or the presence of vegetation cover. But the process is mostly dependent on the ground–thermal gradient, and thus on temperatures at the top of the permafrost (TTOP) and on surface temperatures. This means that permafrost is required and frequency and magnitude of thermal-contraction cracking, which is sensitive to high changes of temperature in short time is increasing with global warming (French 2007).

Initial features of thermal-contraction cracking could be observed in the direct forefield of the present-day tongue of the Weißseeferner, which leaves a vast area of ground moraine during its steady decline. Other thermal-contraction cracks can be found in the ground moraines of the side valley Riffltal and the north facing cirque, Münchner Kar, located just below the Gepatschferner plateau. The three locations have common parameters suitable for cracking: (1) high altitude above 2700 m a.s.l., (2) northern aspect, (3) availability of sediment and (4) availability of moisture.

7.2.3 Permafrost Creep

Due to high relief energy and altitude, creeping permafrost creates the most common periglacial phenomena in alpine environments. The most prominent is rock glaciers (see also Chap. 6) that reach velocities of up to 2 ma^{-1} in the Ötztal Alps (Berger et al. 2004). Some solifluction lobes, although not necessarily associated with permafrost, both bound and unbound can be observed in the Riffltal and the Kühgrube tributary valleys. They all are located within areas of probable or possible permafrost occurrence and close to active rock glaciers. Push moraines can be found at the margins of ground moraines of remnant glaciers, i.e. in the upper Riffltal tributary valley, the upper Ölgrube and the cirque

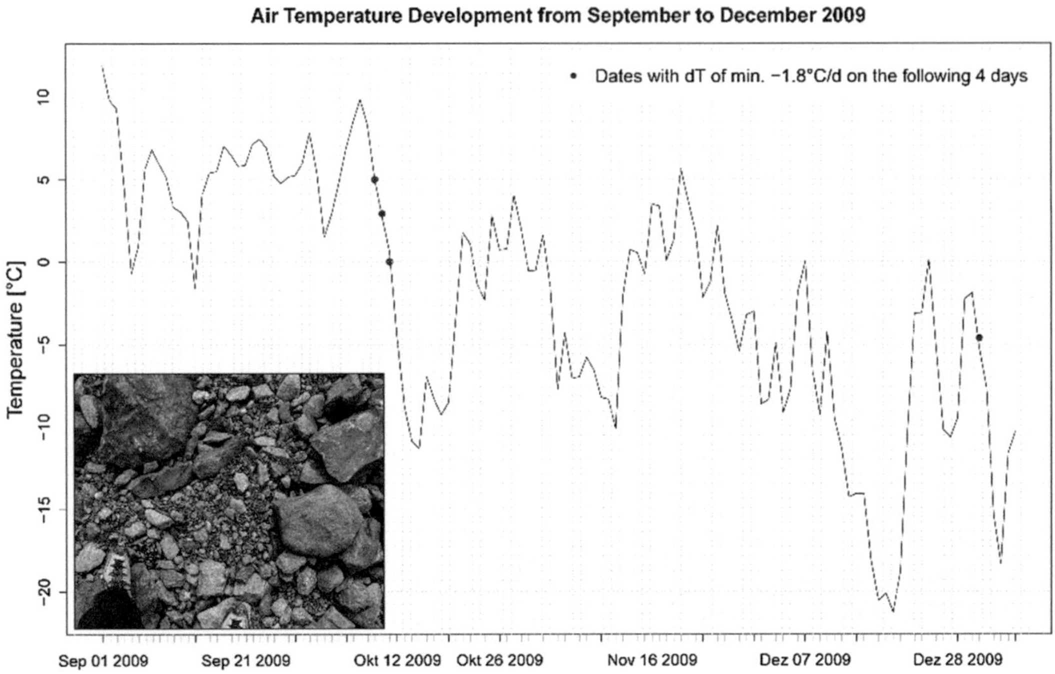

Fig. 7.2 Temperature curve of the meteorological station at lake Weißsee with indication of 4-days series with temperature decreases of 1.8 °C per day; bottom left: stone polygon in the proglacial ground moraine of the Weißseeferner

Münchner Kar. In the Riffltal and the Ölgrube, push moraines are merged with or located slightly above the root zones of active rock glaciers.

Many studies combine geophysical methods and temperature measurements with remote sensing methods for the detection and analysis of creeping mountain permafrost. Aerial photographs and airborne LiDAR data are used for the recognition of surface features on two different recordings, e.g. feature tracking, photogrammetric matching of image points (Kääb and Kneisel 2006; Ishikawa and Hirakawa 2000; Roer and Nyenhuis 2007; Haeberli and Patzelt 1982).

In the study area of the PROSA project, 45 rock glaciers have been identified (Fig. 7.3). 14 of them are classified as active, 11 as inactive, and another 20 are considered fossil. The activity status is either derived by remote sensing techniques or field-based methods. The first group includes feature tracking and mapping of the frontal extent based on multiple orthophotographs, which also allow for the quantification of the activity of single rock glaciers (c.f. Dusik et al. 2015). The latter group includes the investigation of indicators at the rock glacier surface, such as the degree of lichen and vegetation coverage, mobility of surficial blocks, signs of thermokarst and other melt phenomena. Front elevations of active rock glaciers in our study area range from 2450 to 2921 m a.s.l. The highest front is one of the Ölgrube rock glaciers with 90 m, but a comparatively low slope angle of 33°. Active rock glaciers are exposed towards north, northwest or northeast and have diverse geometries. The most common rock glacier class (after Wahrhaftig and Cox 1959 and Barsch 1996) is tongue-shaped rock glaciers. Genetically, most of the rock glaciers cannot be separated in talus or moraine rock glaciers, as different sedimentary processes often coincide; for example, moraine deposits are overlain by rockfall deposits.

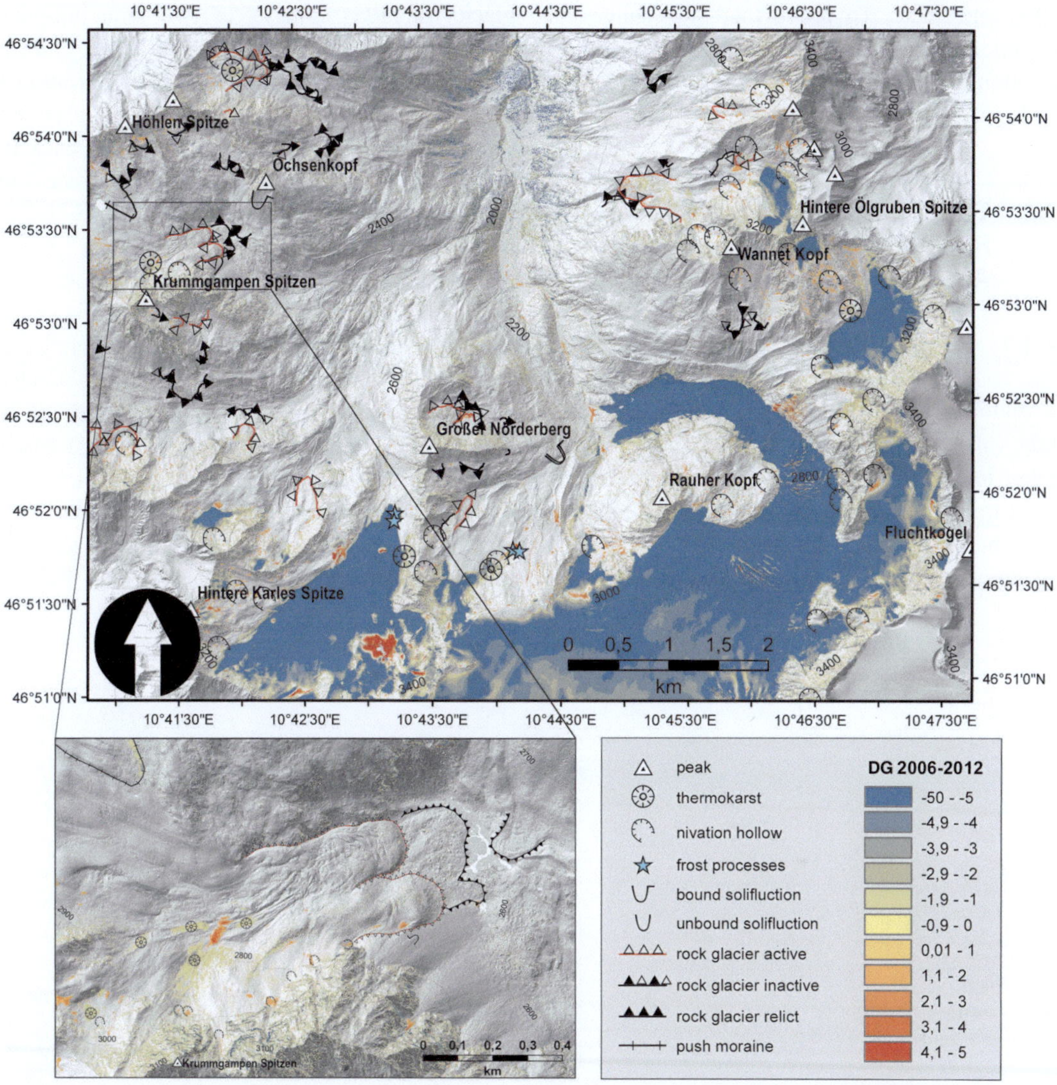

Fig. 7.3 Map of the Upper Kaunertal with changes in surface elevation (DoD = DEM of difference) and geomorphic features related to periglacial morphodynamics and permafrost occurrence; bottom left panel shows the Riffltal rock glacier with thermokarst appearances

7.2.4 Thermokarst

Changes in surface morphology through melting of ground- and dead ice are a common feature in the pro- and periglacial areas of the Upper Kaunertal. Although French (2007) argues "that thermokarst is rarely reported from alpine and mountainous regions" due to the fact that large areas are bedrock surfaces, thermokarst can be found in areas of proglacial ground moraines, rock glaciers and push moraines. Clearly, thermokarst phenomena are not as well developed and characteristic in shape as they are in the discontinuous permafrost belt of Siberia. Nevertheless, subsidence of surfaces that are potentially underlain by ice-rich sediments or contain massive ice can be detected on DEMs of difference (DoD) of the study area (Fig. 7.3). Figure 7.3 (bottom left) shows a part of the Riffltal with clear surface subsidence between

2006 and 2012. To prove that the changes of morphology are not caused by melting snow or nivation hollows, mapping of the latter was done on basis of aerial photographs.

7.3 The Case Study of the Riffeltal Rock Glacier, Central Austrian Alps

7.3.1 Aim of the Study and Study Site

A multipart rock glacier (Fig. 7.4), located in a side valley of the Upper Kaunertal, is investigated to assess its activity and hence the sediment transport rate of the landform to reveal its contribution to the sediment budget of the high alpine valley (c.f. Chap. 17). The Rifflta l rock glacier (RTRG) consists of two main tongues lying side by side and several overlapping lobes. It has a length of 955 m and an overall width of 330 m. Bordered by the Krummgampenspitzen mountain ridge in the south and a whaleback in the north, RTRG forms a barrier between the proglacial area of the remnant Riffl cirque glacier below the Glockturm peak (3355 m a.s.l.) and the valley mouth. The rock glacier fronts are located on flat sedimentary terrain at an elevation of 2623 m a.s.l. In the immediate forefield, a semicircular grassland with unsorted, chaotically distributed boulders of a few decimetres up to several metres in size is interpreted as a fossil rock glacier. It is bordered by a channel, partly fed by sources emanating from the active rock glacier fronts. Another fossil rock glacier is located just opposite the discussed one. Further

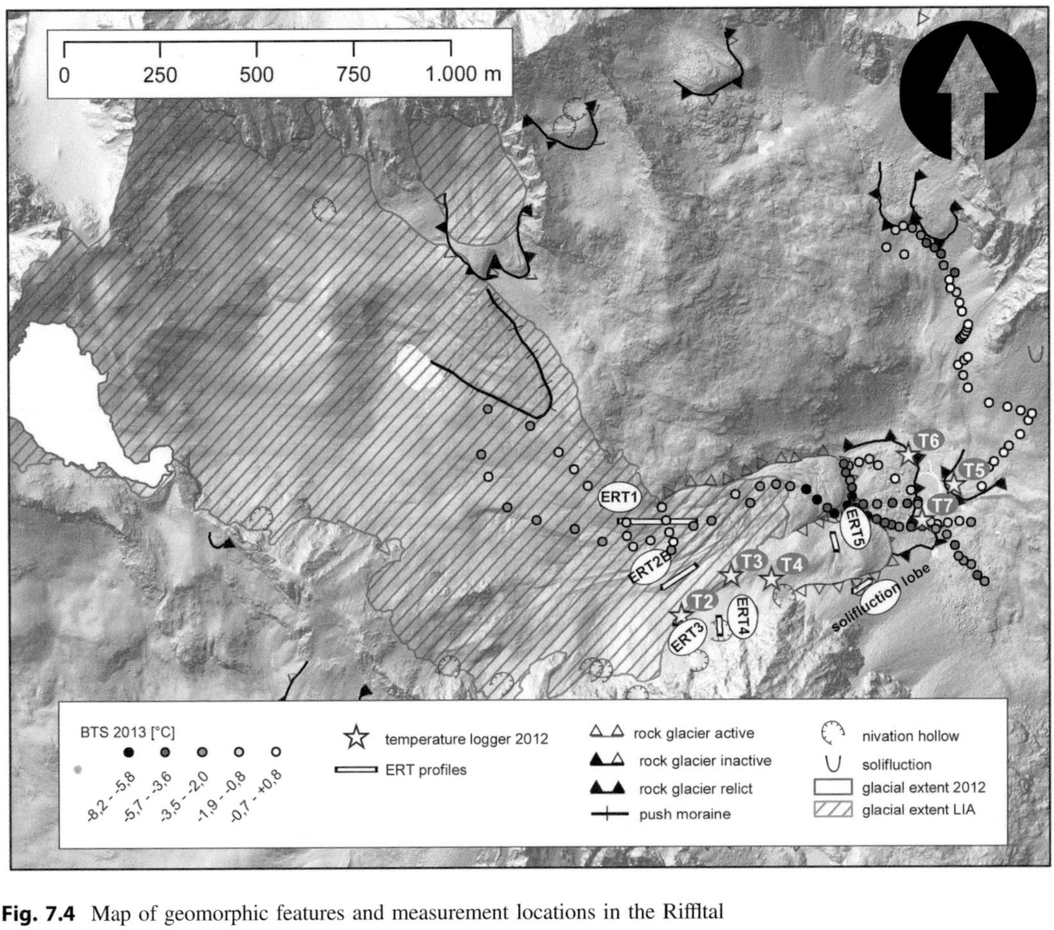

Fig. 7.4 Map of geomorphic features and measurement locations in the Rifflta l

down valley, a huge rock mass has been deposited that consists of angular blocks. It obstructs the valley mouth and hence forms a barrier to sediment transport (aside from suspended material). The root zones of the tongues are different: RT01 arises from the ground moraine landscape below the Glockturm massif in the very back of the valley and thus consists mainly of reddish paragneiss and grey orthogneiss. RT02 developed out of the Little Ice Age (LIA) lateral moraine of the Riffl glacier that is located at the foot of the Krummgampenspitzen peaks. It consists of dark, brittle amphibolite that is delivered by rockfall to the root zone of the rock glacier tongue. Besides the rock composition, both tongues differ in surface elevation, surface homogeneity, mobility of the surficial rocks and lichen cover. The surface of RT01 is flatter than the one of RT02; many blocks, especially in the front part, are covered with lichens, suggesting stability. Fine sediments fill the gaps between blocks of up to a couple of metres in size that are mostly immobile. RT02 shows a more mobile arrangement of surficial boulders ranging from several decimetres to a few metres in size, indicating dynamic activity of the tongue that is fed by the north facing rock slopes of the Krummgampenspitzen mountains.

7.3.2 Methods and Data Acquisition

7.3.2.1 Geophysical Measurements

In order to portray the internal structure with a possible permafrost or ground ice occurrence of the rock glacier, we used three different geophysical methods, electrical resistivity tomography (ERT), ground penetrating radar (GPR) and shallow seismic refraction (SSR), all applied in September 2012.

A total of five two-dimensional ERT profiles (Fig. 7.4) were measured with different arrays (Wenner, dipole–dipole and Wenner half) to ensure maximum penetration depth, best near surface resolution and a stable signal-to-noise ratio. The spacing of the electrodes (n_{max} = 100) was set to 1 or 2 m, respectively, (for details, see Leopold et al. 2013 and Dusik et al. 2015). The start and end points of the profiles were measured with a differential GPS and an uncertainty of up to 2 cm. Based on this and xyz-topography, data from a LiDAR DEM of 2012 were extracted for a topographic correction.

Additionally to the ERT measurement of ERT line 1, the GPR-1 and SSR-1 profile were measured along the same line to validate the ERT results.

The geophones of the seismometer (12-channel Smartseis from Geometrics, San Jose, CA) had a spacing of 8 m. A sledgehammer was used as acoustic source and struck three to five times per shot point in order to reduce the noise and increase acoustic coupling by compressing the surficial material as suggested by Krummel (2005). Shot points were set in between the geophones as well as at 5 and 10 m from the first and last geophone. The calculated travel times were inverted (wavefront inversion combined with network/raytracing) to generate a two-dimensional subsurface velocity model (Dusik et al. 2015; Hoffmann and Schrott 2003; Leopold et al. 2011).

To obtain additional information on the internal structure of the rock glacier, a portable GPR system from Måla Geosystems (RAMAC CU II) with a 50 MHz antennae and 2 m spacing (common offset) was used in step-mode perpendicular to the profile. Pulses were sent out every 0.5 m, and traces were stacked 16 times. An additional common midpoint (CMP) survey with a step interval of 10 cm was applied at 34.5 m of the line. Postprocessing of the data is following the protocol of Leopold et al. (2011), which was then visually interpreted after Neal (2004).

7.3.2.2 Temperature Monitoring

In autumn 2012, six temperature loggers were installed on and around the rock glacier (Fig. 7.4) in order to investigate the ground thermal regime at different locations of the complex landform. They logged temperature data at an hourly resolution until summer 2013. Additionally, 111 point measurements were taken in March 2013 (Fig. 7.4) with a temperature probe of 2 m length (Hannah Instruments, HI 766 TR4) associated with a k-type handheld thermometer (Hannah Instruments). Point locations were taken with a differential GPS (DGPS; accuracy 2 cm) or handheld GPS (accuracy 2–4 m). The measurements are distributed over the Riffltal to validate a linear multivariate regression model of BTS values for the entire catchment (Sect. 7.2.1).

7.3.2.3 Topographic Changes

The analysis of multitemporal DEMs is a convenient tool for the detection of changes in surficial morphology and the description of dynamic landforms. Creep processes can be detected by image correlation-based feature tracking (Scambos et al. 1992; Debella-Gilo and Kääb 2011). One possible method is to apply a normalized cross-covariance algorithm—as provided by the SAGA GIS tool *feature tracking* (*IMCORR*)—to analytical hillshades, resulting in movement vectors of surface features (i.e. rocks). Depending on the resolution and accuracy of the DEMs, seasonal and long-term movements of creeping permafrost can be detected and quantified.

An even more simple method is DEM differencing (c.f. Sect. 7.2.4). Negative values of the DoD display mass wasting processes, like melting of ground ice or extensional flow of rock glaciers. Positive values of the DoD indicate depositional processes, aggradation of permafrost and ground ice, or a rise of rock glacier front heights due to compressional flow (Roer et al. 2005). In terms of error estimation of the DoD, an level of detection (LoD) has to be applied (for details, see Lane et al. 2003; Wheaton et al. 2010).

7.3.3 Results

7.3.3.1 Geophysical Measurements

Two of the five geophysical profiles are chosen to be described in the following, as they represent different important zones of the rock glacier.

ERT line 1 (Fig. 7.5, top left) is located in the upper part of RT01 and starts on blocky terrain that merges to finer surface material from point 45 m on in upward direction. Three iterations with an absolute error of 2.45% result in a subsurface model of specific electric resistivities between 2.2 and 17.3 kΩm. The uppermost metre shows homogenous conditions with low resistivity values of 2 to 4.4 kΩm. Below the first metres, resistivity gets more heterogeneous. Two zones of higher resistivities can be found along the profile. The first one is located between 0 and 55 m with values up to 17 kΩm and a sharp vertical increase that is reversed below 2 m down to 6 m depth with values of 5–6 kΩm. The second but not as distinct zone is found between 130 and 180 m is about 15 m thick and shows values of up to 11 kΩm. According to the depth of investigation (DOI) analysis, the model is sensitive to the physical properties of the subsurface (DOI < 0.2; Hilbich et al. 2009) with exceptions of areas deeper than 20 m and at 70 and 128 m along the profile where DOI is >0.2.

The SSR profile (Fig. 7.5, top right) along the same line is little shorter than the ERT profile. A network ray tracing model was chosen due to a reduced number of shot points. Three layers could be detected, where the first layer shows p-wave velocities between 180 and 330 m s^{-1}, followed by a refractor in 6–12 m depth. P-wave velocity of the second layer was 1149 m s^{-1}. Passing the second refractor at 16–20 m depth, velocities of the third layer increased significantly to between 4438 and 5356 m s^{-1}.

Velocity analysis of the CMP measurement at 34.5 m resulted in a two-layer model with a velocity of 0.108 m ns^{-1} for the upper 7 metres and an increasing velocity of 0.114 m ns^{-1} below. The resulting radar image showed several diffraction hyperbola, most probably caused by point sources such as large boulders and several continuous boundaries that might represent several generations of the rock glacier as well as internal flow structures (Fig. 7.5, middle and bottom right).

With ERT line 3 (Fig. 7.5, bottom left), a small lobe in the root zone of RT02 just below the north exposed talus slopes of the Krumgampenspitzen is investigated. Its surface consists of loose boulders of up to 0.5 m in a sandy-silty matrix and a high moisture content. The two-layer model shows a constant, high sensitivity to the field measurement with DOI

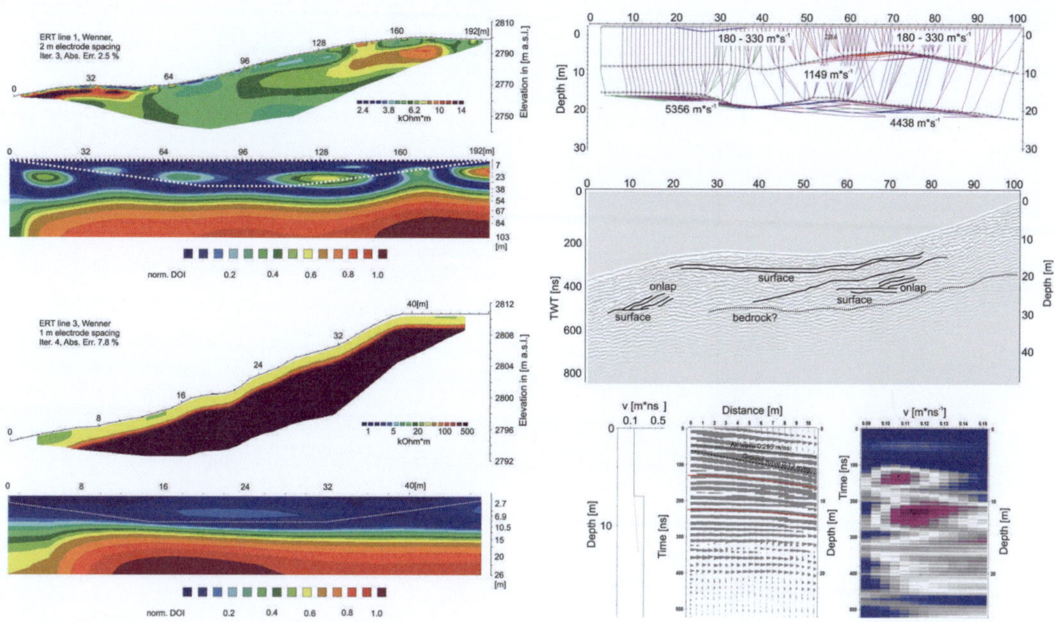

Fig. 7.5 l: ERT models of lines 1 (top) and 3 (bottom); r: SSR model (top) and GPR model (bottom) along line 1

values well below 0.2 (Hilbich et al. 2009). Resistivity values of the upper two metres range from 25 to 50 kΩm and abruptly turn into much higher values of up to 1000 kΩm below 2 m depth.

Combined results from the line 1 (ERT, GPR and SSR) document a mixture of unconsolidated gravel and boulders over bedrock. The geophysical methods could not detect ice-rich permafrost in this part of the rock glacier. This is consistent with the surface geomorphology that does not indicate large heave–subsidence or other cryoturbation forms. The generated specific resistivity values are too low to be interpreted as ice-rich permafrost but could indicate isolated remnants of ice or larger air-filled voids between boulders. This is corroborated by slow p-wave velocities (1149 m s^{-1}) as well as slow electromagnetic wave propagation (0.114 m ns^{-1}) of the upper 15–20 m of the rock glacier body as these values stand for loose and unconsolidated material. The sharp rise to acoustic velocities of >4 km s^{-1} at 20 m depth most likely related to the bedrock at this depth. In summary, the rock glacier seems to have progressed into a more relict stage in the area of line one. However, permafrost is still found in various areas nearby as documented along line 3 for example. Here, the very high electric resistivity values are indicative of an ice-rich subsurface overlain by a coarse boulder matrix.

7.3.3.2 Temperature Monitoring

BTS point measurements (Fig. 7.4) were taken in areas of different elevation, aspect, slope and surface structure with two exceptions: bedrock areas were excluded, as well as foot slopes due to high avalanche risk. Coldest temperatures of −8.2 °C occurred in the direct forefield of the rock glacier where blocky surfaces offer ideal conditions for bulk ventilation (Thompson 1962). Southwest facing slopes and fine-grained surfaces like in the root zone of RT01 exhibit the warmest temperature values that could be measured (0.8 °C). As RT02 is located just below steep avalanche slopes, BTS measurements only exist for RT01. BTS of the tongue increase from the root zone (−0.6 °C) to the front (−7 °C) according to an increase of the grain size of the surficial material (Fig. 7.4).

Temperature loggers TLOG2 and TLOG4 measured temperatures in the root zone of RT02

Fig. 7.6 Temperature curves of six GST loggers on and around the rock glacier for winter and spring 2012/2013

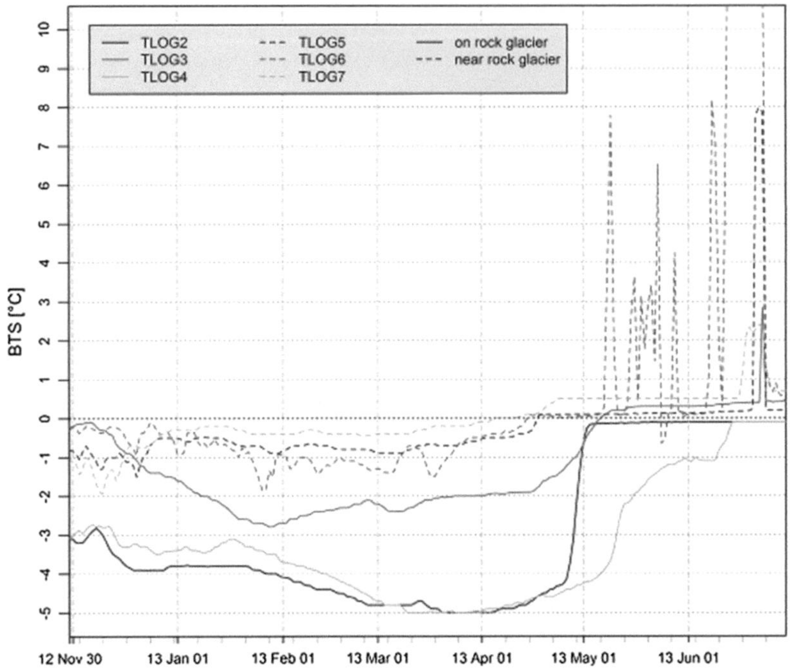

and both of them exhibit a mean temperature of −4.6 °C for late winter (February 1–April 30, 2013). Another temperature logger (TLOG3) was placed on an active lobe of RT02 and measured mean temperatures of −2.1 °C for the same period as TLOG2 and TLOG4. Temperature loggers 5, 6 and 7 (Figs. 7.4 and 7.6) were installed in the more distal apron of the rock glacier and logged mean temperatures ranging from −0.8 to −0.2 °C. The temperature curve of TLOG 6 shows some noise, which might be the consequence of a thin snow cover at that location.

7.3.3.3 Topographic Changes

Horizontal displacements at the surface of RT01 are limited to an area in the south of the tongue neighbouring RT02 (Fig. 7.7*a*). Velocities reach a maximum of 2.6 m in six years for a lobe in the back, while the mean velocities are rather low with only 0.46 m in six years.

The surface of RT02 however plays out to be more dynamic. Highest velocities are reached in the northern central part of the tongue (Fig. 7.7*d*) with 3.9 m from 2006 to 2012. At the front (Fig. 7.7*b*), movement gets as fast as 1.8 m during the investigated timestep. A small lobe in the root zone of RT02 (Fig. 7.7*e*) shows movement patterns in the direction of RT01 reaching 2.8 m in six years.

In accordance with horizontal movement patterns, the DoD analysis for vertical surface changes exhibits little variation for RT01 and clear structures for RT02 (see Fig. 7.7, right).

Positive changes can be found at the front of the tongue RT02 (0.83 m in six years, *b*) and its superposed lobe (1.25 m in six years, *d*). The front of the small lobe in the back reveals positive changes of 0.63. m from 2006 to 2012 (*e*). An area of negative changes is indicated by small letter *h* (Fig. 7.7) with a mean −0.39 m. Unusually, quite a lot of positive values without flow patterns related to rock glacier movement can be found all over RT02. They indicate the high rockfall activity of the Krummgampen mountain ridge.

The surface of RT01 is dominated by negative changes with a mean of 0.44 m for the most dynamic part (Fig. 7.7, indicator *a*). These are supposed to be the consequence of either extensional flow or settling of rock mass due to melting of ground ice. The northern or true left

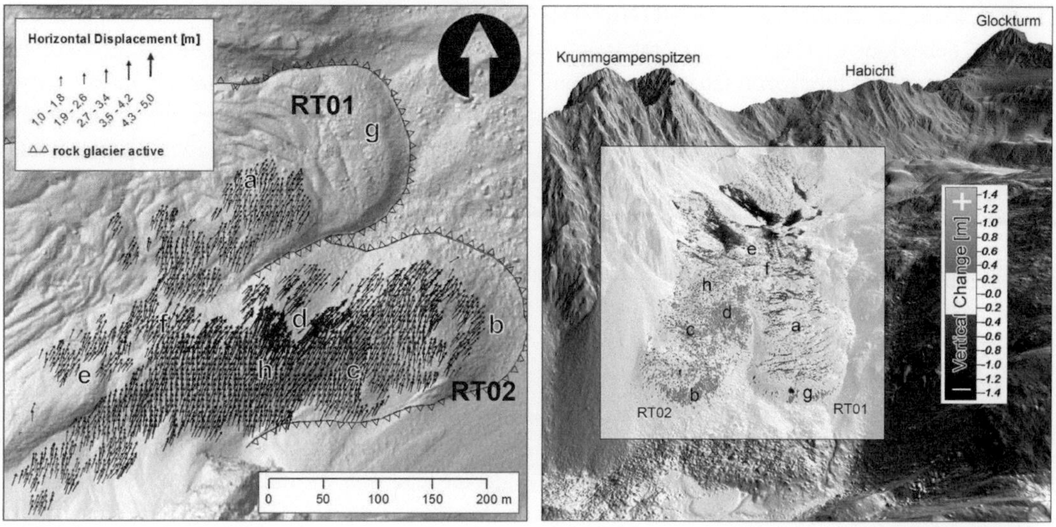

Fig. 7.7 l: Displacements of features on the rock glacier's surface; r: Vertical changes of the rock glacier's surface

part of the tongue is not affected by vertical changes. Some positive values at the front suggest gravitational mass movements of rocks.

7.3.4 Discussion and Conclusions

The applied methods give clear evidence that the two rock glacier tongues are in different states of activity.

RT01 seems to be mainly inactive, aside from some smaller areas of the rock glacier tongue. As proven by the geophysical measurements of profile 1, there is no ground ice occurrence in the back of the sediment body, but evidence for two superposing lobes is given (for details see Dusik et al. 2015). BTS values ranging from −2.0 to −0.6 °C support this finding, as well as feature tracking and difference grid results that do not show specific change patterns in the area of ERT line 1. However, the southern central part of RT01 (indicated by letter *a* in Fig. 7.7) shows regularly arranged velocity vectors, achieved by feature tracking. In the same location, BTS measurements show cold temperatures that relate to the class of probable permafrost occurrence according to Haeberli (1975). The blocky surface material in that location is prone to be well ventilated in early winter when the snow cover is thin or still missing and falling cold air can penetrate easily into the coarse surface (Thompson 1962). The movement of surficial features somehow takes a curve from south over northwest to west orientation, which indicates a pushing from moving masses in the back of RT02 (Fig. 7.7, indicator *h* and *f*).

RT02 in general seems to be active but with varying magnitudes of surfacial motion in different areas as suggested by 2-D resistivity models of three profiles on the tongue and also by temperature logger data. ERT line 4 is characterized by an upper layer of up to 50 kΩm followed by ice-cemented sediments. The same accounts for ERT line 3 but with a solid ice core instead of an ice-sediment matrix. A sharp gradient of resistivity values segregates the 2-m-thick active layer from the frozen ground below. The profiles are representative for the north-shaded areas of RT02. Thus, there is reason to believe that similar subsurface conditions are valid for main parts of the tongue. Enhancing this theory, the temperature logger, TLOG 4, located in the central part of RT02 shows mean temperatures of −4.6 °C for late winter 2012. In this area (Fig. 7.7, indicators *f* and *h*), velocity fields with up to 0.47 m a^{-1} moving towards and parallel to RT01 can be recognized in the feature tracking analysis, which describes the movement of the southern central part of RT01. The dynamic behaviour of RT02 in general is initiated by the creep of the frozen lateral moraine of the LIA extent of the Riff glacier. The moraine material is now superposed by scree slopes built up by rockfall from the Krummgampenspitzen mountains. Those scree slopes consist of sharp-edged blocks and therefore are prone to cold conditions due to bulk ventilation. Avalanche deposits and nivation hollows additionally conserve cold temperatures within the shaded deposits. Although ERT line 5 suggests no ground ice or permafrost in the front of RT02 (Fig. 7.7, indicator *b* and *d*), movement of the sediment body by up to 0.6 m a^{-1} can be detected. Again pushing masses of the creeping scree at the foot of the Krummgampenspitzen mountain range must be responsible (Dusik et al. 2015).

Regarding the physiographic conditions of the rock glaciers surrounding, we may conclude that the landform depicts a sediment sink that blocks sediment pathways from the proglacial area down valley, although at least RT02 at the moment is still an active tongue, both dynamically and climatically.

The tongues RT01 and RT02 are of differing genesis and age. While RT01 developed out of the Riffl glaciers LGM ground moraine, RT02 results from creeping sediments of the Riffl glaciers LIA lateral moraine and is further fed by rockfalls from the Krummgampenspitzen. RT01 is either about to become inactive or already is inactive. The fine sediment deposits in its back and the two-layered GPR model give evidence that the Riffl glacier invaded the tongue during the LIA. This leads to a change of the ground thermal regime and following inactivation of the rock glacier. The existence of a push moraine in the proglacial zone further up from the rock glacier suggests that the retreating Riffl glacier was at least marginally cold based and thus allowed permafrost to aggrade (Waller et al. 2012). This would indicate that RT01 could have been activated after the LIA's maximum extent of the glacier. The actual inactivation is probably caused by an interplay of warming temperatures, missing moisture and missing input of sediments.

The movement of both tongues will someday be limited by the slope, as the fronts arrive on flat ground and opposing a mountain range. Down valley, the way is blocked by the deposits of a rock avalanche that built up a wall. Only solutes are transported by a creek that partially disappears beneath the rockfall masses on its way to the proglacial river Fagge (that drains the Gepatsch glacier).

Furthermore, the rock glacier system itself represents a barrier between the Krummgampenspitzen mountain range in the south and a whaleback and steep stepped terrain in the north. All transport pathways from above the rock glacier lead onto or through the rock glacier.

References

Anderson JG (1906) Solifluction, a component of Subaërial Denudation. J Geol 14:91–112

Arenson LU, Jakob M (2015) Periglacial geohazard risks and ground temperature increases. In: Lollino G (ed) Engineering geology for society and territory. Springer, Cham, pp 233–237

Arenson LU, Colgan W, Marshall HP (2014) Physical, thermal, and mechanical properties of snow, ice, and permafrost. In: Haeberli W, Whiteman C, Shroder JF (eds) Snow and ice-related hazards, risks, and disasters. Elsevier Science, pp 35–75

Barsch D (1969) Studien und Messungen an Blockgletschern in Macun, Unterengadin. Zeitschrift für Geomorphologie. N. F. Supplement 8:11–30

Barsch D (1996) Rockglaciers: indicators for the present and former geoecology in high mountain environments. Springer, Berlin

Berger J, Krainer K, Mostler W (2004) Dynamics of an active rock glacier (Ötztal Alps, Austria). Quat Res 62:233–242

Beylich AA, Warburton J (2007) SEDIFLUX Manual: analysis of source-to-sink fluxes and sediment budgets in changing high-latitude and high-altitude cold environments. Geol Surv Norway Publ 2007:053

Boeckli L, Brenning A, Gruber S, Noetzli J (2012) A statistical approach to modelling permafrost distribution in the European Alps or similar mountain ranges. The Cryosphere 6:125–140. https://doi.org/10.5194/tc-6-125-2012

Debella-Gilo M, Kääb A (2011) Sub-pixel precision image matching for measuring surface displacements on mass movements using normalized cross-correlation. Remote Sens Environ 115:130–142. https://doi.org/10.1016/j.rse.2010.08.012

Dusik J-M, Leopold M, Heckmann T, Haas F, Hilger L, Morche D, Neugirg F, Becht M (2015) Influence of glacier advance on the development of the multipart Riffeltal rock glacier, Central Austrian Alps. Earth Surf Process Land 40:965–980. https://doi.org/10.1002/esp.3695

Ebohon B, Schrott L (2008) Modeling mountain permafrost distribution: a new permafrost map of Austria. In: Ninth international conference on permafrost

Etzelmüller B (2013) Recent advances in mountain permafrost research. Permafrost Periglac Process 24:99–107. https://doi.org/10.1002/ppp.1772

Etzelmüller B, Heggem ESF, Sharkhuu N, Frauenfelder R, Kääb A, Goulden C (2006) Mountain permafrost distribution modelling using a multi-criteria approach in the Hövsgöl area, Northern Mongolia. Permafrost Periglac Process 17:91–104. https://doi.org/10.1002/ppp.554

Etzelmüller B, Farbrot H, Guðmundsson Á, Humlum O, Tveito OE, Björnsson H (2007) The regional distribution of mountain permafrost in Iceland. Permafrost Periglac Process 18:185–199. https://doi.org/10.1002/ppp.583

Farbrot H, Isaksen K, Etzelmüller B, Gisnås K (2013) Ground thermal regime and permafrost distribution under a changing climate in Northern Norway. Permafrost Periglac Process 24:20–38. https://doi.org/10.1002/ppp.1763

French H, Thorn CE (2006) The changing nature of periglacial geomorphology. Géomorphologie: relief, processus, environnement 12. https://doi.org/10.4000/geomorphologie.119

French HM (2007) The periglacial environment, 3rd edn. Wiley, Chichester [u.a.]

Gorbunov AP, Marchenko SS, Seversky EV (2004) The thermal environment of blocky materials in the mountains of Central Asia. Permafrost Periglac Process 15:95–98

Gruber S, Hoelzle M (2001) Statistical modelling of mountain permafrost distribution: local calibration and incorporation of remotely sensed data. Permafrost Periglac Process 12:69–77. https://doi.org/10.1002/ppp.374

Guodong C, Dramis F (1992) Distribution of mountain permafrost and climate. Permafrost Periglac Process 3:83–91. https://doi.org/10.1002/ppp.3430030205

Haeberli W (1973) Die Basis-Temperatur der winterlichen Schneedecke als möglicher Indikator für die Verbreitung von Permafrost in den Alpen. Zeitschrift für Gletscherkunde und Glazialgeologie IX:221–227

Haeberli W (1975) Untersuchungen zur Verbreitung von Permafrost zwischen Flüelapass und Piz Grialetsch (GR). Mitteilungen der Versuchsanstalt für Wasserbau, Hydrologie und Glaziologie, vol 17. Eidgenössische Technische Hochschule Zürich, Zürich

Haeberli W (1985) Creep of mountain permafrost: internal structure and flow of alpine rock glaciers. Mitteilungen der Versuchsanstalt für Wasserbau, Hydrologie und Glaziologie, vol 77. Eidgenössische Technische Hochschule Zürich, Zürich

Haeberli W (2000) Modern research perspectives relating to permafrost creep and rock glaciers: a discussion. Permafrost Periglac Process 11:290–293. https://doi.org/10.1002/1099-1530(200012)11:4%3C290:AID-PPP372%3E3.0.CO;2-0

Haeberli W (2013) Mountain permafrost—research frontiers and a special long-term challenge. Cold Reg Sci Technol 96:71–76. https://doi.org/10.1016/j.coldregions.2013.02.004

Haeberli W, Patzelt G (1982) Permafrostkartierung im Gebiet der Hochebenkar-Blockgletscher, Obergurgl Ötztaler Alpen. Z Gletscherk Glazialgeol 18:127–150

Haeberli W, Hoelzle M, Kääb A, Keller F, Vonder Mühll D (1998) Ten years after drilling through the permafrost of the active rockglacier Murtèl, Eastern Swiss Alps: Answered questions and new perspectives. In: Seventh international permafrost conference: Proceedings, Yellowknife, pp 403–410

Haeberli W, Hallet B, Arenson L, Elconin R, Humlum O, Kääb A, Kaufmann V, Ladanyi B, Matsuoka N, Springman S, Vonder Mühll D (2006) Permafrost creep and rock glacier dynamics. Permafrost Periglac Process 17:189–214. https://doi.org/10.1002/ppp.561

Harris C, Murton JB (2005) Interactions between glaciers and permafrost: an introduction. Geol Soc 242:1–9. https://doi.org/10.1144/GSL.SP.2005.242.01.01 (London, Special Publications)

Harris C, Haeberli W, Vonder Mühll D, King LJ (2001) Permafrost monitoring in the high mountains of Europe: the PACE Project in its global context. Permafrost Periglac Process 12:3–11. https://doi.org/10.1002/ppp.377

Harris SA, Pedersen DE (1998) Thermal regimes beneath coarse blocky materials. Permafrost Periglac Process 9:107–120

Hauck C, Kneisel C (2008) Applied geophysics in periglacial environments. Cambridge University Press, Cambridge

Hauck C, Isaksen K, Vonder Mühll D, Sollid JL (2004) Geophysical surveys designed to delineate the altitudinal limit of mountain permafrost: an example from Jotunheimen, Norway. Permafrost Periglac Process 15:191–205. https://doi.org/10.1002/ppp.493

Hausmann H, Krainer K, Brückl E, Mostler W (2007a) Creep of two alpine rock glaciers: observation and modelling (Ötztal- and Stubai Alps, Austria). Grazer Schriften der Geographie und Raumforschung 43:145–150

Hausmann H, Krainer K, Brückl E, Mostler W (2007b) Creep of two alpine rock glaciers: observation and modelling (Ötztal- and Stubai Alps, Austria). Grazer Schriften der Geographie und Raumforschung 43:145–150

Hausmann H, Krainer K, Brückl E, Mostler W (2007c) Internal structure and ice content of Reichenkar rock glacier (Stubai Alps, Austria) assessed by geophysical investigations. Permafrost Periglac Process 18:351–367. https://doi.org/10.1002/ppp.601

Hausmann H, Krainer K, Brückl E, Ullrich C (2012) Internal structure, ice content and dynamics of Ölgrube and Kaiserberg rock glaciers (Ötztal Alps,

Austria) determined from geophysical surveys. Austrian J Earth Sci 105:12–31
Hilbich C, Hauck C, Hoelzle M, Scherler M, Schudel L, Völksch I, Vonder Mühll D, Mäusbacher R (2008) Monitoring mountain permafrost evolution using electrical resistivity tomography: a 7-year study of seasonal, annual, and long-term variations at Schilthorn, Swiss Alps. J Geophys Res 113. https://doi.org/10.1029/2007jf000799
Hilbich C, Marescot L, Hauck C, Loke MH, Mäusbacher R (2009) Applicability of electrical resistivity tomography monitoring to coarse blocky and ice-rich permafrost landforms. Permafrost Periglac Process 20:269–284. https://doi.org/10.1002/ppp.652
Hoelzle M (1992) Permafrost occurrence from BTS measurements and climatic parameters in the eastern Swiss Alps. Permafrost Periglac Process 3:143–147. https://doi.org/10.1002/ppp.3430030212
Hoelzle M, Wegmann M, Krummenacher B (1999) Miniature temperature dataloggers for mapping and monitoring of permafrost in high mountain areas: first experience from the Swiss Alps. Permafrost Periglac Process 10:113–124. https://doi.org/10.1002/(SICI)1099-1530(199904/06)10:2%3C113:AID-PPP317%3E3.0.CO;2-A
Hoelzle M, Mittaz C, Etzelmüller B, Haeberli W (2001) Surface energy fluxes and distribution models of permafrost in European mountain areas: an overview of current developments. Permafrost Periglac Process 12:53–68. https://doi.org/10.1002/ppp.385
Hoffmann T, Schrott L (2003) Determining sediment thickness of talusslopes and valley fill deposits using seismic refraction: a comparison of 2D interpretation tools. Z Geomorph NF (Suppl):71–87
Ishikawa M (2003) Thermal regimes at the snow–ground interface and their implications for permafrost investigation. Periglac Geomorphol Beginning 21st Century 52:105–120. https://doi.org/10.1016/s0169-555x(02)00251-9
Ishikawa M, Hirakawa K (2000) Mountain permafrost distribution based on BTS measurements and DC resistivity soundings in the Daisetsu Mountains, Hokkaido, Japan. Permafrost Periglac Process 11. https://doi.org/10.1002/1099-1530(200004/06)11:2%3C109:aid-ppp343%3E3.0.co;2-o
Jäckli H (1957) Gegenwartsgeologie des bündnerischen Rheingebietes: Ein Beitrag zur exogenen Dynamik alpiner Gebirgslandschaften. Beiträge zur Geologie der Schweiz: Geotechnische Serie, vol 36. Kümmerly & Frey, Bern
Kääb A, Kneisel C (2006) Permafrost creep within a recently deglaciated glacier forefield: Muragl, Swiss Alps. Permafrost Periglac Process 17:79–85. https://doi.org/10.1002/ppp.540
Kääb A, Frauenfelder R, Roer I (2007) On the response of rockglacier creep to surface temperature increase. Glob Planet Change 56:172–187. https://doi.org/10.1016/j.gloplacha.2006.07.005
Kaufmann V, Ladstädter R (2003) Quantitative analysis of rock glacier creep by means of digital photogrammetry using multi-temporal aerial photographs: two case studies in the Austrian Alps, pp 525–530
Keller F (1992) Automated mapping of mountain permafrost using the program PERMAKART within the geographical information system ARC/INFO. Permafrost Periglac Process 3:133–138. https://doi.org/10.1002/ppp.3430030210
Keller F, Hoelzle M (1996) PERMAKART und PERMAMAP. In: Haeberli W (ed) Simulation der Permafrostverbreitung in den Alpen mit geographischen Informationssystemen: [Arbeitsbericht im Rahmen des Nationalen Forschungsprogrammes "Klimaänderungen und Naturkatastrophen", NFP 31]. vdf, Hochsch.-Verl. an der ETH, Zürich, pp 35–46
King LJ, Gorbunov AP, Evin M (1992) Prospecting and mapping of mountain permafrost and associated phenomena. Permafrost Periglac Process 3:73–81. https://doi.org/10.1002/ppp.3430030204
Kinnard C, Lewkowicz AG (2006) Frontal advance of turf-banked solifluction lobes, Kluane Range, Yukon Territory, Canada. Geomorphology 73:261–276
Kneisel C (2004) New insights into mountain permafrost occurrence and characteristics in glacier forefields at high altitude through the application of 2D resistivity imaging. Permafrost Periglac Process 15:221–227
Krainer K, Mostler W (2006) Flow velocities of active rock glaciers in the austrian Alps. Geogr Ann 88A:267–280
Krainer K, Ribis M (2012) A rock glacier inventory of the tyrolean Alps (Austria). Austrian J Earth Sci 105:32–47
Krautblatter M (2008) Rock permafrost geophysics and its explanatory power for permafrost-induced rockfalls and rock creep: a Perspective. In: Ninth international conference on permafrost. Fairbanks, pp 999–1004
Krautblatter M (2009) Detection and quantification or permafrost change in alpine rock walls and implications for rock instability. Dissertation, University of Bonn
Krummel H (2005) Seismische Quellen. In: Knödel K, Krummel H, Lange G (eds) Geophysik. Springer, Heidelberg, pp 467–526
Lane SN, Westaway RM, Murray Hicks D (2003) Estimation of erosion and deposition volumes in a large, gravel-bed, braided river using synoptic remote sensing. Earth Surf Process Land 28:249–271
Leopold M, Dethier D, Völkel J, Raab T (2008a) Combining sediment analysis and seismic refraction to describe the structure, thickness and distribution of periglacial slope deposits at Niwot Ridge, Rocky Mountains Front Range, Colorado, USA. Z Geomorph NF 52(Suppl):77–94
Leopold M, Dethier D, Völkel J, Raab T, Rikert TC, Caine N (2008b) Using geophysical methods to study the shallow subsurface of a sensitive alpine environment, Niwot Ridge, Colorado Front Range, U.S.A. Arct Antarct Alp Res 40:519–530
Leopold M, Williams M, Caine N, Völkel J, Dethier D (2011) Internal structure of the Green Lake 5 rock glacier, Colorado Front Range, USA. Permafrost Periglac Process 22:107–119. https://doi.org/10.1002/ppp.706

Leopold M, Völkel J, Huber J, Dethier D (2013) Subsurface architecture of the Boulder Creek Critical Zone Observatory from electrical resistivity tomography. Earth Surf Process Land: n/a-n/a. https://doi.org/10.1002/esp.3420

Leopold M, Völkel J, Dethier DP, Williams MW (2014) Changing mountain permafrost from the 1970s to today—comparing two examples from Niwot Ridge, Colorado Front Range, USA. Zeit fur Geo 58 (Supp):137–157. https://doi.org/10.1127/0372-8854/2013/S-00129

Lerjen M, Kääb A, Hoelzle M, Haeberli W (2003) Local distribution of discontinous mountain permafrost. A process study at Flüela Pass, Swiss Alps, pp 667–672

Lewkowicz AG (1992) A solifluction meter for permafrost sites. Permafrost Periglac Process 3:11–18. https://doi.org/10.1002/ppp.3430030103

Lewkowicz AG, Ednie M (2004) Probability mapping of mountain permafrost using the BTS method, Wolf Creek, Yukon Territory, Canada. Permafrost Periglac Process 15:67–80. https://doi.org/10.1002/ppp.480

Lewkowicz AG, Etzelmüller B, Smith SL (2011) Characteristics of discontinuous permafrost based on ground temperature measurements and electrical resistivity tomography, Southern Yukon, Canada. Permafrost Periglac Process 22:320–342. https://doi.org/10.1002/ppp.703

Lozinski W (1909) Über die mechanische Verwitterung der Sandsteine im gemäßigten Klima. Bulletin International de l'Academie de Science de Crocovie class des Sciences Mathématique et Naturalles 1:1–25

Marcott SA, Shakun JD, Clark PU, Mix AC (2013) A reconstruction of regional and global temperature for the past 11,300 years. Science 339:1198–1201. https://doi.org/10.1126/science.1228026

Matsuoka N (2001) Solifluction rates, processes and landforms: a global review. Earth-Sci Rev 55:107–134. https://doi.org/10.1016/S0012-8252(01)00057-5

Matsuoka N, Ikeda A, Date T (2005) Morphometric analysis of solifluction lobes and rock glaciers in the Swiss Alps. Permafrost Periglac Process 16:99–113. https://doi.org/10.1002/ppp.517

Neal A (2004) Ground-penetrating radar and its use in sedimentology: principles, problems and progress. Earth-Sci Rev 66:261–330. https://doi.org/10.1016/j.earscirev.2004.01.004

Noetzli J, Gruber S, Kohl T, Salzmann N, Haeberli W (2007) Three-dimensional distribution and evolution of permafrost temperatures in idealized high-mountain topography. J Geophys Res 112. https://doi.org/10.1029/2006jf000545

Nogués-Bravo D, Araújo MB, Errea MP, Martínez-Rica JP (2007) Exposure of global mountain systems to climate warming during the 21st Century. Glob Environ Change 17:420–428. https://doi.org/10.1016/j.gloenvcha.2006.11.007

Nötzli J (2011) Modeling transient three-dimensional temperature fields in mountain permafrost. Schriftenreihe Physische Geographie, 60: Glaziologie und Geomorphodynamik. Geographisches Institut der Universität Zürich, Zürich

Nötzli J, Hoelzle M, Haeberli W (2003–2004) Mountain permafrost and recent Alpine rock-fall events: a GIS-based approach to determine critical factors. In: Phillips M, Springman S, Arenson LU (eds) Permafrost. Balkema, Lisse, pp 827–832

Otto J-C, Keuschnig M, Götz J, Marbach M, Schrott L (2012) Detection of mountain permafrost by combining high resolution surface and subsurface information—an example from the Glatzbach Catchment, Austrian Alps. Geografiska Annaler: Ser A Phys Geogr 94:43–57

Phillips M, Mutter EZ, Kern-Luetschg M, Lehning M (2009) Rapid degradation of ground ice in a ventilated talus slope: Flüela Pass, Swiss Alps. Permafrost Periglac Process 20:1–14. https://doi.org/10.1002/ppp.638

Rödder T, Kneisel C (2012) Influence of snow cover and grain size on the ground thermal regime in the discontinuous permafrost zone, Swiss Alps. Geomorphology 175–176:176–189. https://doi.org/10.1016/j.geomorph.2012.07.008

Roer I, Nyenhuis M (2007) Rockglacier activity studies on a regional scale: comparison of geomorphological mapping and photogrammetric monitoring. Earth Surf Process Land 32:1747–1758. https://doi.org/10.1002/esp.1496

Roer I, Kääb A, Dikau R (2005) Rockglacier kinematics derived from small-scale aerial photography and digital airborne pushbroom imagery. Zeitschrift für Geomorphologie, NF 49:73–87

Roer I, Kääb A, Kaufmann V, Delaloye R, Avian M, Lambiel C, Haeberli W (2008) Observations and considerations on destabilizing active rock glaciers in the European Alps. In: Ninth international conference on permafrost. Fairbanks, pp 1505–1510

Ruiz L, Trombotto D (2012) Mountain permafrost distribution in the Andes of Chubut (Argentina). In: Proceedings, Tenth International Conference on Permafrost (TICOP), Salekhard, Russia, pp 365–370

Scambos TA, Dutkiewicz MJ, Wilson JC, Bindschadler RA (1992) Application of image cross-correlation to the measurement of glacier velocity using satellite image data. Remote Sens Environ 42:177–186. https://doi.org/10.1016/0034-4257(92)90101-O

Thompson WF (1962) Preliminary notes on the nature and distribution of rock glaciers relative to true glaciers and other effects of the climate on the ground in North America. In: International Association of Scientific Hydrology (ed) International Association of Scientific Hydrology Publication 58: Symposium at Obergurgl, Austria Ward W., pp 212–219

Verleysdonk S, Krautblatter M, Dikau R (2011) Sensitivity and path dependence of mountain permafrost systems. Geografiska Annaler: Ser A Phys Geogr 93:113–135. https://doi.org/10.1111/j.1468-0459.2011.00423.x

Vonder Mühll D, Haeberli W (1990) Thermal characteristics of the permafrost within an active rockglacier (Murtèl/Corvatsch, Grisons, Swiss Alps). J Glaciol 36:151–158

Wahrhaftig C, Cox A (1959) Rock glaciers in the Alaska Range. Geol Soc Am Bull 70:383

Waller RI, Murton JB, Kristensen L (2012) Glacier–permafrost interactions: processes, products and glaciological implications. Sediment Geol 255–256:1–28. https://doi.org/10.1016/j.sedgeo.2012.02.005

Walsh SJ, Bian L, McKnight S, Brown DG, Hammer ES (2003) Solifluction steps and risers, Lee Ridge, Glacier National Park, Montana, USA: a scale and pattern analysis. Geomorphology 55:381–398. https://doi.org/10.1016/S0169-555X(03)00151-X

Washburn AL (1967) Instrumental observations of mass-wasting in the Mesters Vig district, Northeast Greenland. Medd Groenl 166:1–318

Wheaton JM, Brasington J, Darby SE, Sear DA (2010) Accounting for uncertainty in DEMs from repeat topographic surveys: improved sediment budgets. Earth Surf Process Land 35:136–156. https://doi.org/10.1002/esp.1886

World Glacier Monitoring Service (2015) Fluctuations of glaciers database

Part II

Hillslope Processes in the Proglacial Zone

Rock Slope Instability in the Proglacial Zone: State of the Art

8

Samuel T. McColl and Daniel Draebing

Abstract

Rock slope failures are characteristic of mountainous environments. These mass movements produce sediment, alter catchment behaviour and contribute to the dynamics and hazards of high alpine and proglacial areas. This chapter highlights the state of knowledge in the context of proglacial environments and reviews methods of investigating rock slope failure activity and causes. An alignment of extreme conditions and dynamic processes renders proglacial environments exceptionally prone to instability. Glacier retreat and climate change, following the Last Glacial Maximum and more recent stadials, has been a major catalyst for past and ongoing mass movements in alpine areas, and many slopes continue to respond to these legacy perturbations. Rock slope failure activity is preconditioned by rock mass properties and topography, and failures in alpine areas are typically prepared or triggered by: (i) fracture growth and seismicity arising from the unloading of glacial loads over millennial timescales, and (ii) rock fracture growth and loss of strength as a result of hydrological and thermal effects that fluctuate over daily to seasonal timescales, but are superimposed upon long-term trends related to climate change. These insights stem from a growing application of geomorphological, geotechnical, geochemical, geodetic and geophysical techniques that enable the assessment of stability factors and the activity of rock slope failures (both past and contemporary). The rich datasets are being used to inform new understanding of past and ongoing proglacial rock slope instabilities; this understanding will ultimately help to predict process dynamics, environmental change and to mitigate hazards resulting from rock slope failures.

Keywords

Rock slope failure · Rockfall · Deep-seated rock slope deformation · Preconditioning factors · Preparatory factors · Triggers

8.1 Impacts of Proglacial Rock Slope Failures

The failure of rock slopes is a fundamental process of glaciated mountains, strongly influencing the nature and behaviour of their proglacial systems and sediment dynamics. By definition, rock slope failures involve a range of mass movement

S. T. McColl (✉)
Geosciences Group, Massey University, Palmerston North, New Zealand
e-mail: s.t.mccoll@massey.ac.nz

D. Draebing
Chair of Landslide Research, Technische Universität München, Munich, Germany

T. Heckmann and D. Morche (eds.), *Geomorphology of Proglacial Systems*,
Geography of the Physical Environment, https://doi.org/10.1007/978-3-319-94184-4_8

styles that directly displace in situ bedrock, such as falls, topples, slides, flows and deep-seated rock slope deformations. The rock slope failures that affect proglacial environments can range in size from tiny rockfalls through to the largest terrestrial mass movements on the planet (Ivy-Ochs et al. 2009; Hancox and Perrin 2009). For mountainous topography, there is a growing recognition of the importance of rock slope failures for their roles as an agent of erosion (e.g. Korup et al. 2007; Arsenault and Meigs 2005; Larsen et al. 2010; Cave and Ballantyne 2016), and in sediment and solute generation (e.g. Emberson et al. 2015; Korup et al. 2004). In some settings, mass movements (Fig. 8.1a) are the dominant source and mobiliser of clastic sediment to both glacial and proglacial systems (Korup 2005; Arsenault and Meigs 2005). However, rock slope failures (Fig. 8.1b) can also impede sediment transport by changing topography (e.g. Decaulne et al. 2016) and blocking or occluding rivers (Korup 2006). Thus, mass movements play multiple roles in regulating the release, delivery, transport and impedance of sediments in mountainous topography, with a single large landslide being able to persist in these roles for many thousands of years (Fig. 8.1a, b). Their interactions with glacial systems can also drive changes in glacier behaviour (e.g. Deline 2009; Reznichenko et al. 2011) and glacial landform development (e.g. Coquin et al. 2015; Turnbull and Davies 2006), and the potential significance of the mass movements as an origin of much glacial sediment (Fig. 8.1c) has been brought to the fore recently (Dunning et al. 2015; Reznichenko et al. 2012, 2016). The direct (e.g. talus) or indirect (e.g. landslide-dominated moraine) products of rock slope failure can be reworked by surficial mass movement processes, which form another important process in proglacial systems (see Chaps. 10 and 11). In addition to their geomorphic impacts, rock slope failures constitute one of the most significant hazards in proglacial systems, with direct hazards ranging from high-frequency small magnitude events (e.g. Purdie et al. 2015; Ravanel and Deline 2011) through to infrequent but extremely destructive mass movements (Fig. 8.1d, e.g. Huggel et al. 2005; Akçar et al. 2012; Kargel et al. 2016). Indirect, cascading hazards initiated by rock slope failures include the creation and failure of landslide dams (e.g. Schneider et al. 2013), landslide-triggered tsunami (e.g. Chap. 14) and sediment accumulation which can elevate both flood and debris flow activity (e.g. Cox et al. 2014; Fig. 8.1a).

Within mountainous proglacial settings—typified by steep bedrock slopes, a rapidly evolving topography, a harsh, changing climate and often seismically active—the conditions for rock slope instability are rife. Indeed, rock slope failure, along with mass movement of surficial deposits, is a distinguishing feature of most mountainous proglacial systems. Further, glacier retreat and the associated warming of alpine environments since the end of the last global glaciation has been a major catalyst for mass movements in proglacial areas (McColl 2012a), and the present acceleration in warming and glacier retreat appears to be responsible for heightened rock slope failure activity in these environments (Fischer et al. 2012; Deline et al. 2015). Rock slope failure is thus an important tenet of proglacial research, and this chapter serves to highlight the current state of knowledge about rock slope failure in the context of proglacial environments.

The first part of the chapter explores post-glacial to contemporary patterns of rock slope failure activity in proglacial environments and reviews explanations for this activity, updating and extending previous reviews on the subject (namely, McColl 2012a; Ballantyne 2002; Krautblatter and Leith 2015). In the second part, a range of techniques now available for investigating rock slope failures are outlined, in the hope that it will encourage application and further development of these and other techniques.

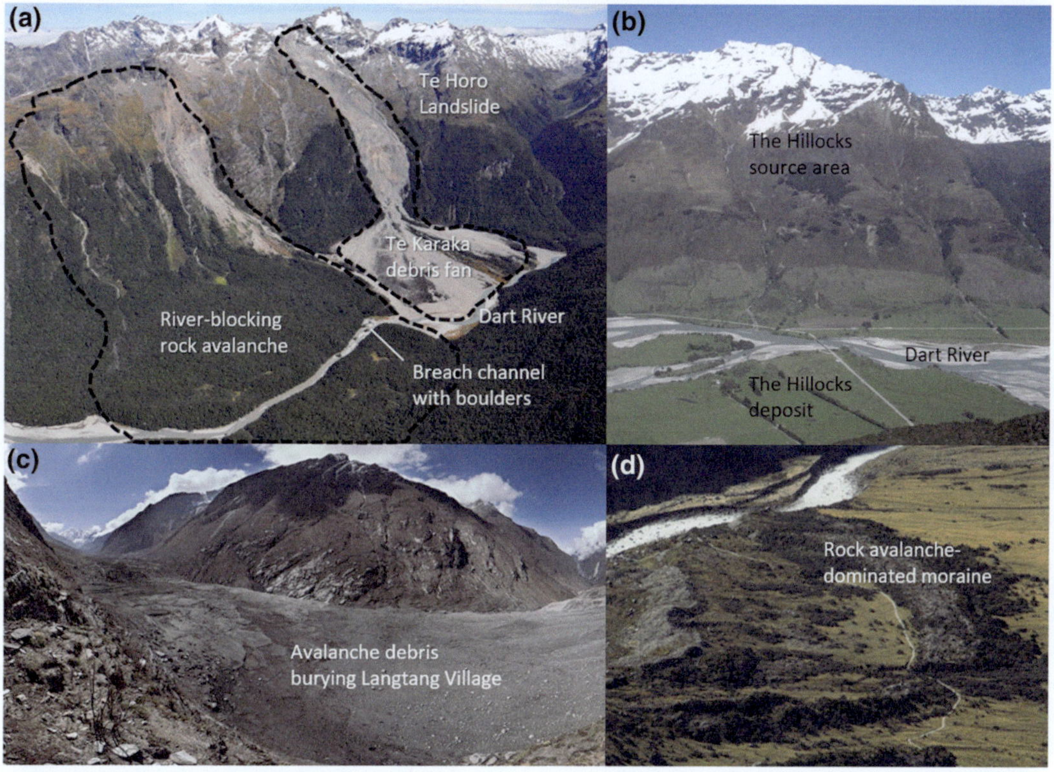

Fig. 8.1 **a** The >150 M m^3 Te Horo landslide in Otago, New Zealand, is likely to be post-glacial in age and has delivered, and continues to deliver millions of tons of sediment to the Dart River and proglacial system of the Dart Glacier (Cox et al. 2014; photo credit: G. Hancox); immediately downstream there is another large post-glacial rock avalanche that would have blocked the Dart River; **b** twenty-two kilometres farther downstream of the previous location, The Hillocks rock avalanche likely blocked the proglacial Dart River in the early Holocene, and the debris continues to occlude the Dart River in the present day, creating one of the narrowest reaches in the ~ 25 km braided floodplain (McColl and Davies 2011); **c** a terminal moraine of the Mueller glacier, in Aoraki/Mt Cook National Park, New Zealand, is thought to be comprised almost exclusively of glacial debris that is of rock avalanche origin (Reznichenko et al. 2016; photo credit: S. Winkler); **d** the site of a Nepalese village tragically destroyed by co-seismic rock slope failure in the proglacial Langtang valley during Nepal's 2015 Gorkha earthquake (Kargel et al. 2016; photo credit: D. F. Breashers/Glacier Works)

8.2 Patterns and Drivers of Proglacial Rock Slope Failure

8.2.1 A Framework for Proglacial Rock Slope Failure

Given the multitude of stability factors influencing proglacial slopes, it is useful to have a framework to contextualise instability and the processes that lead to failure. Rock slope stability depends on the balance of driving and resisting stresses. The mechanical properties of rock slopes are determined largely by gravity-induced normal and shear stresses, the cohesive strength of rock bridges and the frictional properties of discontinuities. The mechanical properties of slope materials, and the balance between driving and resisting stresses change over time. This is a result of changes in water content, the transition between frozen and liquid water phases, and from variations in static, transient or dynamic loads (e.g. from changes in slope mass, snow cover or seismicity). The terms 'stable', 'marginally stable' and 'actively unstable' are used to classify slopes as a function of the likelihood of failure, in recognition of a time-dependent stability condition (Glade and Crozier 2005; McColl 2014). This concept is particularly pertinent for proglacial slopes, which undergo significant changes to their environmental, geomorphic and geotechnical conditions in a relatively short period of time (Fig. 8.2). A range of stability factors can govern how close to the failure threshold a slope is situated, with these factors being divided into preconditioning, preparatory and triggering factors (Glade and Crozier 2005). The preconditioning factors for rock slopes, mostly the inherent rock mass and slope properties, ultimately control where slope failures will occur (e.g. the preferential failure of post-glacial landslides on certain lithologies; Cave and Ballantyne 2016) along with the general time frame of slope response, and thus data on the geotechnical properties and topography of proglacial landscapes are invaluable to have. However, rock mass and slope conditions change with glacial erosion and deglaciation (particularly glacier recession and climate warming), which are preparatory factors that can bring the slopes into a marginally stable state. They do this by reducing the excess strength available to resist driving stresses and/or by increasing the magnitude of potential triggers, both providing greater likelihood for one of a number of possible (internal or external) processes to trigger failure (Fig. 8.2). Thus, to predict slope failures, it is also necessary to also have information about these processes and their effect on rock slopes.

The role that glaciation and deglaciation plays in triggering failures or priming slopes for failure can span timescales of years to millennia, sometimes initiating failure long into subsequent interglacial periods (McColl 2012a) and over multiple glacial cycles (Grämiger et al. 2017). Large rock slope failures in particular are considered to respond to deglaciation over a longer period of time than small rock slope failures or the failure and reworking of surficial debris (e.g. refer to the exhaustion curves of Ballantyne 2003 or the empirical data of Pánek et al. 2016). This legacy effect of glaciation and deglaciation on slopes within proglacial environments is often referred to as a paraglacial response (Ballantyne 2002; McColl 2012a). The paraglacial concept considers that glaciated landscapes undergo a transition in form and process as the landscapes adjust to new major process regimes. Mass movements contribute to this transition, as both an agent of that transition and also as one of the processes undergoing change.

The paraglacial concept combined with a model of how slopes evolve to an unstable state (e.g. Figure 8.2) provides a framework that can be used to contextualise and understand the proglacial slope failures. Using this framework, the following examines major preparatory processes operating during the paraglacial transition: locally induced stress changes in the slope from removal of overburden and glacier loads; seismic triggering of mass movements associated with glacio-isostatic adjustment; and rock mass changes driven by variations in temperature and moisture. These preparatory processes, and the consideration we give to them, are not exhaustive, but they have received the most attention and at the present time provide the greatest potential in explaining and predicting proglacial slope failures.

8.2.2 Glacier Loads and Local Stress Changes

Glacier recession lowers slope stability through the removal of lateral ice support, exposure of slopes to a new microclimate and weathering regime, and by enhancing a process of rock mass weakening initiated by unloading. The exposed slopes are typically in a state of disequilibrium, having endured a long history of glacial erosion which can steepen, heighten and unload slopes of their rock overburden. The removal of support as a glacier recedes, often referred to as glacial debuttressing and frequently cited as a major driver of proglacial slope instability, may act as a trigger or a factor that prepares the slope for failure at a later time (McColl 2012). Spatial and temporal relationships between pre-historical (Cossart et al. 2008) and historical (e.g.

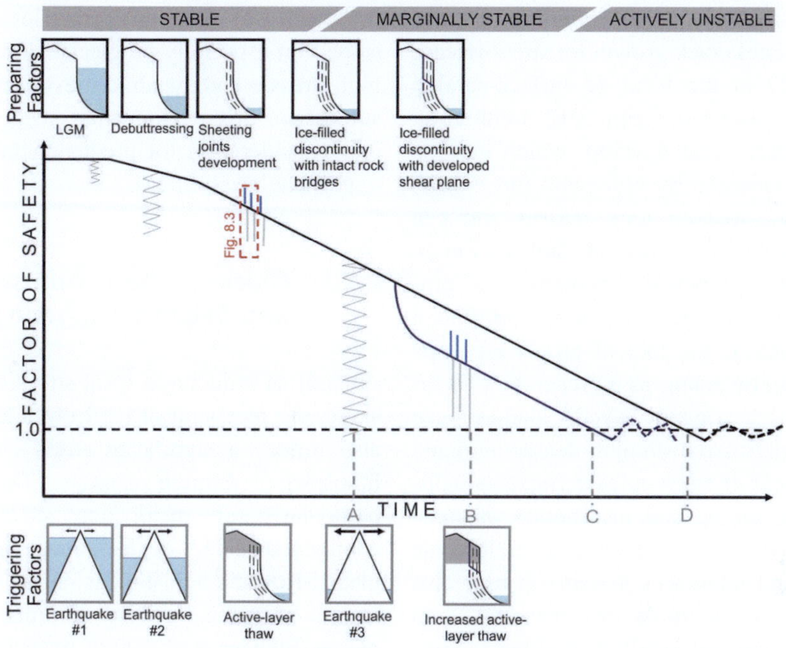

Fig. 8.2 A conceptual model for the reduction in stability of a slope over time as several internal and external conditions change in a proglacial environment. The stability is represented here as the Factor of Safety (FoS), with a value below unity (i.e. failure) when the resisting forces in the slope are exceeded by the destabilising forces. The FoS is shown here to reduce through time along different potential pathways, with new preparatory factors initiating more rapid reductions in stability as deglaciation and climate warming progress. Potential triggers are represented underneath the line and in the corresponding schematic diagrams below the graph. A system-state change from rock-dominant state (black line) to an ice-dominant state (dark blue line) occurs when an ice-filled discontinuity with continuous shear plane develops, and the system is more sensitive to failure. The letters along the x-axis represent external thresholds being crossed at A and B, and internal thresholds being crossed at C (sensitive rock state dominated by ice-filled fractures) and D (e.g. weathering-triggered). Explanations for the preparing factors and triggers are provided in the following sections and Fig. 8.3

Sigurdsson and Williams 1991) glacier ice limits and rock slope failures lend support to the role of debuttressing. However, its importance as a major trigger of slope instability following glacier recession has been called into question in the light of recent landslide dating information that suggests a delayed reaction in the timing of post-glacial rock slope failures (e.g. Prager et al. 2008; Ballantyne et al. 2014), and on theoretical grounds because of the limited ability of ductile glaciers for providing completely effective slope support, especially in temperate glacier systems (McColl and Davies 2013; McColl et al. 2010). But there is a paucity of empirical data for rock slope failure in the initial stages of major deglaciation events, making the assessment of debuttressing difficult; this paucity is hypothesised to arise from a lack of preservation of landslide debris, which is rapidly reworked during early deglaciation (Dunning et al. 2015; McColl and Davies 2013; Ballantyne et al. 2014; Ballantyne and Stone 2013).

The steep slopes exposed by glacier recession become less stable over time due to progressive strength degradation. High static self-weight stresses enhanced by glacial erosion and cyclic stresses from seismicity, thermal changes and fluctuations in groundwater, snow and ice loads, induce rock fatigue through sub-critical fracture growth (Leith 2012; McColl 2012a; Grämiger et al. 2017). Unloading of both rock overburden during glacial activity and ice from glacier retreat also relieves rock slopes of confining stresses, which too initiates crack growth (or stress-release joints), usually in the form of surface-parallel fractures (Nichols 1980; Leith 2012; Leith et al. 2014). The cracks that develop, which may be subsequently enlarged by cryogenic forces (discussed below), weaken rock masses and can form potential failure surfaces (Krautblatter et al. 2013). The time-dependent response, or progressive failure, of rock slopes is therefore a factor that controls the rate of proglacial slope failure activity, by acting as a preparatory factor or possibly as a trigger. Several studies have argued that progressive strength degradation and the development of stress-release fractures helps to explain the timing and distribution of some post-glacial rock slope failures (e.g. Ballantyne et al. 2014, and references therein). It may also help explain a multitude of historical (and therefore also pre-historical) rock slope failures that have occurred with no obvious trigger. For example, in New Zealand many of the largest alpine rock slope failures in the Southern Alps in the past 25 years have not been associated with any notable trigger such as rainfall, seismicity or an extreme temperature (e.g. Hancox and Thomson 2013; Hancox et al. 2005; Cox et al. 2008, 2015; McSaveney and Massey 2013; McSaveney 2002; Allen and Huggel 2013), with many of these being best explained (for now) as being prepared and possibly triggered by strength degradation.

The pace of strength degradation in itself is controlled by the rock mass properties, slope geometry, slope history, modes of failure and the stress changes experienced by the slope in the period following deglaciation. For example, frequent small-scale rockfalls from very steep, rapidly exposed rock slopes with abundant slope-parallel fractures in steep temperate maritime glaciers of New Zealand appear to be responding to historical glacier retreat (Purdie et al. 2015). In contrast, in a catchment farther south the failure of an 11 million m^3 deep-seated landslide, thought to have been primed by fracture propagation, occurred thousands of years after its glacier had abandoned the slope (McSaveney and Massey 2013). Given the importance of strength degradation for any rock slope, but especially proglacial slopes due to the high stresses and dynamic stress changes they are subject to, this phenomenon needs to be taken into consideration for predicting failure activity (e.g. Stead et al. 2004).

8.2.3 Glacio-Isostatic Adjustment and Seismic Triggering

As well as inducing a local stress response to a slope, the recession of ice from a landscape can also induce a significant stress change in the lithosphere, through isostasy. The unloading, particularly of major ice sheets following continental-scale deglaciation, has induced glacial-isostatic adjustments on the order tens of metres of vertical crustal deformation in some places (Stewart et al. 2000). Associated with this response are stress changes on pre-existing faults that can enhance or trigger seismic activity (e.g. Hetzel and Hampel 2005), even in places of low tectonic activity. The earthquakes produced by glacial-isostatic adjustment may be sufficient to trigger landslides, and because of the very slow response times of glacio-isostatic adjustment, the earthquakes and landslides produced may occur many thousands of years after ice has retreated. Further, the high, steep relief typical of alpine environments is likely to amplify seismic shaking (Murphy 2006), especially once glacier thinning has increased the surface relief of ridges and peaks (McColl et al. 2012), making proglacial slopes even more prone to co-seismic landsliding.

The idea that post-glacial earthquake activity can trigger rock slope failure activity in glaciated environments is supported by recent empirical datasets. For example, there is a temporal correlation between the maximum rates of post-glacial uplift in the Scottish Highlands and landslide activity (Ballantyne et al. 2014). A spatial correlation has been identified between the magnitude of post-glacial uplift and landslide density in a fiord in Iceland (Cossart et al. 2014). In the French Alps, post-glacial fault and landslide activity are shown to cluster in space and time (Sanchez et al. 2010). In the central European Alps, an abundance of very large, ancient rock slope failure deposits, and some evidence of temporal and spatial clustering of these deposits (Ostermann and Prager 2014; Ostermann and Sanders 2017), may point towards seismic origins and seismic magnitudes not expected for the current tectonic regime. Thus, enhanced seismic activity in the post-glacial period appears to have played a role in triggering slope failures. The corollary of this wolud be that in many places (i.e. particularly tectonically inactive mountains), the peak rates of post-glacial-isostatic adjustment have long since passed, and thus the probability of large-scale co-seismic rock slope failure is hopefully diminished.

8.2.4 Hydrological Effects on Slope Stability

Water within slopes can trigger rock slope failures, from small rockfalls to larger-scale rock slope failures. The efficient interception of precipitation by the mountainous topography typical of many proglacial settings (e.g. rates that can exceed 10 m per year in the New Zealand Southern Alps; Whitehouse 1988) creates an abundant potential source of groundwater. Infiltration and percolation of water into the rock (which may only be a few per cent of the precipitation landing on a slope, e.g. Sims et al. 2015) depend on the rock mass permeability and storage capacity. Permeability is enhanced by high relief and preferential paths of vertical water flow along open fractures; but percolation is affected by aperture, roughness and filling material (Dietrich et al. 2005). Depending on rock mass permeability, groundwater fluctuations may occur rapidly with individual rainfall events or be insensitive to short-term changes in precipitation and instead respond to longer-term, seasonal changes in water infiltration. Shallower rock slope failures can be expected to be more sensitive to individual rainfall events than deeper-seated failures because deep-seated failures will require larger rises in the groundwater table to trigger instability and consequently greater antecedent precipitation (Van Asch et al. 1999). The main impact of rising groundwater (i.e. porewater pressure) for rock slope stability is the reduction in effective normal stresses, and therefore frictional strength of potential or active failure surfaces. The build-up of hydrostatic water pressure within vertical fractures—if for example, an aquiclude prevents seepage—can also reduce the stability of a slope by transmitting shear forces (Terzaghi 1962), for example the Young River and Tschierva landslides (discussed further below).

Rainfall can trigger rockfalls especially when the intensity is extremely high (Krautblatter and Moser 2009; Wieczorek and Jäger 1996), but in alpine areas thermal effects (e.g. Stoffel et al. 2005) tend to be more important at triggering rockfall. However, Krautblatter and Moser (2009) observed and quantified another type of rainfall-triggered mass movement from rockwalls, which may be of greater significance for the supply of sediment in the proglacial area. Referred to as 'secondary rockfall events', these are typically fine-grained rock particles, stored in a rockfall source area, that are dislodged by flowing water or debris-saturated flows during rainfall events. These processes might be significant for reworking earlier rockfall debris that build up on the rockwalls during periods without high-intensity rainfall. Rainfall triggering of larger-scale failures may occur if the slopes are critically stable. In Aoraki/Mt Cook National Park in New Zealand, a 400-mm rainfall event over three days likely triggered the 2013 Ball Ridge landslide which was primed for failure by a long period of rock mass weakening associated with glacier recession (Hancox and Thomson 2013). Large, deep-seated slow-moving landslides can also respond to changes in groundwater, for example the 100–200 M m^3 Mueller Rockslide in New Zealand appears to undergo periods of faster movement following heavy or prolonged rainfall (Deline et al. 2015; McColl 2012b). Rainfall has also been cited as a potential control on the timing of alpine rock slope failure activity over both historical and Holocene timescales (e.g. Grove 1972; Dortch et al. 2009; Porter and Orombelli 1981; Borgatti and Soldati 2010; Johnson et al. 2017).

8.2.5 Thermal Effects on Slope Stability

Proglacial environments are especially sensitive to climate change, and of the factors affecting alpine rock slope stability, thermal effects have received the greatest attention. Operating at a range of temporal scales, thermal changes in both ice and rock can induce mechanical changes in a slope sufficient to cause instability (Krautblatter et al. 2013; Viles 2013). Permafrost degradation can increase regional rockfall activity on multi-annual scales (Ravanel and Deline 2011; Fischer et al. 2012), while seasonal active-layer thaw drives patterns of seasonal rockfall activity (e.g. in the European Alps in 2003; Gruber et al. 2004). In addition, frost weathering and thermo-mechanical stresses due to temperature fluctuations affect long-term (i.e. decadal to millennial) rock slope stability and can act in conjunction with permafrost as preparatory factors and triggers (McColl 2012a; Krautblatter et al. 2013). Within proglacial environments, the influences of thermal effects have been investigated for a range of rock slope failure types from rockfall processes (Ravanel et al. 2010; Fischer et al. 2012) to rock avalanches (Allen and Huggel 2013; Huggel et al. 2005; Fischer et al. 2010).

8.2.5.1 Temperature Variability and Its Effects on Rock Slopes

Temperatures and ice within rock slopes are influenced by the heat flux between the atmosphere, snow cover and the rock. The mechanisms of heat flux can be distinguished into (1) conductive heat transport along a temperature gradient in intact rock, (2) advective heat flux by percolating water along fractures (Gruber and Haeberli 2007) and (3) convective heat flux due to air movement along fractures (Moore et al. 2011). Snow cover strongly influences heat fluxes into and out of the rock (Draebing et al. 2014), by increasing albedo, long-wave emissivity and absorptivity and, decreasing thermal conductivity (Zhang 2005). Due to topographic and meteorological factors, snow cover is highly variable in steep rockwalls (Haberkorn et al. 2015; Wirz et al. 2011). Consequently, rock slope temperatures and meltwater effects, both of which have a strong bearing on stability, are highly heterogeneous over space and time (Draebing et al. 2017a; Hasler et al. 2011b).

Rock warming results in thermal expansion and crack closure, whereas cooling results in rock contraction and crack opening (Draebing et al. 2017b). Crackmeter data from a steep alpine rock slope (Steintaelli, Switzerland) revealed frequent (daily) short-term crack expansions of up to 1 cm and seasonal crack deformation of up to 2 cm (Fig. 4.4c; Draebing et al. 2017b). High-frequency cyclic expansion and contraction can produce thermo-mechanical stresses sufficient to induce fracture growth or displacements (Hall and Thorn 2014; Eppes et al. 2016). Daily thermal changes are effective up to 0.5 m depth in intact rock (Gunzburger and Merrien-Soukatchoff 2011) and can prepare or trigger rockfalls as recently demonstrated by Collins and Stock (2016). In contrast, seasonal thermal changes can propagate to depths of up to 100 m due to fracture air convection (Gischig et al. 2011b). At such depths, the thermally induced stress changes are small (on the order of 10–100 kPa), but they may be sufficient to induce displacement on critically stressed failure surfaces (Gischig et al. 2011c).

The freezing and thawing of ice within rock slopes also influence stability. When water freezes to ice, volumetric expansion of ice can develop cryostatic pressure up to 207 MPa (Matsuoka and Murton 2008) while cryostatic pressure due to ice segregation is in the range of 20–30 MPa (Hallet et al. 1991). Acting on fracture walls, these pressures can induce crack propagation or transmit shearing stresses into the surrounding rock or onto potential sliding surfaces. However, ice within cracks adds cohesion, increasing crack (tensile) strength, and the freezing of the rock itself increases the uniaxial compressive and tensile strength of intact-rock bridges and rock–rock contacts (Krautblatter et al. 2013). All of these effects reverse with warming and melting, with the occurrence or melting of permafrost altering the driving and resisting stresses in rock slopes (Krautblatter et al. 2013; Davies et al. 2001); but as shown below these effects vary temporally and spatially (Fig. 8.3).

8.2.5.2 Seasonal Time Windows of Instability

Based on the interaction between snow cover, ground heat transport and mechanical processes, Draebing et al. (2014) identified two seasonal time windows of rock instability (Fig. 8.3). Snowmelt in summer can amplify active-layer thawing along fractures (Gruber and Haeberli 2007) and cause ice erosion in fractures (Hasler et al. 2011a). Therefore, shear resistance is decreased due to active-layer thawing while shear force increases due to hydrostatic pressure development, which in conjuction leads to potential instabilities within time window of instability #1 (Fig. 8.3). The existence of time window #1 is supported by the timing of rock slope deformation during snowmelt periods in the permafrost-affected Jettan rock slide (Blikra and Christiansen 2014) and an increase in rockfall activity revealed by rockfall inventories in the European Alps (Ravanel et al. 2010; Luethi et al. 2015). However, time window #1 may not develop in all conditions. Crackmeter measurements at Steintaelli, Switzerland, demonstrate crack closure at N-facing positions during snowmelt (Fig. 8.4a). The data suggest that hydrostatic pressure was insufficient to counter crack closure (Draebing et al. 2017a), possibly because refreezing of meltwater at the rock surface prevented sufficient water from being able to infiltrate into fractures (Phillips et al. 2016). High-magnitude precipitation events as described in the chapter on hydrological effects on slope stability can hydrostatically trigger rock slope failures similar to time window #1. Permafrost or seasonal frozen ground can enhance hydrostatic pressure build-up by preventing seepage (e.g. McSaveney and Massey 2013; Fischer et al. 2010).

A second time window of instability (#2 in Fig. 8.3) may develop in autumn when cooling of rock due to conductive heat transport may be amplified by convective heat transport in fractures (Moore et al. 2011). Cooling causes refreezing of water in cracks and the resulting volumetric expansion (Matsuoka and Murton 2008) can lead to the development of cryostatic

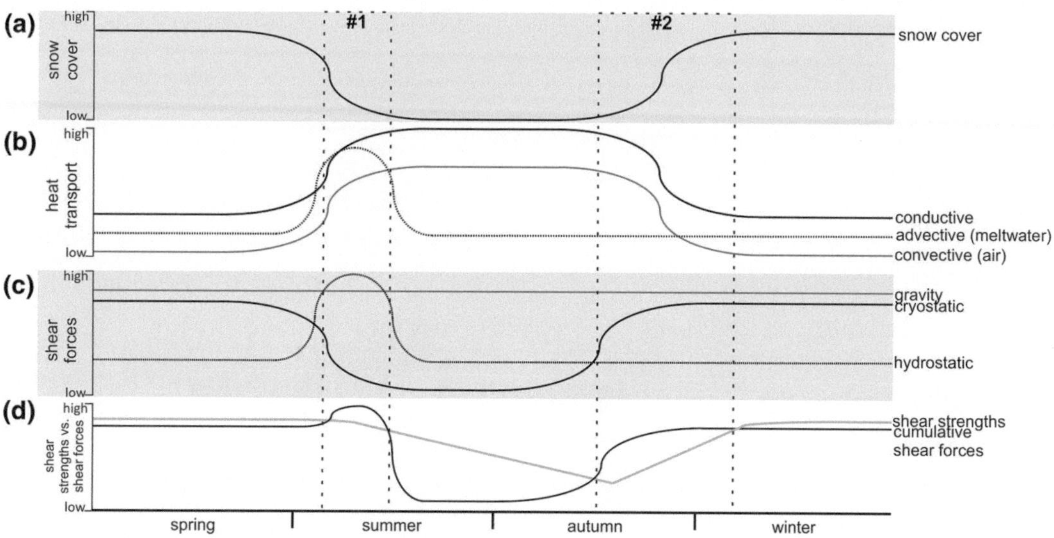

Fig. 8.3 Seasonal evolution of **a** snow cover, **b** thermal heat transfer, **c** shear forces such as gravity, hydrostatic and cryostatic pressures and **d** shear resistances (adapted from Draebing et al. 2014). Dotted rectangles indicate time windows of instability. Printed with permission by Elsevier

pressure (i.e. increased shear force) that exceeds shear resistance (Draebing et al. 2014). The timing of increased rockfall events during periods of cooling in late summer and early autumn observed by Ravanel et al. (2010) in the Mt. Blanc area was probably driven by volumetric expansion and supports the existence of time window of instability #2. Draebing et al. (2017a) observed the occurrence of high cooling rates necessary for volume expansion (arrows in Fig. 8.4) in the Steintaelli, however, little or no crack dilation was observed. This observation may be supported by a recent modelling study by Rode et al. (2016) which suggests that the moisture preconditions required for volumetric expansion are rarely fulfilled and that ice segregation is a more probable frost weathering process in poleward-facing rockwalls. If unfrozen water is present, cryosuction enables water migration and increases crack aperture due to ice segregation. Draebing et al. (2017b) observed a cryogenic-induced crack opening during winter of 1 cm that continued to widen from January until the end of the snow-covered period (Fig. 8.4b); although fracture opening was reversed in summer by ice relaxation without any net change of fracture aperture, repeated cycles or sustained cold may fatigue fractures. At Piz Kesch, Phillips et al. (2017) observed a 150,000 m^3 failure of rock during stable cold

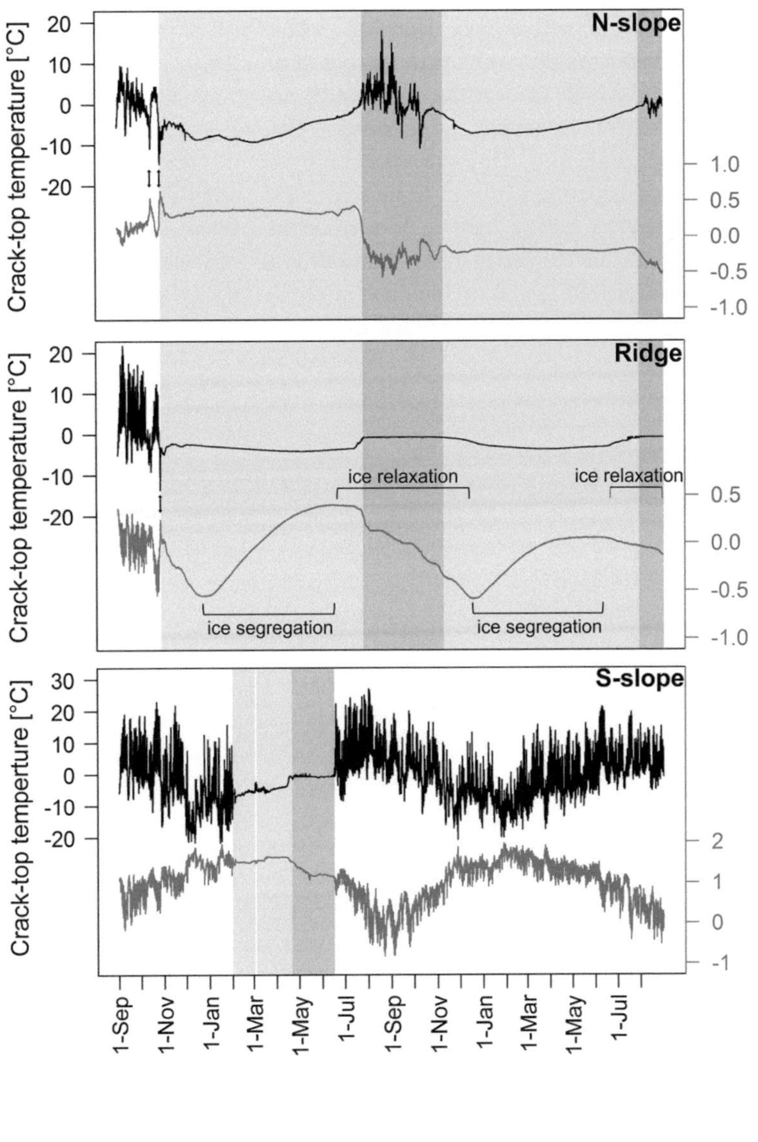

Fig. 8.4 Crack-top temperature and crack deformation for crackmeters on the N slope, ridge and S slope at Steintaelli, Switzerland, between September 2012 and August 2014. Snow cover is indicated as light-grey rectangles and zero-curtain periods which highlight melting periods are indicated as dark-grey rectangles. Arrows indicate temperature drops and potential occurrence of volumetric expansion. For information on crackmeter location, see Draebing et al. (2017a, b). The figure is adapted from Draebing et al. (2017b) and printed with permission by Wiley & Sons

temperatures in mid-winter. Therefore, we suggest that ice segregation can prolong the time window of instability #2 and cryogenic stresses can trigger rock slope failures even in winter.

8.2.5.3 Longer-Term Thermal Effects

In the long-term, permafrost and weathering processes can prepare rock slope failures. The cyclic cryostatic pressures due to ice segregation and volumetric expansion, along with thermo-mechanical stresses, can lead to ice-(Jia et al. 2015; Murton et al. 2006) or thermal-induced fatigue (Hall and Thorn 2014; Eppes et al. 2016). The fatigue reduces the resistance, bringing a slope to a marginally stable state (Fig. 8.2) more sensitive to triggers (Draebing et al. 2014). In addition, as proglacial environments undergo the transition from cold glacial to warmer interglacial conditions, changes to air temperatures and snow-cover alter the spatial and temporal patterns of rock temperatures and permafrost. This in turn affects the efficiency of thermal-induced effects to rock slopes relative to other processes driving proglacial rock slope failure patterns. For example, rock slope back-weathering rates derived from talus slopes are often inconsistent with modern rockfall rates. This is suggested to be a result of initially increased rockfall directly after deglaciation, driven by stress-release and joint propagation (Ballantyne 2002), but followed by a second peak in talus slope accumulation when enhanced frost cracking developed under a periglacial climate (in the Isle of Skye; Hinchliffe and Ballantyne 1999); and with both mechanisms being diminished now.

8.2.6 Discussion

While instability may be a common condition of proglacial rock slopes, the pace of landslide activity and therefore the causes vary over time and space. In many proglacial valleys, there may have been a peak in the frequency or probability of rock slope failures shortly after deglaciation that diminished over time (through easing glacio-isostatic adjustment, stress relaxation and thermal effects). In some locations, the subsequent pace of landslide activity may have remained fairly stable over the Holocene (Ballantyne et al. 2014), while in other places there may not have been a notable peak in post-glacial failure activity (Johnson et al. 2017). Nonetheless, many proglacial zones are likely still responding to localised stress perturbations (e.g. ongoing glacier retreat, historical climate change), and regardless of paraglacial factors, the extreme conditions of proglacial rock slopes mean that landslide activity continues to be a dominant process. Within some catchments, the focus of hotspot landslide activity might be changing. In presently glaciated or permafrost-affected valleys for instance, the zones of maximal slope failure activity may be increasing in altitude as temperatures rise (e.g. the increased zonal elevation of historical landslide activity vs. pre-historical activity in the NZ Southern Alps; Allen and Huggel 2013) or moving farther up into the catchments as glaciers continue to recede, rather than reducing completely. Further, although some contemporary proglacial systems may have lower rates of slope failure activity today, many are still experiencing the legacy effect of earlier periods of more enhanced landslide activity, particularly from the very large failures which can persist in the landscape for many thousands of years; whether these fell onto ice or onto the proglacial zone they can continue to contribute sediment, influence river form, or impede sediment transport, for many thousands of years.

8.3 Advances in Rock Slope Investigation and Monitoring

Current understanding of rock slope failure processes, as presented in the preceding sections of this chapter, has been informed by a rich combination of local and regional data on landslide activity and controlling factors. The remainder of the chapter provides an overview of the range of techniques used to generate these datasets.

Understanding and predicting rock slope failure activity require knowledge of geotechnical rock mass properties, the geomorphic history of the slopes, the nature (frequency and magnitude) of potential triggering processes and an abundance of data on the slope failure patterns over a range of spatial and temporal scales. All of these data are challenging to collect because the environments are usually difficult to access and evidence is readily eroded, buried or modified by a range of dynamic processes. Further, it requires being able to tease out the multitude of potential factors and processes responsible for the slope failures, many of which themselves are not well understood and are not constant. Nonetheless, these data are essential for understanding rock slope failure processes and a range of techniques are available to investigate and monitor rock slope failures in proglacial environments. These investigations provide an opportunity for deepening understanding of rock slope activity and causes, are the basis of warning systems for risk reduction (e.g. Stead et al. 2004; Eberhardt 2012) and are useful for assessing the impacts of slope failure to proglacial systems. In this section, we provide a short (non-exhaustive) overview of the main tools used for the investigation of rock slope stability and failure activity (see Table 8.1), which we separate into techniques to measure preconditioning, preparing and triggering factors, monitor rock slope kinematics, and reconstruct past rock slope failures.

8.3.1 Techniques to Measure Preconditioning Factors of Proglacial Rock Slope Failures

Preconditioning factors encompass lithology and structure, intact-rock strength, and rock mass quality. Geomorphological mapping provides qualitative information of these preconditioning factors and can be used to qualitatively evaluate process activity and provide information on source areas (Otto and Dikau 2004). Furthermore, preconditioning factors can be quantitatively assessed by using geotechnical mapping approaches (Moore et al. 2009; Vehling et al. 2016; Chap. 9). To calculate rock slope stability, the resulting data can be quantitatively analysed using the Joint Roughness Coefficient (Barton and Choubey 1977) or semi-quantitatively using the Rock Mass Rating (Bieniawski 1989), Rock Mass Strength (Selby 1980) or Geological Strength Index (Hoek and Brown 1997). If rock slopes are not accessible, discontinuity properties and structures exceeding 0.2–0.6 m (Sturzenegger and Stead 2009) can be derived from Terrestrial Laser Scanning (TLS; Abellán et al. 2014) or photogrammetry (Brideau et al. 2012). The fracture distribution within rock slopes or rockslide boundaries can be mapped by using geophysical techniques such as ground-penetrating radar (Willenberg et al. 2008; Heincke et al. 2006) or refraction seismics (Phillips et al. 2016; Heincke et al. 2010).

8.3.2 Techniques to Measure Preparing and Triggering Factors of Proglacial Rock Slope Failures

Preparing and triggering factors include water conditions, glacier thinning, permafrost and weathering processes. Groundwater conditions in rock slopes or rockslides can be assessed using piezometers (Preisig et al. 2016) or geoelectric techniques (Sass 2004, 2005a; Heincke et al. 2010). To investigate permafrost occurrence and monitor active-layer thawing, temperature boreholes (Phillips et al. 2016) and temperature loggers (Haberkorn et al. 2015; Draebing et al. 2017a) provide high temporal point information on the thermal regime and can be used to establish regional permafrost probability maps (Hipp et al. 2014) or to validate thermal rock slope models (Myhra et al. 2017; Noetzli et al. 2007). Spatial distribution of permafrost in rock slopes can be derived using Electrical Resistivity Tomography (ERT; Krautblatter and Hauck 2007; Krautblatter et al. 2010; Chap. 7) or Seismic Refraction Tomography (SRT; Krautblatter and Draebing 2014; Draebing et al. 2017a). Weathering processes are difficult to detect due to low process frequency or

Table 8.1 Remote sensing, in situ measurement and dating techniques are used to measure preconditioning, preparing and triggering factors, monitor rock slope kinematics or reconstruct past landslides

Method	Nature of measurement	References
Preconditioning factors		
Geomorphological mapping	Localisation of landslide processes and related sources	Otto and Dikau (2004)
Geotechnical mapping	Discontinuity mapping	Moore et al. (2009), Vehling (Ch. 4.2)
Terrestrial Laserscanning	Rock slope characterisation of discontinuities and structure	Abellán et al. (2014)
Photogrammetry	Rock slope characterisation of discontinuities and structure	Brideau et al. (2012)
Ground-penetrating radar	Fracture mapping	Willenberg et al. (2008)
Seismic Refraction	Detection of fracture systems	Phillips et al. (2016), Willenberg et al. (2008)
	Determination of rockslide body	Heincke et al. (2010)
Preparing and triggering factors		
Piezometers	Groundwater and hydrostatic pressure measurements	Preisig et al. (2016)
Temperature boreholes	Monitoring of rock slope thermal regime	Phillips et al. (2016)
Temperature data loggers	Monitoring of rock surface thermal regime	Draebing et al. (2017a), Haberkorn et al. (2015)
Electrical resistivity	Monitoring of rock slope moisture	Sass (2004, 2005)
	Determination hydrogeological regime of rockslides	Heincke et al. (2010)
	Detection of permafrost occurrence and active-layer thawing in rock slopes	Krautblatter and Hauck (2007), Krautblatter et al. (2010)
Seismic Refraction	Detection of permafrost occurrence and active-layer thawing in rock slopes	Draebing et al. (2017a), Krautblatter and Draebing (2014)
Acoustic emission	Temporal and spatial distribution of frost cracking in rock slopes	Amitrano et al. (2012), Girard et al. (2013)
Crackmeters	Differentiating weathering processes and timing of backweathering	Draebing et al. (2017a, b), Hasler et al. (2012), Vehling (Ch. 4.2)
Schmidt hammer	Detection of rock strength reduction due to weathering processes	Viles et al. (2011)
Rock slope kinematics		
Terrestrial Laserscanning	Detailed volumetric inventory of rockfalls	Oppikofer et al. (2008), Strunden et al. (2015)
Airborne LiDAR	Detailed DTM, calculation of erosion rates	Fischer et al. (2011), Gischig et al. (2011a)
InSAR	Large spatial and temporal rock slope displacement	Strozzi et al. (2010)
GB InSAR	Large spatial and temporal rock slope displacement	Gischig et al. (2009, 2011a), Kos et al. (2016)

(continued)

Table 8.1 (continued)

Method	Nature of measurement	References
Time-lapse photogrammetry	Estimation of rockfall volume and rockfall timing	Ravanel et al. (2010)
	Construction of DTM to calculate of erosion rates or slope displacement	Clayton et al. (2017), Fischer et al. (2011), Micheletti et al. (2015)
Total station	Spatial and temporal rock slope displacement	Nishii and Matsuoka (2012)
dGPS	Spatial and temporal rock slope displacement	McColl et al. (2012), Nishii and Matsuoka (2012)
Real-Time GPS	High-resolution temporal rock slope displacement	Blikra and Christiansen (2014), Wirz et al. (2016)
Extensometer, crackmeter	Spatial and temporal rock slope displacement	Blikra and Christiansen (2014), Draebing et al. (2017a, b)
Inclinometers	spatial and temporal rock slope displacement	Blikra and Christiansen (2014), Gischig et al. (2011b)
Cosmogenic nuclides	Dating of rockslide scars and calculation of palaeo-movement rates	Grämiger et al. (2017), Hermanns et al. (2012)
Height difference of quartz veins	Holocene rock slope retreat rate	André 1997, Dahl (1967)
Reconstruction of past landslides using landslide deposits		
Geomorphological mapping	Landform distribution and process activity	Otto and Dikau (2004)
Painted rock slopes	Timing and magnitude of rockfall from restricted area	Matsuoka and Sakai (1999), Prick (2003)
Rockfall collector measurements	Short-term rockfall rates	Krautblatter and Moser (2009), Sass (2005b), Vehling et al. (Ch. 4.2)
Rockfall deposits on snow	Integral values of winter rockfall activity	Matsuoka and Sakai (1999), Rapp (1960)
Terrestrial Laserscanning	Talus slope accumulation rates	Heckmann et al. (2012)
Time-lapse photogrammetry	Construction of DTM to calculate rockfall accumulation rates	Heckmann et al. (2008)
Electrical resistivity	Detection of permafrost occurrence and sediment thickness of rockfall deposits	Scapozza et al. (2011), Siewert et al. (2012)
Seismic Refraction	Detection of permafrost occurrence and sediment thickness of rockfall deposits	Draebing (2016), Sass (2006)
Ground-penetrating radar	Detection of thickness and structure of landslide deposits	Dunning et al. (2015), Sass et al. (2007)
Lichenometric dating	Rockfall frequency in response to climatic patterns	André (1997), McCarroll et al. (2001)
14C	Dating of organic soil in rockfall or landslide deposits	Hinchliffe and Ballantyne (2009), Hancox et al. (2013)
Cosmogenic nuclides	Dating of landslide deposits	Ballantyne et al. (2014), Hermanns et al. (2015)

magnitude. Acoustic emission such as microseismic measurements is applied to monitor and locate frost cracking processes (Amitrano et al. 2012; Girard et al. 2013). Crackmeters are used to monitor fracture movement and infer frost cracking processes such as volumetric expansion or ice segregation (Draebing et al. 2017a, b; Hasler et al. 2012; Chap. 9). Repeated Schmidt hammer measurements enable the detection of rock strength reduction which can be used as a proxy for rock weathering (Viles et al. 2011).

8.3.3 Techniques to Monitor Rock Slope Kinematics

Rock slope kinematics (i.e. movement) can be assessed with remote sensing techniques and in situ measurements. LiDAR techniques such as Terrestrial Laserscanning (TLS; Strunden et al. 2015; Oppikofer et al. 2008; Chap. 9) or Airborne LiDAR (Fischer et al. 2011; Gischig et al. 2011a) enable the construction of detailed Digital Terrain Models (DTM) and with repeat surveys allows the calculation of DTMs of Difference (DoDs) which provide a detailed volumetric inventory on rockfall processes or rates of rock slope displacement. Alternatively, rock slope displacement can be derived from repetitive measurements using radar techniques such as InSAR (Strozzi et al. 2010) or ground-based InSAR (Gischig et al. 2009, 2011a; Kos et al. 2016). Photogrammetric approaches are also used to produce Digital Elevation Models (DEMs), with the Structure-for-Motion method becoming popular (James and Robson 2012). With photogrammetric techniques, rock slope displacements or erosion can be derived in the same way as LiDAR processing (Fischer et al. 2011), but with the added benefit that the time series can be prolonged with archival aerial photography (Micheletti et al. 2015; Clayton et al. 2017). To reconstruct rockfall volumes in a semi-qualitative way, historical photos can be used to identify rockfall scars and develop historical rockfall inventories (Ravanel and Deline 2011).

Alternatives to remote sensing techniques are repetitive point measurements of movement or deformation. These techniques require access to the slope and practical constraints usually result in low density of spatial information, but they have the advantage of directly measuring slope movement, they can be more accurate and precise than some remote sensing techniques, and some of them can be developed into continuous or near-continuous systems. Total stations (Nishii and Matsuoka 2012) and differential GPS systems (Nishii and Matsuoka 2012; McColl and Davies 2013) are two popular point measurement techniques, that both can be used to collect near-continuous data, which help to link movement with preparing or triggering factors such as hydrostatic pressure due to snowmelt (Blikra and Christiansen 2014; Wirz et al. 2016). Other, more traditional, methods are manual extensometers, automatic crackmeters or inclinometers (Eberhardt 2012). Manual extensometer measurements enable monitoring of large transects with low temporal resolution and, therefore provide information on seasonal or annual displacement of rock slopes (Draebing 2015). Temporal resolution can be increased using crackmeters (Draebing et al. 2017b; Hasler et al. 2012; Vehling et al. 2016; Chap. 9); however, the application of crackmeters is connected to a decrease of spatial resolution due to higher costs and smaller size than extensometers. Inclinometer measurements in boreholes provide information on rock slope deformation and are suited for the monitoring of rotational block movement or identifying the location of a failure surface. The measurements can be combined with thermal borehole measurements for permafrost monitoring; however, boreholes in alpine areas are expensive and their use should legitimate the costs. Therefore, they are often integrated in early warning systems monitoring hazardous rock slopes (Blikra and Christiansen 2014; Gischig et al. 2011a).

To reconstruct rock slope deformation or rock slope erosion on Holocene timescales, exposure dating techniques, such as measurement of quartz vein relief (Dahl 1967; André 1997) or cosmogenic nuclide dating, can be used to assess the development of failure scarps (Grämiger et al. 2017; Hermanns et al. 2012). As a consequence of slow rock slope deformation, failure scarp height and area increases, gradually exposing rock to the atmosphere. Hermanns et al. (2012) and Pánek et al. (2017) have both dated rock surface exposure and used age differences to calculate palaeo-rock slope deformation rates.

8.3.4 Reconstruction of Past Landslides Using Landslide Deposits

Geomorphological mapping provides spatial information on landslide deposits, size and process activity (Otto and Dikau 2004). Past rockfall or landslide activity can be reconstructed based on the analysis of these rock slope failure deposits. To estimate the timing and magnitude of rockfall from small restricted areas, rock slopes can be painted and coloured rockfall deposits can be collected and analysed (Matsuoka and Sakai 1999; Prick 2003). Snow patches provide natural seasonal sediment traps and rockfall accumulation gives insight on winter rockfall activity (Matsuoka and Sakai 1999; Rapp 1960). Temporal resolution of rockfall activity can be increased using rockfall collectors (Krautblatter et al. 2012; Sass 2005b; Chap. 9) or DoDs derived from Terrestrial Laserscanning (Heckmann et al. 2012). These data enable an analysis of rockfall activity and correlation to rock slope properties at different sites. If the time periods between observations are small, the rockfall events can be linked to triggers (Sass 2005b) such as rainfall (Krautblatter and Moser 2009).

Rockfall and landslide deposit structure and thickness can be quantified using single or a multiple geophysical techniques (Sass 2006, 2007; Otto and Sass 2006) including ERT (Siewert et al. 2012), refraction seismics and ground-penetrating radar (Dunning et al. 2015; Sass et al. 2007). In addition, geophysical measurements can provide information on permafrost occurrence (Scapozza et al. 2011; Draebing 2016; Chap. 7). Rockfall deposit thickness can be included in sediment budget approaches (Otto et al. 2009) and can be used to establish Holocene rock slope backweathering rates (Otto and Sass 2006; Sass 2006, 2007). Dating techniques such as ^{14}C, lichenometry and cosmogenic nuclides can be used to establish long-term rockfall rates or reconstruct landslide activity. In proglacial areas, the application of dendrologic measurements is seldom possible due to the lack of datable trees. Lichenometric dating enables the reconstruction of rockfall frequencies in response to climatic changes (André 1997; McCarroll et al. 2001); however, the existence or establishment of a calibration curve for lichen growth is a precondition for the technique's application. Organic soil in rockfall deposits can be used to establish dates and rates of rockfall accumulation (Hinchliffe and Ballantyne 2009). Large rock slope failures have traditionally been dated using radiocarbon measurement of organic debris buried under, incorporated within or accumulated upon the deposit (e.g. Hancox et al. 2013). However, a technique being more commonly applied to alpine landslides is cosmogenic nuclide exposure age dating (e.g. Hermanns et al. 2015; Ballantyne et al. 2014). For a database of very large dated paraglacial landslide events, see McColl (2012).

8.4 Conclusion

Rock slope failure is an integral process and part of proglacial systems. Steep topography, deformed rock masses, seismic activity and extreme climate conditions make proglacial slopes inherently unstable. These slopes are therefore particularly sensitive to climate change and glacier recession which can induce localised stress changes, influence seismicity, and set off a chain of processes that degrade the strength of the rock masses. The resultant instability and rock slope failures that develop can range from small-scale rockfall through to some of the largest terrestrial slope failures on the planet. These factors arguably make proglacial environments some of the most unstable and rapidly evolving parts of the terrestrial environment. The impacts of rock slope failures in the proglacial environment include the large-scale and sometimes rapid erosional modification of landscapes, the production and impedance of sediment, and the considerable threat of landslides to communities living within and downstream of mountainous topography.

The causes of proglacial rock slope failure activity include the preconditioning factors of geology (often tectonically deformed rocks) and high relief, and the preparatory factors of glacial erosion, debuttressing and stress-release, and progressive strength degradation from weathering and gravitational stresses. Dominant triggers include rainfall or groundwater changes, but appear to be more strongly affected by thermal changes and seismic triggers. There is a general pattern of increased landslide activity (both small-scale rockfall and large, catastrophic failures) in the several millennia immediately following the onset of deglaciation, but rock slopes continue to respond to the antecedent (paraglacial) conditions and fresh perturbations provided by ongoing glacier retreat and climate change. Predicting rock slope failure within proglacial settings requires knowledge of how slopes have responded in the past, the alignment of data on regional and local-scale stability factors, and projections of how these stability factors may continue to change in the future. The final section of the chapter provided a brief review of major techniques that allow these data to be collected, and that offer promise for further growing understanding of proglacial rock slope behaviour.

References

Abellán A, Oppikofer T, Jaboyedoff M et al (2014) Terrestrial laser scanning of rock slope instabilities: state of science. Earth Surf Process Land 39:80–97. https://doi.org/10.1002/esp.3493

Akçar N, Deline P, Ivy-Ochs S et al (2012) The AD 1717 rock avalanche deposits in the upper Ferret Valley (Italy): a dating approach with cosmogenic 10Be. J Quat Sci 27:383–392

Allen S, Huggel C (2013) Extremely warm temperatures as a potential cause of recent high mountain rockfall. Glob Planet Change 107:59–69

Amitrano D, Gruber S, Girard L (2012) Evidence of frost-cracking inferred from acoustic emissions in a high-alpine rock-wall. Earth Planet Sci Lett 341–344:86–93. https://doi.org/10.1016/j.epsl.2012.06.014

André M-F (1997) Holocene Rockwall Retreat in Svalbard: a triple-rate evolution. Earth Surf Process Land 22:423–440. https://doi.org/10.1002/(SICI)1096-9837(199705)22:5%3C423:AID-ESP706%3E3.0.CO;2-6

Arsenault AM, Meigs AJ (2005) Contribution of deep-seated bedrock landslides to erosion of a glaciated basin in southern Alaska. Earth Surf Process Land 30:1111–1125

Ballantyne C (2003) Paraglacial landform succession and sediment storage in deglaciated mountain valleys: theory and approaches to calibration (with 6 figures). Zeitschrift für Geomorphologie Supplement 1–18

Ballantyne CK (2002) Paraglacial geomorphology. Quat Sci Rev 21:1935–2017. https://doi.org/10.1016/S0277-3791(02)00005-7

Ballantyne CK, Sandeman GF, Stone JO, Wilson P (2014) Rock-slope failure following Late Pleistocene deglaciation on tectonically stable mountainous terrain. Quat Sci Rev 86:144–157. https://doi.org/10.1016/j.quascirev.2013.12.021

Ballantyne CK, Stone JO (2013) Timing and periodicity of paraglacial rock-slope failures in the Scottish highlands. Geomorphology 186:150–161. https://doi.org/10.1016/j.geomorph.2012.12.030

Barton N, Choubey V (1977) The shear strength of rock joints in theory and practice. Rock Mechanics Felsmechanik Mécanique des Roches 10:1–54. https://doi.org/10.1007/BF01261801

Bieniawski ZT (1989) Engineering rock mass classifications: a complete manual for engineers and geologists in mining, civil, and petroleum engineering. Wiley, Hoboken

Blikra LH, Christiansen HH (2014) A field-based model of permafrost-controlled rockslide deformation in northern Norway. Geomorphology 208:34–49. https://doi.org/10.1016/j.geomorph.2013.11.014

Borgatti L, Soldati M (2010) Landslides as a geomorphological proxy for climate change: a record from the Dolomites (Northern Italy). Geomorphology 120:56–64

Brideau M-A, Sturzenegger M, Stead D et al (2012) Stability analysis of the 2007 Chehalis lake landslide based on long-range terrestrial photogrammetry and airborne LiDAR data. Landslides 9:75–91. https://doi.org/10.1007/s10346-011-0286-4

Cave JAS, Ballantyne CK (2016) Catastrophic Rock-Slope failures in NW Scotland: quantitative analysis and implications. Scott Geogr J 132:185–209. https://doi.org/10.1080/14702541.2016.1156148

Clayton A, Stead D, Kinakin D, Wolter A (2017) Engineering geomorphological interpretation of the Mitchell Creek Landslide, British Columbia, Canada. Landslides. https://doi.org/10.1007/s10346-017-0811-1

Collins BD, Stock GM (2016) Rockfall triggering by cyclic thermal stressing of exfoliation fractures. Nat Geosci 9:395–400. https://doi.org/10.1038/ngeo2686
Cossart E, Braucher R, Fort M, Bourlés DL, Carcaillet J (2008) Slope instability in relation to glacial debuttressing in alpine areas (Upper Durance catchment, southeastern France): evidence from field data and 10Be cosmic ray exposure ages. Geomorphology 95:3–26
Coquin J, Mercier D, Bourgeois O et al (2015) Gravitational spreading of mountain ridges coeval with Late Weichselian deglaciation: impact on glacial landscapes in Tröllaskagi, Northern Iceland. Quat Sci Rev 107:197–213. https://doi.org/10.1016/j.quascirev.2014.10.023
Cossart E, Mercier D, Decaulne A et al (2014) Impacts of post-glacial rebound on landslide spatial distribution at a regional scale in Northern Iceland (Skagafjörður). Earth Surf Process Land 39:336–350. https://doi.org/10.1002/esp.3450
Cox SC, Allen SK, Ferris BG (2008) Vampire rock avalanches, Aoraki/Mount Cook National Park, New Zealand. GNS Science
Cox SC, McSaveney MJ, Rattenbury MS, Hamling IJ (2014) Activity of the landslide Te Horo and Te Koroka fan, Dart River, New Zealand during January 2014
Cox SC, McSaveney MJ, Spencer J et al (2015) Rock avalanche on 14 July 2014 from Hillary Ridge, Aoraki/Mount Cook, New Zealand. Landslides 12:395–402. https://doi.org/10.1007/s10346-015-0556-7
Crozier MJ (2005) Management frameworks for landslide hazard and risk: issues and options. Wiley, Chichester
Dahl R (1967) Post-glacial micro-weathering of bedrock surfaces in the Narvik District of Norway. Geografiska Annaler Series A, Phys Geogr 49:155. https://doi.org/10.2307/520884
Davies MC, Hamza O, Harris C (2001) The effect of rise in mean annual temperature on the stability of rock slopes containing ice-filled discontinuities. Permafrost Periglac Process 12:137–144
Decaulne A, Cossart E, Mercier D et al (2016) An early Holocene age for the Vatn landslide (Skagafjörḫur, central northern Iceland): insights into the role of postglacial landsliding on slope development. The Holocene 26:1304–1318
Deline P (2009) Interactions between rock avalanches and glaciers in the Mont Blanc massif during the late Holocene. Quat Sci Rev 28:1070–1083
Deline P, Gruber S, Delaloye R et al (2015) Ice loss and slope stability in high-mountain regions. In: Haeberli W, Whiteman C (eds) Snow and ice-related hazards, risks, and disasters. Elsevier, Amsterdam, pp 521–561
Dietrich P, Helmig R, Sauter M, et al (eds) (2005) Flow and transport in fractured porous media. Springer, Berlin
Dortch JM, Owen LA, Haneberg WC et al (2009) Nature and timing of large landslides in the Himalaya and Transhimalaya of Northern India. Quat Sci Rev 28:1037–1054
Draebing D (2015) Influences of snow cover on thermal and mechanical processes in steep permafrost rock walls. PhD Thesis, University of Bonn
Draebing D (2016) Application of refraction seismics in alpine permafrost studies: a review. Earth-Sci Rev 155:136–152. https://doi.org/10.1016/j.earscirev.2016.02.006
Draebing D, Haberkorn A, Krautblatter M et al (2017a) Thermal and mechanical responses resulting from spatial and temporal snow cover variability in permafrost rock slopes, Steintaelli, Swiss Alps. Permafrost Periglac Process 28:140–157. https://doi.org/10.1002/ppp.1921
Draebing D, Krautblatter M, Hoffmann T (2017b) Thermo-cryogenic controls of fracture kinematics in permafrost rockwalls. Geophys Res Lett 44:3535–3544. https://doi.org/10.1002/2016GL072050
Draebing D, Krautblatter M, Dikau R (2014) Interaction of thermal and mechanical processes in steep permafrost rock walls: a conceptual approach. Geomorphology 226:226–235
Dunning SA, Rosser NJ, McColl ST, Reznichenko NV (2015) Rapid sequestration of rock avalanche deposits within glaciers. Nat Commun 6:7964. https://doi.org/10.1038/ncomms8964
Eberhardt E (2012) Landslide monitoring: the role of investigative monitoring to improve understanding and early warning of failure. Cambridge University Press, Cambridge, UK
Emberson R, Hovius N, Galy A, Marc O (2015) Chemical weathering in active mountain belts controlled by stochastic bedrock landsliding. Nat Geosci 9:42–45. https://doi.org/10.1038/ngeo2600
Eppes MC, Magi B, Hallet B et al (2016) Deciphering the role of solar-induced thermal stresses in rock weathering. Geol Soc Am Bull 128:1315–1338. https://doi.org/10.1130/B31422.1
Fischer L, Amann F, Moore JR, Huggel C (2010) Assessment of periglacial slope stability for the 1988 Tschierva rock avalanche (Piz Morteratsch, Switzerland). Eng Geol 116:32–43. https://doi.org/10.1016/j.enggeo.2010.07.005
Fischer L, Eisenbeiss H, Kääb A et al (2011) Monitoring topographic changes in a periglacial high-mountain face using high-resolution DTMs, Monte Rosa East Face, Italian Alps. Permafrost Periglac Process 22:140–152. https://doi.org/10.1002/ppp.717
Fischer L, Purves RS, Huggel C et al (2012) On the influence of topographic, geological and cryospheric factors on rock avalanches and rockfalls in high-mountain areas. Nat Hazards Earth Syst Sci 12:241
Girard L, Gruber S, Weber S, Beutel J (2013) Environmental controls of frost cracking revealed through in situ acoustic emission measurements in steep bedrock. Geophys Res Lett 40:1748–1753. https://doi.org/10.1002/grl.50384

Gischig V, Loew S, Kos A et al (2009) Identification of active release planes using ground-based differential InSAR at the Randa rock slope instability, Switzerland. Nat Hazards Earth Syst Sci 9:2027–2038

Gischig V, Amann F, Moore JR et al (2011a) Composite rock slope kinematics at the current Randa instability, Switzerland, based on remote sensing and numerical modeling. Eng Geol 118:37–53. https://doi.org/10.1016/j.enggeo.2010.11.006

Gischig VS, Moore JR, Evans KF et al (2011b) Thermomechanical forcing of deep rock slope deformation: 1. Conceptual study of a simplified slope. J Geophys Res https://doi.org/10.1029/2011jf002006

Gischig VS, Moore JR, Evans KF et al (2011c) Thermomechanical forcing of deep rock slope deformation: 2. The Randa rock slope instability

Glade T, Crozier MJ (2005) The nature of landslide hazard impact. In: Glade T, Anderson MG, Crozier MJ (eds) Landslide hazard and risk. Wiley, London, pp 43–74

Grämiger LM, Moore JR, Gischig VS et al (2017) Beyond debuttressing: Mechanics of paraglacial rock slope damage during repeat glacial cycles. J Geophys Res Earth Surf 122:1004–1036. https://doi.org/10.1002/2016JF003967

Grove JM (1972) The incidence of landslides, avalanches, and floods in western Norway during the Little Ice Age. Arctic Alpine Res 131–138

Gruber S, Haeberli W (2007) Permafrost in steep bedrock slopes and its temperature-related destabilization following climate change. J Geophys Res. https://doi.org/10.1029/2006JF000547

Gruber S, Hoelzle M, Haeberli W (2004) Permafrost thaw and destabilization of Alpine rock walls in the hot summer of 2003. Geophys Res Lett 31:n/a-n/a. https://doi.org/10.1029/2004gl020051

Gunzburger Y, Merrien-Soukatchoff V (2011) Near-surface temperatures and heat balance of bare outcrops exposed to solar radiation. Earth Surf Process Land 36:1577–1589. https://doi.org/10.1002/esp.2167

Haberkorn A, Hoelzle M, Phillips M, Kenner R (2015) Snow as a driving factor of rock surface temperatures in steep rough rock walls. Cold Reg Sci Technol 118:64–75. https://doi.org/10.1016/j.coldregions.2015.06.013

Hall K, Thorn CE (2014) Thermal fatigue and thermal shock in bedrock: an attempt to unravel the geomorphic processes and products. Geomorphology 206:1–13. https://doi.org/10.1016/j.geomorph.2013.09.022

Hallet B, Walder JS, Stubbs CW (1991) Weathering by segregation ice growth in microcracks at sustained subzero temperatures: verification from an experimental study using acoustic emissions. Permafrost Periglac Process 2:283–300. https://doi.org/10.1002/ppp.3430020404

Hancox GT, Langridge RM, Perrin ND et al (2013) Recent mapping and radiocarbon dating of three giant landslides in Northern Fiordland, New Zealand. GNS Science

Hancox GT, McSaveney MJ, Manville VR, Davies TR (2005) The October 1999 Mt Adams rock avalanche and subsequent landslide dam-break flood and effects in Poerua River, Westland, New Zealand. NZ J Geol Geophys 48:683–705

Hancox GT, Perrin ND (2009) Green Lake Landslide and other giant and very large postglacial landslides in Fiordland, New Zealand. Quat Sci Rev 28:1020–1036

Hancox GT, Thomson R (2013) The January 2013 Mt Haast Rock Avalanche and Ball Ridge Rock Fall in Aoraki/Mt Cook National Park, New Zealand. GNS Science

Hasler A, Gruber S, Beutel J (2012) Kinematics of steep bedrock permafrost. J Geophys Res Earth Surf 117: n/a-n/a. https://doi.org/10.1029/2011jf001981

Hasler A, Gruber S, Font M, Dubois A (2011a) Advective heat transport in Frozen Rock Clefts: conceptual model, laboratory experiments and numerical simulation. Permafrost Periglac Process 22:378–389. https://doi.org/10.1002/ppp.737

Hasler A, Gruber S, Haeberli W (2011b) Temperature variability and offset in steep alpine rock and ice faces. The Cryosphere 5:977–988. https://doi.org/10.5194/tc-5-977-2011

Heckmann T, Bimböse M, Krautblatter M et al (2012) From geotechnical analysis to quantification and modelling using LiDAR data: a study on rockfall in the Reintal catchment, Bavarian Alps, Germany. Earth Surf Process Land 37:119–133. https://doi.org/10.1002/esp.2250

Heckmann T, Haas F, Wichmann V, Morche D (2008) Sediment budget and morphodynamics of an alpine talus cone on different timescales. Zeitschrift für Geomorphologie, Supplementary Issues 52:103–121. https://doi.org/10.1127/0372-8854/2008/0052S1-0103

Heincke B, Green A, Van Der Kruk J, Willenberg H (2006) Semblance-based topographic migration (SBTM): a method for identifying fracture zones in 3D georadar data. Near Surf Geophys 4:79–88

Heincke B, Günther T, Dalsegg E et al (2010) Combined three-dimensional electric and seismic tomography study on the Åknes rockslide in Western Norway. J Appl Geophys 70:292–306. https://doi.org/10.1016/j.jappgeo.2009.12.004

Hermanns RL, Fauqué L, Wilson CGJ (2015) 36Cl terrestrial cosmogenic nuclide dating suggests late pleistocene to early holocene mass movements on the south face of Aconcagua mountain and in the Las Cuevas–Horcones valleys, Central Andes, Argentina, Geological Society, London, Special Publications, 399 (1):345–368. https://doi.org/10.1144/sp399.19

Hermanns R, Redfield T, Bunkholt H et al (2012) Cosmogenic nuclide dating of slow moving rockslides in Norway in order to assess long-term slide velocities. Landslides and engineered slopes: protecting society through improved understanding. Taylor & Francis Group, London, pp 849–854

Hetzel R, Hampel A (2005) Slip rate variations on normal faults during glacial-interglacial changes in surface loads. Nature 435:81–84

Hinchliffe S, Ballantyne CK (1999) Talus accumulation and Rockwall retreat, Trotternish, isle of Skye, Scotland. Scott Geogr J 115:53–70. https://doi.org/10.1080/00369229918737057

Hinchliffe S, Ballantyne CK (2009) Talus structure and evolution on sandstone mountains in NW Scotland. The Holocene 19:477–486. https://doi.org/10.1177/0959683608101396

Hipp T, Etzelmüller B, Westermann S (2014) Permafrost in Alpine Rock Faces from Jotunheimen and Hurrungane, Southern Norway. Permafrost Periglac Process 25:1–13. https://doi.org/10.1002/ppp.1799

Hoek E, Brown ET (1997) Practical estimates of rock mass strength. Int J Rock Mech Min Sci 34:1165–1186. https://doi.org/10.1016/S1365-1609(97)80069-X

Huggel C, Zgraggen-Oswald S, Haeberli W et al (2005) The 2002 rock/ice avalanche at Kolka/Karmadon, Russian Caucasus: assessment of extraordinary avalanche formation and mobility, and application of QuickBird satellite imagery. Nat Hazards Earth Syst Sci 5:173–187

Ivy-Ochs S, Poschinger A, Synal H-A, Maisch M (2009) Surface exposure dating of the Flims landslide, Graubünden, Switzerland. Geomorphology 103:104–112

James MR, Robson S (2012) Straightforward reconstruction of 3D surfaces and topography with a camera: accuracy and geoscience application. J Geophys Res Earth Surf 117:n/a-n/a. https://doi.org/10.1029/2011jf002289

Jia H, Xiang W, Krautblatter M (2015) Quantifying rock fatigue and decreasing compressive and tensile strength after repeated Freeze-Thaw cycles: rock fatigue model. Permafrost Periglac Process 26:368–377. https://doi.org/10.1002/ppp.1857

Johnson BG, Smith JA, Diemer JA (2017) A chronology of post-glacial landslides suggests that slight increases in precipitation could trigger a disproportionate geomorphic response: precipitation increase may be disproportationate to landslide response. Earth Surf Process Land. https://doi.org/10.1002/esp.4168

Kargel JS, Leonard GJ, Shugar DH et al (2016) Geomorphic and geologic controls of geohazards induced by Nepals 2015 Gorkha Earthquake. Science 351:aac8353-aac8353. https://doi.org/10.1126/science.aac8353

Korup O (2005) Large landslides and their effect on sediment flux in South Westland, New Zealand. Earth Surf Process Land 30:305–323

Korup O (2006) Rock-slope failure and the river long profile. Geology 34:45. https://doi.org/10.1130/G21959.1

Korup O, Clague JJ, Hermanns RL et al (2007) Giant landslides, topography, and erosion. Earth Planet Sci Lett 261:578–589

Korup O, McSaveney MJ, Davies TR (2004) Sediment generation and delivery from large historic landslides in the Southern Alps, New Zealand. Geomorphology 61:189–207

Kos A, Amann F, Strozzi T et al (2016) Contemporary glacier retreat triggers a rapid landslide response, Great Aletsch Glacier, Switzerland. Geophys Res Lett 43:12466–12474. https://doi.org/10.1002/2016gl071708

Krautblatter M, Draebing D (2014) Pseudo 3-D *P* wave refraction seismic monitoring of permafrost in steep unstable bedrock. J Geophys Res Earth Surf 119:287–299. https://doi.org/10.1002/2012JF002638

Krautblatter M, Funk D, Günzel FK (2013) Why permafrost rocks become unstable: a rock-ice-mechanical model in time and space. Earth Surf Process Land 38:876–887. https://doi.org/10.1002/esp.3374

Krautblatter M, Hauck C (2007) Electrical resistivity tomography monitoring of permafrost in solid rock walls. J Geophys Res. https://doi.org/10.1029/2006JF000546

Krautblatter M, Leith K (2015) Glacier- and permafrost-related slope instabilities. In: Huggel C, Carey M, Clague JJ, Kaab A (eds) The high-mountain cryosphere. Cambridge University Press, Cambridge, pp 147–165

Krautblatter M, Moser M (2009) A nonlinear model coupling rockfall and rainfall intensity based on a four year measurement in a high Alpine rock wall (Reintal, German Alps). Nat Hazards Earth Syst Sci 9:1425

Krautblatter M, Moser M, Schrott L et al (2012) Significance of rockfall magnitude and carbonate dissolution for rock slope erosion and geomorphic work on Alpine limestone cliffs (Reintal, German Alps). Geomorphology 167:21–34

Krautblatter M, Verleysdonk S, Flores-Orozco A, Kemna A (2010) Temperature-calibrated imaging of seasonal changes in permafrost rock walls by quantitative electrical resistivity tomography (Zugspitze, German/Austrian Alps). J Geophys Res Earth Surf 115:n/a-n/a. https://doi.org/10.1029/2008jf001209

Larsen IJ, Montgomery DR, Korup O (2010) Landslide erosion controlled by hillslope material. Nat Geosci 3:247–251. https://doi.org/10.1038/ngeo776

Leith K, Moore JR, Amann F, Loew S (2014) In situ stress control on microcrack generation and macroscopic extensional fracture in exhuming bedrock. J Geophys Res Solid Earth 119:594–615

Leith KJ (2012) Stress development and geomechanical controls on the geomorphic evolution of alpine valleys

Luethi R, Gruber S, Ravanel L (2015) Modelling transient ground surface temperatures of past rockfall events: towards a better understanding of failure mechanisms in changing periglacial environments. Geografiska Annaler: Series A, Phys Geogr 97:753–767. https://doi.org/10.1111/geoa.12114

Matsuoka N, Murton J (2008) Frost weathering: recent advances and future directions. Permafrost Periglac Process 19:195–210. https://doi.org/10.1002/ppp.620

Matsuoka N, Sakai H (1999) Rockfall activity from an alpine cliff during thawing periods. Geomorphology 28:309–328

McCarroll D, Shakesby RA, Matthews JA (2001) Enhanced rockfall activity during the Little Ice Age: further lichenometric evidence from a Norwegian talus. Permafrost Periglac Process 12:157–164
McColl S, Davies T (2011) Evidence for a rock-avalanche origin for 'The Hillocks'"moraine", Otago, New Zealand. Geomorphology 127:216–224
McColl S, Davies T, McSaveney M (2010) Glacier retreat and rock-slope stability: debunking debuttressing. In: Geologically active: delegate papers 11th congress of the international association for engineering geology and the environment, Auckland, Aotearoa, pp 5–10
McColl ST (2012a) Paraglacial rock-slope stability. Geomorphology 153:1–16
McColl ST (2012b) Paraglacial rockslope stability. PhD Thesis. University of Canterbury, Christchurch, New Zealand
McColl ST (2014) Landslide causes and triggers. In: Shroder JF, Davies TRH (eds) Landslide hazards, risks, and disasters. Academic Press, pp 17–42
McColl ST, Davies TR (2013) Large ice-contact slope movements: glacial buttressing, deformation and erosion. Earth Surf Process Land 38:1102–1115
McColl ST, Davies TR, McSaveney MJ (2012) The effect of glaciation on the intensity of seismic ground motion. Earth Surf Process Land 37:1290–1301
McSaveney M (2002) Recent rockfalls and rock avalanches in Mount Cook National Park, New Zealand. Rev Eng Geol 15:35–70
McSaveney M, Massey C (2013) Did radiative cooling trigger New Zealand's 2007 Young River Landslide? In: Margottini C, Canuti P, Sassa K (eds) Landslide science and practice. Springer, Heidelberg, pp 347–353
Micheletti N, Lambiel C, Lane SN (2015) Investigating decadal-scale geomorphic dynamics in an alpine mountain setting. J Geophys Res Earth Surf 120:2155–2175. https://doi.org/10.1002/2015JF003656
Moore JR, Gischig V, Katterbach M, Loew S (2011) Air circulation in deep fractures and the temperature field of an alpine rock slope. Earth Surf Process Land 36:1985–1996. https://doi.org/10.1002/esp.2217
Moore JR, Sanders JW, Dietrich WE, Glaser SD (2009) Influence of rock mass strength on the erosion rate of alpine cliffs. Earth Surf Process Land 34:1339–1352. https://doi.org/10.1002/esp.1821
Murphy W (2006) The role of topographic amplification on the initiation of rock slopes failures during earthquakes. In: Evans SG, Mugnozza GS, Strom A, Hermanns RL (eds) Landslides from Massive Rock Slope Failure. Springer, Dordrecht, pp 139–154
Murton JB, Peterson R, Ozouf J-C (2006) Bedrock fracture by ice segregation in cold regions. Science 314:1127–1129. https://doi.org/10.1126/science.1132127
Myhra KS, Westermann S, Etzelmüller B (2017) Modelled distribution and temporal evolution of permafrost in steep rock walls along a latitudinal transect in Norway by CryoGrid 2D: permafrost in steep rock walls. Permafrost Periglac Process 28:172–182. https://doi.org/10.1002/ppp.1884
Nichols T (1980) Rebound, its nature and effect on engineering works. Q J Eng Geol Hydrogeol 13:133–152
Nishii R, Matsuoka N (2012) Kinematics of an alpine retrogressive rockslide in the Japanese Alps. Earth Surf Process Land 37:1641–1650. https://doi.org/10.1002/esp.3298
Noetzli J, Gruber S, Kohl T et al (2007) Three-dimensional distribution and evolution of permafrost temperatures in idealized high-mountain topography. J Geophys Res. https://doi.org/10.1029/2006JF000545
Oppikofer T, Jaboyedoff M, Keusen H-R (2008) Collapse at the eastern Eiger flank in the Swiss Alps. Nat Geosci 1:531–535. https://doi.org/10.1038/ngeo258
Ostermann M, Prager C (2014) Major Holocene rock slope failures in the Upper Inn-and Ötz valley region (Tyrol, Austria). In: Kerschner H, Krainer K, Spötl C (eds) From the Foreland to the Central Alps. DEUQUA Excursions Geozon, Berlin, pp 116–126
Ostermann M, Sanders D (2017) The Benner pass rock avalanche cluster suggests a close relation between long-term slope deformation (DSGSDs and translational rock slides) and catastrophic failure. Geomorphology 289:44–59
Otto J, Sass O (2006) Comparing geophysical methods for talus slope investigations in the Turtmann valley (Swiss Alps). Geomorphology 76:257–272
Otto J-C, Dikau R (2004) Geomorphologic system analysis of a high mountain valley in the Swiss Alps. Zeitschrift für Geomorphologie, NF 323–341
Otto J-C, Schrott L, Jaboyedoff M, Dikau R (2009) Quantifying sediment storage in a high alpine valley (Turtmanntal, Switzerland). Earth Surf Process Land 34:1726–1742. https://doi.org/10.1002/esp.1856
Pánek T, Engel Z, Mentlík P et al (2016) Cosmogenic age constraints on post-LGM catastrophic rock slope failures in the Tatra Mountains (Western Carpathians). CATENA 138:52–67. https://doi.org/10.1016/j.catena.2015.11.005
Pánek T, Mentlík P, Engel Z et al (2017) Late Quaternary sackungen in the highest mountains of the Carpathians. Quat Sci Rev 159:47–62. https://doi.org/10.1016/j.quascirev.2017.01.008
Phillips M, Haberkorn A, Draebing D et al (2016) Seasonally intermittent water flow through deep fractures in an Alpine Rock Ridge: Gemsstock, Central Swiss Alps. Cold Reg Sci Technol 125:117–127. https://doi.org/10.1016/j.coldregions.2016.02.010
Phillips M, Wolter A, Lüthi R et al (2017) Rock slope failure in a recently deglaciated permafrost rock wall at Piz Kesch (Eastern Swiss Alps), February 2014. Earth Surf Process Land 42:426–438. https://doi.org/10.1002/esp.3992
Porter SC, Orombelli G (1981) Alpine Rockfall Hazards: recognition and dating of rockfall deposits in the western Italian Alps lead to an understanding of the

potential hazards of giant rockfalls in mountainous regions. Am Sci 69:67–75
Prager C, Zangerl C, Patzelt G, Brandner R (2008) Age distribution of fossil landslides in the Tyrol (Austria) and its surrounding areas. Nat Hazards Earth Syst Sci 8:377–407
Preisig G, Eberhardt E, Smithyman M et al (2016) Hydromechanical rock mass fatigue in deep-seated landslides accompanying seasonal variations in pore pressures. Rock Mech Rock Eng 49:2333–2351. https://doi.org/10.1007/s00603-016-0912-5
Prick A (2003) La désagrégation des roches et les chutes de pierres en milieu de montagne polaire (Longyearbyen, Spitsberg)(Rock weathering and rock falls in polar mountain environment). Bulletin de l'Association de géographes français 80:73–85
Purdie H, Gomez C, Espiner S (2015) Glacier recession and the changing rockfall hazard: implications for glacier tourism. NZ Geogr 71:189–202
Rapp A (1960) Recent development of mountain slopes in Kärkevagge and surroundings, Northern Scandinavia. Geogr Ann 42:65. https://doi.org/10.2307/520126
Ravanel L, Allignol F, Deline P et al (2010) Rock falls in the Mont Blanc Massif in 2007 and 2008. Landslides 7:493–501. https://doi.org/10.1007/s10346-010-0206-z
Ravanel L, Deline P (2011) Climate influence on rockfalls in high-Alpine steep rockwalls: the north side of the Aiguilles de Chamonix (Mont Blanc massif) since the end of the 'Little Ice Age'. The Holocene 21:357–365. https://doi.org/10.1177/0959683610374887
Reznichenko NV, Davies TR, Alexander DJ (2011) Effects of rock avalanches on glacier behaviour and moraine formation. Geomorphology 132:327–338
Reznichenko NV, Davies TR, Shulmeister J, Larsen SH (2012) A new technique for identifying rock avalanche–sourced sediment in moraines and some paleoclimatic implications. Geology 40:319–322
Reznichenko NV, Davies TRH, Winkler S (2016) Revised palaeoclimatic significance of Mueller Glacier moraines, Southern Alps, New Zealand: revised interpretation of Mueller Glacier Moraines, Southern Alps. Earth Surf Process Land 41:196–207. https://doi.org/10.1002/esp.3848
Rode M, Schnepfleitner H, Sass O (2016) Simulation of moisture content in alpine rockwalls during freeze-thaw events: simulation of moisture content in Alpine Rock Walls. Earth Surf Process Land 41:1937–1950. https://doi.org/10.1002/esp.3961
Sanchez G, Rolland Y, Corsini M et al (2010) Relationships between tectonics, slope instability and climate change: cosmic ray exposure dating of active faults, landslides and glacial surfaces in the SW Alps. Geomorphology 117:1–13. https://doi.org/10.1016/j.geomorph.2009.10.019
Sass O (2004) Rock moisture fluctuations during freeze-thaw cycles: preliminary results from electrical resistivity measurements. Polar Geogr 28:13–31. https://doi.org/10.1080/789610157
Sass O (2005a) Rock moisture measurements: techniques, results, and implications for weathering. Earth Surf Process Land 30:359–374. https://doi.org/10.1002/esp.1214
Sass O (2005b) Temporal variability of rockfall in the Bavarian Alps, Germany. Arct Antarct Alp Res 37:564–573. https://doi.org/10.1657/1523-0430(2005)037[0564:TVORIT]2.0.CO;2
Sass O (2006) Determination of the internal structure of alpine talus deposits using different geophysical methods (Lechtaler Alps, Austria). Geomorphology 80:45–58. https://doi.org/10.1016/j.geomorph.2005.09.006
Sass O (2007) Bedrock detection and talus thickness assessment in the European Alps using geophysical methods. J Appl Geophys 62:254–269. https://doi.org/10.1016/j.jappgeo.2006.12.003
Sass O, Krautblatter M, Morche D (2007) Rapid lake infill following major rockfall (bergsturz) events revealed by ground-penetrating radar (GPR) measurements, Reintal, German Alps. The Holocene 17:965–976
Scapozza C, Lambiel C, Baron L et al (2011) Internal structure and permafrost distribution in two alpine periglacial talus slopes, Valais, Swiss Alps. Geomorphology 132:208–221. https://doi.org/10.1016/j.geomorph.2011.05.010
Schneider JF, Gruber FE, Mergili M (2013) Recent cases and geomorphic evidence of landslide-dammed lakes and related hazards in the mountains of Central Asia. In: Landslide science and practice. Springer, Heidelberg, pp 57–64
Selby M (1980) A rock mass strength classification for geomorphic purposes, with tests from Antarctica and New Zealand. Zeit Geomorph, NF 24:31–51
Siewert MB, Krautblatter M, Christiansen HH, Eckerstorfer M (2012) Arctic rockwall retreat rates estimated using laboratory-calibrated ERT measurements of talus cones in Longyeardalen, Svalbard. Earth Surf Process Land 37:1542–1555. https://doi.org/10.1002/esp.3297
Sigurdsson O, Williams RS (1991) Rockslides on the Terminus of "Jokulsargilsjokull", Southern Iceland. Geografiska Annaler Series A, Phys Geogr 73:129. https://doi.org/10.2307/521018
Sims A, Cox SC, Fitzsimons S, Holland P (2015) Seasonal infiltration and groundwater movement in schist bedrock, Southern Alps, New Zealand. J Hydrol 54:33
Stead D, Coggan J, Eberhardt E (2004) Realistic simulation of rock slope failure mechanisms: the need to incorporate principles of fracture mechanics. Int J Rock Mech Min Sci 41:563–568. https://doi.org/10.1016/j.ijrmms.2004.03.100
Stewart IS, Sauber J, Rose J (2000) Glacio-seismotectonics: ice sheets, crustal deformation and seismicity. Quat Sci Rev 19:1367–1389
Stoffel M, Lièvre I, Monbaron M, Perret S (2005) Seasonal timing of rockfall activity on a forested slope at Täschgufer (Swiss Alps)–a dendrochronological approach. Zeitschrift für Geomorphologie 89–106

Strozzi T, Delaloye R, Kääb A et al (2010) Combined observations of rock mass movements using satellite SAR interferometry, differential GPS, airborne digital photogrammetry, and airborne photography interpretation. J Geophys Res. https://doi.org/10.1029/2009JF001311

Strunden J, Ehlers TA, Brehm D, Nettesheim M (2015) Spatial and temporal variations in rockfall determined from TLS measurements in a deglaciated valley, Switzerland. J Geophys Res Earth Surf 120:1251–1273. https://doi.org/10.1002/2014JF003274

Sturzenegger M, Stead D (2009) Close-range terrestrial digital photogrammetry and terrestrial laser scanning for discontinuity characterization on rock cuts. Eng Geol 106:163–182. https://doi.org/10.1016/j.enggeo.2009.03.004

Terzaghi K (1962) Stability of steep slopes on hard unweathered rock. Geotechnique 12:251–270

Turnbull JM, Davies TR (2006) A mass movement origin for cirques. Earth Surf Process Land 31:1129–1148

Van Asch TW, Buma J, Van Beek LP (1999) A view on some hydrological triggering systems in landslides. Geomorphology 30:25–32. https://doi.org/10.1016/S0169-555X(99)00042-2

Vehling L, Rohn J, Moser M (2016) Quantification of small magnitude rockfall processes at a proglacial high mountain site, Gepatsch glacier (Tyrol, Austria). Zeitschrift für Geomorphologie, Supplementary Issues 60:93–108

Viles H, Goudie A, Grab S, Lalley J (2011) The use of the Schmidt Hammer and Equotip for rock hardness assessment in geomorphology and heritage science: a comparative analysis. Earth Surf Process Land 36:320–333. https://doi.org/10.1002/esp.2040

Viles HA (2013) Linking weathering and rock slope instability: non-linear perspectives. Earth Surf Process Land 38:62–70. https://doi.org/10.1002/esp.3294

Whitehouse I (1988) Geomorphology of the central Southern Alps, New Zealand: the interaction of plate collision and atmospheric circulation. Zeitschrift für Geomorphologie NF 69:105–116

Wieczorek GF, Jäger S (1996) Triggering mechanisms and depositional rates of postglacial slope-movement processes in the Yosemite Valley, California. Geomorphology 15:17–31

Willenberg H, Loew S, Eberhardt E et al (2008) Internal structure and deformation of an unstable crystalline rock mass above Randa (Switzerland): Part I—Internal structure from integrated geological and geophysical investigations. Eng Geol 101:1–14

Wirz V, Geertsema M, Gruber S, Purves RS (2016) Temporal variability of diverse mountain permafrost slope movements derived from multi-year daily GPS data, Mattertal, Switzerland. Landslides 13:67–83. https://doi.org/10.1007/s10346-014-0544-3

Wirz V, Schirmer M, Gruber S, Lehning M (2011) Spatio-temporal measurements and analysis of snow depth in a rock face. The Cryosphere 5:893–905. https://doi.org/10.5194/tc-5-893-2011

Zhang T (2005) Influence of the seasonal snow cover on the ground thermal regime: an overview. Rev Geophys. https://doi.org/10.1029/2004RG000157

9 Rockfall at Proglacial Rockwalls—A Case Study from the Kaunertal, Austria

Lucas Vehling, Joachim Rohn and Michael Moser

Abstract

Since the Little Ice Age, high alpine regions have faced rapid glacier melting that has contributed to enhanced rockfall activity at recently deglaciated rockwalls. At the Gepatschferner, Kaunertal (Austria), rockfall activity has been quantified for the past several years using rockfall collector nets, 'natural' rockfall traps and multi-temporal LiDAR. Toppling and sliding activity of large unstable rock blocks, considered as precursors of rockfalls, were monitored by steel tape measurements and electrical crackmeters. The highest rockfall activity was measured at recently deglaciated rockwalls with low rock mass quality, where rockwall back-weathering rates locally exceeded 10 mm/a. Those rates are among the highest ever published. 108 mid- and high-magnitude rockfalls with volumes between 100 and 30,000 m^3 were released between 2006 and 2012. Their scars are clustered in the proglacial high-altitude parts of the Kaunertal. As well, rockwall activity was concentrated in the autumn and winter months.

Keywords

PROSA project · Rockfall rates · Displacement rates · Crackmeter · Rock mass strength

9.1 Introduction and Study Area

Rockfall is an important geomorphic process in high mountains that may be enhanced because of the ongoing environmental changes in the European Alps (Noetzli et al. 2003; Jaboyedoff et al. 2012). Rockfalls and other types of rock slope instabilities are the result of the interaction of complex processes which have been discussed in Chap. 8: Glacio-isostatic adjustment after glacier melting causes stress redistribution in rock masses that leads to rock slope instabilities on different temporal and spatial scales (Ballantyne 2002). As a consequence of global warming, which particularly affect high alpine regions, also the degradation of rock permafrost is an important trigger of rockfalls (Chap. 8; Jaboyedoff et al. 2012). The rapid deglaciation also provokes rockfall activity because rockwalls are freshly exposed to subaerial weathering processes (Matsuoka 2008).

Rockfall volumes may range from small-magnitude single rockfalls that encompass

L. Vehling (✉) · J. Rohn · M. Moser
Friedrich-Alexander University Erlangen-Nürnberg, Erlangen, Germany
e-mail: vehlingl@googlemail.com

T. Heckmann and D. Morche (eds.), *Geomorphology of Proglacial Systems*, Geography of the Physical Environment, https://doi.org/10.1007/978-3-319-94184-4_9

a few kilograms of rock to high-magnitude rockfalls whose volumes exceed 100,000 m^3. There are numerous examples for large rock avalanches and rockslides that thought to be triggered by the late Pleistocene deglaciation (for a compilation of these examples, see Abele 1974; Prager et al. 2008). However, there has been less quantitative data for small-magnitude rockfall volumes at recently deglaciated rock slopes because systematic quantitative investigations are only available for few high alpine areas (Sass 2005; Krautblatter et al. 2012). Consequently, the recent work at the Kaunertal which is presented in this chapter aims to extend the quantitative database for post-Little Ice Age rockfall activity in recently deglaciated areas. It summarizes the most recent results of the rockfall measurements carried out in the framework of the PROSA project (Chap. 1). For these investigations, quantitative rockfall data were gathered by the combination of several methods: rockfall collector nets (Krautblatter and Moser 2009), 'natural' rockfall traps (Vehling et al. 2016a), the evaluation of laserscanning data (Abellán et al. 2014), and field mapping of rockfall deposits.

The upper Kaunertal encompasses 62 km^2 planar area (Fig. 9.1). General information of geology and morphology of the study area is provided in Chap. 1, while here the focus is on bedrock areas: The planar rock face area is 12.9 km^2 (18.6 km^2 real surface area, calculated from a 5 m resolution digital elevation model). From all rock slopes, around 50% are directly adjacent to glaciers or have been covered by glacier ice during the Little Ice Age maximum in 1850. These rock slopes are termed proglacial rockwalls, here. The rockwalls are composed of metamorphic rocks, mainly paragneiss, orthogneiss and amphibolite (Fig. 9.1). Mica schists and magmatic dykes occur in minor spatial extension. All rock faces have between two and four joint sets that originate from the complex geologic history of the Ötztal crystalline unit where the upper Kaunertal belongs to. Furthermore, a dense set of tectonic lineaments crosses the study area.

9.2 Methods

Mass movements at rockwalls were identified and accounted by the combination of several methods: Rockfall rates of small-magnitude rockfalls were determined at 22 locations (Fig. 9.1) that differ in lithology, rock mass quality, deglaciation time, altitude and aspect so that the calculated rockfall rates could be employed for the regionalization of rockfall activity in the whole upper Kaunertal (Chap. 17; Heckmann et al. 2016; Vehling 2016). Mid- and high-magnitude rockfalls were identified catchment wide by using high-resolution airborne laserscanning data (Chap. 1). At some large unstable rock blocks, the displacement patterns were investigated by extensometer measurements to tackle short-term variations of rockwall activity.

9.2.1 Quantification of Small-Magnitude Rockfalls

Small-magnitude rockfall volumes were determined on seven rockfall collector nets (Fig. 9.2). The collector nets are placed directly below the rockwall on the talus slope and fixed by iron hooks. The collector nets were between 84 and 230 m^2 large, gathering rockfall from eight different rock faces (one collector net gathers rockfall from two different rock faces simultaneously) of several hundred to three thousand square meters, each. A 60 cm high steel fence or an artificial sediment rampart of equal height prevented rockfall particles from leaving the net. Furthermore, the net itself damped the bouncing of bigger particles. The gathered particles were mechanical weighted in the field and their volume was calculated assuming a density of 2.7 g/cm^3 for gneiss and 3.1 g/cm^3 for amphibolite (Vehling 2016).

Furthermore, rockfall rates were determined below recently deglaciated rockwalls at 'natural' rockfall traps. In these locations, sharp-angular rockfall particles could be distinguished

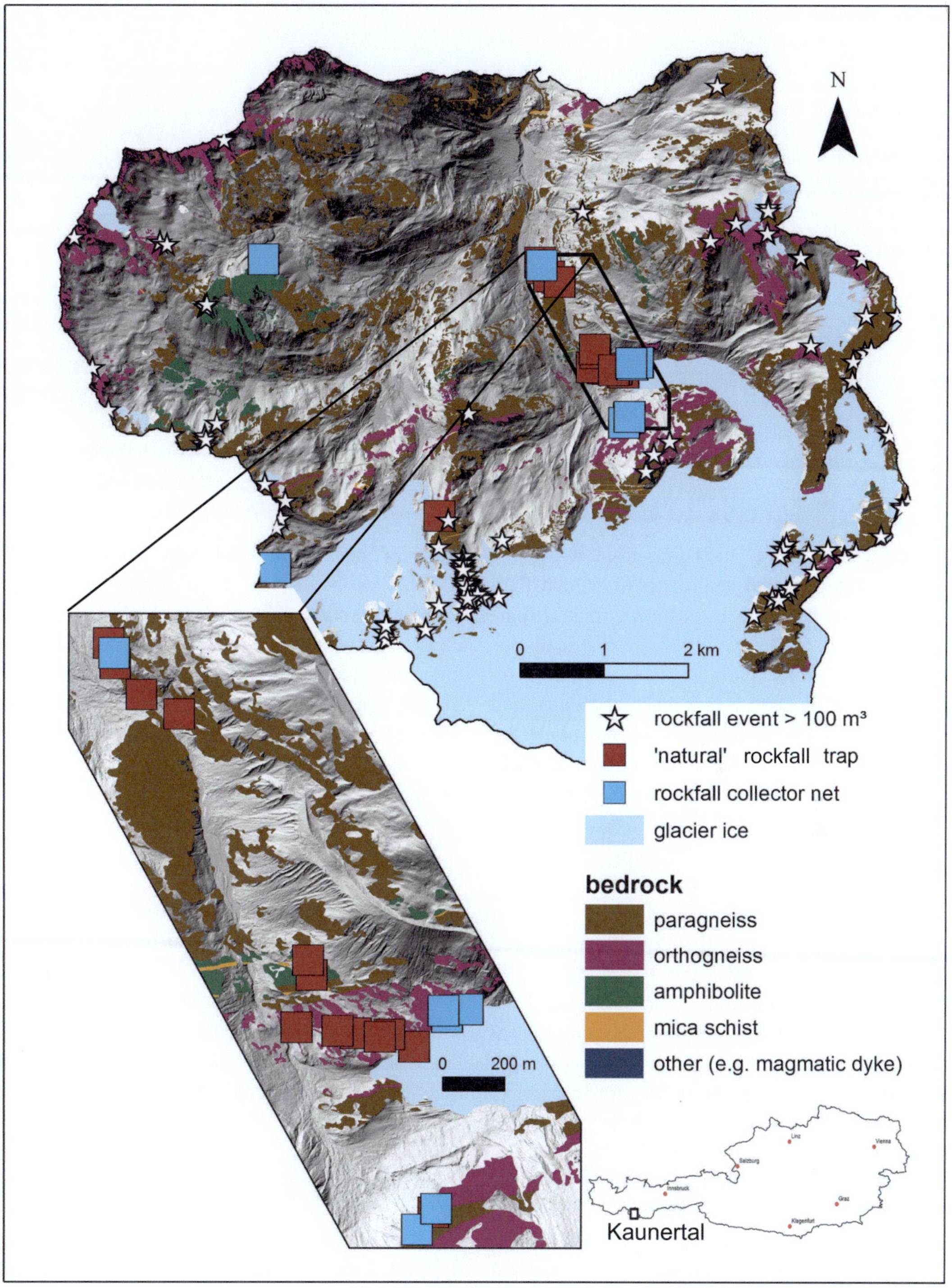

Fig. 9.1 The location and geology of the upper Kaunertal. Information on the rockfall collectors and rockfall traps follow in Sect. 9.2

Fig. 9.2 Rockfall collector net in the proglacial zone (2250 m a.s.l.). The contact-line of net and rockwall is 20 m. The lower part of the rockwall is build up by paragneiss, while in the upper part granitic orthogneiss prevails. So, particles from two different rock faces ('rockwall contributing areas') could be accounted with one collector net

unequivocally from the fine-grained, partly rounded fluvio-glacial sediments on which they were deposited. At those locations, rockfall particles were likewise weighted mechanically or the mass of larger particles was estimated by measuring a-, b-, and c-axis of the particles and employing shape coefficients to calculate the volume (Luckman 1973). The volume of small talus cones was measured by a ruler, taking into account a pore volume of 25% (Sass and Wollny 2001). The 'natural' rockfall traps bridged the gap between the small-magnitude rockfalls, measured at the collector nets and the mid- and high-magnitude rockfalls that could be quantified catchment wide by the evaluation of multi-temporal airborne laser scans (see Sect. 9.2.2). Furthermore, by employing 'natural' rockfall traps, rockfall activity could be quantified at rockwalls where the construction of collectors was not feasible due to dangerous rockfall activity or steep talus slopes. Thus, the rockfall traps were necessary to avoid an inadmissible pre-selection of certain study sites just because of technical reasons.

To determine time- and source area-dependent rockfall rates, the rockfall 'contributing rockwall area' of each collector net and rockfall trap was derived from a 1 m digital elevation model (DEM1) employing the D8-algorithm to the collector net or rockfall trap in SAGA GIS (Conrad et al. 2015). In a further step, only rockwall sections above 40° slope angle were considered as rockfall contributing areas. Due to the well-known deglaciation history of the proglacial zone (e.g. from aerial photographs), the time span of the deposition of the rockfall particles at the rockfall traps could be determined at an annual resolution so that back-weathering rates could be calculated (in mm/a).

9.2.2 Quantification of Mid- and High-Magnitude Rockfalls

Mid- and high-magnitude rockfalls were identified catchment wide by field mapping and quantified by the analysis of multi-temporal DEMs. As those rockfalls left distinct depositional landforms (i.e. big boulders, tongue-shaped talus sheets, irregular surface, low shadow angle), they could be distinguished from the deposits formed by continuous small-magnitude rockfall activity. The deposits of the latter are dominated by smaller particles; talus cones have regular surfaces and a higher shadow angle.

Two airborne LiDAR datasets were employed for the quantification of rockfalls, one from

September 2006 and one from September 2012. While the former one was provided by the Tyrolean government, the latter one was acquired and processed by the PROSA project partners (TU Vienna). For each visually identified rockfall scar or rockfall deposit, a cut–fill analysis between the DEMs from September 2006 and September 2012 was carried out after Wheaton et al. (2010) in ESRI ArcGIS, using the GCD-software package (www.gcd.joewheaton.org). The point clouds had a high point density (up to 14 pt/m^2) so that the volumetric analysis could be carried out on a DEM1. Steep topography at the rock faces and differential snow cover in high altitudes made it difficult to reliable quantify rockfall events below a volume of 100 m^3 catchment wide by LiDAR; despite of the general high accuracy of the LiDAR point clouds.

9.2.3 Extensometer Measurements

The displacement of large unstable rock blocks (up to some thousand m^3) in the proglacial zone was repeatedly measured by a precision steel tape and continuously by electrical crackmeters. The precision steel tape was hooked between two dowels so that the distance between the two dowels could be measured (Moser et al. 2009). The crackmeters were installed at large joints that are caused by slope movements to monitor displacements of unstable rock blocks (Fig. 9.3). Simultaneously, rock temperature was measured by a probe at the crackmeters. The crackmeters, fabricated by GEOKON Inc., were calibrated against thermal dilatation of the steel wire employing the calibration factors provided by GEOKON Inc. (for further information, see Vehling 2016). To rule out that direct sun radiation distorts the distance measurements, only the 4:00 a.m. values were obtained for the analysis of the rock block displacements. Under rough environmental conditions, the crackmeter measurements reached an accuracy of 0.1 mm.

9.2.4 Rock Mass Mapping

Rock mass mapping was carried out by determining the most important discontinuity properties along scanlines after Priest (2012). Uniaxial compressive strength (UCS) of each rock type was estimated by more than 400 Schmidthammer measurements (Barton and Choubey 1977). For rock mass mapping of large and inaccessible rock slopes, it was necessary to sum up several rock outcrops to larger homogenous sections where rock mass properties did not differ noticeably. The geotechnical parameters were rated by rock mass classification schemes (e.g. RMR—Bieniawski 1973; RMS—Selby 1980) in order to characterize the rock masses in a numerical way.

9.3 Results and Discussion

9.3.1 Precursory Processes of Rockfalls

To identify rockfall precursors, this chapter focusses on small-scale rockslides and rock topples affecting proglacial rockwalls. However, large rockslides play also an important role for the initiation of high-magnitude rockfalls or rock avalanches. One prominent example from the Kaunertal is the 'Schwarze Wand,' where 1,200,000 m^3 of rock move with an average velocity of one meter per year toward the glacier (Vehling et al. 2016b).

Examples of small-scale rock topples are some unstable rock blocks, located directly in front of the glacier tongue. The location has been deglaciated 15 years ago. Additionally, the rock slope was formerly undercut by the proglacial Fagge River that drains the main glacier tongue and formed a steep gorge. The unstable rock blocks are located at the shoulders of this gorge (Fig. 9.3). Those boulders encompass some 100 m^3 up to a few 1000 m^3 and are prone to failure because they are almost fully detached

Fig. 9.3 Outer part of an unstable rock block, instrumented by electrical crackmeters. The rock mass body is located at the shoulder of a 15 m deep gorge in the proglacial area of the Gepatsch glacier. Note for scale: The crackmeters are 0.5 m long

from the rockwall. Their dynamics were detected by repeated steel tape measurements. Further, one unstable frontal block of the rock mass body was instrumented by two electric extensometers (Fig. 9.3).

The crackmeter measurements revealed quite slow displacements of 4 mm (crackmeter 1) during 20 months (Fig. 9.4). The displacement rate reached a maximum during the autumn and the early winter months while between December and March, the rock was frozen and no significant movement occurred.

Some of the precision steel tape measurement results at the gorge and additionally from another location nearby are presented in Fig. 9.5: While 1B, 1C, and 3 hold quite low displacement rates, 1A and 2 showed more pronounced dynamics.

Summing up the displacement patterns of all monitored rock topples (in total 10 measurement tracks) in the proglacial area and combining those with the crackmeter measurements, two main tendencies were revealed:

First, the highest displacement rates were reached during the autumn and early winter months (Fig. 9.4). This was also valid for some of the steel tape measurements (Fig. 9.5). The exact time of the displacement at the steel tape

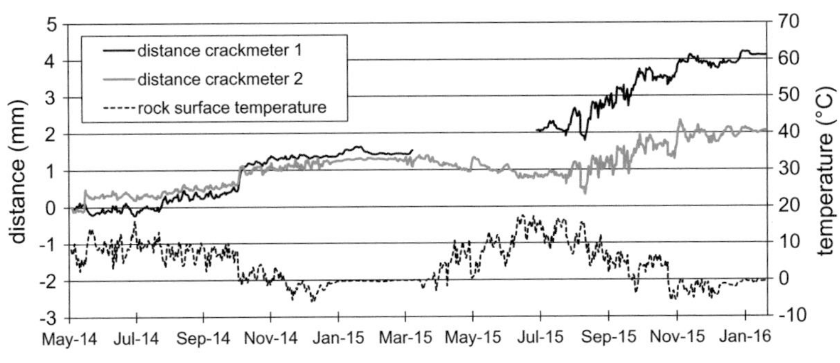

Fig. 9.4 Displacement of the unstable rock block (Fig. 9.3) measured by two electrical crackmeters (left *y*-axis). Rock surface temperature is indicated on the right y-axis

Fig. 9.5 Precision steel tape measurement tracks from five different rock blocks

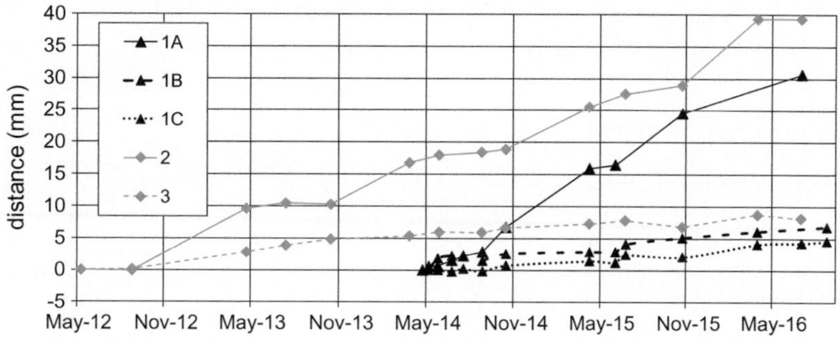

measurement tracks was not resolvable because the locations where inaccessible during the winter months because of a 2 m high snow cover of the valley. The displacement at the steel tape measurement tracks might also took place entirely during the short snowmelt period in the spring as it was reported from comparable studies (Moser et al. 2009). We attribute the autumn and early winter maximum to the first snow cover that leads to pressure on the unstable blocks and in cracks between the blocks and the rockwall. Further, cold climate weathering processes are especially efficient during the cold wet autumn climate (Matsuoka 2008). Second, the overall displacement rates were rather slow (mainly some mm per year) so that mid-magnitude rockfalls will need a considerable time span of several years to initiate at the monitored locations. The exception may be track 1A that shows the highest displacement rate (to the end more than 15 mm per year) and is certainly right before failure so that the location was not approachable in summer 2016 anymore.

9.3.2 Small-Magnitude Rockfalls

Rockfall intensities and corresponding back-weathering rates were calculated for 22 different rockwalls above the collector nets and 'natural' rockfall traps. The results are summarized in Fig. 9.6, where altitude, slope aspect, rock type and back-weathering rate of each investigated rockwall are indicated.

Rockfall intensity varied over four orders of magnitude (back-weathering rate: 0.01–14.0 mm/a). The highest back-weathering rates approached more than 10 mm/a at some rockwalls in orthogneiss, directly in front of the Gepatsch glacier tongue. These locations showed extraordinary high rockfall intensity compared to previous studies (see Sect. 9.3.4). However, altitude and slope aspect did not determine rockfall intensity at the 22 investigated rockwalls (Fig. 9.6), as there can be seen no relationship between those factors. The highest values were measured at locations with extremely unfavorable rock mass properties that were located close to a tectonic fault (south- and north-facing orthogneiss rockwalls, mica schist rockwalls in Fig. 9.6). So there was a clear correlation between rockfall intensity and rock mass properties (Fig. 9.7a–c).

Figure 9.7a shows that there is an inverse exponential relationship between rock mass strength (RMS) and back-weathering rate. Only some locations were far outside the regression model, mainly because of special conditions at those locations (e.g. fluvial undercutting at S6). Similar results yielded the RMR classification (Bieniawski 1973). The correlation between GSI and back-weathering (Fig. 9.7b) was quite weak as the state of weathering of the rock mass is one of only two factors in the GSI. At the recently deglaciated rockwalls, the state of weathering was generally low ('not weathered'), so that GSI was quite high. Thus, the parameter weathering was not suitable to classify rock masses in our study area. Basically, the methodology of the

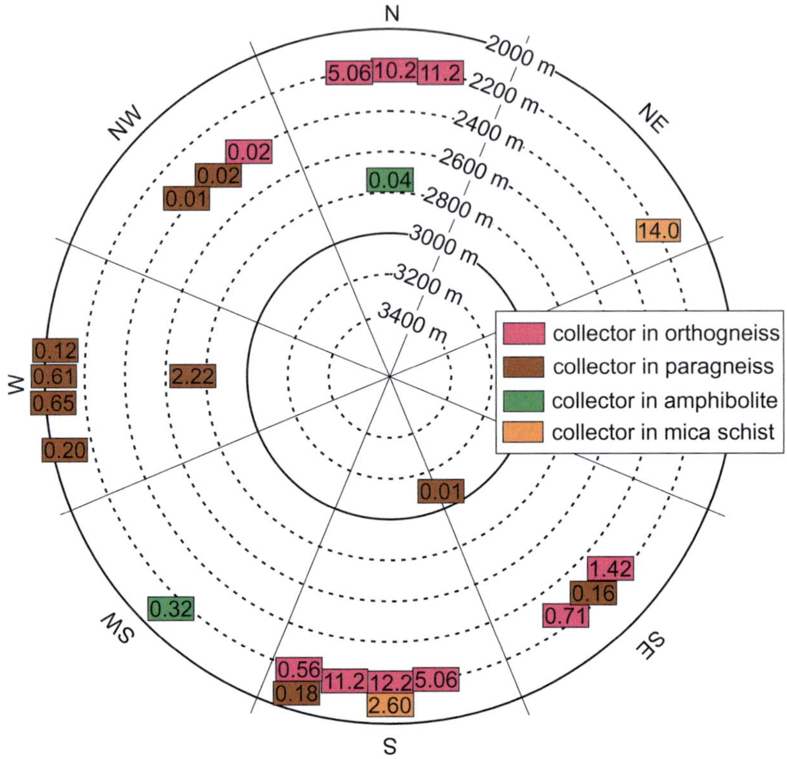

Fig. 9.6 Location (slope aspect, altitude) and rock type of the rockwalls at the 22 rockfall collectors. The number inside the colored rectangles is the average back-weathering rate in mm/a. Two collectors (5.06, 11.2) are inside a steep gorge so that rockfall from both, the north- and the opposite south-facing rockwall is deposited in one rockfall trap

rock mass classifications prevented a better correlation, as common classifications also imply parameters that are of minor importance for contemporary rockfall intensity. For example, rockfall intensity was more or less insensitive to the uniaxial compressive strength (UCS—Fig. 9.7d) of intact rock surfaces. UCS contributes 20% to the RMS and RMR rock mass classification. In contrast, joint set parameters were the most essential drivers of rockfall intensity (Moore et al. 2009), but they contribute less than two thirds to the points in the classification systems. Joint spacing is thought to be the most relevant geotechnical factor for slope stability in previous studies (e.g. Selby 1980; Sass 2005) and is thought to be especially important in the polymetamorphic rocks of the upper Kaunertal. Consequently, joint spacing reaches the best correlation of the single geotechnical joint properties in the data set (Fig. 9.7c).

While there was surprisingly no correlation between slope inclination (calculated from one meter DEM) and back-weathering (Fig. 9.7e), we found a clear connection between back-weathering rate and the time of the deglaciation of the rockwall (Fig. 9.7f). Indeed, only rank order correlation was possible to analyze the classified dataset, but definitely the most recent deglaciated rockwalls delivered the largest amount of rockfall, while at non-proglacial sites the quantities were least, as the non-proglacial locations showed low rates between 0.01 and 0.04 mm/a. We explain this by the dynamic equilibrium between rock mass quality and rockwall inclination on the one hand (Selby 1982) and rockfall activity at the non-proglacial rockwalls on the other: All parts of rock that were prone to failure were mobilized during the Holocene, so that only 'base level' rockfall activity takes place at the non-proglacial sites ('exhaustion effect'—Ballantyne 2002). On the opposite, the proglacial rockwalls showed rates entirely above 0.1 mm/a and approached more than 10 mm/a at some locations that became ice-free in the recent years. Those rockwalls are still adjusting to the recent environmental

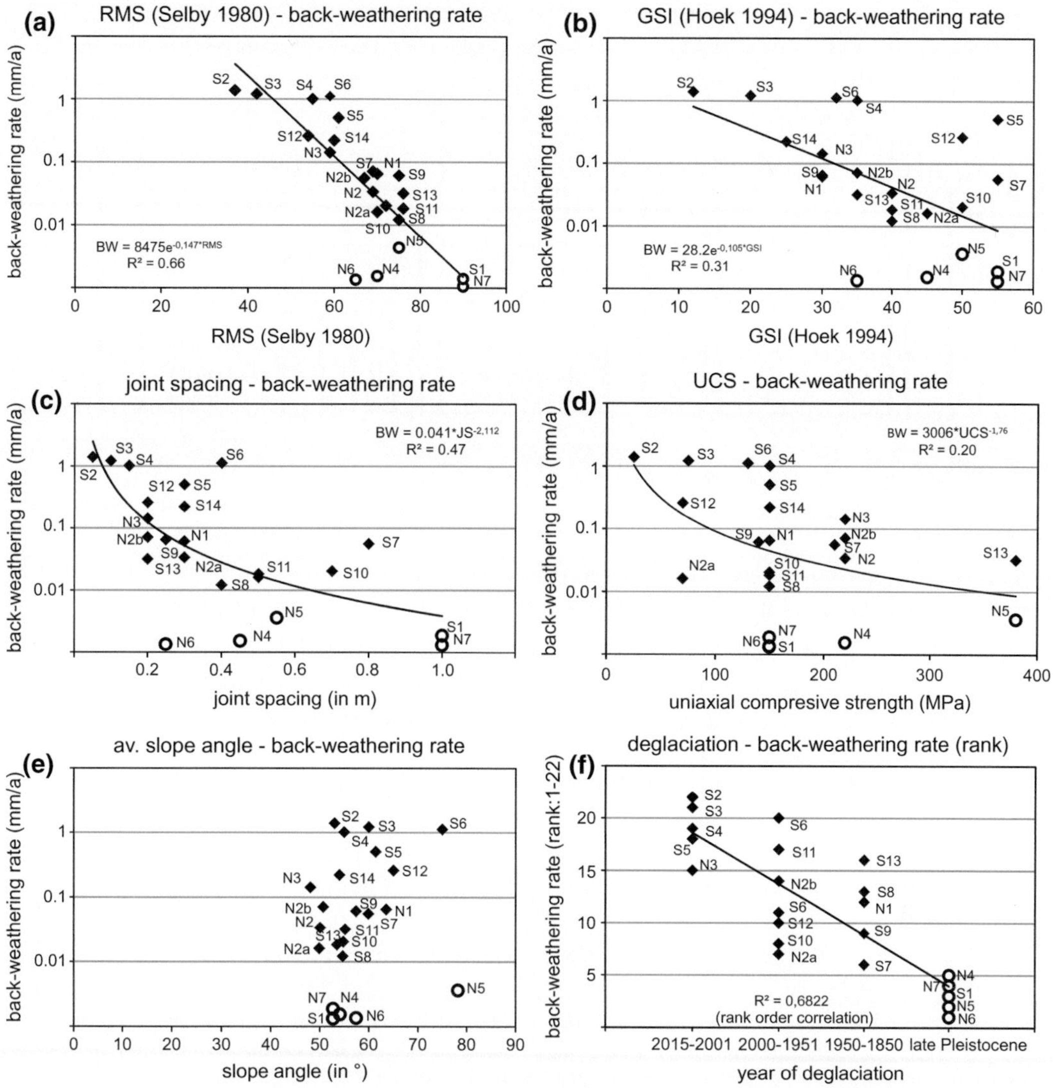

Fig. 9.7 Correlations between back-weathering rate and RMS (**a**), GSI (**b**), joint spacing (**c**), UCS (**d**), slope angle (**e**) and the year of deglaciation (**f**). Squares: Proglacial locations; circles: non-proglacial locations

conditions that are dominated by glacier melting so that instable parts of the rockwall were removed soon after the deglaciation by gravitational activity.

Beside the huge spatial variations, our measurements traced also the intra-annual fluctuations of rockfall intensity over more than three years (Fig. 9.8). Figure 9.8 shows that the back-weathering rate was highest during the winter months, while in the summer months a declining back-weathering rate prevailed—except for summer 2015. However, the nets were inaccessible during the winter because of snow cover so that we could not reveal rockfall activity on a monthly basis throughout the entire year. The unequal distribution of rockfall intensity is related to weather conditions, similar to the observations that were made in Sect. 9.3.1: The winter maximum is certainly because of cold climate weathering processes that disintegrate rock particles from intact rock (Matsuoka 2008). These particles are subsequently mobilized by

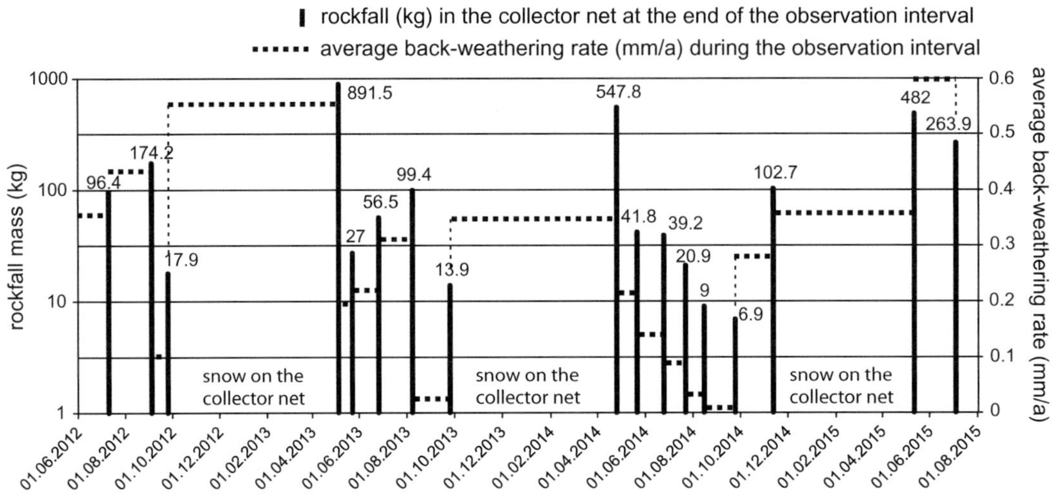

Fig. 9.8 Intra-annual rockfall variability on collector net 2 during three years (modified from Vehling et al. 2016a). Black columns indicate the rockfall masses that are collected from the net at the end of each observation timespan (left *y*-axis). The horizontal dotted lines represent the average back-weathering rate during each observation time (right *y*-axis)

small snow avalanches during thawing episodes (Vehling 2016). This process is most efficient during the spring months March and April. Evidences could be witnessed on talus cones during field work in the late April and similar observations were reported by Matsuoka and Sakai (1999). The declining back-weathering rate during the summer months was measured despite of several rainstorms in July and August (Vehling et al. 2016a). However, the rainstorms never exceeded 5 mm/15 min and thus might be incapable to mobilize rockfall particles from the rockwall (Krautblatter and Moser 2009). Furthermore, there are no ledges on the rockwall that could have served as intermediate storages which could be depleted during rainfalls. We conclude that there is also an 'exhaustion effect' on the intra-annual timescale: During the course of the summer, no rockfall particles remain in unstable positions that could be mobilized by moderate rainfall events in the late summer before unstable pebbles and blocks are provided for failure during the next winter.

9.3.3 Mid- and High-Magnitude Rockfalls

Overall, 108 mid- and high-magnitude rockfall events with a volume greater than 100 m^3 were released in the study area in six years between September 2006 and September 2012 (Fig. 9.9). More than 80% of the events took place at proglacial rockwalls (see also Fig. 9.1 for the spatial distribution of mid- and high-magnitude rockfalls). The rockfalls clustered in high altitudes above 2800 m (Fig. 9.9). This suggests that permafrost degradation is a significant factor for mid- and high-magnitude rockfalls in the upper Kaunertal as those altitudes are subject to rock permafrost (Boeckli et al. 2012). Nearly, every event was released at rockwalls that are at least possibly underlain by permafrost. Furthermore, north-east to west facing rockwalls were especially affected by mid- and high-magnitude rockfalls, mainly because of the enhanced presence of permafrost at these slopes. Most frequently, rockfalls were triggered in paragneiss (brown stars in Fig. 9.9): Normalized to

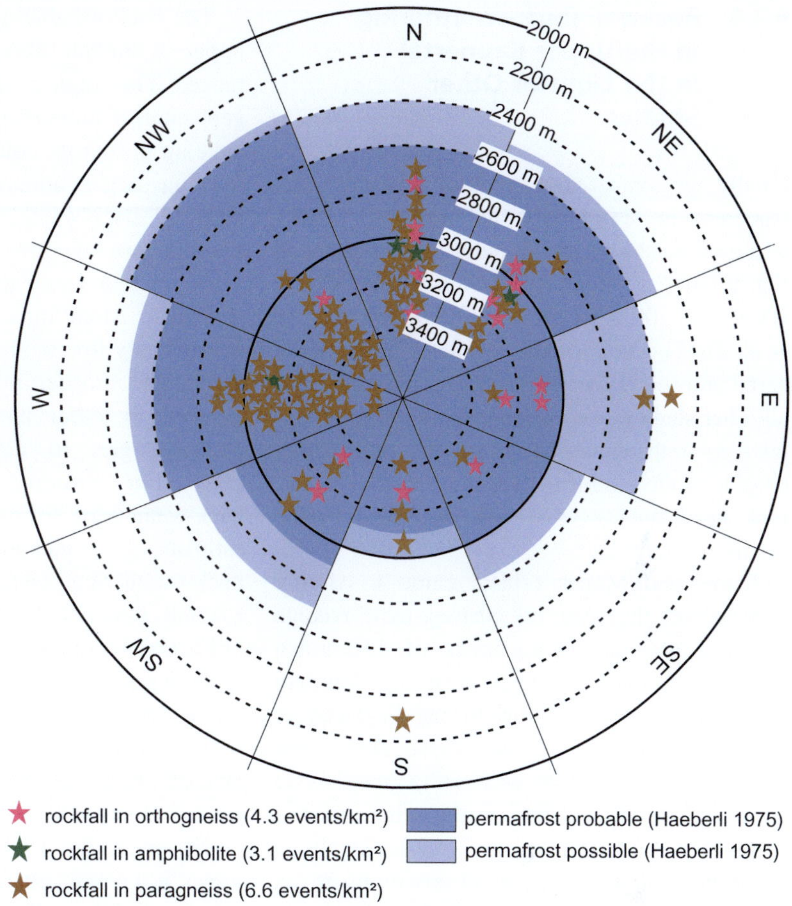

Fig. 9.9 Rockfall events (>100 m^3) in the upper Kaunertal between 2006 and 2012 respect to altitude, aspect and rock type of their scars

the rockwall area, there is a density of 6.6 events per km^2 in paragneiss. The event density in orthogneiss (4.3 events/km^2) and amphibolite (3.1 events/km^2) is smaller. The reason for the unequal distribution is that especially the paragneisses are dissected by large-spacing lineaments which serve as failure planes for bigger rock blocks. Those lineaments frequently hold thin zones of mica schist where shear strength is quite low (Engl et al. 2008). Further, the joint set that evolved parallel to the foliation is most pronounced in the paragneisses and thus favors mid-magnitude rockfalls. However, this tendency was not visible at the small-magnitude rockfalls so that this factor might be more relevant for larger rockfalls (Abele 1974).

The biggest single rockfall event between 2006 and 2012 had a volume of 30,000 m^3, but the majority of the events were between 100 and 500 m^3. The dominance of mid-magnitude rockfalls and the absence of large rockfalls >30,000 m^3 is on the one hand caused by the short observation timespan. There is also no trace of a rockfall event larger than 30,000 m^3 since the Little Ice Age (Vehling 2016). So, on the other hand it was concluded that the dense joint sets in the polymetamorphic gneisses are responsible for the absence of higher magnitude rockfalls, because unstable rock masses rather tend to fail more frequently as mid-magnitude rockfalls (Abele 1974).

9.3.4 Rockwall Back-Weathering in the Upper Kaunertal in the Light of Other Studies

Finally, we compared our rockfall measurements (squares in Fig. 9.10) to rockwall back-weathering rates from 43 other studies (bars in Fig. 9.10) (e.g. Rapp 1960; Curry and Morris 2004; Sass 2005; Moore et al. 2009; Krautblatter et al. 2012). Some of the locations in the Kaunertal showed higher back-weathering rates than all other studies. We attribute the higher rockfall intensity to the enhanced morphodynamics in the proglacial environment that significantly reduce rock mass quality as it was discussed in the sections before.

Curry and Morris (2004) came to similar conclusions; they determined long-term average rockwall back-weathering rates of 1.23 mm/a (1.01–2.44 mm/a) at steep cliffs in Southern Wales during the late Pleistocene deglaciation from talus volumes. Today, those cliffs were merely active as back-weathering rates were about 0.01 mm/a and thus in the same order of magnitude as at the non-proglacial rockwalls in the Kaunertal. Comparable observations in an alpine environment were also made by Luckman and Fiske (1995) who noticed that contemporary rockfall rates are in a clear disproportion to talus volumes in formerly glaciated valleys. Our study provides new data from numerous freshly deglaciated rockwalls that explains the disproportion.

The back-weathering rates determined in the upper Kaunertal also cover a much broader range: The higher range is caused by the geotechnical properties of the rock masses that have up to four pronounced joint sets: Instead of forming larger homogenous zones of similar rockwall back-weathering rates, the polymetamorphic gneisses are subject to weathering and mass wasting along predefined linear zones of weakness. Along those zones, rock mass quality is especially low so that rockfall is concentrated there. The broader range is also methodically induced as other authors employed different methodologies to determine back-weathering rates. For example studies that calculate back-weathering from talus cones tend to overestimate catchment-wide rockwall back-weathering because they only measure rockfall intensity below generally active rock cliffs. On the other hand, those studies integrate rockfall rates over thousands of years so that they fail to tackle rockfall peaks that happen under rockfall-favoring conditions (for a comprehensive discussion of these problems see Krautblatter and Dikau 2007).

However, the back-weathering rates in the upper Kaunertal are certainly even higher than measured at the collector nets and rockfall traps: Those measurements do not incorporate the numerous mid- and high-magnitude rockfalls. Adding mid- and high-magnitude rockfall volumes the catchment-wide denudation rate is even much higher (Chap. 17; Krautblatter et al. 2012; Vehling 2016).

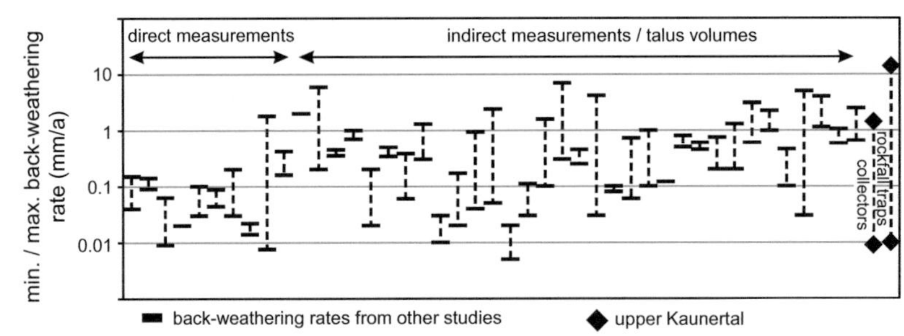

Fig. 9.10 Rockwall back-weathering rates from the literature in comparison to this study. The top and bottom of the bars indicate the maximum and minimum value reported in each publication

9.4 Conclusions

This study provides new quantitative insights into rockwall back-weathering in proglacial environments. It showed that back-weathering rates by small-magnitude rockfalls on recently deglaciated slopes are locally much higher than reported from other environments in other studies. The highest values approach more than 10 mm/a and rates cover a huge range (0.01–14.0 mm/a) that is mainly caused by the small-scale variations of rock mass quality. Consequently, Rock Mass Strength (Selby 1980) correlates quite well with rockfall intensity ($r^2 = 0.68$).

Mid- and high-magnitude rockfalls (100–30,000 m^3) cluster in high-altitude regions that are affected by degrading permafrost. Paragneisses are especially prone to mid-magnitude rockfalls because of their marked foliation and continuous large-spacing lineaments.

Extensometer measurements and short-term observations over three years at collector nets revealed that rockwall activity is concentrated during the early winter and early spring months, frequently triggered by small snow avalanches in steep couloirs and on small ledges.

Acknowledgements The research work was carried out in the PROSA-joint (High-resolution measurements of morphodynamics in rapidly changing PROglacial Systems of the Alps) project funded by the Deutsche Forschungsgemeinschaft (DFG; grant numbers RO 2211/5-1, RO 2211/5-2, RO 2211/5-3).We thank Philipp Glira (TU Vienna) and Ludwig Hilger (University of Eichstätt-Ingolstadt) for the excellent preparation of the airborne LiDAR data and Harald Meier, Markus Schleier, Renneng Bi, Johannes Wiedenmann and Ingvar Krieger for their help during the construction of the rockfall collector nets. Further, we acknowledge Tobias Heckmann and Samuel McColl for the suggestions to revise the manuscript and the good ideas to improve it. Finally, we acknowledge the TiWAG hydropower company for granting free access to the mountain road and the Tyrolean government for generously providing airborne LiDAR data of the year 2006.

References

Abele G (1974) Bergstürze in den Alpen. Wissenschaftliche Alpenvereinshefte, Number 25, Munich

Abellán A, Oppikofer T, Jaboyedoff M et al (2014) Terrestrial laser scanning of rock slope instabilities: state of science. Earth Surf Process Land 39:80–97. https://doi.org/10.1002/esp.3493

Ballantyne CK (2002) Paraglacial geomorphology. Quat Sci Rev 21:1935–2017. https://doi.org/10.1016/S0277-3791(02)00005-7

Barton N, Choubey V (1977) The shear strength of rock joints in theory and practice. Rock Mech Felsmechanik Méchanique des Roches 10:1–54. https://doi.org/10.1007/BF01261801

Bieniawski Z (1973) Engineering classification of jointed rock masses

Boeckli L, Brenning A, Gruber S, Noetzli J (2012) Permafrost distribution in the European Alps: calculation and evaluation of an index map and summary statistics. The Cryosphere 6:807–820. https://doi.org/10.5194/tc-6-807-2012

Conrad O, Bechtel B, Bock M et al (2015) System for Automated Geoscientific Analyses (SAGA) v. 2.1.4. Geoscientific Model Dev 8:1991–2007. https://doi.org/10.5194/gmd-8-1991-2015

Curry AM, Morris CJ (2004) Lateglacial and Holocene talus slope development and rockwall retreat on Mynydd Du, UK. Geomorphology 58:85–106. https://doi.org/10.1016/S0169-555X(03)00226-5

Engl D, Fellin W, Zangerl C (2008) Scherfestigkeiten von Scherzonengesteinen-Ein Beitrag zur geotechnischen Bewertung von tektonischen Störungszonen und Gleitzonen von Massenbewegungen. Bulletin für angewandte Geologie 13:63–81

Haeberli W (1975) Untersuchungen zur Verbreitung von Permafrost zwischen Flüelapass und Piz Grialetsch (Graubünden). Versuchsanstalt für Wasserbau, Hydrologie und Glaziologie an der ETH

Heckmann T, Hilger L, Vehling L, Becht M (2016) Integrating field measurements, a geomorphological map and stochastic modelling to estimate the spatially distributed rockfall sediment budget of the Upper Kauner Valley, Austrian Central Alps. Geomorphology 260:16–31. https://doi.org/10.1016/j.geomorph.2015.07.003

Jaboyedoff M, Derron M-H, Jakubowski J et al (2012) The 2006 Eiger rockslide, European Alps. In: Clague JJ, Stead D (eds) Landslides. Cambridge University Press, Cambridge, pp 282–296

Krautblatter M, Dikau R (2007) Towards a uniform concept for the comparison and extrapolation of rockwall retreat and rockfall supply. Geografiska Annaler, Ser A Phys Geogr 89:21–40. https://doi.org/10.1111/j.1468-0459.2007.00305.x

Krautblatter M, Moser M (2009) A nonlinear model coupling rockfall and rainfall intensity based on a four year measurement in a high Alpine rock wall (Reintal, German Alps). Nat Hazards Earth Syst Sci 9:1425

Krautblatter M, Moser M, Schrott L, Wolf J, Morche D (2012) Significance of rockfall magnitude and carbonate dissolution for rock slope erosion and geomorphic work on Alpine limestone cliffs (Reintal, German Alps). Geomorphology 167:21–34

Luckman B, Fiske C (1995) Estimating long-term rockfall accretion rates by lichenometry. In: Slaymaker O (ed) Steepland geomorphology. Wiley, Hoboken, pp 221–254

Luckman BH (1973) Scree slope characteristics and associated geomorphic processes in Surprise Valley, Jasper National Park. McMaster University, Alberta

Matsuoka N (2008) Frost weathering and rockwall erosion in the southeastern Swiss Alps: Long-term (1994–2006) observations. Geomorphology 99:353–368. https://doi.org/10.1016/j.geomorph.2007.11.013

Matsuoka N, Sakai H (1999) Rockfall activity from an alpine cliff during thawing periods. Geomorphology 28:309–328

Moore JR, Sanders JW, Dietrich WE, Glaser SD (2009) Influence of rock mass strength on the erosion rate of alpine cliffs. Earth Surf Process Land 34:1339–1352. https://doi.org/10.1002/esp.1821

Moser M, Wunderlich T, Meier H (2009) Kinematische Analyse der Bergzerreißung Hornbergl–Reutte (Tirol). Jahrbuch der Geol. Bundesanstalt 149.1. Wien

Noetzli J, Hoelzle M, Haeberli W (2003) Mountain permafrost and recent Alpine rock-fall events: a GIS-based approach to determine critical factors. In: Proceedings of the 8th international conference on permafrost, pp 827–832

Prager C, Zangerl C, Patzelt G, Brandner R (2008) Age distribution of fossil landslides in the Tyrol (Austria) and its surrounding areas. Nat Hazards Earth Syst Sci 8:377–407

Priest SD (2012) Discontinuity analysis for rock engineering. Springer Science & Business Media, Berlin

Rapp A (1960) Recent development of mountain slopes in Kärkevagge and surroundings, Northern Scandinavia. Geogr Ann 42:65. https://doi.org/10.2307/520126

Sass O (2005) Spatial patterns of rockfall intensity in the Northern Alps

Sass O, Wollny K (2001) Investigations regarding Alpine talus slopes using ground-penetrating radar (GPR) in the Bavarian Alps, Germany. Earth Surf Process Land 26:1071–1086. https://doi.org/10.1002/esp.254

Selby M (1980) A rock mass strength classification for geomorphic purposes, with tests from Antarctica and New Zealand. Zeit Geomorph, NF 24:31–51

Selby M (1982) Controls on the stability and inclinations of hillslopes formed on hard rock. Earth Surf Process Land 7:449–467

Vehling L (2016) Gravitative Massenbewegungen an alpinen Felshängen-Quantitative Bedeutung in der Sedimentkaskade proglazialer Geosysteme

Vehling L, Rohn J, Moser M (2016a) Quantification of small magnitude rockfall processes at a proglacial high mountain site, Gepatsch glacier (Tyrol, Austria). Zeitschrift für Geomorphologie, Supplementary Issues 60:93–108

Vehling L, Baewert H, Glira P, Moser M, Rohn J, Morche D (2016b) Quantification of sediment transport by rockfall and rockslide processes on a proglacial rock slope. Kauner Valley, Austria

Wheaton JM, Brasington J, Darby SE, Sear DA (2010) Accounting for uncertainty in DEMs from repeat topographic surveys: improved sediment budgets. Earth Surf Process Land 35:136–156

GCD 4.0—Geomorphic change detection software. http://gcd.joewheaton.org/downloads/older-versions/gcd-4-0. Accessed 5 Sept 2017

Glacial Sediment Stores and Their Reworking

10

Philip R. Porter, Martin J. Smart
and Tristram D. L. Irvine-Fynn

Abstract

In the light of heightened geomorphological activity associated with progressive deglaciation in alpine regions, the storage of sediments within and flux of sediments through the proglacial zone represents an increasingly important area for contemporary geomorphological and sedimentological study. The term 'paraglacial' is used to describe this heightened geomorphological activity, and the general model of paraglacial basin sediment yield is one of initial increase as deglaciation commences, followed by progressive decline during and after deglaciation. Against the backdrop of this paraglacial model, in this chapter we consider storage, release and reworking of sediments from lateral and forefield slopes in the proglacial zone. The propensity of lateral slope units to stand stable at steep angles might indicate that such units represent a long-term store of sediments. However, their genetic complexity and geomorphological evidence of sediment reworking, in the form of deep gullies and associated debris cones and fans, would suggest that such units can yield sediments dependant on multiple climatic, geomorphical and biogeomorphical factors. Similarly, the extent to which the generally lower-angled forefield slopes act as a source or sink of sediments is subject to conjecture. Although the forefield is regarded as the most dynamic component of the alpine sediment flux system, largely due to the efficacy of fluvial action, the great diversity of forms and associated genetic complexity, together with the operation of time-variant processes, will likely add great variability to long-term patterns of basin sediment yield. In addition to fluvial activity and associated sediment mobilisation and redistribution, processes such as permafrost degradation, melt of forefield buried ice and associated slumping and debris flowage offer additional sediment release mechanisms that may punctuate the often-assumed uni-directional decline in sediment yield associated with progressive deglaciation. As meltwater discharge declines, after the so-called 'deglaciation discharge dividend' peaks, progressive eluviation of fines and consequent armouring of previously abundant sediment supply areas will likely lead to overall declining sediment yields with time, enhanced by progressive vegetation colonisation. However,

P. R. Porter (✉) · M. J. Smart
School of Life and Medical Sciences, University of Hertfordshire, Hatfield, Hertfordshire AL10 9AB, UK
e-mail: p.r.porter@herts.ac.uk

T. D. L. Irvine-Fynn
Department of Geography and Earth Sciences, Aberystwyth University, Aberystwyth SY23 3DB, UK

T. Heckmann and D. Morche (eds.), *Geomorphology of Proglacial Systems*,
Geography of the Physical Environment, https://doi.org/10.1007/978-3-319-94184-4_10

this net decline will inevitably be punctuated by stochastic geomorphological events. While uncertainty therefore exists concerning the detailed timescales of sediment release associated with deglaciation, contemporary progressive deglaciation offers an unparalleled opportunity to directly observe the genesis of deglaciation landforms, their modification and associated sediment fluxes and fluctuations in basin-scale sediment storage and release.

Keywords
Sediment storage · Sediment supply
Lateral slopes · Forefield · Moraine
Paraglacial

10.1 Introduction

Understanding the storage and flux of sediments through the proglacial zone of deglaciating catchments is an important topic within glacial geomorphology, primarily as there is great uncertainty surrounding the timescales that sediments are stored and released, which has implications for fully understanding landform genesis and the response of landforms and glaciated catchments to current and forecast environmental changes. The steep slopes found in alpine glacial environments, combined with associated potential for intense geomorphological activity, offer favourable conditions for the release, reworking and storage of glacial sediments and play an important part in the geomorphic coupling and overall connectivity of the down-valley sediment cascade (Cavalli et al. 2013; Heckmann and Schwanghart 2013; Carrivick and Heckmann 2017; Chap. 16). It is no surprise, therefore, that processes acting on hillslopes have formed the focus for much research investigating the sedimentary response to deglaciation.

The concept of a so-called paraglacial period of enhanced geomorphological activity conditioned by glaciation (or indeed deglaciation) originally detailed by Church and Ryder (1972) and subsequently developed by several authors, most notably Ballantyne (2002a), has received much attention in recent years, as the geomorphological consequences of deglaciation become ever more apparent. Hillslopes and associated slope processes in glaciated regions are likely to play a key role in that heightened geomorphological activity and associated release, reworking and subsequent redeposition, or storage, of glacial sediments. However, any sedimentary or 'paraglacial', response to deglaciation is complex, and different landsystems will respond at differing rates and spatial scales (Ballantyne 2002b) and this is certainly the case with hillslopes, which can both store and release sediments at the catchment scale, with potentially significant spatial and temporal variability. Hillslopes in deglaciating environments also demonstrate the attributes of both primary and secondary paraglacial systems. Primary paraglacial systems are those where sediment release is directly conditioned by glaciation (e.g. sediment-mantled slopes), while secondary paraglacial systems are those where rates of sediment release are additionally controlled by reworking of paraglacial sediment stores [e.g. debris cones and fans at the base of drift-mantled slopes (Ballantye 2002b)]. The presence, operation and interaction of these primary and secondary systems inevitably add complexity to the deglaciation response.

Direct glacial influence on sediment release and reworking is likely to decrease as distance down valley increases, but smaller glaciers have been observed to play an important role in efficiently coupling slope and fluvial processes (e.g. Lukas et al. 2005), ensuring that sediment fluxes from glaciated alpine catchments are dominated by fluvial processes (e.g. Orwin et al. 2010; Carrivick et al. 2013; Chaps. 12 and 13). However, critical to that fluvial reworking and redistribution is the supply of sediments from adjacent slope units (Chap. 11) and the degree of connectivity between slope and fluvial systems, both of which will also vary in space and time (e.g. Cavalli et al. 2013; Heckmann and Schwanghart

2013; Lane et al. 2017; Chap. 16). Potential supply limitation from those slope units to fluvial systems therefore becomes an important parameter partially dictating the extent to which fluvial systems act as an effective mechanism of sediment redistribution into the down-valley sediment cascade (Cavalli et al. 2013). Processes acting upon hillslopes and associated storage, release and reworking of sediments in the hillslope domain therefore play a significant role in controlling the basin-scale flux and yield of sediments during deglaciation.

It could be argued, however, that in comparison with the wealth of research conducted in high-latitude environments, investigation of the paraglacial response to deglaciation in alpine environments is relatively understudied. This is perhaps unsurprising, given the very obvious and dynamic geomorphological activity and landscape modification associated with processes of thermo-erosion and mass movement at high latitudes (e.g. Etzelmüller 2000; Lyså and Lønne 2001; Irvine-Fynn et al. 2005, 2011; Porter et al. 2010; Ewertowski and Tomczyk 2015). In comparison, many deglaciating alpine environments, at least superficially, appear to exhibit relative stability following Little Ice Age glacier retreat and subsequent stabilisation of many proglacial areas through, for example, relatively undisturbed vegetation colonisation well beyond pioneer stages (e.g. Eichel et al. 2013, 2016; Chap. 19). However, as deglaciation gathers pace in alpine regions (e.g. Barry 2006; Radic and Hock 2011; Klaar et al. 2015), the requirement for a fuller understanding of sediment fluxes becomes enhanced, both due to a need to understand the complex geomorphological and sedimentological responses to deglaciation and to assess any potential impacts on, or from, human activities in alpine areas (e.g. Moore et al. 2009; Otto et al. 2009; Carrivick et al. 2013).

10.2 Overview

Perhaps the most visually obvious manifestation of hillslope glacigenic sediment storage, release and redistribution within deglaciating alpine environments are the large, often dissected, lateral moraines that flank many systems throughout alpine regions (e.g. Curry et al. 2006, 2009a, b; Lukas et al. 2012; Chap. 11; Fig. 10.1). Although perhaps less visually obvious, the extensive suites of sedimentary ice-marginal landforms and features found in proglacial zones also represent an important store and source of glacial and related sediments associated with deglaciation, likely subject to extensive sediment redistribution and reworking that may lead to both modification of landscape morphology and impacts on sediment fluxes through the down-valley cascade during deglaciation.

In this chapter, we consider storage within, and supply of sediments from, slopes within ice-marginal and proglacial environments. Inevitably, however, there is the potential for overlap with the content of other contributions in this volume, as hillslopes represent a fundamental and pivotal connection between geomorphic systems, controlling both the supply and storage of sediments and being subject to the full range of geomorphological processes (Chaps. 11, 15 and 17). We limit our consideration to landforms in the proglacial zone, as these features not only represent sizeable and important stores of glaciogenic and other sediments, but represent a highly dynamic geomorphological environment that impacts upon and interacts with the multiple geomorphical and biogeomorphical (Chap. 19) systems within the basin sediment cascade (Chaps. 15 and 17) that are considered elsewhere in this volume.

Although subglacial and supraglacial sediments also represent an important source from which the sediments contained within ice-marginal landforms and features may ultimately be derived (Barr and Lovell 2014), these systems and sediments are considered in Sects. 10.3 and 10.4, respectively, and are therefore not discussed in detail here. Similarly, the reduction in lateral slope support associated with deglaciation, known as debuttressing, is likely to enhance the likelihood of rockfall activity (e.g. Stoffel et al. 2014; Vehling et al. 2016) and associated delivery of predominantly larger calibre sedimentary debris to moraine

Fig. 10.1 Deglaciating proglacial alpine landscapes, Feegletscher Nord, Valais Switzerland. **a** Aerial view showing recent rockfall and slope debris to the left of the image, proglacial lake dammed by moraines and rockfall debris and heavily dissected lateral moraine sequence to the top right of the image. **b** Proximal face of lateral moraines showing extensive dissection and gullying, with debris cone build-up progressively burying slope units with increasing distance from the glacier snout (out of shot behind the photographer). **c** Gullied lateral moraine slopes showing distal dipping fabric and slope-foot debris accumulation. Note the mature vegetation at the slope crest, which extends down the distal slope away from the camera

sequences (e.g. Shulmeister et al. 2009; Cossart et al. 2008; Reznichenko et al. 2011), while periglacial systems such as rock glaciers also represent an effective store and potential source of slope sediments (Stoffel and Huggel 2012; Müller et al. 2014) that may be subsequently reworked. Rockfall activity, however, is discussed in detail in Chaps. 8 and 9, while rock glaciers and other periglacial features are considered in Chaps. 6 and 7.

Therefore, rather than considering the broader scale catastrophic and/or slow mass movements which may arise as a result of the same climatic forcing that drives deglaciation and contribute additional sediments for paraglacial reworking and redistribution, here we restrict our consideration to those sedimentary slope units directly related to glacial activity and subject to paraglacial modification. These comprise lateral and forefield glacial and glaciofluvial sedimentary landforms, their component slope units and processes operating thereon.

10.3 Lateral Slopes

10.3.1 Formation and Fluxes

Although large suites of lateral moraines commonly exhibiting numerous, narrow, parallel ephemeral channels, hereafter referred to as gullies, are a characteristic of many deglaciating alpine systems (Fig. 10.1), the details of lateral moraine formation and internal structure remain incompletely understood. This situation has arisen in part, due to moraine spatial location with respect to the contemporary glacier front and patterns of glacier fluctuation based thereon, being the focus for much research to date (Lukas and Sass 2011; Lukas et al. 2012). Early theories of lateral moraine formation generally assumed that sub-aerial weathering and resultant erosion were primarily responsible for accumulation of sediment at the ice margin, with additional contribution of sediment from englacial sources (e.g. Eyles and Rogerson 1978; Röthlisberger and Schneebeli 1979; Eyles 1983). Latter theories invoke processes of repeated 'stacking' of ice-derived debris flows at the glacier margins, with sedimentary stratification, gently sloping distal morphology and limited dating providing evidential support for such incremental formation (e.g. Small 1983, 1987). However, substantial subglacial and glaciofluvial sediments observed within lateral moraines at Findelengletscher, Switzerland, contrast with observations made elsewhere and highlight the potential importance of sediment transfer via englacial pathways (Lukas et al. 2012), while work conducted in geomorphologically active alpine areas has highlighted the rapid rate at which rockfall debris may become incorporated in the englacial environment and advected towards the glacier margin for subsequent deposition (e.g. Dunning et al. 2015). More recent geophysical investigations indicate complex ice-marginal moraine depositional history and evidence of polygenesis (e.g. Midgley et al. 2013; Tonkin et al. 2017). These observations combined, highlight the potential genetic complexity of lateral moraine features. Irrespective of the precise modes of formation, it is clear that lateral moraine complexes represent a substantial store of glacial and/or glaciofluvial sediment (Otto et al. 2009) that has the potential to be extensively reworked and redistributed during deglaciation (Chap. 11).

The extent to which any given slope unit will yield sediments which are then incorporated into the down-valley transfer of sediments, or sediment 'cascade', will depend on multiple factors, including, but not restricted to, slope geotechnical properties such as lithology and degree of consolidation (e.g. Curry et al. 2009a, b; Lukas et al. 2012), topographic setting (e.g. Barr and Lovell 2014) and the consequent nature and efficacy of geomorphological processes (e.g. Curry 1999; Curry et al. 2006). These factors will clearly vary in space and time (Ballantyne 2002b; Orwin and Smart 2004a) with the result that the detailed sedimentological consequences of paraglacial activity are uncertain (Curry and Ballantyne 1999), with relatively few studies to date directed towards identifying a sedimentological signature associated with paraglacial reworking (e.g. Benn and Ballantyne 2005; Curry et al. 2009a, b). This uncertainty is in no small part due to the fact that a given slope unit can act as both a store and a source of sediment, with secondary paraglacial activity furthering the complexity of the landscape response. Lateral moraines represent an interesting example of this dual 'sink' and 'source' role. Being commonly located away from highly dynamic proglacial fluvial systems, lateral moraines are more usually dissected, reworked and modified on an episodic basis through the operation of debris flow activity, facilitated by, for example, extreme precipitation events, spring thaw (e.g. Blair 1994; Kellerer-Pirklbauer et al. 2010) and thermo-erosion of dead ice bodies (e.g. Kjær and Krüger 2001; Schomacker and Kjær 2008; Lukas et al. 2012).

10.3.2 Lateral Slope Stability

The extent to which any slope unit releases sediments for reworking into the down-valley sediment cascade will relate in part to the overall geomorphic stability of that unit. Lateral moraines may be particularly important in this respect, due to their ability in many cases to stand stably at extreme angles. Many lateral moraines exhibit very steep (>60°) proximal slopes that appear to retain stable form, despite ongoing paraglacial reworking (e.g. Curry et al. 2006, 2009a, b) and are indicative of the storage of glaciogenic sediments in a quasi-stable state. Where moraines exist in such a steep, stable form, any role as a source of sediment is likely to become diminished. Slope stability, or lack thereof, then becomes an important determinant of the delivery and storage of sediments within a deglaciating environment, as enhanced paraglacial activity and associated sediment delivery from current or former ice-marginal areas is not an inevitable and immediate consequence of deglaciation where factors such as time, climatic setting and geotechnical properties may well induce long-term slope stability. The extent to which the stability of a given slope unit is compromised, such that the unit might act as a source of sediment, will be dictated by multiple factors. These may include processes such as debuttressing and resultant gravitational deformation (e.g. Hugenholtz et al. 2008), the action of fluvial processes, which may also be exacerbated by extreme events such as glacier lake drainage (e.g. Iturrizaga 2008; Chap. 14), antecedent saturation and soil suction levels (e.g. Springman et al. 2003; Hürlimann et al. 2012) and complex combinations and interactions of geomorphic processes, such as freeze–thaw, snowmelt, water infiltration and consequent sub-surface flow, seepage and outflow (Hürlimann et al. 2012). Clearly therefore, detailed genesis and long-term stability of moraines will be dictated by multiple elements, with glacial conditioning and local geomorphological factors being of particular importance (Hugenholtz et al. 2008).

Given the likely complex genesis of lateral moraines, combined with relative uncertainty over their stability, preservation and reworking potential and formative mechanisms (Lukas et al. 2012), they present something of a conundrum, having the potential to act as both a dynamic sediment source where paraglacial modification and reworking is taking place (e.g. Irvine-Fynn et al. 2011) and a relative sediment sink where stability is evident (e.g. Otto et al. 2009; Blair 1994).

The material properties and genetic mechanisms that permit some lateral moraines to stand stably at angles that in some cases exceed 70° (e.g. Whalley 1975; Lebourg et al. 2004; Curry et al. 2006; Lukas et al. 2012) remain incompletely understood, although recent work conducted in the European Alps has highlighted the importance of processes such as overconsolidation (e.g. Lukas et al. 2012) and the steepness of underlying bedrock (Lukas, pers. comm.). This characteristic of many lateral moraines is important in the context of sedimentary deglaciation dynamics as, irrespective of the source of sediment (subglacial, englacial, glaciofluvial, sub-aerial), slope stability necessarily means that sediments are held in transient storage, but are also potentially available for release, reworking and subsequent redeposition, as evidenced by the presence of gullies and associated slope-foot deposits (Fig. 10.1b, c). This ability of steep lateral moraines to hold sediments in quasi-stable form has implications for the correct interpretation of paraglacial sediment dynamics. Models of paraglacial system behaviour (e.g. Church and Ryder 1972; Matthews 1992) often propose maximum sediment production at or soon after deglaciation, followed by a simple and continuous, uni-directional decline. However, transient storage in and stochastic release of sediments from sources such as quasi-stable lateral moraine sequences has the capacity to disrupt this interpretation and allow release and reworking of sediments for many decades, if not centuries following deglaciation (Ballantyne 2002b; Curry et al. 2009a, b) with enhanced availability of sediments associated with deglaciation not always resulting in a consistent increase in sediment release and reworking (e.g. Cossart et al. 2008). This 'interrupted sediment cascade' (Curry et al. 2006; Fig. 10.2) continues in an episodic manner, such that paraglacial slope adjustment following ice recession and wasting since the Last Glacial Maximum may still impact upon erosional and mass movement processes operating today (e.g. Kellerer-Pirklbauer et al. 2010; Vehling et al. 2016).

The geotechnical properties of lateral moraine sediments offer one avenue of investigation towards a fuller understanding of lateral moraine stability. Whalley (1975) was able to characterise various mechanical aspects of lateral moraines at the Feegletscher, Switzerland, and calculated a friction angle of 45°, while suggesting that soil suction may have a role to play in enhancing slope strength and therefore stability. Distally dipping fabric, evident with clastic material >0.5 m long, results in imbrication of the Feegletscher moraines and might be expected to enhance stability as a result. However, the presence of steep, stable moraines with proximally dipping fabric observed elsewhere suggests that a distal dipping fabric is not a requirement for stability and extreme slope angles (Whalley 1975). In situ studies carried out at the same location, combined with geotechnical laboratory tests of extracted samples from upper high-angle (<80°) slope units, indicate that imbricated and

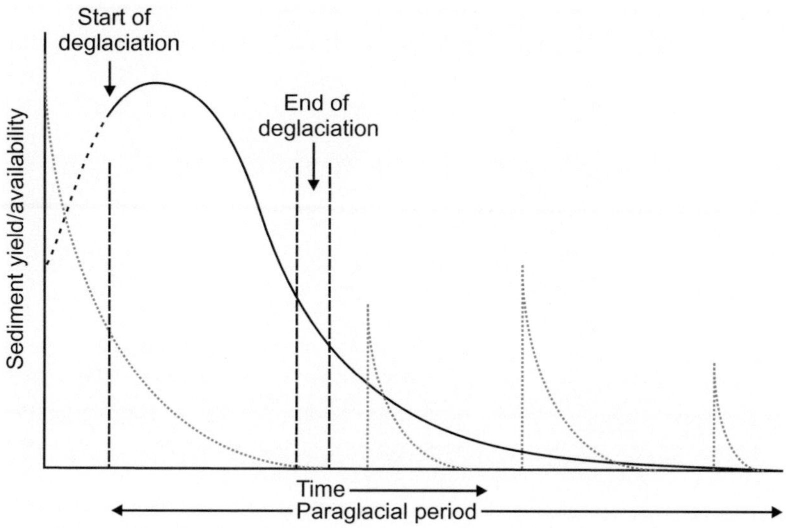

Fig. 10.2 Paraglacial models illustrating catchment sediment responses to deglaciation. The solid black curve illustrates the classic model proposed by Church and Ryder (1972), whereby basin sediment yields progressively decline with time since deglaciation. The dashed grey curves illustrate responses to deglaciation conditioned by sediment availability/exhaustion as proposed by Ballantyne (2002b), with episodic events taking place during and after deglaciation conditioned by, for example, extreme rainfall events (adapted from Ballantyne 2002a)

proximally dipping mica–schist clasts inhibit shallow translational shear on the proximal slopes, which may assist in the retention of stable form at this site (Curry et al. 2009a, b; Fig. 10.1c). A 'buttressing' effect provided by the inter-gully slopes is also speculated to assist in slope stabilisation, with suction effects potentially enhancing surface crust formation, but with limited impact on long-term stability (Curry et al. 2009a, b).

A comprehensive sedimentological study of lateral moraine morphology and properties carried out at Findelengletscher, Switzerland (Lukas et al. 2012), provides further information on lateral moraine genesis and resultant factors that may influence slope stability and resultant potential for sediment release and reworking. Rather than a dominant supraglacial source for lateral moraine sediments, subglacial and glaciofluvial sediments dominate, deposited by debris flows once material has been transferred from the bed to the ice surface via englacial pathways. Key, however, to proximal slope stability is the process of overconsolidation. Lukas et al. (2012) argue that this overconsolidation arises through a combination of glaciotectonisation of pre-existing sediments and incremental 'plastering' of till onto the proximal slopes of lateral moraines by moving ice. The resulting overconsolidation is suggested to be an important factor in enhancing slope stability, retarding paraglacial slope modification and consequently enhancing preservation potential. It is regarded as a potentially widespread process that may offer an explanation for slope stability and the extreme slope angles of up to 80°, as observed at Findelengletscher, without having to invoke the presence of a distally dipping fabric, a feature not ubiquitous to all lateral moraines (Curry et al. 2009a, b; Lukas et al. 2012).

The stability and steepness of proximal lateral moraine slopes often gives rise to cross-sectional asymmetry, with distal slopes exhibiting lower slope angles, although it should be noted that cross-sectional asymmetry is not a ubiquitous observation (Lukas and Sass 2011). In contrast to steep proximal slopes, lower-angled distal slopes, that usually represent depositional fan surfaces resting at the angle of repose, are generally less prone to intense paraglacial modification and the

Fig. 10.3 Lateral-frontal moraine of the Feegletscher Nord, Valais, Switzerland. **a** Dashed line indicates approximate location of the glacier front during the Little Ice Age maximum. Note the heavily vegetated distal slopes of the moraine. **b** View towards the frontal moraine from the proximal side. Note the mature vegetation on the stable proximal slopes of the moraine to the far right of the image and the potential for slope-foot fluvial transport where the proglacial meltwater stream impinges on the moraine in the middle distance

formation of gullies and, as such, tend to become more readily stabilised by vegetation cover (Fig. 10.3). During deglaciation therefore, paraglacial modification of slope form may act to release sediments for reworking, while vegetation colonisation acts to decrease geomorphological activity and stabilise slope form where conditions are conducive to that colonisation (Matthews 1992; Ballantyne 2002a; Eichel et al. 2016; Chap. 19). Conversely, at sites in Norway and the central Swiss Alps, Curry (1999) and Curry et al. (2006) suggest that progressive vegetation colonisation on geomorphologically active proximal moraine slopes is thought to be a response to slope stabilisation, rather than a causal factor. The role of vegetation acting to stabilise, or arising as a result of stabilisation and enhancing longer-term stability, therefore adds further complexity to the functioning of lateral moraines as both stores and sources of sediment. Feedback between vegetation and geomorphic process is likely to strongly condition slope form, but the details of this feedback are incompletely understood (Eichel et al. 2016; Chap. 19) and represent an important area for further research, as deglaciation and vegetation colonisation gather pace in many locations.

10.3.3 Reworking of Lateral Slopes

Where paraglacial reworking does take place on lateral moraines, it frequently gives rise to dissection of moraine slopes, formation of gullies and redistribution of sedimentary materials, with associated colluvial debris cones and fans accumulating at the slope foot (e.g. Curry and Ballantyne 1999; Curry et al. 2006, 2009a, b; Lukas et al. 2012; Fig. 10.1b, c). Clearly this dissection of moraines provides evidence of mass movement and resultant redistribution and transfer of sediment to the glacier surface or, where glacier retreat has left the moraine unit wholly exposed, the proglacial zone. Despite their apparent large-scale overall stability, steep lateral moraines are therefore potentially important conditioning factors for paraglacial slope modification (Curry et al. 2009a, b) and represent a dynamic environment that may provide an efficient transport link between hillslopes and other geomorphic systems such as channelised flow in forefield areas (Cavalli et al. 2013; Eichel et al. 2016). However, in common with studies of lateral moraine formation and internal structure, paraglacial modification of slope form generally (Curry 1999, 2000) and lateral moraine form in particular (Lukas et al. 2012) has received relatively scant research attention to date.

Paraglacial modification of lateral moraine slope form can take place through a variety of processes, including debris flowage, debris slides, stream action, solifluction, snow avalanching and stream action (Ballantyne 2002a; Curry et al. 2006; Chap. 11). Such processes are largely responsible for the characteristic paraglacial landscapes of gullied lateral moraines and valley sides, slope-foot debris cones, fans and valley-floor deposits (Ballantyne and Benn 1994; Ballantyne 1995, Curry 1999; Curry et al. 2006; Fig. 10.1a). It is clear that extensive paraglacial modification of slope form through, for example, gullying, can take place within short timescales of the order of a few decades (e.g. Ballantyne and Benn 1994) and that debris flow activity is one of the prime agents responsible for the redistribution of glacial sediments contained within hillslopes or moraines. Obvious flow tracks, levées and debris cones are often visible (e.g. Eyles and Kocsis 1988; Owen 1991; Ballantyne and Benn 1994; Curry 2000), and rainfall (e.g. Chiarle et al. 2007) and rapid snowmelt and resultant liquefaction (e.g. Ballantyne and Benn 1994; Palacios et al. 1999; Curry 2000) are key triggers in many debris flow events. However, the precise controls on the extent and efficacy of debris flowage and other paraglacial hillslope activity represent a relatively understudied aspect of deglaciation terrain relaxation (Curry 2000). Studying paraglacial modification of glaciogenic slope sediments in western Norway, Curry (2000) utilises gully density as a surrogate indicator of paraglacial modification of sediments within lateral moraines. In common with other studies of paraglacial slope modification (e.g. Curry et al. 2006, 2009a, b; Cavalli et al. 2013), debris flow activity emerges as a dominant mechanism, with snow avalanches representing a secondary paraglacial process. Of several intrinsic and extrinsic conditioning factors studied that may influence paraglacial debris flow activity, slope gradient, sediment availability and water supply emerge as important factors. However, a relative lack of paraglacial slope modification even in high relief areas demonstrates a clear need for a fuller understanding of the detailed constraints on paraglacial activity (Curry 2000).

One obvious constraint is sediment supply, and this aspect of paraglacial activity warrants a re-evaluation of the original Church and Ryder model that emphasises time as the key variable dictating paraglacial sediment supply, release and reworking, with maximum sediment supply occurring shortly after deglaciation and then steadily declining as glacier shrinkage continues (Church and Ryder 1972). Ballantyne (2002b) therefore suggests that an exponentially declining exhaustion model might be more appropriate, with sediment yield being dictated by sediment availability and perturbations in yield being dictated by factors such as changing base level or episodic sediment release from, for example, hillslopes during extreme rainfall events (Fig. 10.2; Ballantyne 2002b). This concept of sediment availability as a key controlling factor has significance when considering the role that hillslopes may play as stores and sources of sediment in a deglaciating landscape. Indeed, Curry (1999) suggests that upslope sediment availability is a likely control on slope stabilisation and to a certain extent, the delivery of sediments from ice-marginal hillslopes itself ensures a continuously diminishing supply of sediment and longer-term stability, as debris flowage delivers material to the slope foot, progressively burying upper slope units, reducing slope gradient, reducing upslope sediment supply and potentially facilitating slope stability, with vegetation colonisation furthering stability (Fig. 10.1b).

Considering lateral moraine materials further up-valley, the supraglacial accumulation of sediments sourced from adjacent hillslopes clearly represents an important input to lateral moraines. Although there is an extensive literature on debris-covered glaciers (see Chap. 4), the extent to which ice-marginal sediment delivery to a glacier surface may be regarded as strictly paraglacial in nature is open to question. Although we do not consider rockfall delivery in this section, it is thought that enhanced rockfall activity is a likely consequence of deglaciation (e.g. Fischer et al. 2006; Stoffel et al. 2014; Chaps. 8 and 9), and it seems not unreasonable to assume that ice-marginal sediment delivery to a glacier surface is also likely to increase as deglaciation continues, through processes of debuttressing (in instances where sedimentary slope units are being supported by the glacier), melt of permafrost that is acting to stabilise sedimentary units and through a general increase in water supply, associated with ice melt and shifting in the relative proportions of precipitation from snow to rainfall. Potential links between deglaciation and delivery of ice-marginal sediments to the glacier surface have been demonstrated in Arctic settings (e.g. Porter et al. 2010), and there is a clear need for further research to assess the links between slope sedimentary systems and supraglacial systems in alpine settings, particularly considering the impact that surface debris cover can have on glacier melt, behaviour and ultimately moraine formation (e.g. Reznichenko et al. 2011). It is also becoming increasingly clear that a genetic origin for lateral moraines based wholly upon supraglacial accumulation of sediments derived from valley-side slopes may not appropriately explain formative mechanisms. The presence of sub- and englacial sediments and glaciofluvial deposits within lateral moraines highlights the potential importance of alternative mechanisms for supraglacial sediment accumulation and subsequent reworking through processes such as thermo-erosion and debris flow activity (e.g. Etzelmüller 2000; Irvine-Fynn et al. 2011; Lukas et al. 2012; Porter et al. 2010; Tonkin et al. 2017).

10.4 Forefield Slopes

10.4.1 Formation and Fluxes

Although often less visually obvious than large lateral moraine and valley-side glaciogenic sediment sequences, the often extensive suites of moraine landforms commonly observed in deglaciating forefields represent a dynamic source and store of glacial sediments and exhibit characteristics of primary and secondary paraglacial activity. Although forefield slope gradients are typically lower than those observed on lateral moraine slopes, the presence of often spatially extensive and high-energy fluvial activity in the proglacial area (Chaps. 12 and 13) offers an important means of sediment transport and redistribution, making the forefield area the most dynamic part of the alpine sediment flux system (Fenn and Gurnell 1987; Warburton 1990; Maizels 1993; Marren 2005; Otto et al. 2009). The presence of ice cores within proglacial moraine structures adds additional interest, as not only do ice cores potentially influence sub-surface drainage pathways with potential slope stability implications (e.g. Langston et al. 2011; Muir et al. 2011), but thermo-erosion offers an additional mechanism of sediment release from moraine slope units, aided by associated meltwater supply. There is extensive literature detailing mechanisms and processes of glaciofluvial entrainment, transport and deposition within proglacial environments and useful overviews can be found in Gurnell and Clark (1987), Warburton (1990), Marren (2005) and Lane et al. (2017). Likewise, the genesis, morphology and structure of ice-marginal moraines is an area that has received much research attention, and we would direct readers to Bennett (2001) and Barr and Lovell (2014) as useful starting points for discussion of these aspects of glacial geomorphology. As with the previous section concerning lateral slope forms and processes, in this section we confine our consideration to discussion of the role of moraine proglacial landforms as potential stores or sources of sediment and consider the mechanisms and extent of paraglacial activity operating on moraine slope units to further the redistribution of glaciogenic sediments in deglaciating environments.

Glacier forefields are necessarily comprised of deposits that reflect multiple episodes of erosion, transport and deposition, having inevitably been subject to the operation of multiple geomorphological processes and can be viewed as transitioning from glacial to non-glacial conditions. The diversity and morphological characteristics of forefield deposits will broadly reflect processes of sediment production and modification beneath the glacier, forefield geomorphological activity and glacier dynamics and in particular, the glacial response to deglaciation and the extent of any reworking and redistribution by meltwater (Orwin and Smart 2004a). Exposure through glacier retreat and thinning makes deposits potentially available for remobilisation, transport and deposition by non-glacial processes, the presence of flowing water in the form of proglacial meltwaters being a particularly effective agent of erosion in a geomorphological context of abundant sediment supply. Observations indicate that up to 80% of total suspended sediment yield in proglacial rivers may be sourced from the forefield area (Orwin and Smart 2004b). However, given the likely complexity of the genetic history of forefield sediments, they will comprise a wide range of deposits of varying ages and therefore varying stability and vulnerability to reworking (Orwin and Smart 2004a). For example, negative feedback in some areas of deglaciating basins (e.g. transient storage of sediment in slope-foot debris cones) can act to reduce, rather than increase, basin sediment yields, through a decline in sediment system connectivity, despite increases in that connectivity elsewhere in the basin in response to deglaciation (e.g. Lane et al. 2017). Indeed, the extent to which the recently deglaciated forefield might represent a source or a sink for sediments is open to conjecture. Observed in an Arctic setting, Hodgkins et al. (2003) found that the extent to which the forefield zone may act as a net sink or a source of sediments may depend upon factors such as the nature of the temporally variant run-off regime, while age since exposure also likely represents an important factor in dictating vulnerability to reworking (e.g. Orwin and Smart 2004a), reflecting the generalised form of the paraglacial model of declining sediment yield as time since deglaciation advances (e.g. Church and Ryder 1972; Fig. 10.2).

At the basin scale, it is common for the forefield to be dominated by a relatively flat glaciofluvial plain that extends from the glacier margin (Marren 2005). In alpine environments, the forefield environment is typically delineated laterally by steep valley-side slopes and potentially high and steep lateral moraines, while the frontal margin is often marked by a prominent moraine (e.g. Fig. 10.3), invariably breached by flowing meltwaters. This frontal moraine is often the most morphologically distinctive forefield moraine form (Benediktsson et al. 2009), and in alpine environments frontal moraines are usually associated with maximum recent glacier advance during the Little Ice Age (Fig. 10.3a). Dependent on age, spatial extent and intensity of geomorphological activity, frontal moraines may support vegetation cover, particularly on distal slopes (Fig. 10.3a), but equally, exposed sediments will be vulnerable to processes of reworking and redistribution. The stability of these moraines is a topic that has attracted much recent interest in the light of deglaciation and resultant formation of proglacial water bodies dammed behind frontal moraines, with consequent risk of glacier lake outburst floods (GLOFs) should the moraine dam be breached (e.g. Stoffel and Huggel 2012; Westoby et al. 2014; Ashraf et al. 2015. See also Chap. 14).

The sediments stored within the forefield area are typically characterised by substantial volumes of the products of glacier erosion and are usually dissected by meltwater streams with resultant capacity for extensive glaciofluvial reworking and redistribution (e.g. Warburton 1990; Hodgkins et al. 2003; Orwin and Smart 2004a; Leggat et al. 2015). It is not uncommon for deglaciating forefields to be further characterised by a diverse range of glacial, fluvial and non-glacial landforms, features and deposits. Setting aside the glaciofluvial landforms, stores and processes that are considered elsewhere in this volume (Chap. 12), it is the diverse groups of moraines that are particularly important when considering paraglacial activity operating within the forefield area, although clearly relief and slope angles are both likely to be lower on average than that observed in recently deglaciated lateral locations.

Forefield moraines are an important source of information about the extent and dynamics of glaciers and ice masses. Consequently, they have been extensively studied and, despite the fact that mechanisms of formation remain manifold (Hiemstra et al. 2015), they have been widely used as an indirect proxy for climatic variation and past glacier dynamics and extents (e.g. Benn and Ballantyne 2005; Beedle et al. 2009; Evans et al. 1999). Numerous moraine types and classifications exist, leading to a complex picture in part due to 'inconsistent terminology' (Winkler and Matthews 2010, p 87) with sometimes 'incompatible' interpretations of the same landform being made (Evans et al. 1999, p 673) and confusion being heightened by restricting moraine analysis to geomorphology alone, without a detailed consideration of the detailed sedimentology of moraines that is evident in more recent work (e.g. Lukas et al. 2012; Reinardy et al. 2013; Chandler et al. 2016). Given the dynamic nature of proglacial environments and associated potential for moraine modification and even potential eradication by a range of post-depositional processes (Kirkbride and Brazier 1998; Kirkbride and Winkler 2012), preservation potential is variable, as will be the role that moraines play in the storage and supply of sediment. The extent to which moraines may provide an accessible source of sediment for reworking and redistribution will depend upon myriad factors including, but not limited to: time since exposure, slope angle, degree of glaciotectonisation and compaction, the presence of an ice core and associated de-icing and thermo-erosion, the nature and extent of vegetation colonisation and the degree of transport and supply/weathering limitations.

10.4.2 Reworking of Forefield Slopes

10.4.2.1 Buried Ice

The presence and subsequent prolonged degradation of buried remnant glacier ice within moraine landforms may influence moraine formation, morphology, slope activity and flow of meltwaters and groundwater in the forefield area (e.g. Kjær and Krüger 2001; Langston et al. 2011; Chaps. 6 and 7). Retreating debris-covered glaciers can leave forefields of ice-cored forms and ice-rich debris (e.g. Bosson et al. 2015), and this is particularly true close to the ice margins, where dead ice can be incorporated during the process of moraine formation (Lukas et al. 2012). Ice cores may survive decades to millennia and at considerable distances from the active ice margins (Barr and Lovell 2014). Their degradation is typically evident as backwasting (i.e. lateral retreat) and down-wasting (i.e. thinning; Krüger and Kjær 2000), commonly giving rise to inverted topography and the reworking of sediments due to slumping, fall-sorting and the formation of features such as sinkholes, extension fractures, kettles, cracks, slips and mud/debris flows (Johnson 1971; Kjær and Krüger 2001). The redistribution and remobilisation of sediments can also be facilitated further by ice-core melt and consequent sediment mobilisation, where removal of slope sediments through, for example, slumping or other slope mass movement exposes ice which is then vulnerable to thermo-erosion and backwasting (e.g. Kjær and Krüger 2001; Schomacker and Kjaer 2008). Similarly, slow melt of buried ice can initiate sinkhole formation and associated collapse of slope units through undermining, and over-steepening of adjacent slopes that may initiate processes such as sliding and backslumping (Kjær and Krüger 2001).

These processes of slope readjustment in response to a melting ice core and consequent sediment redistribution may be enhanced through fluvial-induced thermo-erosion where meltwaters are present. It is well established that fluvial activity plays an important role in shaping the detailed morphology of forefield landscapes (e.g. Carrivick et al. 2013), primarily through the mechanical processes of entrainment, transport and deposition and where flowing water is able to remove surficial deposits and expose buried ice, melt will progress rapidly. However, the presence of buried ice may also modify and partially dictate the location and efficacy of sub-surface water flow paths, and it has been established that moraines can contain complex sub-surface hydrological systems associated with the presence and interaction of sediments, buried and ground ice and bedrock (Langston et al. 2011).

Buried ice can act in two key ways to modify sub-surface drainage. Firstly, ice can act as an aquiclude, presenting a relatively impermeable barrier to sub-surface waters, controlling the routing of water. Secondly, buried ice may act as a source of water through the ongoing process of slow, sub-surface melting, or through rapid melting where ice becomes exposed at the surface. The accumulation of water in response to any sub-surface ice melt also offers the potential for saturation of sediments, with resultant implications for slope stability should that saturation destabilise slope units or indeed penetrate to surface sediment horizons. In Arctic settings, collapse of moraine sediments through melt of buried ice provides a potentially important component of the sediment cascade as a means of releasing sediments for redistribution, with resultant impact on moraine slope form and basin sediment yield (e.g. Etzelmüller 2000; Etzelmüller et al. 2000; Lyså, and Lønne 2001; Lukas et al. 2005; Porter et al. 2010; Irvine-Fynn et al. 2011). Where sub-surface ice and water are in contact, there is also the potential for effective thermo-erosion. However, largely due to a paucity of field observations and the practical difficulties of assessing both sub-surface water flow and ice presence and characteristics, the detailed role of forefield moraine landforms in controlling sub-surface water storage and flow remains poorly constrained (Langston et al. 2011) and by implication, so do the resultant influences upon surface slope processes and stability.

10.4.2.2 Hydrological Interactions

It is well established that sediment transfer within, and from, deglaciated forefields reflects the competing controls of sediment supply and the efficacy of transport processes (e.g. Warburton 1990; Hodgkins et al. 2003). Networks of high-energy streams and rivers typically dominate the downstream transfer of glaciogenic sediments from forefield environments, and fluvial processes are one of the primary agents of within-catchment landscape change due to their pivotal role in forefield sediment reworking and redistribution (Warburton 1990; Marren 2005; Carrivick et al. 2013; Lane et al. 2017).

The impact of fluvial–slope interactions has its most obvious morphological expression where forefield moraines have been dissected by meltwater channels. Such dissection will arise through 'normal' processes of glaciofluvial erosion, but more extreme examples of dissection and associated slope modification have been associated with outburst floods (e.g. Staines et al. 2015) or the failure of moraine dams in mountain environments (e.g. Korup and Tweed 2007). However, there are limited quantitative data available to establish the overall importance of such large-scale sediment redistribution events in remobilising and redistributing sediments as deglaciation terrain relaxation takes place, nor to establish the likely recurrence of such events (Ballantyne 2002a; Beylich and Warburton 2007). More commonly, there will be 'normal' or 'continuous' post-depositional modification of forefield landforms through the action of flowing meltwater, interspersed with discrete, extreme events (e.g. Cossart et al. 2008; Carrivick et al. 2013).

The impact of connections between slope and glaciofluvial sedimentary systems and resultant landform modification are most evident at the margins of proglacial streams, where fluvial processes may directly interact with hillslope units, presenting the most obvious sign that sediment is being delivered from the slopes to the valley floor (Beylich and Warburton 2007; Cavalli et al. 2013; Fig. 10.3b). Any resulting erosion not only modifies slope form, but may potentially generate slope instability through processes of undercutting and resultant collapse (Iturrizaga 2008; Winkler and Matthews 2010). Direct proglacial channel margin erosion will be limited to the valley floor, however, upslope instability may be induced and only become apparent when slopes are destabilised and mass movements occur (Orwin and Smart 2004a, b; Schrott et al. 2006).

It is clear that fluvial activity plays an important role in moraine morphological modification and associated release of sediments and that forefield sediments generally can contribute significantly to total basin sediment yield (e.g. Warburton 1990; Orwin and Smart 2004a; Leggat et al. 2015) and in an episodic fashion in the case of mass movements (Schrott et al. 2006; Carrivick et al. 2013). However, over time the hydrological efficacy of flowing water in mobilising forefield sediments is likely to decline. This arises from the eventual decline of the 'deglaciation dividend' (Kaser et al. 2010), whereby the initial enhanced melt, or dividend, associated with deglaciation reduces. Aside from this dwindling supply of meltwaters in the latter stages of deglaciation, surface sediments and slopes are likely to stabilise because of eluviation of fines and progressive armouring of surface layers caused by processes such as overland flow and rainsplash. This process can take place rapidly, and forefield surfaces may cease to function as significant sources of sediment within decades of deglaciation and exhibit stability to all but extreme events (Orwin and Smart 2004a).

The reworking and redistribution by meltwater of sediments within moraines may eventually lead to a situation where much of the forefield will comprise reworked glacigenic material (Carrivick et al. 2013) such that relief becomes subdued and the area of meltwater river channels increases (Staines et al. 2015). Forefield slope stability will likely be restricted to areas unaffected by fluvial activity or ice-core degradation (Ewertowski et al. 2010). This increase in slope stability and consequent decrease in sediment availability and mobilisation with time have been evident from studies of the patterns of aggradation and degradation in glacier forefields, where geomorphological activity decreases in both spatial extent and in intensity with distance from the glacier and by implication, time since deglaciation (e.g. Orwin and Smart 2004a). Hillslope activity will then likely become restricted to inherently unstable slopes, or those unaffected by or resistant to fluvial activity. As deglaciation continues, fluvial activity in general will progressively become more important. However, the apparent stability and persistence of moraine landforms and sediments observed in contemporary forefield areas suggest that the reworking and redistribution of deposits by fluvial activity may not be spatially extensive and that geomorphological activity is increasingly dominated by the interactions with specific processes (e.g. fluvial) over relatively small areas of catchments (Carrivick et al. 2013) or by the occurrence of extreme high-magnitude, episodic, low-frequency events such as those induced by changes in temperature or precipitation (Stoffel and Huggel 2012; Blair 1994).

The importance of episodic rainfall events in mobilising sediments from forefield slopes has been noted in a number of studies (e.g. Richards 1984; Kellerer-Pirklbauer et al. 2010; Cavalli et al. 2013; Leggat et al. 2015; Chap. 11) and can be responsible for both destabilising slopes (Blair 1994; Deline et al. 2015) and contributing to the exhaustion of supplies of fine sediments, with resultant reduction in overall sediment mobilisation (Orwin and Smart 2004b). The role of precipitation as a sediment transfer mechanism may become increasingly important during deglaciation due to the increasing proportion of total run-off comprising liquid precipitation as stores of snow and ice deplete (Collins 2008), while rainfall-induced extreme discharge events clearly have the capacity to mobilise even well-armoured and stabilised forefield sediments (e.g. Luckman 1981), interrupting the theoretical uni-directional decline in basin sediment yield as deglaciation progresses. Given forecast changes in precipitation frequency and intensity associated with climate change (Stoffel and Huggel 2012), the potential for precipitation events to further affect basin sediment yields is only likely to increase.

10.5 Environmental Change and Alpine Paraglacial Activity

As the impacts of climate change are becoming more obvious in alpine regions, an increased interest in interactions and feedbacks between paraglacial activity and associated biogeographical and hydrological processes has developed. In the European Alps for example, larger magnitude debris flow events may become increasingly common due to predicted increases in autumn and spring rainfall, permafrost melt and enhanced sediment delivery (Stoffel et al. 2014). It is also hypothesised that, given the combination of enhanced sediment availability, permafrost degradation and changes in rainfall, debris flow events with 'little or no historical precedent' could be facilitated (Stoffel and Huggel 2012, p 430).

To date, catchment-scale studies of paraglacial activity have largely focussed upon the storage, release and reworking of sediments. However, as climate warms, there is also a need to consider the related issues of enhanced biological and hydrological activity that develop in tandem with enhanced storage and transfer of sedimentary materials. It is becoming increasingly clear that any consideration of the landscape response to deglaciation cannot be considered in isolation from the closely associated hydrological and biological responses. In addition to enhancing the vulnerability of sediments to processes of reworking and redistribution, glacier retreat and down-wasting make those same sediments available for microbial colonisation, with resultant build-up of nutrient pools that encourage pedogenesis and increased organic matter development (e.g. Bernasconi et al. 2008; Schurig et al. 2013). In response to increased soil, organic matter and nutrient availability (Chap. 18), vegetation succession is initialised, leading ultimately to the growth of higher-order plants that may play an important role in conditioning slope stability (Bradley et al. 2014; Eichel et al. 2016; Chap. 19). Indeed, vegetation–landform interactions are becoming increasingly recognised as an important, albeit understudied, aspect of the landscape response to deglaciation, with the result that the discipline of 'biogeomorphology' is currently seen to be an actively developing and useful means of enhancing understanding of forefield dynamics (Eichel et al. 2013, 2016). Rapid deglaciation and revelation of ice-marginal and forefield sediments also offer opportunities to study in detail the mechanisms of and factors determining the efficacy of vegetation colonisation and succession (e.g. Jumpponen et al. 1999).

It has been established that the extent and stage of vegetation colonisation is a likely strong determinant of sediment availability and catchment sediment yield (Ballantyne 2002a; Klaar et al. 2015) and, in general terms, vegetation succession and associated ecosystem complexity will develop on a similar timescale to the progressively declining sediment yield assumed in many paraglacial models (Ballantyne 2002a; Klaar et al. 2015; Fig. 10.2). For slope units that are not subject to regular disturbance from geomorphological activity, vegetation colonisation impacts paraglacial adjustment through the stabilisation of landforms and a resultant decline in sediment availability as the paraglacial 'period' progresses (Ballantyne 2002a; Orwin and Smart 2004b; Marston 2010; Klaar et al. 2015). As vegetation colonisation progresses, sediment cohesion and shear strength increase and rainfall interception and infiltration become enhanced, thereby reducing surface run-off and likelihood of surface sediment mobilisation (Klaar et al. 2015).

However, in much the same way that a simple uni-directional decline in sediment yield may not adequately describe the reality of the landscape response to deglaciation (e.g. Ballantyne 2002b; Fig. 10.2), vegetation succession may also be interrupted by geomorphic processes (Matthews 1992) dependent on magnitude and frequency (Eichel et al. 2013), resulting in the persistence of younger successional stages of vegetation colonisation (e.g. Eichel et al. 2013) with resultant impacts on sediment availability. Feedbacks and relationships between geomorphic systems and vegetation ecosystems are therefore potentially complex, and the details of microbial and subsequent vegetation colonisation and consequent impacts on slope stability and sediment redistribution (or stability-induced lack thereof) are presently obscure (Klaar et al. 2015) and represent a current research priority.

In terms of hydrological impacts, glacial and associated sediments and moraines play an important role in controlling the timing and quantity of water release from glaciated catchments (e.g. Langston et al. 2011; Cook et al. 2013), with any paraglacial modification potentially altering the nature and functioning of water flow paths, with resultant impacts on the quantity and timing of basin water yields and the ecology of meltwater-fed ecosystems (e.g. Brown et al. 2006; Milner et al. 2009). Forefield surface glacial sediments, such as those contained within moraines, can play an important role in temporarily storing and thereby buffering basin meltwater discharge and sustaining baseflow, but the processes of water storage and transfer through moraines and other depositional glacial landforms remain obscure (Langston et al. 2011). Given the likely increasing intensity of sediment redistribution associated with deglaciation, understanding the associated impacts on meltwater flow specifically and basin hydrology generally represents an important area for further research, especially considering the significant reliance on meltwater as a resource in many alpine regions.

10.6 Conclusion

As deglaciation in alpine regions continues, research interest in the resultant impacts on sediment storage, release and reworking is likely to become further enhanced, adding to the growth of interest in paraglacial geomorphology identified by Ballantyne (2002a). Great uncertainty still exists, however, concerning the timescales over which the geomorphological response to deglaciation persists (Dadson and Church 2005) and the subsequent patterns of sediment release from basins during deglaciation. The basic notion of maximal sediment delivery immediately following deglaciation, followed by a slow, uni-directional decline as slope units and other terrain surfaces stabilise, may be broadly appropriate over large timescales. However, over shorter timescales, it is clear that enhanced delivery of sediments from slope units is not an inevitable or immediate consequence of deglaciation and that a variety of geological, geomorphological, hydrological and biogeomorphological factors will inevitably add a somewhat stochastic element to patterns of sediment release. Understanding these patterns of sediment release and indeed spatial and temporal variations in sediment storage represents an important area of current glaciological research, as contemporary deglaciation offers an unparalleled, albeit ultimately time-limited, opportunity to directly observe the genesis of deglaciation landforms, their modification and associated sediment fluxes and fluctuations in basin-scale sediment storage. This not only permits a fuller understanding of the complexities of the geomorphological processes during deglaciation, but contributes to enhanced understanding of glacial depositional landform genesis. Furthermore, with rapid warming evident in alpine regions, melt of permafrost and a predicted greater frequency of extreme precipitation events have the potential to remobilise the substantial glaciogenic sediment stores that are present in many alpine regions. It is therefore critical, both from a geomorphological and human impact point of view, that a fuller and detailed understanding of the processes that drive that remobilisation is gained, such that greater accuracy can be afforded to predictions of landscape development and associated potential impacts on human activity.

References

Ashraf A, Naz R, Iqbal MB (2015) Heterogeneous expansion of end-moraine dammed lakes in the Hindukush-Karakoram-Himalaya ranges of Pakistan during 2001–2013. J Mt Sci 12:1113–1124

Ballantyne CK (1995) Paraglacial debris-cone formation on recently deglaciated terrain, western Norway. Holocene 5:25–33. https://doi.org/10.1177/095968369500500104

Ballantyne CK (2002a) Paraglacial geomorphology. Quatern Sci Rev 21:1935–2017. https://doi.org/10.1016/S0277-3791(02)00005-7

Ballantyne CK (2002b) A general model of paraglacial landscape response. Holocene 12:371–376. https://doi.org/10.1191/0959683602hl553fa

Ballantyne CK, Benn DI (1994) Paraglacial slope adjustment and resedimentation following recent glacier retreat, Fåbergstølsdalen, Norway. Arct Alp Res 26(3): 255–269

Barr ID, Lovell H (2014) A review of topographic controls on moraine distribution. Geomorphology 226:44–64

Barry RG (2006) The status of research on glaciers and global glacier recession: a review. Prog Phys Geogr 30:285–306

Beedle MJ, Menounos B, Luckman BH, Wheate R (2009) Annual push moraines as climate proxy. Geophys Res Lett 36(20)

Benediktsson ÍÖ, Ingolfsson O, Schomacker A, Kjaer KH (2009) Formation of submarginal and proglacial end moraines: implications of ice-flow mechanism during the 1963–64 surge of Brúarjökull, Iceland. Boreas 38:440–457

Benn DI, Ballantyne CK (2005) Palaeoclimatic reconstruction from Loch Lomond Readvance glaciers in the West Drumochter Hills, Scotland. J Quat Sci 20:577–592

Bennett MR (2001) The morphology, structural evolution and significance of push moraines. Earth Sci Rev 53:197–236

Bernasconi SM, Christi I, Hajdas I, Abbaspour K (2008) Weathering, soil formation and initial ecosystem evolution on a glacier forefield: a case study from the Damma Glacier, Switzerland. Mineral Mag 72:19–22

Beylich A, Warburton J (2007) Analysis of source-to-sink-fluxes and sediment budgets in changing high-latitude and high-altitude cold environments. SEDIFLUX manual. NGU report

Blair R Jr (1994) Moraine and valley wall collapse due to rapid deglaciation in Mount Cook National Park, New Zealand. Mt Res Dev 347–358

Bosson J-B, Deline P, Bodin X et al (2015) The influence of ground ice distribution on geomorphic dynamics since the Little Ice Age in proglacial areas of two cirque glacier systems. Earth Surf Proc Land 40:666–680. https://doi.org/10.1002/esp.3666

Bradley JA, Singarayer JS, Anesio AM (2014) Microbial community dynamics in the forefield of glaciers. Proc R Soc B 281(1795):20140882

Brown LE, Milner AM, Hannah DM (2006) Stability and persistence of alpine stream macroinvertebrate communities and the role of physicochemical habitat variables. Hydrobiologia 560:159–173

Carrivick JL, Heckmann T (2017) Short-term geomorphological evolution of proglacial systems. Geomorphology 287:3–28. https://doi.org/10.1016/j.geomorph.2017.01.037

Carrivick JL, Geilhausen M, Warburton J et al (2013) Contemporary geomorphological activity throughout the proglacial area of an alpine catchment. Geomorphology 188:83–95. https://doi.org/10.1016/j.geomorph.2012.03.029

Cavalli M, Trevisani S, Comiti F, Marchi L (2013) Geomorphometric assessment of spatial sediment connectivity in small Alpine catchments. Geomorphology 188:31–41. https://doi.org/10.1016/j.geomorph.2012.05.007

Chandler BM, Evans DJ, Roberts DH (2016) Characteristics of recessional moraines at a temperate glacier in SE Iceland: insights into patterns, rates and drivers of glacier retreat. Quatern Sci Rev 135:171–205

Chiarle M, Iannotti S, Mortara G, Deline P (2007) Recent debris flow occurrences associated with glaciers in the Alps. Glob Planet Change 56:123–136. https://doi.org/10.1016/j.gloplacha.2006.07.003

Church M, Ryder JM (1972) Paraglacial sedimentation: a consideration of fluvial processes conditioned by glaciation. Geol Soc Am Bull 83:3059. https://doi.org/10.1130/0016-7606(1972)83%5b3059:PSACOF%5d2.0.CO;2

Collins DN (2008) Climatic warming, glacier recession and runoff from Alpine basins after the Little Ice Age maximum. Ann Glaciol 48:119–124

Cook SJ, Porter PR, Bendall CA (2013) Geomorphological consequences of a glacier advance across a paraglacial rock avalanche deposit. Geomorphology 189:109–120

Cossart E, Braucher R, Fort M Bourlès DL Carcaillet J (2008) Slope instability in relation to glacial debuttressing in alpine areas (Upper Durance catchment, southeastern France): evidence from field data and ^{10}Be cosmic ray exposure ages. Geomorphology 95(1–2): 3–26

Curry AM (1999) Paraglacial modification of slope form. Earth Surf Proc Land 24:1213–1228

Curry AM (2000) Observations on the distribution of paraglacial reworking of glacigenic drift in western Norway. Nor Geogr Tidsskr 54:139–147

Curry AM, Ballantyne CK (1999) Paraglacial modification of glacigenic sediment. Geogr Ann Ser A Phys Geogr 81:409–419

Curry AM, Cleasby V, Zukowskyj P (2006) Paraglacial response of steep, sediment-mantled slopes to post-'Little Ice Age' glacier recession in the central Swiss Alps. J Quat Sci 21:211–225. https://doi.org/10.1002/jqs.954

Curry AM, Sands TB, Porter PR (2009a) Geotechnical controls on a steep lateral moraine undergoing paraglacial slope adjustment. Geol Soc Lond Spec Publ 320:181–197

Curry A, Porter P, Irvine-Fynn T et al (2009b) Quantitative particle size, microtextural and outline shape analyses of glacigenic sediment reworked by paraglacial debris flows. Earth Surf Proc Land 34:48–62

Dadson SJ, Church M (2005) Postglacial topographic evolution of glaciated valleys: a stochastic landscape evolution model. Earth Surf Proc Land 30:1387–1403

Deline P, Hewitt K, Reznichenko N, Shugar D (2015) Rock avalanches onto glaciers. Landslide hazards, risks and disasters. Elsevier, Amsterdam, pp 263–319

Dunning SA, Rosser NJ, McColl ST, Reznichenko NV (2015) Rapid sequestration of rock avalanche deposits within glaciers. Nat Commun 6:7964. https://doi.org/10.1038/ncomms8964

Eichel J, Krautblatter M, Schmidtlein S, Dikau R (2013) Biogeomorphic interactions in the Turtmann glacier forefield, Switzerland. Geomorphology 201:98–110. https://doi.org/10.1016/j.geomorph.2013.06.012

Eichel J, Corenblit D, Dikau R (2016) Conditions for feedbacks between geomorphic and vegetation dynamics on lateral moraine slopes: a biogeomorphic feedback window: conditions for biogeomorphic feedbacks on lateral moraine slopes. Earth Surf Proc Land 41:406–419. https://doi.org/10.1002/esp.3859

Etzelmüller B (2000) Quantification of thermo-erosion in pro-glacial areas-examples from Svalbard. Zeitschrift für Geomorphol, NF 343–361

Etzelmüller B, Ødegård RS, Vatne G et al (2000) Glacier characteristics and sediment transfer system of Longyearbreen and Larsbreen, western Spitsbergen. Nor Geogr Tidsskr 54:157–168

Evans DJ, Lemmen DS, Rea BR (1999) Glacial landsystems of the southwest Laurentide Ice Sheet: modern Icelandic analogues. J Quat Sci 14:673–691

Ewertowski MW, Tomczyk AM (2015) Quantification of the ice-cored moraines' short-term dynamics in the high-Arctic glaciers Ebbabreen and Ragnarbreen, Petuniabukta, Svalbard. Geomorphology 234:211–227

Ewertowski M, Kasprzak L, Szuman I, Tomczyk A (2010) Depositional processes within the frontal ice-cored moraine system, Ragnar glacier, Svalbard. Quaestiones Geographicae 29:27

Eyles N (1983) Chapter 4—the glaciated valley landsystem. In: Eyles N (ed) Glacial geology. Pergamon, Amsterdam, pp 91–110

Eyles N, Kocsis S (1988) Sedimentology and clast fabric of subaerial debris flow facies in a glacially-influenced alluvial fan. Sed Geol 59:15–28

Eyles N, Rogerson R (1978) Sedimentology of medial moraines on Berendon Glacier, British Columbia, Canada: implications for debris transport in a glacierized basin. Geol Soc Am Bull 89:1688–1693

Fenn CR & Gurnell AM (1987) Proglacial channel processes. In: Gurnell AM & Clark MJ (eds) Glaciofluvial sediment transfer: an alpine perspective, Wiley, Chichester, UK, pp 423–472

Fischer L, Kääb A, Huggel C, Noetzli J (2006) Geology, glacier retreat and permafrost degradation as controlling factors of slope instabilities in a high-mountain rock wall: the Monte Rosa east face. Nat Hazards Earth Syst Sci 6:761–772

Gurnell AM, Clark MJ (1987) Glacio-fluvial sediment transfer. Wiley, USA

Heckmann T, Schwanghart W (2013) Geomorphic coupling and sediment connectivity in an alpine catchment—exploring sediment cascades using graph theory. Geomorphology 182:89–103. https://doi.org/10.1016/j.geomorph.2012.10.033

Hiemstra JF, Matthews JA, Evans DJ, Owen G (2015) Sediment fingerprinting and the mode of formation of singular and composite annual moraine ridges at two glacier margins, Jotunheimen, southern Norway. Holocene 25:1772–1785

Hodgkins R, Cooper R, Wadham J, Tranter M (2003) Suspended sediment fluxes in a high-Arctic glacierised catchment: implications for fluvial sediment storage. Sed Geol 162:105–117

Hugenholtz CH, Moorman BJ, Barlow J, Wainstein PA (2008) Large-scale moraine deformation at the Athabasca Glacier, Jasper National Park, Alberta, Canada. Landslides 5:251–260

Hürlimann M, Abancó C, Moya J (2012) Rockfalls detached from a lateral moraine during spring season.

2010 and 2011 events observed at the Rebaixader debris-flow monitoring site (Central Pyrenees, Spain). Landslides 9:385–393

Irvine-Fynn T, Moorman B, Sjogren D et al (2005) Cryological processes implied in Arctic proglacial stream sediment dynamics using principal components analysis and regression. Geol Soc Lond Spec Publ 242:83–98

Irvine-Fynn TDL, Barrand NE, Porter PR et al (2011) Recent high-arctic glacial sediment redistribution: a process perspective using airborne lidar. Geomorphology 125:27–39. https://doi.org/10.1016/j.geomorph.2010.08.012

Iturrizaga L (2008) Paraglacial landform assemblages in the Hindukush and Karakoram Mountains. Geomorphology 95:27–47

Johnson P (1971) Ice cored moraine formation and degradation, Donjek glacier, Yukon Territory, Canada. Geogr Ann Ser A Phys Geogr 53(3–4):198–202

Jumpponen A, Väre H, Mattson KG et al (1999) Characterization of 'safe sites' for pioneers in primary succession on recently deglaciated terrain. J Ecol 87:98–105

Kaser G, Großshauser M, Marzeion B (2010) Contribution potential of glaciers to water availability in different climate regimes. Proc Nat Acad Sci 107: 20223–20227

Kellerer-Pirklbauer A, Proske H, Strasser V (2010) Paraglacial slope adjustment since the end of the last glacial maximum and its long-lasting effects on secondary mass wasting processes: Hauser Kaibling, Austria. Geomorphology 120:65–76

Kirkbride MP, Brazier V (1998) A critical evaluation of the use of glacier chronologies in climatic reconstruction, with reference to New Zealand. J Quat Sci 13:55–64

Kirkbride M, Winkler S (2012) Correlation of Late Quaternary moraines: impact of climate variability, glacier response, and chronological resolution. Quatern Sci Rev 46:1–29

Kjær KH, Krüger J (2001) The final phase of dead-ice moraine development: processes and sediment architecture, Kötlujökull, Iceland. Sedimentology 48:935–952

Klaar MJ, Kidd C, Malone E et al (2015) Vegetation succession in deglaciated landscapes: implications for sediment and landscape stability. Earth Surf Proc Land 40:1088–1100

Korup O, Tweed F (2007) Ice, moraine, and landslide dams in mountainous terrain. Quatern Sci Rev 26: 3406–3422. https://doi.org/10.1016/j.quascirev.2007.10.012

Krüger J, Kjær KH (2000) De-icing progression of ice-cored moraines in a humid, subpolar climate, Kötlujökull, Iceland. Holocene 10:737–747

Lane SN, Bakker M, Gabbud C et al (2017) Sediment export, transient landscape response and catchment-scale connectivity following rapid climate warming and Alpine glacier recession. Geomorphology 277:210–227. https://doi.org/10.1016/j.geomorph.2016.02.015

Langston G, Bentley LR, Hayashi M et al (2011) Internal structure and hydrological functions of an alpine proglacial moraine. Hydrol Process 25:2967–2982

Lebourg T, Riss J, Pirard E (2004) Influence of morphological characteristics of heterogeneous moraine formations on their mechanical behaviour using image and statistical analysis. Eng Geol 73:37–50

Leggat MS, Owens PN, Stott TA et al (2015) Hydro-meteorological drivers and sources of suspended sediment flux in the pro-glacial zone of the retreating Castle Creek Glacier, Cariboo Mountains, British Columbia, Canada: suspended sediment fluxes in the pro-glacial zone. Earth Surf Proc Land 40:1542–1559. https://doi.org/10.1002/esp.3755

Luckman B (1981) The geomorphology of the Alberta Rocky Mountains: a review and commentary. Zeitschrift für Geomophol 91–119

Lukas S, Nicholson LI, Ross FH and Humlum, O (2005) Formation, Meltout Processes and Landscape Alteration of high-arctic ice-cored moraines-examples from Nordenskiold Land, Central Spitsbergen. Polar Geography 29(3):157–187

Lukas S, Sass O (2011) The formation of Alpine lateral moraines inferred from sedimentology and radar reflection patterns: a case study from Gornergletscher, Switzerland. Geol Soc Lond Spec Publ 354:77–92

Lukas S, Graf A, Coray S, Schlüchter C (2012) Genesis, stability and preservation potential of large lateral moraines of Alpine valley glaciers—towards a unifying theory based on Findelengletscher, Switzerland. Quatern Sci Rev 38:27–48

Lyså A, Lønne I (2001) Moraine development at a small High-Arctic valley glacier: Rieperbreen, Svalbard. J Quat Sci 16:519–529

Maizels J (1993) Lithofacies variations within sandur deposits: the role of runoff regime, flow dynamics and sediment supply characteristics. Sed Geol 85:299–325

Marren PM (2005) Magnitude and frequency in proglacial rivers: a geomorphological and sedimentological perspective. Earth Sci Rev 70:203–251. https://doi.org/10.1016/j.earscirev.2004.12.002

Marston RA (2010) Geomorphology and vegetation on hillslopes: interactions, dependencies, and feedback loops. Geomorphology 116:206–217

Matthews JA (1992) The ecology of recently-deglaciated terrain: a geoecological approach to glacier forelands. Cambridge University Press, Cambridge

Midgley NG, Cook SJ, Graham DJ, Tonkin TN (2013) Origin, evolution and dynamic context of a Neoglacial lateral–frontal moraine at Austre Lovénbreen, Svalbard. Geomorphology 198:96–106

Milner AM, Brown LE, Hannah DM (2009) Hydroecological response of river systems to shrinking glaciers. Hydrol Process 23:62–77

Moore R, Fleming S, Menounos B et al (2009) Glacier change in western North America: influences on hydrology, geomorphic hazards and water quality. Hydrol Process 23:42–61

Muir DL, Hayashi M, McClymont AF (2011) Hydrological storage and transmission characteristics of an alpine talus. Hydrol Process 25:2954–2966

Müller J, Gärtner-Roer I, Kenner R et al (2014) Sediment storage and transfer on a periglacial mountain slope (Corvatsch, Switzerland). Geomorphology 218:35–44

Orwin JF, Smart CC (2004a) The evidence for paraglacial sedimentation and its temporal scale in the deglacierizing basin of Small River Glacier, Canada. Geomorphology 58:175–202

Orwin JF, Smart CC (2004b) Short-term spatial and temporal patterns of suspended sediment transfer in proglacial channels, Small River Glacier, Canada. Hydrol Process 18:1521–1542

Orwin JF, Lamoureux SF, Warburton J, Beylich A (2010) A framework for characterizing fluvial sediment fluxes from source to sink in cold environments. Geogr Ann Ser A Phys Geogr 92:155–176

Otto J-C, Schrott L, Jaboyedoff M, Dikau R (2009) Quantifying sediment storage in a high alpine valley (Turtmanntal, Switzerland). Earth Surf Proc Land 34:1726–1742. https://doi.org/10.1002/esp.1856

Owen LA (1991) Mass movement deposits in the Karakoram Mountains: their sedimentary characteristics, recognition and role in Karakoram landform evolution. Zeitschrift für Geomorphol 35:401–424

Palacios D, Parrilla G, Zamorano JJ (1999) Paraglacial and postglacial debris flows on a Little Ice Age terminal moraine: Jamapa Glacier, Pico de Orizaba (Mexico). Geomorphology 28:95–118

Porter PR, Vatne G, Ng F, Irvine-fynn TD (2010) Ice-marginal sediment delivery to the surface of a high-Arctic glacier: Austre Brøggerbreen, Svalbard. Geogr Ann Ser A Phys Geogr 92:437–449

Radic V, Hock R (2011) Regionally differentiated contribution of mountain glaciers and ice caps to future sea-level rise. Nat Geosci 4:91

Reinardy BT, Leighton I, Marx PJ (2013) Glacier thermal regime linked to processes of annual moraine formation at Midtdalsbreen, southern Norway. Boreas 42:896–911

Reznichenko NV, Davies TR, Alexander DJ (2011) Effects of rock avalanches on glacier behaviour and moraine formation. Geomorphology 132:327–338

Richards K (1984) Some observations on suspended sediment dynamics in Storbregrova, Jotunheimen. Earth Surf Proc Land 9:101–112

Röthlisberger F, Schneebeli W (1979) Genesis of lateral moraine complexes, demonstrated by fossil soils and trunks: indicators of postglacial climatic fluctuations

Schomacker A, Kjær KH (2008) Quantification of dead-ice melting in ice-cored moraines at the high-Arctic glacier Holmströmbreen, Svalbard. Boreas 37:211–225

Schrott L, Götz J, Geilhausen M, Morche D (2006) Spatial and temporal variability of sediment transfer and storage in an Alpine basin (Reintal valley, Bavarian Alps, Germany). Geogr Helv 61:191–200. https://doi.org/10.5194/gh-61-191-2006

Schurig C, Smittenberg RH, Berger J et al (2013) Microbial cell-envelope fragments and the formation of soil organic matter: a case study from a glacier forefield. Biogeochemistry 113:595–612

Shulmeister J, Davies TR, Evans DJ et al (2009) Catastrophic landslides, glacier behaviour and moraine formation—a view from an active plate margin. Quatern Sci Rev 28:1085–1096

Small R (1983) Lateral moraines of glacier de Tsidjiore Nouve: form, development, and implications. J Glaciol 29:250–259

Small, RJ (1987) Moraine sediment budgets. In: Gurnell AM & Clark MJ (eds) Glaciofluvial sediment transfer: an alpine perspective, Wiley, Chichester, UK, pp 165–197

Springman S, Jommi C, Teysseire P (2003) Instabilities on moraine slopes induced by loss of suction: a case history. Geotechnique 53:3–10

Staines KE, Carrivick JL, Tweed FS et al (2015) A multi-dimensional analysis of pro-glacial landscape change at Sólheimajökull, southern Iceland. Earth Surf Proc Land 40:809–822

Stoffel M, Huggel C (2012) Effects of climate change on mass movements in mountain environments. Prog Phys Geogr 36:421–439

Stoffel M, Tiranti D, Huggel C (2014) Climate change impacts on mass movements—case studies from the European Alps. Sci Total Environ 493:1255–1266

Tonkin TN, Midgley NG, Graham DJ, Labadz JC (2017) Internal structure and significance of ice-marginal moraine in the Kebnekaise Mountains, northern Sweden. Boreas 46:199–211

Vehling L, Rohn J, Moser M (2016) Quantification of small magnitude rockfall processes at a proglacial high mountain site, Gepatsch glacier (Tyrol, Austria). Zeitschrift für Geomorphol 60(Supplementary Issues): 93–108

Warburton J (1990) An alpine proglacial fluvial sediment budget. Geogr Ann Ser A Phys Geogr 72:261–272. https://doi.org/10.1080/04353676.1990.11880322

Westoby MJ, Glasser NF, Brasington J et al (2014) Modelling outburst floods from moraine-dammed glacial lakes. Earth Sci Rev 134:137–159. https://doi.org/10.1016/j.earscirev.2014.03.009

Whalley W (1975) Abnormally steep slopes on moraines constructed by valley glaciers. Eng Behav Glacial Mater 60–66

Winkler S, Matthews JA (2010) Observations on terminal moraine-ridge formation during recent advances of southern Norwegian glaciers. Geomorphology 116: 87–106

Slope Wash, Gully Erosion and Debris Flows on Lateral Moraines in the Upper Kaunertal, Austria

11

Jana-Marie Dusik, Fabian Neugirg and Florian Haas

Abstract

The science on high-mountain proglacial systems is a rather young field of study, but has gained importance since a few years ago. In this context, we investigated a part of the lateral Little Ice Age (LIA) moraine that was built up by the second longest glacier of the Eastern Alps. The Gepatschferner glacier in the Upper Kaunertal, Central Austrian Alps, created up to 150-m-high moraine slopes during its LIA advance that are now prone to paraglacial reworking. To analyse the degree of reworking and to identify the driving forces behind the slope development with both their spatial and temporal variations, we conducted several case studies, mainly based on data acquired by remote sensing techniques (multitemporal TLS and aerial photographs) and their derivatives (DEMs, DoD, orthophotos). First, a medium-term (5 years) overall balance of erosion and deposition of the studied slope is calculated. Second, seasonal variations of the process dynamics are uncovered based on short-term TLS measurements within c. one year. Third, the sediment contributing area (SCA) is delineated to estimate fluvial reworking. Fourth, following paraglacial adjustment studies by Curry et al. (2006), we measured and analysed gully development with time since deglaciation. The case studies lead to the conclusions that paraglacial adjustment of the study slope is still in progress. Extreme events in summer play a dominant role for morphodynamics, followed by processes during winter, whereas springtime offers important preparatory conditions for sediment transport. Fluvial transport is considered to have minor effects on the moraine development compared to gravitational processes. And last but not least, the development of gullies depends more on natural boundary conditions than on time since deglaciation.

Keywords

PROSA project · Lateral moraine · Gullying Morphological budget · Sediment contributing area · Seasonal morphodynamics

J.-M. Dusik (✉)
Bavarian State Office for Environment (LfU), Geological Survey, Hof/Saale, Germany
e-mail: jana.dusik@gmx.de

F. Neugirg
Municipal administration, Strömsund, Sweden

F. Haas
Catholic University of Eichstätt-Ingolstadt, Eichstätt, Germany

T. Heckmann and D. Morche (eds.), *Geomorphology of Proglacial Systems*, Geography of the Physical Environment, https://doi.org/10.1007/978-3-319-94184-4_11

11.1 Introduction

Many glaciated areas in the European Alps are characterised by a rapid deglaciation since the end of the Little Ice Age (LIA). Klok and Oerlemans (2003) reported a rise of the equilibrium line altitude by 54 m on average for 17 European glaciers in the period 1920–1950. Thus, many bedrock areas and sediment stores, formed and protected by glacier ice, now become vulnerable to geomorphic processes due to deglaciation (Holm et al. 2004; Matthews and Shakesby 2004; Cossart et al. 2008; Hugenholtz et al. 2008; Kellerer-Pirklbauer et al. 2010; McColl 2012). One kind of such sediment stores are steep lateral moraines with unconsolidated glacial till (see Chap. 10). The combination of steep slopes, loose material and missing or only sparse vegetation cover facilitates the massive reworking of the sediments by geomorphic processes like snow avalanches, debris flows and slope wash. Their geomorphic forms are clearly detectable from terrestrial and aerial observations and are characterised especially by incised gullies, debris cones and levées (Haas et al. 2012) and a subsequent reworking by adjacent rivers (Morche et al. 2012).

Studies on reworking rates of moraine slopes were carried out by Ballantyne and Benn (1994, 1996), and Curry (1999), in western Norway. Curry et al. (2006) examined erosion rates on lateral moraines at 17 locations in the central Swiss Alps, and Schiefer and Gilbert (2007) conducted a morphometric analysis of gully systems in a proglacial area in Canada. Curry et al. (2006) concluded that gullies on lateral moraines reached their maximum dimension after approx. 50 years, and a stabilisation of the gully system was reached 80–140 years after deglaciation.

A lot of studies exist on fluvial processes on alpine slopes consisting of loose moraine material (e.g. Becht 1995; Wetzel 1992; Haas 2008). Most of them have been carried out by using classical methods as sediment traps or erosion pins. During the last years, terrestrial laser scanning (TLS) and Structure from Motion (SfM) emerged as powerful tools to measure geomorphic processes with low magnitudes in great detail (c.f. a review by Carrivick and Heckmann 2017).

Within the proglacial area of the upper Kaunertal (Kauner valley) with its retreating glaciers Gepatschferner and Weißseeferner (see Chap. 1), similar conditions as described in the aforementioned studies can be found. However, the Kaunertal features areas with very high and very low reworking dynamics in direct vicinity. It may therefore be assumed that the time since deglaciation is not the only influencing parameter controlling the geomorphic dynamics on these moraine slopes. Thus, one goal of the PROSA project (Chap. 1) was to analyse the geomorphic activity of steep and unconsolidated lateral moraines in this area and to identify additional factors that influence the reworking rates, the landforms and the temporal variability. This is achieved by using geodetic methods with different temporal and spatial resolution in order to observe the geomorphic activity on different time scales. The following sections give an insight into the results of this project.

11.2 Materials, Methods and Study Site

Besides airborne laserscanning (ALS; 2006, summer 2012, autumn 2012 and 2014) and terrestrial laserscanning (TLS; multiple epochs between 2011 and 2015), aerial photographs (beginning in 1954), series of photographs taken from an unmanned aerial vehicle (UAV, since 2013) and terrestrial photographs (since May 2015) have been used to generate digital elevation models (DEMs) for different moraine test sites distributed over the whole valley. The postprocessing of the laserscanning data (Fig. 11.1) includes the filtering of vegetation in order to produce bare ground DEMs and thus to exclude the detection of surface changes reflecting, e.g. plant growth. Photographical data are processed by a digital Structure-from-Motion (SfM) approach to obtain three-dimensional point clouds. The single DEMs are referenced in one geodetic reference system to be usable for

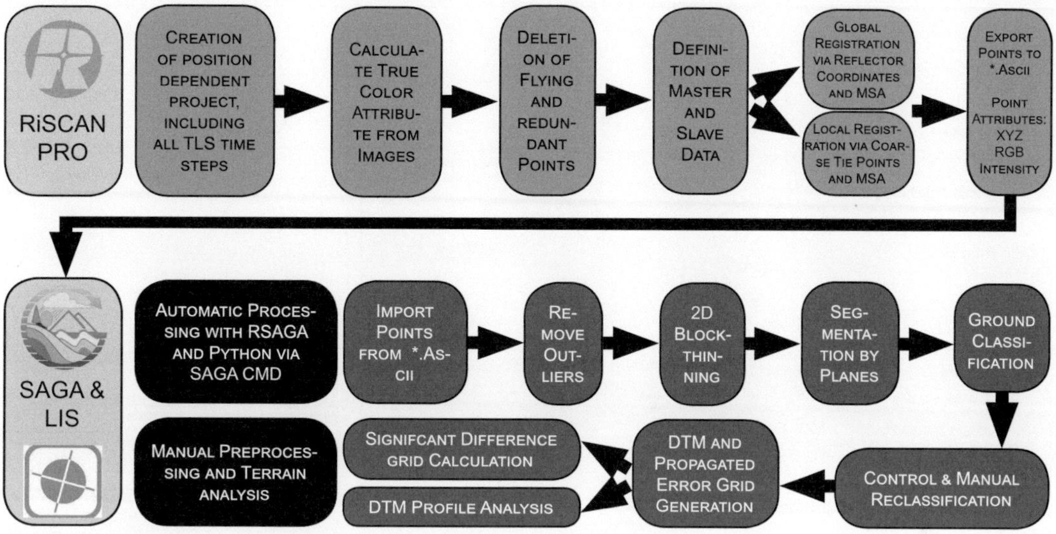

Fig. 11.1 Postprocessing work flow for laserscanning data (from ALS, TLS)

quantitative analyses on different temporal and spatial scales.

The following case studies focus exemplarily on morphometric analyses of landforms and the quantification of morphodynamics within a part of the proglacial lateral moraine (Fig. 11.2) of the Gepatschferner on different timescales. For the investigation of morphodynamics, we used the morphological budgeting approach that is associated with DEMs of difference (DoD). Furthermore, we demonstrate a modelling approach used for regionalising the test site data, both based on multitemporal TLS data and available multitemporal aerial images.

11.3 Results and Discussion

11.3.1 TLS-Based Morphological Budgeting of Slope Wash, Linear Erosion and Debris Flows

During the PROSA project period, a lot of locations have been surveyed several times per year. Figure 11.3 shows the results of multitemporal TLS surveys between 2010 and 2015 for a section of the lateral moraine of the Gepatschferner. A total of 2967.4 m^3 of sediments were transferred from the moraine to the proglacial river Fagge, which represents the base level of the slope. Erosion mainly took place in the upper third of the slope that is characterised by numerous gullies with steep sidewalls that have developed in (over)consolidated, unsorted and thick glacial sediments. 4122.2 m^3 of the 7089.6 m^3 of eroded material were re-deposited at the lower slope.

Due to the superposition of geomorphic changes, the recognition of erosion and deposition patterns of specific processes in DoDs that span several seasonal cycles is not an easy task. It can be concluded though that denudational processes associated with very wet conditions dominantly drive the development of the moraine crest. Further downslope, erosion patterns become more linear; with decreasing slope angles, transport pathways follow existing channels that were created by debris flows and are reworked by fluvial processes. Below the angle of repose (c. 30°), depositional processes prevail; diagnostic features include accumulation areas, but also linear structures recognised as debris flow levées. Due to the short measuring period, the corresponding DoD (Fig. 11.3) does not show depositional patterns in the shape of fans or cones, but the morphology of the slope clearly indicates debris cones/alluvial fans that emanate from the

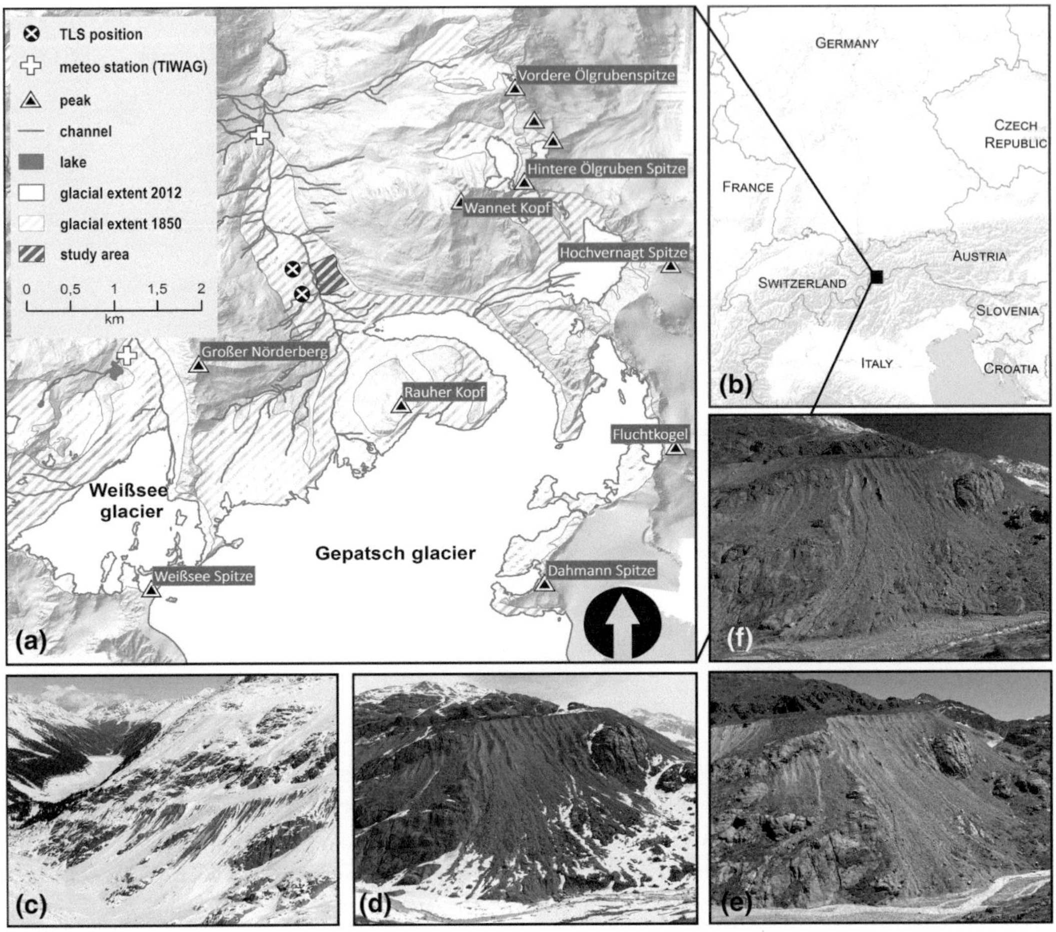

Fig. 11.2 **a** Study area with indication of measurement locations, test sites and morphological properties; **b** the PROSA study area within the European Alps; **c** winter photography of the test site; **d** spring photography of the test site; **e** summer photography of the test site; **f** autumn photography of the test site

gullies. In order to detect (and single out) surface changes caused by processes of even smaller magnitude (e.g. frost heave, slope wash), measurements would have to be taken at a higher spatial and temporal resolution, and with higher accuracy than in the present study.

11.3.2 Seasonal Variability of Erosional Processes

Meteorological parameters as influencing and triggering factors of geomorphic processes are often discussed in the context of gravitational and fluvial processes (e.g. Hagg and Becht 2000; Wetzel 1992; Haas 2008) and process-response systems. Summertime is considered to be the most dynamic season regarding erosion processes in high alpine areas (e.g. Oostwoud Wijdenes and Ergenzinger 1998). Therefore, most studies focused on the influence of local weather conditions on hillslope processes during the snow-free period. The flushing of channels and related fluvial sediment transport triggered by extreme rainfall events dominates erosion during summer (Becht 1995). As summer debris flows can be directly related to extreme rainfall events, Zimmermann (1990) and Hagg and Becht (2000) introduced local rainfall intensity thresholds for the triggering of debris flows in various parts of

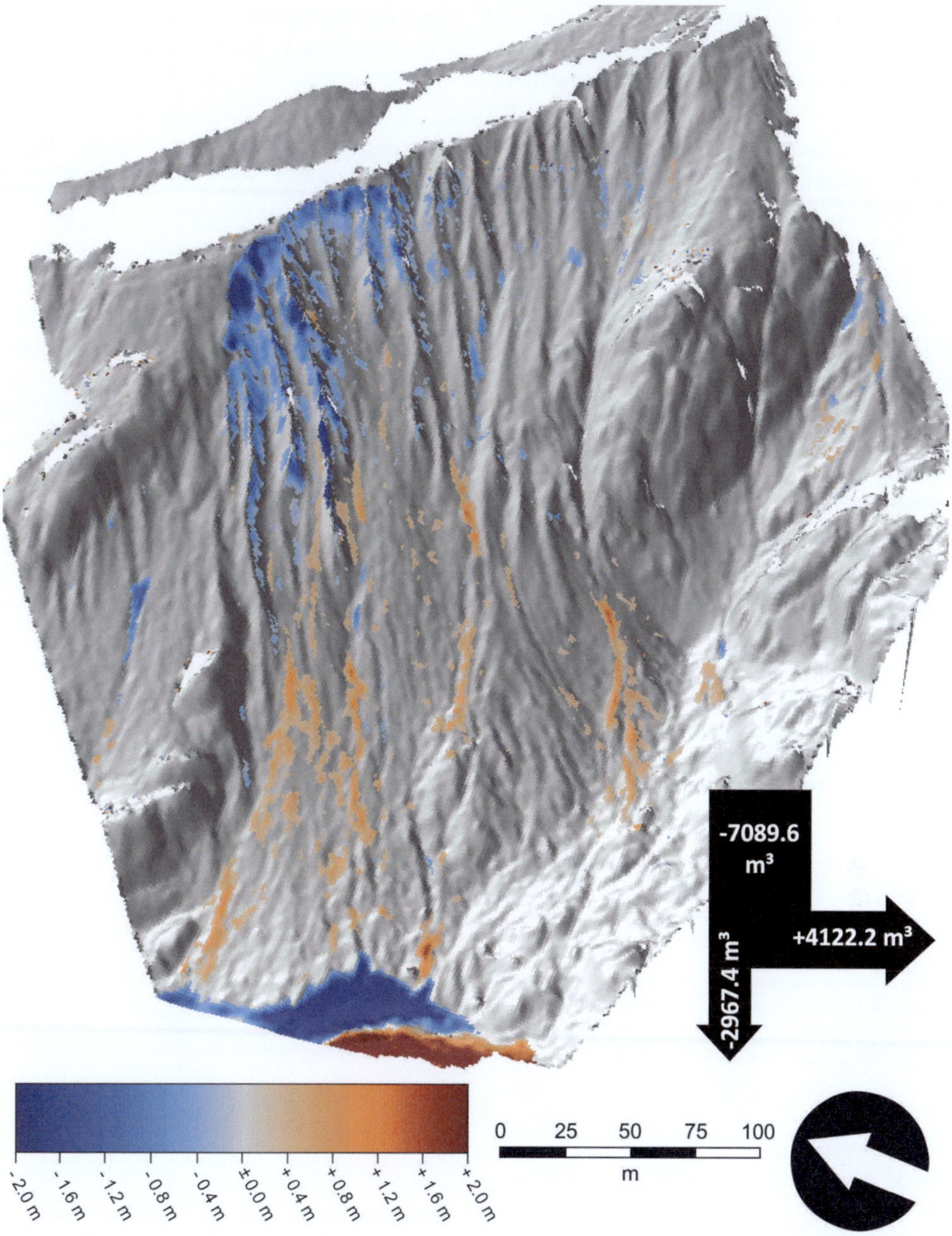

Fig. 11.3 DEM of difference (DoD) and inferred sediment budget representing the longest measurement period (July 2010 to September 2015)

the European Alps. Up to now, only few studies have addressed hydrological processes during the cold season in relation to sediment transport (Wetzel 1992; Oostwoud Wijdenes and Ergenzinger 1998; Schindewolf et al. 2016; Neugirg 2016). Winter erosion is mostly considered to be negligible due to the snow-covered conditions and the fact that precipitation falls as snow, minimising erosion by surface run-off. This may be the case for flat terrain and gently dipping and vegetated slopes with constant snow cover over the entire winter period. On steep, unvegetated slopes, however, the moderate or slow movement of snow (e.g. sliding, creeping) increases due to a loss in cohesion and decreasing shear strength (Gude and Scherer 1995). Conditions of high solar radiation or warm air masses with or without rainfall on the snow cover favour the process. Wetzel (1992), Becht (1995) and Haas (2008) additionally describe processes during the winter season triggered by, e.g. freezing and thawing with a filling of channel beds by the mobilised material during the winter. Schauer (1999) investigated small-scale snow slides of thin snow packs in steep terrain under the influence of insolation and high moisture content. Recent studies by Schindewolf et al. (2016) and Neugirg et al. (2016a) measured winter denudation rates that are four times higher than those in the summer. While winter denudation—i.e. nival and freeze–thaw processes—is driven by gravitational mass transports, summer denudation is dominated by slope wash or fluvial transport.

Within this case study, a seasonal differentiation of the geomorphic activity can be performed based on measurements with a high temporal and spatial resolution, geared to the process magnitude and the specific conditions of the different seasons. The high spatial resolution is achieved by using DEMs generated from TLS data (Riegl VZ4000 Laser scanner; Table 11.1) in order to detect even small-scale geomorphic processes.

Surveying dates were chosen in order to reveal the seasonal activity of different processes (gullying, debris flows, sliding, freeze–thaw-induced rock and debris falls) based on the spatial patterns of erosion and deposition. Therefore, DEMs of difference (DoD) were calculated under consideration of a constant Level of Detection (LoD). The LoD is derived from the DoD uncertainty (which is the root sum of squared standard deviations of error σ_{DEM1} and σ_{DEM2}, pertaining to the pre- and post-event DEMs), multiplied by t = 1.96 in order to account for a 95% confidence interval (c.f. Lane et al. 2003; Table 11.2).

Table 11.1 Overview of used TLS data, and the error of co-registration

Date	σ of pointcloud co-registration [m]
28 September 2013	0.005
28 May 2014	0.026 (Master, global registration)
10 July 2014	0.005
23 September 2014	0.011
20 October 2014	0.014

Data acquisition was performed with a Riegl VZ-4000

Table 11.2 Properties of the LiDAR-based DEMs used for the seasonal analysis of morphodynamics at the test site within the proglacial moraine of the Gepatschferner

Season	Date 1	Date 2	Range of days	Value range of significant changes [m]	Range of grid cell-based propagated errors [m]	LoD [m]
Winter	28 September 2013	28 May 2014	242	−1.94–1.77	0–3.99	0.20
Spring	28 May 2014	10 July 2014	43	−1.78–2.00	0–4.23	0.21
Summer	10 July 2014	23 September 2014	75	−1.93–1.86	0–4.26	0.21
Autumn	23 September 2014	20 October 2014	27	−1.82–1.96	0–4.25	0.21

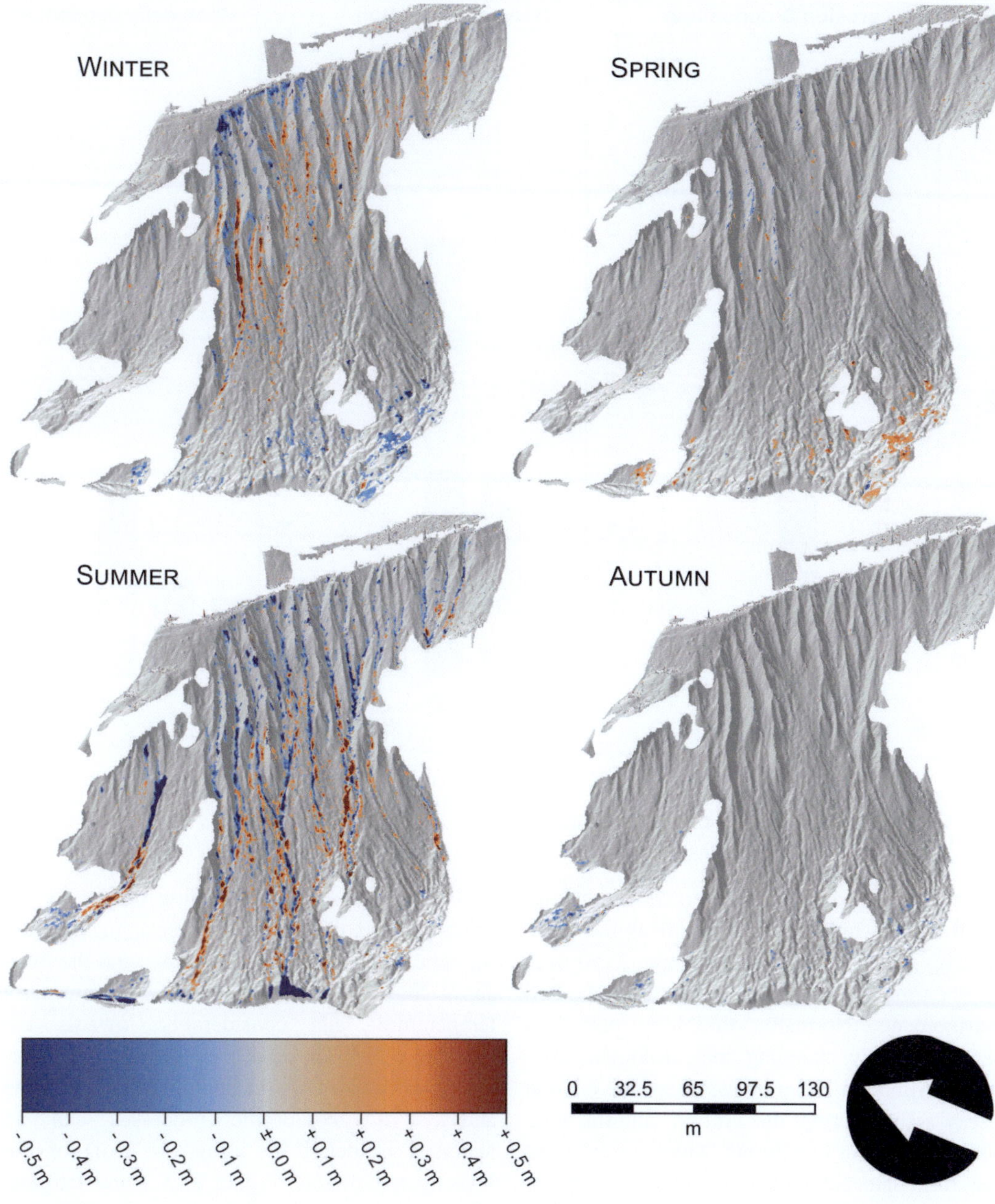

Fig. 11.4 DoD representing winter 2013/2014 (top left), spring 2014 (top right), summer 2014 (bottom left) and autumn 2014 (bottom right)

$$\mathrm{LOD} = 1.96\sqrt{\sigma^2_{\mathrm{DEM1}} + \sigma^2_{\mathrm{DEM2}}}$$

Values of $0 \pm \mathrm{LoD}$ are masked from the DoD, and hence are not taken into account. The interpretation of the results is performed by evaluating precipitation and temperature data within the inter-survey period.

The results (Fig. 11.4) show surface changes and thus reflect geomorphic activity during four seasons, Winter 2013/2014, Spring 2014, Summer 2014 and Autumn 2014, on the

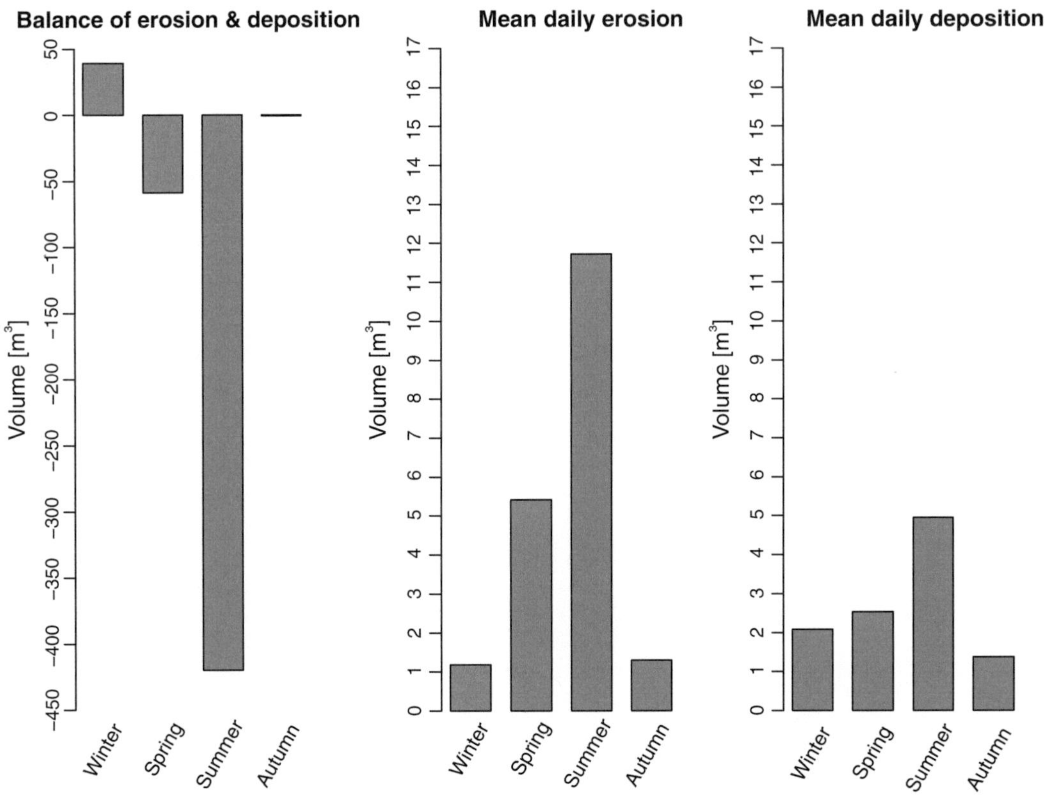

Fig. 11.5 Overall balance (left) and mean daily erosion (middle) and deposition (right) for the seasons of winter 2013/14, spring 2014, summer 2014 and autumn 2014

aforementioned lateral moraine. As the winter period can be rather long in high alpine environments (mean altitude of the slope: 2150 m a. s.l.), the time span is quite long between the two measurements. Normally, this long period is characterised by changing meteorological conditions. Although it can be assumed that erosion is less active during the coldest months with stable snow cover conditions, which would be at least from late November to late February for this location. Springtime begins with the first melting of the snow cover, lasting till the onset of summer in June/beginning of July. Air temperature increases noticeably with clear skies during daytime and most of the winter snow has melted, aside from north exposed, avalanche-fed foot slopes and permafrost or glaciated terrain. Precipitation in summer usually appears as convective heavy rainfall during thunderstorms. In autumn, air temperature and rainfall are decreasing again and, depending on the meteorological conditions, winter sets in quickly. Snowfall is not unusual but sparse, and the fresh snow mostly melts again unless mean daily temperature falls considerably towards the freezing point. The effects of the specific meteorological conditions on the appearance and the activity of geomorphic processes can be analysed in detail by using the DoD of the investigated period. Figure 11.5 show barplots with information on the overall balances (left) and daily mean erosion (middle) and deposition (right) of the respective seasonal timestep DoDs.

11.3.2.1 Winter Morphodynamics (Fig. 11.4, Top Left)

The end of September and October 2013 were comparably humid with some higher rainfall events above 20 mm in 24 h (Fig. 11.6). Daily

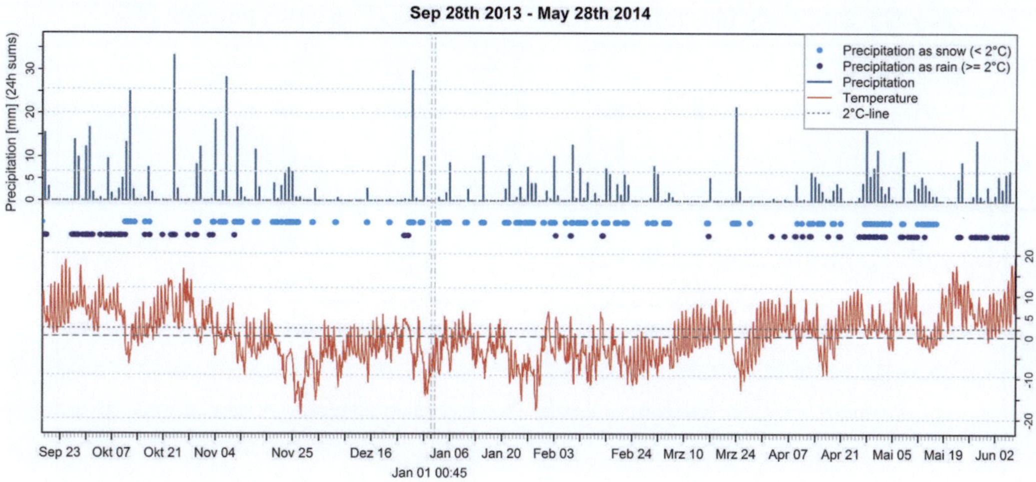

Fig. 11.6 Meteorological data for winter 2013/2014

sums above 19 mm/day count as extraordinary precipitation, due to the dry conditions in the study area. At the same time, temperatures stayed above 0 °C with one exception around the 12th of October. That time already belongs to autumn in high-mountain environments, which is normally a time of dry and sunny weather conditions in the Upper Kaunertal. The unusually high precipitation values in 2013 preserve conditions favourable for fluvial processes and debris flows all around the catchment. Between the beginning of November 2013 and the beginning of March 2014, air temperatures stayed below or slightly above 0 °C and precipitation appears in the solid phase, forming a constant snow cover. Those four months can be considered as a period without geomorphic activity. Beginning with the 7th of March, day temperatures frequently cross the 0 °C line with a constant increase until the 21st of March. From then on, two consecutive periods of several weeks starting with low temperatures (minimum of −12 °C) and a rise to 13 °C and a month of changing positive temperatures follow. In the first half of May 2014, another temperature drop below 0 °C with relatively dry conditions and snow occurred. The erosion pattern for the whole winter period indicates high dynamics for Winter 2013/14 especially on the upper slope. Negative surface changes took place in the channel heads and on the gully sidewalls. The eroded material was deposited and further transported downslope within the channels.

11.3.2.2 Spring Morphodynamics (Fig. 11.4, Top Right)

At the beginning of the spring season, rainfall increased, followed by dry conditions with warm temperatures up to 23 °C and ending with the highest rainfall intensities of this period in the last week of June (7 mm h^{-1} and a daily sum of 37 mm) (Fig. 11.7). During these rainfall events, temperatures are slightly above 0 °C, which indicate predominant wet snowfall. In the first days of July, rainfall of lower magnitude followed, and day–night temperature differences increased. According to the meteorologically stable situation, the DoD of Spring 2014 does not show erosion and deposition patterns outside the range of the LoD. Slight geomorphic activity is found in the upper northern part of the study site. Small displacement values capture gully sidewalls, indicating sliding, and according patchy deposition in rills below. Positive values are also found at the foot slope. However, it is hard to say if they arise from geomorphic activity or remaining snow patches.

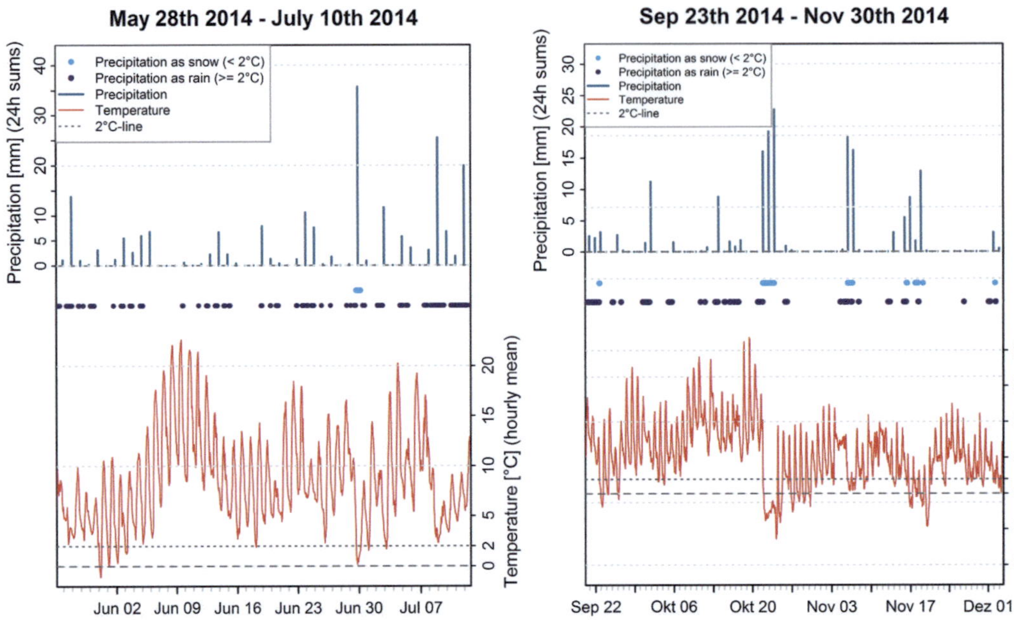

Fig. 11.7 Meteorological data for spring 2014 (left) and autumn 2014 (right)

11.3.2.3 Summer Morphodynamics (Fig. 11.4, Bottom Left)

July 2014 is characterised by almost continuous rainfalls with up to 32 mm in 24 h and temperatures ranging from 3 to 23 °C (Fig. 11.8). The highest daily precipitation sum was 40.5 mm just before both precipitation and temperatures decreased by the middle of August. In September, rainfall intensities decreased steadily, as well as temperature that fell below 0 °C by the 23rd September.

The peak of 13th August triggered several mass movements that reached the main channel. Well prepared by the moisture of constant rainfall in the previous 21 days, the moraine material was easily oversaturated and prone to erosion. As a consequence, patterns of sheet erosion are found along the channel heads and sidewalls. Linear erosion took place within the channel beds of the upper slope. In the northern part of the moraine, some clusters of deposition are visible within a very pronounced channel bed and further below to the north and to the south accumulation took place just below the angle of repose.

11.3.2.4 Autumn Morphodynamics (Fig. 11.4, Bottom Right)

The end of September and the first three weeks of October continued with dry but warm conditions (Fig. 11.7, right). By the 21st of October, temperature dropped significantly with upcoming three days of medium rainfall intensities of 17–22 mm per day.

The DoD of autumn 2014 reflects the meteorological situation with balanced erosion and deposition values around 1 m^3, indicating only a slight geomorphic activity.

Winter patterns reveal deposition in the channels above the angle of repose, and erosion at the channel headcuts and sidewalls. This pattern can be explained by freeze–thaw processes on gully sidewalls and ridges that loosen particles, which are then transported by gravitational processes, slope wash or snow gliding under melting conditions (Wetzel 1992; Haas 2008). Oostwoud Wijdenes and Ergenzinger (1998) describe the same phenomenon for bedrock (marls) in the French Alps and call them 'miniature debris flows' (MDF), triggered by oversaturation of the surface material due to

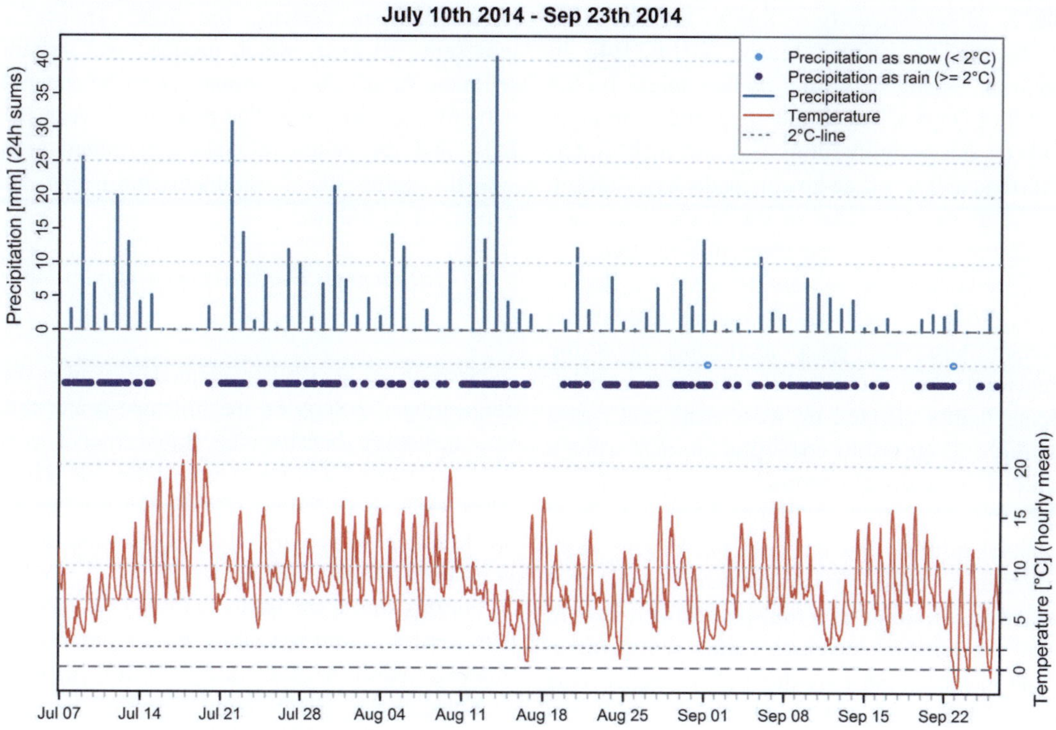

Fig. 11.8 Meteorological data for summer 2014

infiltration of snow melt water and a subsequent movement of the low-viscosity debris-water-matrix (Oostwoud Wijdenes and Ergenzinger 1998; Wetzel 1992). Sheet erosion of the channel sidewalls and a corresponding filling of the channel are typical for the late winter season. Also induced by melting snow and infiltration of the meltwater into the softened moraine sediments, the shear strength of the steep channel sidewalls is reduced, causing slides. Conspicuous deposition along the channel sidewalls at the upper slope is related to slides or collapses of the same (Curry et al. 2006).

The summer season is the most dynamic period; it is dominated by debris flows within channels, fluvial erosion from the channel heads downwards and erosion of the channel sidewalls by sliding. One debris flow channel that was inactive in previous years was reactivated and shows major surface changes.

11.3.2.5 Conclusion

The absolute volumes of erosion and deposition (Fig. 11.5) clearly demonstrate seasonal variation. Processes like frost heave and creep, rill erosion or any other change below the LoD are not (or only cumulatively) represented in the balance. Freeze–thaw cycles, especially in spring, facilitate small magnitude rock or debris falls with subsequent channel infilling, sidewall and headcut erosion. The measurement of these is restricted by the detection limits of the method and thus has an influence on the relative volumes of erosion and deposition that can hardly be estimated. In the summer of 2014, erosion exceeds deposition, which means that the investigated lateral moraine section is coupled to the channel network during summer time. In winter, deposition values exceed erosion. Because of a steady snow cover below the angle of repose, surface processes are restricted to the upper slope, which is clearly visible in the DoD that represents the winter season 2013/2014. In addition, higher deposition values might be the result of (1) swelling of the upper sediment layer due to steady infiltration of snow melt water, (2) deposition of sediment pathways starting above the upper border of the moraine part and (3) a bias related to measurement limitations.

Similar observations are reported by Bechet et al. (2016) in the Draix-Bléone area, Southern French Alps. The black marls that build the landscape there behave like moraine material, being highly affected by weathering and easily erodible. Their results correspond to ours when a shift of the seasonal patterns is considered due to differing climatic conditions. They observed channel infill during the cold but mostly snow cover-free period by frost-driven loosening of sediments, leading to a transport-limited system until precipitation increases during summertime. High-intensity and long-duration rainfall events flush the channels until erodible sediment sources are exhausted, and the system switches to a supply-limited state later in the year (Bechet et al. 2016).

11.3.3 Modelling Hillslope Sediment Yield

Since the measurement of linear erosion and slope wash is only possible on a test site basis, Haas et al. (2011b) developed a statistical model to regionalise measured sediment output from the plot to the catchment scale. In their study, bedload sediment yield was measured by sediment traps installed in selected channels. The measured sediment yield was then related to the sediment contributing area (SCA) of each sediment trap; the SCA is delineated using a set of rules, a DEM and a raster representation of the channel network (see Haas et al. 2011b). The model is based on the assumption that the SCA is a subset of the hydrological catchment, since not all areas contribute sediment to the traps. Some areas that are hindering the sediment mobility, like flat areas or areas with vegetation cover are excluded when deriving the SCA. The SCA represents all areas which produce and sustain sediment supply and are connected to the channel network sediment trap. The relation between the SCA and the sediment yield determined at a specific point (here: sediment trap) can be expressed as follows:

$$\boldsymbol{log \cdot sediment\,yield = intercept + slope * log \cdot SCA}$$

Neugirg et al. (2015) adapted the rules for delineating the SCA on the hillslope scale; this was necessary because the initial model was developed and tested for entire catchments (Haas 2008; Haas et al. 2011a). The model was developed further by Neugirg et al. (2016a) and tested in different climatic settings.

For this study, we use DoD from TLS surveys that were accumulated along flow directions in order to obtain the sediment yield (SY [m^3]) of each raster cell's contributing area (Pelletier and Orem 2014; Neugirg et al. 2016b). The sum of the DoD of any area (here, the contributing area of a particular raster cell) represents the net balance, with negative values indicating net erosion and sediment export, and zero indicating a balance of erosion and deposition. Within the contributing area of a raster cell, this balance has to be either negative (=sediment export) or zero (balanced). The SCA was delineated using the rules described above, with a channel gradient threshold of min. 5° was chosen to assure continuous connectivity along the flow path. The size of the SCA was computed using the real surface area of each cell in order to obtain realistic values. Following the methodology proposed by Neugirg et al. (2016a), the SCA and corresponding SY were sampled at one random location per channel, thus generating an independent sample of 'virtual sediment traps'. This random sampling was repeated 100 times in order to obtain the variability of the regression parameters (Fig. 11.9). Where the accumulated DoD showed spurious positive values (c.f. Fig. 11.9, bottom right), they were excluded from the sample. Possible reasons for positive

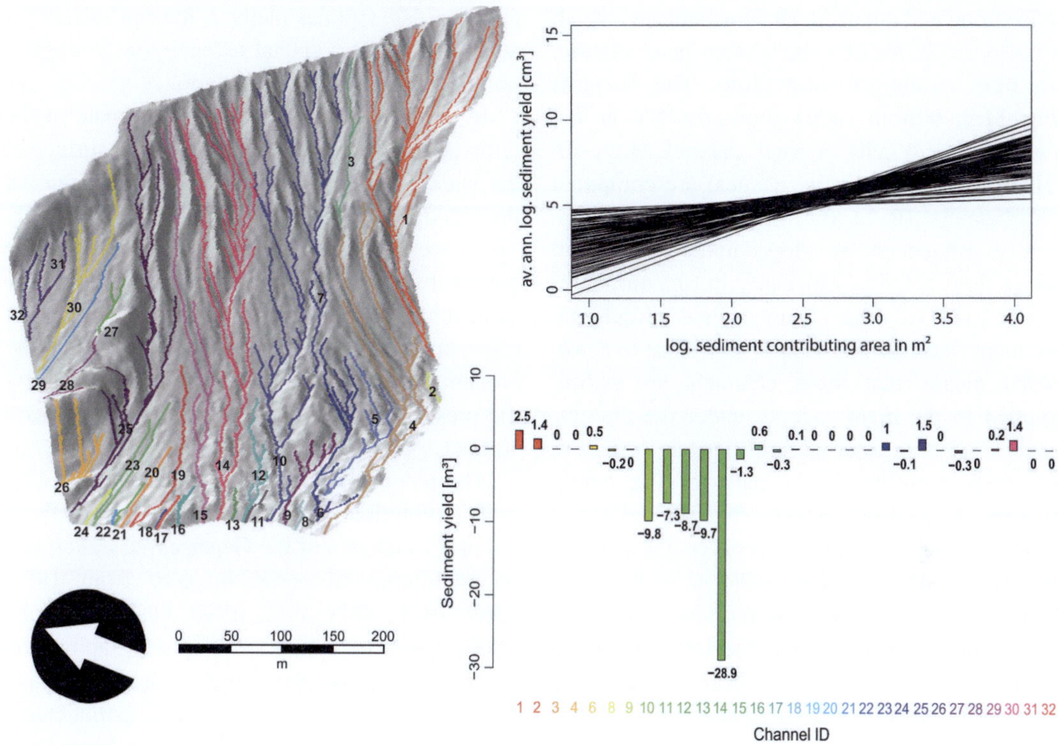

Fig. 11.9 Moraine section and channels (numbered on the map and in bottom right diagram) used for the SCA model. Top right diagram shows linear regressions between log sediment yield and log sediment contributing area, and barplots (bottom right) show the SY at each channel outlet

values include DEM uncertainty and changes in flow routing between the two epochs (e.g., due to debris flows).

Generally, channels with a greater SCA show higher sediment yield at the foot of the slope. Log SCA explains, on average, 21% (up to 61%, compare Table 11.3) of the observed variance in sediment yield at the virtual sediment traps. The model was validated by a tenfold cross-validation of 4121 observations, resulting in a root mean square error (RMSE) of 0.195 m^3.

Compared to model results of study sights described in Neugirg et al. (2016a), the ranges of intercept and slope values vary in range. This can be explained by substantially longer slope lengths and a higher variability of channel lengths within the slope. Best correlations are found for very short channel lengths (IDs 8, 16, 17, 18, 21 and 22, cf. Fig. 11.9) that have their starting points below the angle of repose. We suggest two reasons to explain this behaviour: first, channel heads of the lateral moraine in the Upper Kaunertal are not prone to extensive fluvial erosion, but are heavily influenced by gravitational processes; and second, discharge needs a certain minimum catchment area to

Table 11.3 Summary statistics for the SCA model

	Intercept	Slope	R^2	p-Value
Min.	−1.132	0.1203	0.00	1.457 e-7
Mean	1.9214	1.3541	-/-	-/-
Median	1.956	1.3298	0.20	-/-
Max.	4.683	2.5994	0.61	0.84

concentrate and create linear flow patterns, which is not possible for extremely steep headcuts that are most prone to sheet flow. The barplots (Fig. 11.9, bottom right) show the SY at the lowermost grid cells of each channel. High SY values (i.e. large negative values) are computed for channels 10–14 (cf. Fig. 11.9). These are clearly influenced by slope undercutting and debris flow activity, especially during the summer of 2014. All other channels yield virtually no sediment (indicated by values very close to zero), which means that these channels are poorly coupled to the main channel under the circumstances of the inter-survey period; sediment delivered from the upper parts of the lateral moraine is re-deposited on the alluvial/debris fans. Only high-magnitude events or debris flows are capable of delivering sediment to the main channel on large parts of the moraines (see Haas et al. 2012). Positive values (actually indicating spurious results) could have been caused by a decrease in density and associated heave caused by frost action.

11.3.4 Paraglacial Adjustment of Proglacial Moraines

We use an ALS-based DEM (1 m resolution) acquired in 2012 for a morphometric analysis of the lateral moraine under the assumption that morphometric parameters of moraine sections (e.g. gullies) of different ages since deglaciation reflects different stages of paraglacial adjustment (c.f. Curry et al. 2006). A total of 90 single gullies, distributed over eleven more or less homogenous gully zones, were digitised from the DEM and orthophotos.

An age surface estimating the time since deglaciation for each gully site was interpolated from glacial extents using glacier inventories dating back to 1850 (Patzelt 1980; Groß 1987; Lambrecht and Kuhn 2007; Hartl 2010). This approach relies on the assumption of constant recession rates between the mapped glacier extents, ignoring intermediate re-advances. Points sampled from the dated glacier outlines were used to interpolate (splines method) the age surface; in areas with similar glacial extents, the 'youngest' points were used for interpolation.

In order to obtain a minimum estimate of the gully erosion volumes, the pre-incision surface of the lateral moraine was interpolated between the present-day inter-gully ridges. The inter-gully ridges were digitised to mask the gullies, and the gaps between inter-gully ridges were closed by local thin plate spline interpolation. The amount (elevation differences and volumes) of gully erosion was obtained by computing a DoD using the present-day and the interpolated pre-incision surface.

The reconstruction of landform (specifically: gully) evolution on the lateral moraine walls in the proglacial area of the Gepatschferner is based on multitemporal aerial imagery from 1953, 1969, 1971, 1979, 1982, 1990, 1997, 2003 and 2009. Where necessary, aerial photographs were processed by bundle block adjustment and orthorectification. Gully shapes and spatial extent were digitised for each of the ten time steps covered by aerial photos (Fig. 11.10). Gullies in the outermost regions, and gullies that touch bedrock areas in the early years of the aerial photos were excluded from the analysis. The extent of each gully within the selected moraine sections was defined by inter-gully sidewall ridges, the upper moraine ridge or by interspersed bedrock outcrops that were uncovered by erosion over the years. Lower limits of the gullies were set where inter-gully sidewalls disappeared, touching the boundary to alluvial and debris cones forming below the gullies. We follow Curry et al. (2006) in using gully density as a measure of the state of moraine reworking. Where possible, the mean frequency of gullies per kilometre was determined. In some aerial photos, the distinction of initial gullies and small rills was difficult; therefore, gully frequency should be interpreted with respect to spatial trends only. In addition, we analysed gully floor slope profiles from the channel head to the end of the flow path (on alluvial cones or at the junction with the channel network or the glacier) that were obtained from the (present-day) DEM.

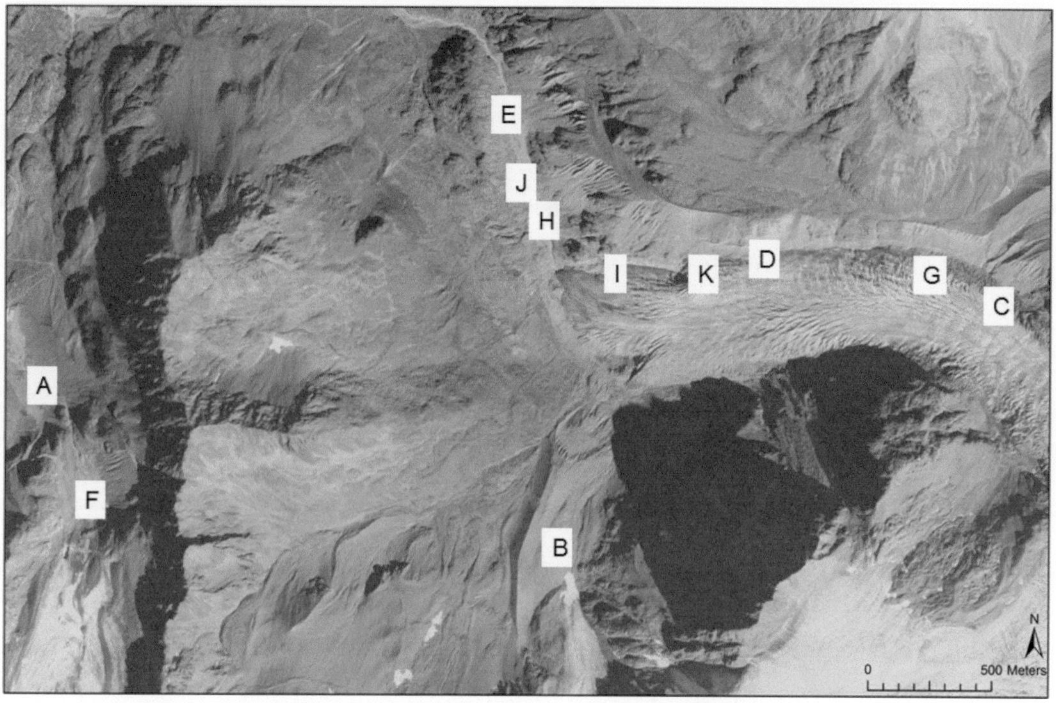

Fig. 11.10 Overview of eleven gully locations (A to K) analysed on basis of the DEM

11.3.4.1 Volumetric and Spatial Extent Changes

The eleven gully sites show great differences in estimated minimum gully volumes, both between and within groups. In contrast, the two oldest (A and B) and the two youngest age groups (J and K) are characterised by very similar reconstructed gully volumes. These four gully locations appear to be inactive nowadays, as no significant changes were detected on recent DoD. For the seven age groups between them, the gully volumes are much more variable. Only two pairs of age groups show significant differences in gully volumes (A–B and J–K). All other differences between age groups are not significant at the 5% significance level (Mann-Whitney-U two sample test). Furthermore, Fig. 11.11 indicates that there is no correlation between the time since deglaciation and the estimated gully volume.

The planimetric areas of the currently existing gullies show a similar pattern. The two oldest and youngest age groups are dissimilar to the seven other age groups in between: very homogenous planimetric gully areas in the two youngest and oldest age groups and a great scatter of planimetric areas in the seven other age groups. There are also only two age group pairs (the same that are different with respect to volumes) that are significantly different in their planimetric area at the 5% significance level (Fig. 11.12).

The volumetric change and the spatial extent of the gullies on the upper moraine slopes represent surrogate measures for the reworking intensity. Both analyses show that a strict correlation between time since deglaciation and intensity of reworking is not given. As concluded by Curry et al. (2006), gullies appear to reach their maximum dimensions approx. 50 years after deglaciation. If this conclusion is also valid for our studied gullies, the two youngest age groups should show a much greater amount of reworked material and greater spatial extents. The existence of differently shaped and sized gullies indicates that a general trend of gully initiation, formation and degradation cannot be inferred from the measurements in our study

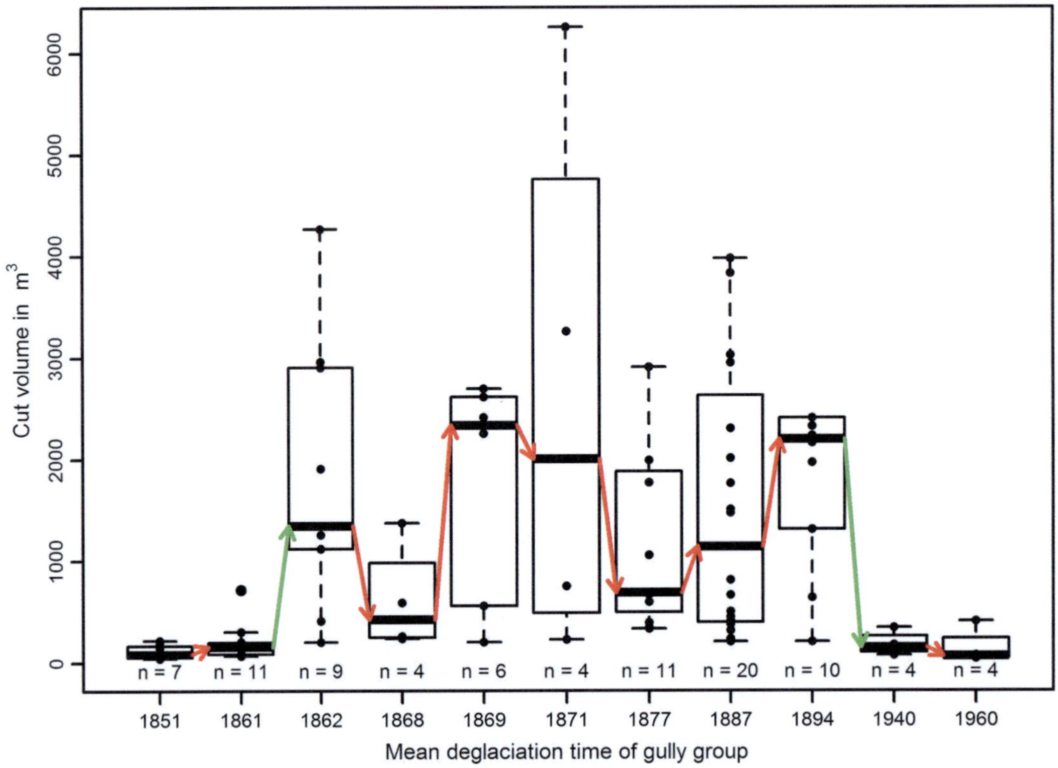

Fig. 11.11 Estimated volumes of gullies on moraine sections of different age since deglaciation. Green arrows: significant differences, red arrows: non-significant differences

area; this is probably due to alternative influencing factors in addition to the time since deglaciation.

11.3.4.2 Influence of Slope Length on Gully Development

The slope length is measured as the overland flow length starting from the headcut and ending at the foot of the accumulation zone of the respective gully. The measured flow paths had lengths between 87 and 489 m. There is no statistical correlation between the slope length and the volumetric loss or the planimetric gully area in our study area (Fig. 11.13).

The gullies investigated by Curry et al. (2006) show a stabilisation 80–140 years after deglaciation. In our study area, we found several groups of gullies that have been free of ice for more than 140 years but turned out as not stabilised yet. There are gully locations with a short time since deglaciation with only small planimetric areas and small cut volumes, and likewise there are many gullies with a longer time since deglaciation which show off much greater spatial extents and cut volumes. The time since deglaciation and the analysed variables show no clear trend in contrast to the outcomes of the study by Curry et al. (2006) at the Fee glacier. Nevertheless, several aspects have to be kept in mind when comparing the results of these two studies: (1) the Fagge river is—from time to time—undercutting parts of the slope during high-discharge events. The natural slope gradient is disturbed by these flood events in the lower lying slope parts, which possibly leads to increased erosion of the moraine wall. (2) Erosion at the slope foot by the Gepatsch glacier is also leading to increased slope erosion and gully incision at locations C, D, G, I and K. (3) In contrast to the Gepatschferner area, the moraine walls in the Fee glacier region are (nearly) free of bedrock outcrops. These existing bedrock

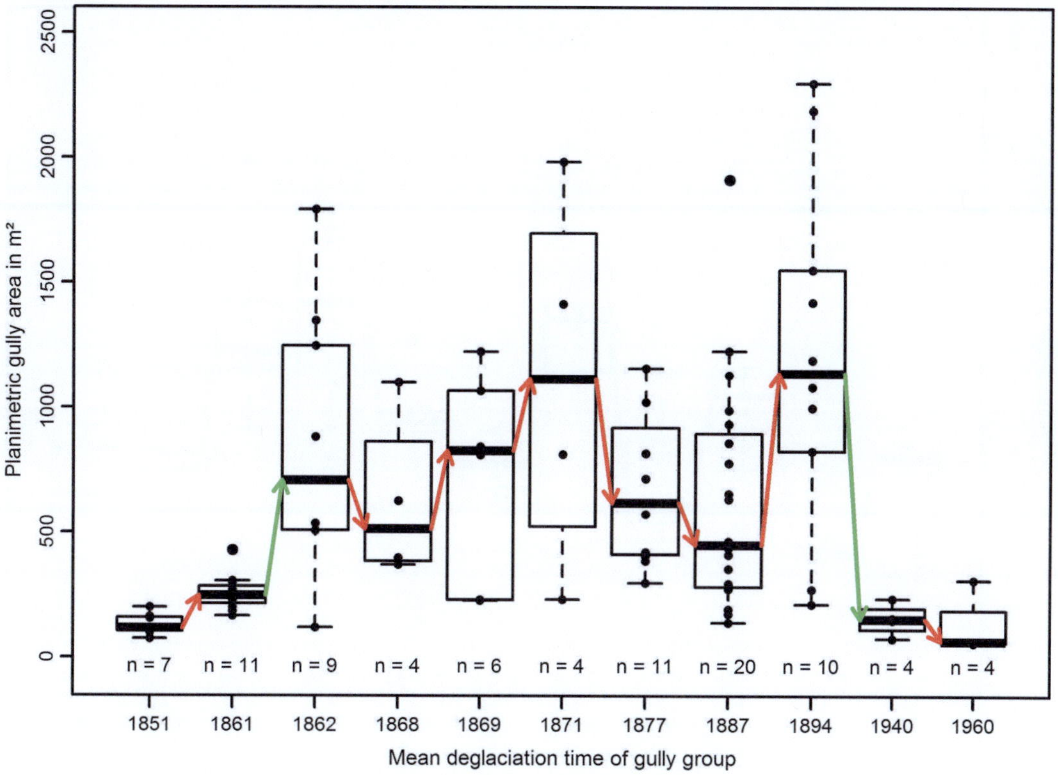

Fig. 11.12 Planimetric gully areas of different age group gullies. Green arrows: significant differences, red arrows: non-significant differences

outcrops in the Gepatsch area lead to a disturbance in gully development.

11.4 Conclusions

The investigated part of the Gepatschferner's lateral LIA moraine is prone to reworking by non-glacial processes following deglaciation. The medium-term measurement of change detection clearly shows that erosion by sheetflow, sliding along gully walls and collapsing of gully ridges dominates the headcuts, while deposition by slope wash and debris flows takes place below the angle of repose. Only about 40% of the mobilised sediments left the slope system, while nearly 60% were re-deposited within the slope. Undercutting of the moraine is another visible pattern on the DoD and is associated with flooding of the channel bed by the main proglacial river Fagge during times of extraordinary discharge and high sediment loads.

As shown by the case study on seasonal patterns of reworking, winter erosion plays an important role in hillslope development. Spatial patterns of surface changes observed in the cold season reflect mass movements, related to gravitational processes, temperature and moisture-induced snow sliding as well as freeze–thaw processes. Sheet erosion in channel heads and on the sidewalls of steep channels is the dominant process with deposition downslope in the channel bed. Rill erosion is only taking place if it comes to meltwater or rainfall-induced fluvial processes that flush the channel. They are spatially limited to the upper slope above the angle of repose where snow cover thickness and persistence are minimised due to the steepness and solar insolation. Springtime denudation was expected to be more effective, as day–night temperature

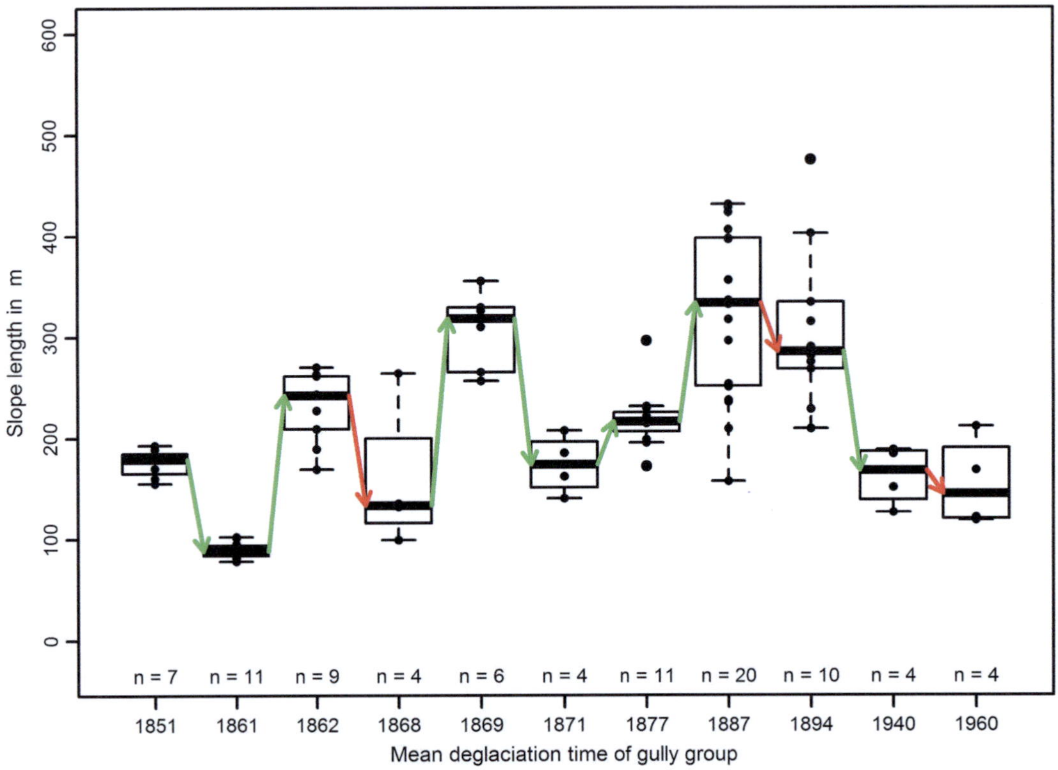

Fig. 11.13 Mean slope length of different age group gullies. Green arrows: significant differences, red arrows: non-significant differences

gradients are steep at this time of the year and freeze–thaw cycles are common as well as changing phases of precipitation. Thus, loosening of particles and softening of the moraine material by the infiltration of snowmelt and rain prepare its mobilisation. March 2014 already had some very warm days of up to 13 °C. This probably led to an earlier occurrence of transport processes. Summer erosion patterns are driven by rainfall duration, intensity and the availability of sediment. Mass movements and linear erosion mostly took place within existing channel beds after several days of intense rain with directly following rainfall events. The highest rates of morphodynamics, with large volumes of transported sediment and extensive gully modification, were caused by high-magnitude precipitation events that triggered debris flows, e.g. in the summer of 2014. Mass movements like debris flows in the Upper Kaunertal are related to comparatively low intensities of max. 12 mm h^{-1} or 41 mm in 24 h by comparison to other regions (c.f. Hagg and Becht 2000). Autumn does not show any conspicuous patterns as climate conditions are normally dry and steady with generally still warm temperatures above the freezing point until late October and only few freeze–thaw cycles.

In order to account for linear fluvial erosion, we computed sediment yield by cumulating DoD (summer 2014) along flow paths (Pelletier and Orem 2014; Neugirg et al. 2016a, b) in 32 channels. The resulting regression models are in good agreement with other studies in research areas with unsorted and compacted material (Haas et al. 2011a; Neugirg et al. 2016a), both in terms of model parameters and R^2. But following the statement: '[…] lower intercept values indicate lower erosion values at the channel heads, higher slope values indicate a stronger increase of erosion with increasing SCA' (Neugirg et al. 2016a), it can be concluded that fluvial processes account for little erosion on the studied part of the lateral moraine, compared to the amount of sediments transported by especially debris flows.

Regarding the case study on paraglacial adjustment by fluvial incision, we conclude that the intensity of moraine wall reworking is not only based on the time since deglaciation. There is no clear association between time since deglaciation and planimetric areas or cut volume of the gully system along the LIA moraine crest. Hence, natural factors like bedrock outcrops, connected rivers and glaciers act as controlling parameters for morphodynamics, as they influence slope gradient and length and are able to hamper the degree of gully incision.

References

Ballantyne CK, Benn DI (1994) Paraglacial slope adjustment and resedimentation following recent glacier retreat, Faabergstolsdalen, Norway. Arct Antarct Alp Res 26:255–269

Ballantyne CK, Benn DI (1996) Paraglacial slope adjustment during recent deglaciation and its implications for slope evolution in formerly glaciated environments. Brooks S, Anderson MG (Eds.) Advances in Hillslope Processes, pp 1173–1195. Wiley, Chichester

Bechet J, Duc J, Loye A, Jaboyedoff M, Mathys N, Malet J-P, Klotz S, Le Bouteiller C, Rudaz B, Travelletti J (2016) Detection of seasonal cycles of erosion processes in a black marl gully from a time series of high-resolution digital elevation models (DEMs). Earth Surf Dynam 4:781–798. https://doi.org/10.5194/esurf-4-781-2016

Becht M (1995) Untersuchungen zur aktuellen Reliefentwicklung in alpinen Einzugsgebieten: Mit 40 Tabellen. Münchener Universitätsschriften, vol 47. Geobuch-Verl., München

Cossart E, Braucher R, Fort M, Bourlès D, Carcaillet J (2008) Slope instability in relation to glacial debuttressing in alpine areas (Upper Durance catchment, southeastern France): evidence from field data and 10Be cosmic ray exposure ages. Geomorphology 95:3–26. https://doi.org/10.1016/j.geomorph.2006.12.022

Curry AM (1999) Paraglacial modification of slope form. Earth Surf Proc Land 24:1213–1228

Curry AM, Cleasby V, Zukowskyj P (2006) Paraglacial response of steep, sediment-mantled slopes to post-Little Ice Age glacier recession in the central Swiss Alps. J Quat Sci 21:211–225. https://doi.org/10.1002/jqs.954

Groß G (1987) Der Flächenverlust der Gletscher in Österreich 1850-1920-1969. Z Gletscherk Glazialgeol 23:131–141

Gude M, Scherer D (1995) Snowmelt and slush torrents: preliminary report from a field campaign in Kärkevagge, Swedish Lappland. Geogr Annaler A 77:199–206. https://doi.org/10.2307/521329

Haas F (2008) Fluviale Hangprozesse in Alpinen Einzugsgebieten der Nördlichen Kalkalpen: Quantifizierung und Modellierungsansätze. Dissertation

Haas F, Heckmann T, Becht M, Cyffka B (2011a) Ground-based laserscanning—a new method for measuring fluvial erosion on steep slopes. In: Hafeez MM (ed) GRACE, remote sensing and ground-based methods in multi-scale hydrology: Proceedings of the symposium JHS01 [entitled: GRACE, remote sensing and ground-based methods in multi-scale hydrology] held during the IUGG GA in Melbourne (28 June–7 July 2011). IAHS Publ, Wallingford, pp 163–168

Haas F, Heckmann T, Wichmann V, Becht M (2011b) Quantification and modeling of fluvial bedload discharge from hillslope channels in two alpine catchments (Bavarian Alps, Germany). Z Geomorph NF 55 (Suppl):147–168

Haas F, Heckmann T, Hilger L, Becht M (2012) Quantification and modelling of debris flows in the proglacial area of the Gepatschferner, Austria, using ground-based LiDAR. In: Collins A (ed) Erosion and sediment yields in the changing environment: proceedings of an IAHS International Commission on continental erosion symposium held at the institute of mountain hazards and environment, CAS-Chengdu, China, 11–15 Oct 2012. IAHS Press, Wallingford, pp 293–302

Hagg W, Becht M (2000) Einflüsse von Niederschlag und Substrat auf die Auslösung von Hangmuren in Beispielgebieten der Ostalpen. In: Becht M, Schmidt K-H (eds) Angewandte und vernetzte geomorphologische Prozeßforschung, vol 123. Borntraeger, Berlin, pp 79–92

Hartl L (2010) The Gepatschferner from 1850–2006: changes in length, area and volume in relation to climate. University of Innsbruck

Holm K, Bovis M, Jakob M (2004) The landslide response of alpine basins to post-Little Ice Age glacial thinning and retreat in southwestern British Columbia. Geomorphology 57:201–216

Hugenholtz CH, Moorman B, Barlow J, Wainstein P (2008) Large-scale moraine deformation at the Athabasca Glacier, Jasper National Park, Alberta, Canada. Landslides 5:251–260. https://doi.org/10.1007/s10346-008-0116-5

Carrivick JL, Heckmann T (2017) Short-term geomorphological evolution of proglacial systems. Geomorphology 287:3–28

Kellerer-Pirklbauer A, Proske H, Strasser V (2010) Paraglacial slope adjustment since the end of the last glacial maximum and its long-lasting effects on secondary mass wasting processes: Hauser Kaibling, Austria. Geomorphology 120:65–76. https://doi.org/10.1016/j.geomorph.2009.09.016

Klok E, Oerlemans J (2003) Deriving historical equilibrium-line altitudes from a glacier length record by linear inverse modelling. Holocene 13:343–351. https://doi.org/10.1191/0959683603hl627rp

Lambrecht A, Kuhn M (2007) Glacier changes in the Austrian Alps during the last three decades derived from the new Austrian glacier inventory. Ann Glaciol 46:177–184. https://doi.org/10.3189/172756407782871341

Lane SN, Westaway RM, Murray Hicks D (2003) Estimation of erosion and deposition volumes in a large, gravel-bed, braided river using synoptic remote sensing. Earth Surf Proc Land 28:249–271. https://doi.org/10.1002/esp.483

Matthews JA, Shakesby RA (2004) A twentieth-century neoparaglacial rock topple on a glacier foreland, Ötztal Alps, Austria. Holocene 14:454–458. https://doi.org/10.1191/0959683604hl706rr

McColl ST (2012) Paraglacial rock-slope stability. Geomorphology 153–154:1–16. https://doi.org/10.1016/j.geomorph.2012.02.015

Morche D, Haas F, Baewert H, Heckmann T, Schmidt K-H, Becht M (2012) Sediment transport in the proglacial Fagge River (Kaunertal/Austria). In: Collins A (ed) Erosion and sediment yields in the changing environment: proceedings of an IAHS International Commission on continental erosion symposium held at the institute of mountain hazards and environment, CAS-Chengdu, China, 11–15 Oct 2012. IAHS Press, Wallingford, pp 72–80

Neugirg F (2016) Quantifizierung, Analyse und Modellierung von Erosionsprozessen auf Steilhängen in unterschiedlichen Klimaten durch hochaufgelöste Geländemodellen. Dissertation, Katholische Universität Eichstätt-Ingolstadt

Neugirg F, Kaiser A, Schindewolf M, Becht M, Schmidt J, Haas F (2015) Monitoring and modelling slope dynamics in an Alpine watershed—a combined approach of soil science, remote sensing and geomorphology. Proc IAHS 371:181–187. https://doi.org/10.5194/piahs-371-181-2015

Neugirg F, Stark M, Kaiser A, Vlacilova M, Della Seta M, Vergari F, Schmidt J, Becht M, Haas F (2016a) Erosion processes in calanchi in the Upper Orcia Valley, Southern Tuscany, Italy based on multitemporal high-resolution terrestrial LiDAR and UAV surveys. Geomorphology 269:8–22. https://doi.org/10.1016/j.geomorph.2016.06.027

Neugirg F, Kaiser A, Huber A, Heckmann T, Schindewolf M, Schmidt J, Becht M, Haas F (2016b) Using terrestrial LiDAR data to analyse morphodynamics on steep unvegetated slopes driven by different geomorphic processes. CATENA 142:269–280. https://doi.org/10.1016/j.catena.2016.03.021

Oostwoud Wijdenes DJ, Ergenzinger P (1998) Erosion and sediment transport on steep marly hillslopes, Draix, Haute-Provence, France: an experimental field study. CATENA 33:179–200. https://doi.org/10.1016/S0341-8162(98)00076-9

Patzelt G (1980) The Austrian glacier inventory: status and first results. IAHS Publication 126

Pelletier JD, Orem CA (2014) How do sediment yields from post-wildfire debris-laden flows depend on terrain slope, soil burn severity class, and drainage basin area?: Insights from airborne-LiDAR change detection. Earth Surf Proc Land 39:1822–1832. https://doi.org/10.1002/esp.3570

Schauer T (1999) Beispiele von Erosionsprozessen in Zusammenhang mit den Standortfaktoren Nutzung und Vegetation im Bayerischen Alpenraum. Relief Boden Pal'aoklima 14:117–128

Schiefer E, Gilbert R (2007) Reconstructing morphometric change in a proglacial landscape using historical aerial photography and automated DEM generation. Geomorphology 88:167–178

Schindewolf M, Kaiser A, Neugirg F, Richter C, Haas F, Schmidt J (2016) Seasonal erosion patterns under alpine conditions: benefits and challenges of a novel approach in physically based soil erosion modeling. Zeit fur Geo 60(Supp):109–123. https://doi.org/10.1127/zfg_suppl/2015/s-00185

Wetzel K-F (1992) Abtragsprozesse an Hängen und Feststoffführung der Gewässer. Dargestelllt am Beispiel der pleistozänen Lockergesteine des Lainbachgebietes (Benediktbeuern/ Obb.). Münchener Geographische Abhandlungen B, vol 17. Geobuch-Verlag, München

Zimmermann M (1990) Debris flows 1987 in Switzerland: geomorphological and meteorological aspects. In: Sinniger RO, Monbaron M (eds) Hydrology in mountainous regions II—artificial reservoirs; water and slopes, vol 194, pp 387–393

Part III
Proglacial Rivers and Lakes

12 Sediment Transport in Proglacial Rivers

Luca Mao, Francesco Comiti, Ricardo Carrillo and Daniele Penna

Abstract

Suspended and bedload transport in mountain rivers are notoriously difficult to monitor and thus to quantify. Even more challenging are proglacial streams that are generally characterized by high flow velocities, high sediment concentrations and transport rates, large and abrupt temporal variability of water discharge and transport rates, as well as by frequent morphological changes. However, the quantification of suspended and bedload transport rates is crucial to understand and predict short- and long-term morphodynamics in such sensitive systems. In this chapter, we present an updated summary on the current knowledge on the monitoring methods and on the observed dynamics of sediment transport in proglacial rivers, including both suspended and bedload fractions.

Keywords

Sediment transport · Sediment yield Proglacial river · Measurement · Rating curves · Hysteresis

12.1 Introduction

Suspended and bedload fluxes are crucial physical processes that are key in determining morphology and morphodynamics of proglacial rivers. A proper understanding of sediment fluxes can allow assessing the current dynamics of proglacial rivers and also inferring geomorphic responses to climate changes in glacierized environments. However, measuring sediment fluxes in the channel network is highly challenging (Gray et al. 2010), both in lowland and mountain rivers. Particularly, mountain rivers have the advantage of being smaller in size (in terms of width and depth) and thus potentially easier to "handle", but their flows are faster and more turbulent, the particles transported at the bed can reach the boulder size range, and sediment-transporting flow events can be short-lasting and limited during the year compared to lowland rivers

L. Mao (✉) · R. Carrillo
Department of Ecosystems and Environment, Pontificia Universidad Católica de Chile, Santiago, Chile
e-mail: LuMao@lincoln.ac.uk

F. Comiti
Faculty of Science and Technology, Free University of Bozen-Bolzano, Bozen-Bolzano, Italy

D. Penna
Department of Agricultural, Food and Forestry Systems, University of Florence, Florence, Italy

L. Mao
School of Geography, University of Lincoln, Lincoln, UK

T. Heckmann and D. Morche (eds.), *Geomorphology of Proglacial Systems*, Geography of the Physical Environment, https://doi.org/10.1007/978-3-319-94184-4_12

(Comiti and Mao 2012). Therefore, activities and instruments required for sediment transport monitoring—especially relative to the bedload fraction—differ in lowlands and mountain river systems.

Within the latter, proglacial streams feature distinct characteristics that affect—both positively and negatively—the approach for sediment transport measurements. On the one hand, their nivo-glacial hydrological regime facilitates monitoring activities, as sediment-transporting flows are frequent during the summer months and highly predictable, being mostly linked to air temperature dynamics rather than to intense precipitation events (Mao et al. 2014; Srivastava et al. 2014). On the other hand, such frequent (roughly every day during summers) high and turbid flows in proglacial channels make the survey of their channel beds, the characterization and tracking of bed sediments possible only during a limited time in summer months, i.e., when melt (snow and glacier alike) flows are inhibited by low air temperatures, and thus, runoff is dominated by groundwater sources (Penna et al. 2014; Engel et al. 2016; Penna et al. 2017). In addition, the low flow conditions ideal for such activities are typical of winter and early spring months, when accessibility of proglacial areas can be very problematic or even dangerous (e.g., snow avalanche hazard). Monitoring methods for both suspended and bedload fractions can be either direct or indirect (also called surrogate, Gray et al. 2010; Aigner et al., 2017), with the former measuring the actual sediment transport flux (in terms of volume or weight per unit of time), whereas the latter measure a variable which is correlated to the sediment transport intensity. With indirect methods, the actual transported sediment volume or weight can be obtained only after calibrating the variable against direct measurements.

The aim of this chapter is to present an updated summary of the current knowledge on sediment transport in proglacial rivers, including both suspended and bedload fractions. The methodological issues regarding their monitoring will be treated first, followed by sections discussing the actual dynamics of sediment transfer in such systems. Our perspective for future investigations will close the chapter.

12.2 Suspended Sediment Transport Measurements and Monitoring

Suspended sediment transport can be monitored using direct or indirect (surrogate) methods. Direct sampling involves the use of bottles to collect—manually or automatically—samples of stream water on which sediment concentration has to be determined. Manual sampling consists generally in using a sampler with a rigid bottle with a 48–79 mm (inner diameter) nozzle on the open end, which allows for the entrance of stream flow and the exit of air from the bottle. An automatic pumping water sampler allows collecting water samples at a certain interval. The collected water samples have to be filtered, dried, and then analyzed by gravimetric procedures to quantify the suspended sediment concentration (SSC). For more details, see Gray and Landers (2014).

In proglacial rivers, the shallow and highly turbulent flows typical of these channels make acceptable to take water samples from one point only along a cross section, as the effective mixing makes cross-sectional—and to a less extent vertical—variations in sediment concentration relatively small, in contrast to deeper and wider lowland rivers. However, the sampling point should be placed within the main flow and actually on a site where well-mixed conditions are ensured.

The temporal resolution of suspended sediment monitoring in proglacial streams has been reported in the literature as quite variable. For a period of 26 days (July–August 1992) within the ablation season in Svalbard, Hodgkins (1996) took discrete suspended sediment samples every 1–2 h. Sawada and Johnson (2000) collected water samples every hour during two ablation seasons with an Alpha Sigma automatic pump-sampler. The same number of ablation seasons was investigated in the Place Glacier by Richards and Moore (2003), who collected a total of 687 samples, but with intervals ranging between 1 and 12 h depending on stream flow condition. A more extensive monitoring was carried out by Arora et al. (2014), who covered four ablation seasons in the Himalaya but with longer sampling interval; only two samples per day during the low and high daily flow conditions. The same sampling interval, at low and high water discharges, was used by Wulf et al. (2012) to average a single value of daily suspended sediment concentration. O'Farrell et al. (2009) monitored the Matanuska River in New Zealand from 1994 to 2004, deploying an automatic pumping water sampler with a 1–2 h interval. In Norway, a water discharge and sediment transport monitoring program in the glacier-fed Bayelva River was carried out for 11 years with a frequency of 2–4 water samples per day, by an automated pumping sampler (Bogen and Bonsnes 2003). Twenty years of data (1973–1992) from a station located 4 km downstream the Jôkulsâ â Solheimasandi glacier snout was used by Lawler (1994) to examine changes in suspended sediment transport. Despite the long period of monitoring, the low frequency (monthly) of suspended sediment sampling did not allow him to detect transient and diurnal sediment pulses.

As mentioned above, the temporal frequency of suspended sediment sampling has been rarely lower than 1 h, thus not allowing to recognize shorter-term fluctuations in SSC. Capturing these short-term variations is instead ensured by surrogate devices that allow for continuous records of flow parameters related to SSC (see Gray and Landers 2014), such as turbidity (an optical characteristic of the flow, measured in nephelometric units, NTU) or its acoustic backscatter (measured by acoustic Doppler instruments as ADCP). In mountain rivers, the most widely used sensors for such continuous monitoring are turbidimeters. A very important step in their application—sometimes underrated—is the need to establish robust calibration curves for the NTU—SSC relationship by means of numerous direct samples, as these curves strongly depend on sediment lithology and are thus site-specific (Gray and Landers 2014). Additionally, the calibration relationships for turbidity-sediment concentration should be derived for each season as separate in proglacial streams (i.e., runoff-origin specific, Fig. 12.1), as it is very likely that the fine sediment transported out of glaciers has a markedly different optical signature compared to that mobilized from other sediment sources. Furthermore, the NTU-SSC relationship can also depend on the grain size of sediments transported in suspension. Also, turbidity associated with intense glacier melt periods may exceed 3000 NTU in proglacial streams (unpublished data from the upper Saldur/Saldura River, South Tyrol, Italy), thus making necessary to use top quality, expensive turbidimeters able to measure beyond this range, at least up to 4000 NTU. Chen et al. (2016) installed a turbidity meter 3 km downstream from the Zhadang glacier, collecting data during 15 days at a 5-min temporal resolution with near-infrared sediment sensor (940-nm wavelength photodiode). The same interval was used by Stott et al. (2014), during almost one month of the ablation season in a glacierized basin in south–west Greenland. For calibration, 335 water samples were taken by integrating portable and manual samplers.

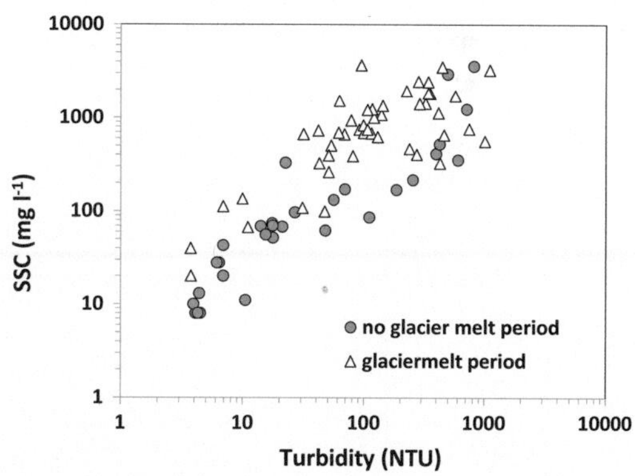

Fig. 12.1 Preliminary relationship between suspended sediment concentration (SSC) and turbidity in the Sulden River (South Tyrol, Italy). Data relative to no glacier melt refer to May and October, whereas data from August and September are grouped into the glacier melt period. The difference in the relationships is due to different color and grain size of sediments transported during the glacier melt period. The recording station is located at about 1100 m a.s.l. and drains a catchment of about 130 km^2, of which about 11% is covered by glaciers

12.3 Suspended Sediment Transport Dynamics in Proglacial Rivers

In glacierized rivers, not only the NTU-SSC curve varies through the year, but also—and to a much larger extent—the relation between water discharge (Q) and SSC can vary dramatically overtime. Indeed, the Q-SSC curve can vary from one year to another depending on climatic and sediment delivery conditions (Fig. 12.2a), and also during the ablation season, depending on the origin of meltwaters (i.e., snow versus glacier melt, Fig. 12.2b). During the same ablation season, the relationship between discharge and suspended sediment transport can also change (Fig. 12.3), depending on the activity and connectivity of the main sources of sediment (channel bed, slopes, tributaries, glaciers).

In proglacial rivers, the continuous monitoring of suspended sediment transport by turbidimeters permits to observe short-scale fluctuations associated to daily discharge cycles during the melting season. At the scale of a single daily event, the Q-SSC relationship is often nonlinear, as SSC during the rising limb of the hydrograph can be considerably different than during the falling limb of the same hydrograph, resulting in hysteresis loops. Hysteresis loops can be detected visually (Fig. 12.4), but recent metrics also permit quantitative comparisons among events and sites (e.g., Langlois et al. 2005; Lawler et al. 2006; Aich et al. 2014; Zuecco et al. 2016).

Hysteresis is clockwise when the peak of SSC occurs before the peak of Q, whereas the opposite (i.e., SSC peaking after Q) depicts a counter-clockwise hysteretic loop. Hysteresis loops at the scale of single flood events have been detected and reported since the 1980s worldwide (e.g., Wood 1977; Williams 1989; Jeje et al. 1991; Lenzi and Marchi 2000; Gao and Josefson 2012; Yeshaneh et al. 2013). In non-glacial rivers, the direction and magnitude of hysteretic loops have been mostly related to the activity and proximity of sediment sources. In particular, clockwise and counter-clockwise hystereses have been attributed to unlimited and limited sediment supply, respectively. Clockwise hysteresis has been associated with early exhaustion of sediment sources and closeness of the main sources of sediments (e.g., Asselman 1999; Rovira and Batalla 2006; Gao and Pasternack 2007). Conversely, counter-clockwise

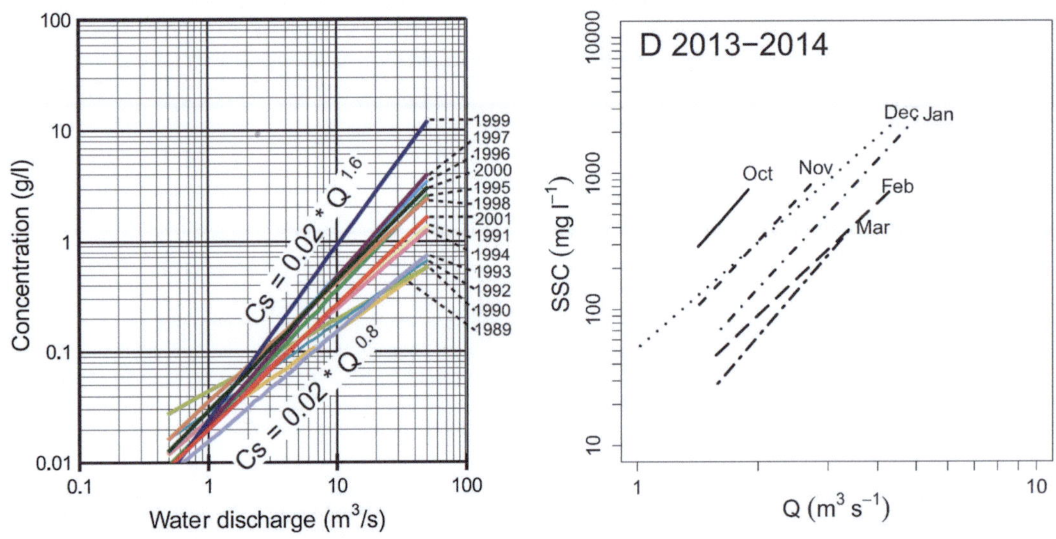

Fig. 12.2 Examples of variability of the discharge–suspended sediment concentration relationships between years (figure reproduced from Bogen and Bonsnes 2003, on the left) and months (figure reproduced from Mao and Carrillo, 2017, with permission from Elsevier, on the right)

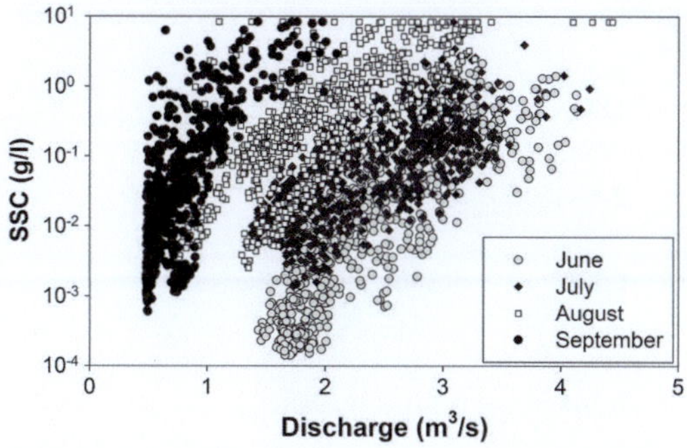

Fig. 12.3 Plot of water discharge versus SSC measured in the upper Saldur River (South Tyrol, Italy). The drainage area at the monitoring station is 19 km^2, of which about 12% currently covered by glaciers. More details on the catchment can be found in Mao et al. (2014)

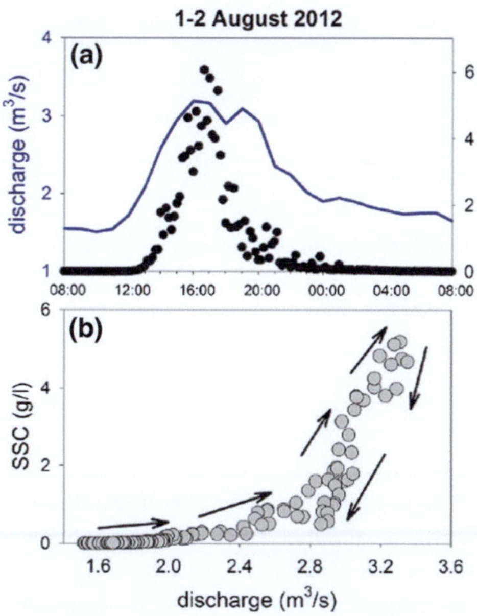

Fig. 12.4 Clockwise loop between suspended sediment concentration and water discharge during a glacier melt-induced daily event (August 1–2, 2012) in the Saldur River (Italy)

hysteresis has been associated with (i) morphological changes in the channel after the occurrence of the hydrograph peak, (ii) a main source of sediment being far from the monitoring point, or (iii) to general inactivity of sediment sources (e.g., Klein 1984; Walling and Webb 1982). More recently, it was demonstrated how hysteretic loops may change in magnitude and shape along a river for the same flood, and therefore, the distance of the recording station with respect to sediment sources seems to be crucial for loops characteristics (Tena et al. 2014).

In proglacial rivers, there is little evidence of both clockwise and counter-clockwise loops relative to suspended sediment during the ablation season (Table 12.1). For the proglacial river below the Gangotri Glacier snout (Indian Himalayas), Singh et al. (2005) reported clockwise hysteresis during the entire ablation season. Clockwise hysteresis was also reported by Bogen (1980), Sawada and Johnson (2000), and Chen et al. (2016), and has been related to high sediment supply from sources close to the monitoring station or abundance of fine sediments stored in channel bed. Orwin and Smart (2004) also observed dominance of clockwise hysteresis due to ready mobilization of fine sediments within the channel near the Small River Glacier (Canada). However, these authors also reported that a monitoring station located more downstream registered mostly counter-clockwise hysteresis, again revealing that the distance of sediment sources from the recording station can play a major role in determining the type of loops. Interestingly, changes in dominant hysteresis loops during the ablation season have been observed. For instance, Hodgkins (1996), Stott et al. (2014) and Mao and Carrillo (2017) reported on a progressive shift from clockwise to counter-clockwise hysteresis due to a progressive exhaustion of sediment supply in glacierized basins. For the Estero Morales (Chile), Mao and Carrillo (2017) showed that during snowmelt, a large supply of fine sediments is provided in the lower and middle part of the basin and hysteresis patterns tend to be clockwise. Instead, during glacier melting the source of fine sediments is the proglacial area, producing counter-clockwise hysteresis. In contrast, the event-scale analysis in the Saldur River (unpublished data, see Fig. 12.4) highlights the dominance of counter-clockwise loops in June and July (snowmelt-related daily fluctuations), whereas August and September (glacier melt-induced daily fluctuations) are dominated by clockwise loops. Interestingly, such pattern does not reflect

Table 12.1 List of proglacial rivers for which hysteresis loops of suspended sediment transport has been analysed

Reference	Location	Basin	Basin (km^2)	Glacier	Range SSC ($mg\ l^{-1}$)	Range of Q ($m^3\ s^{-1}$)	Timescale	Hysteresis[a]
Singh et al. (2005)	Himalayas	Bhagirathi	556	Gangotri	200–4500	38–170	Daily	C
Sawada and Johnson (2000)	North America	Slim	2456	Kaskawulsh	0.1–9	0–340	Daily/season	C
Chen et al. (2016)	Himalayas	Nam Co		Zhadang	0.2–4	0.2–1.7	Daily	C
Orwin and Smart (2004)	North America	Small River	6.86	Small River	50–450	0.04–2.3	Daily/season	C and CC
Hodgkins (1996)	Svalvard	Scott Turnerbreen	12.8	Scott Turnerbreen	100–6000	0.1–3.5	Daily/season	C–CC
Stott et al. (2014)	Greenland	Sermitsiaq	55	Sermitsiaq	4–110	0.3–6.7	Daily/season	C–CC
Mao and Carrillo, 2017	Andes	Estero morales	27	San Francisco	63–903	1.6–3.4	Daily/season	C–CC
O'Farrell et al. (2009)	North America	Matanuska	600	Matanuska			Annual	CC
Wulf et al. (2012)	Himalayas				0–151,000	up to 4000	Annual	CC
Andermann et al. (2012)	Himalayas	Narayani			0–40,000	up to 10	Annual	CC
Riihimaki et al. (2005)	Alaska	Bench	11	Bench	0–13,000	1–6	Daily/season	C

[a]*C* clockwise; *CC* counter-clockwise

bedload transport dynamics in the same stream (see Sect. 12.5), where counter-clockwise loops were mostly observed during the glacier melting period. A tributary draining a small glacierized sub-catchment delivering only fine sediment to the main Saldur channel—just upstream of the monitoring station—could likely be responsible for this discrepancy between bedload and suspended hysteresis. Finally, in the same river, most of the rainfall events were characterized by counter-clockwise loops, likely as a consequence of most sediment sources being farther up in the catchment relative to runoff generating areas.

Because suspended sediment transport depends on water discharge but also on sediment availability, annual scale hysteresis can be observed (O'Farrell et al. 2009). Wulf et al. (2012) showed that suspended sediment transport in the proglacial Sutlej River (Himalaya) featured annual counter-clockwise hysteresis (Fig. 12.5), due to enhanced glacial sediment evacuation during monsoonal rainfall events in the late glacial melt period. Wulf et al. (2012) used this hysteretic trend to interpret the relative activity of sediment sources, further suggesting that future glacial retreat and increased monsoonal frequency could increase suspended sediment transport. Andermann et al. (2012) observed similar hysteretic effects during the ablation season in the Nepal Himalayas, but they attributed this to groundwater dilution effects rather than changes in sediment supply effects. Instead, Riihimaki et al. (2005) reported strong clockwise hysteresis in the relationship between sediment

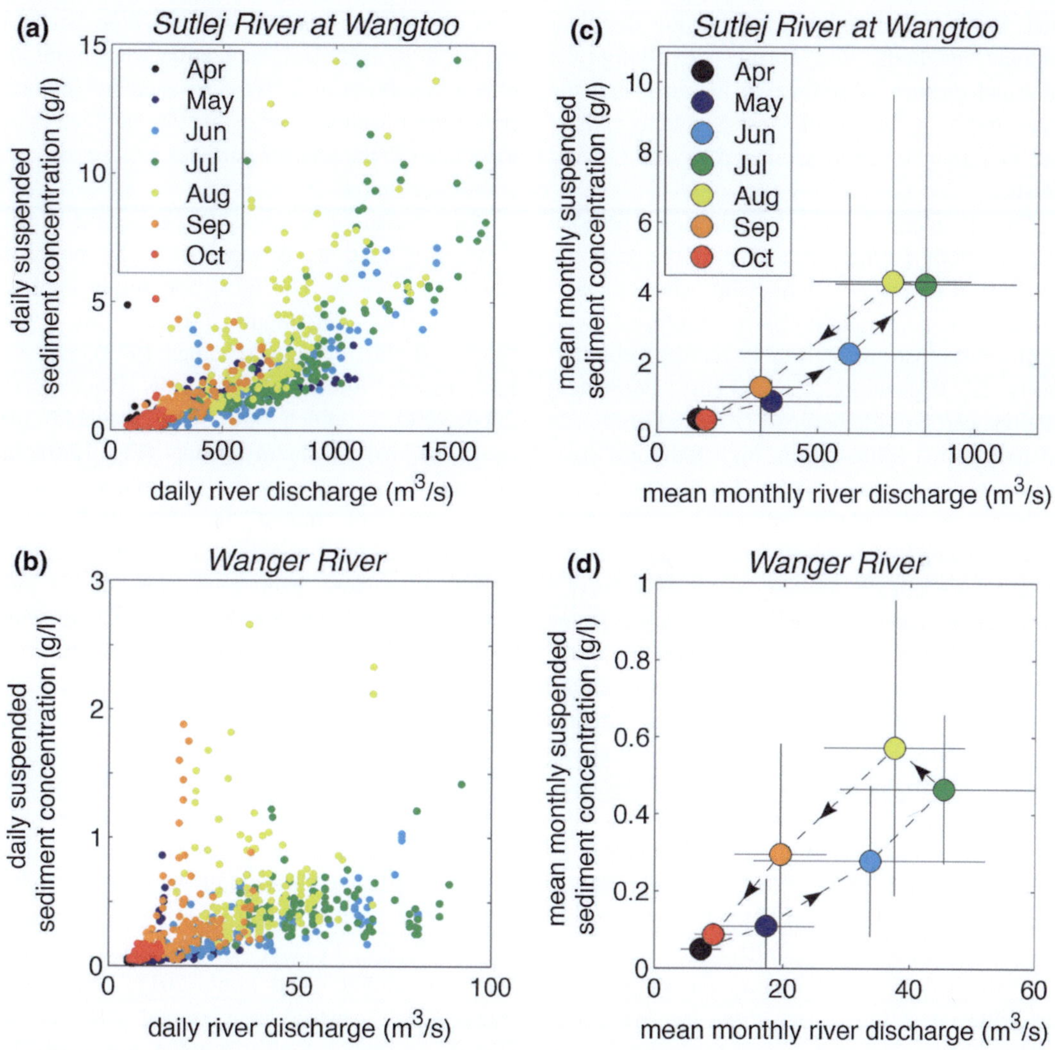

Fig. 12.5 Time series of daily river discharge and suspended sediment concentration data, and annual hysteresis loops of the mean monthly suspended sediment flux in the Sutlej and Wanger Rivers (figure reproduced from Wulf et al. 2012)

and water discharge during the glacier ablation season of the Bench Glacier (Alaska) , arguing that this was due to sediment exhaustion from the subglacial system at both the seasonal and flood event timescales. Rainfall events can strongly influence suspended sediment yield in proglacial rivers. In the Castle Creek Glacier (Canada), Leggat et al. (2015) showed that during rainfall events the main source of sediment changed from the proximal deposits to the glacier to more diffuse sources at the basin scale, and also that suspended sediment load during these rainfall events was 600% higher than the average of the ablation season.

12.4 Bedload Measurements and Monitoring

In rivers, bedload transport comprises relatively coarse (from sand to boulders) sediment particles that roll, slide, and saltate close to the channel bed. Bedload can be monitored using direct or indirect methods, the former involving the physical capture of sediments in traps, and the latter implying the use of sensors able to register the vibration or sound produced by the moving clasts.

Direct methods for monitoring bedload include both permanent and portable trapping devices. Permanent monitoring sites usually consist in a retention basin equipped with appropriate instrumentation to monitor continuously the deposition of sediments. Permanent stations can provide bedload rates and grain size of transported sediments at high temporal resolution, as in the case of the Rio Cordon (Italy; Rainato et al. 2017) or of the Erlenbach (Switzerland; Rickenmann et al. 2012), where a basket automatically moving below the crest of a check dam catches the sediment falling into the retention basin located just downstream. Such systems are rather expensive in terms of initial installation and are feasible in small streams only. We are not aware of permanent trapping stations in proglacial streams, and smaller trapping solutions as Reid (or Birkbeck) slot samplers (see, e.g., Lucía et al. 2013; Liébault et al. 2016; Habersack et al., 2017) are not ideal in such systems due to prolonged high bedload rates and poor site accessibility.

Permanent direct monitoring can alternatively be achieved at intake structures for hydroelectricity plants. The volume of transported sediments can be derived by the frequency of flushing or purging operation of intake structures, provided that some corrections on the type of purging and sediment packing density can be made with confidence (Bezinge et al. 1989; Lane et al. 2016). Later, bedload transport has been assessed in proglacial streams using data provided by flushing of sediments from hydropower structures in the Bas Glacier d'Arolla (Switzerland; Warburton 1990), Gråelva River (Norway; Bogen and Møen 2003), and the Pitzbach stream (Austria; Rickenmann and McArdell 2008; Turowski and Rickenmann 2009). Unfortunately, intake structures are generally equipped with metal grids which let only the smaller fraction of bedload transport fall into settling basins, and thus do not trap and record the whole bedload transport. Nonetheless, such structures permit an effective calibration of sensors used for indirect monitoring, which allow estimating bedload rates associated to the coarser particles moving at high flows (see below).

An alternative solution to permanent direct monitoring stations is represented by portable sampling devices. The most commonly applied solution in mountain rivers are the "Helley-Smith" sampler (Emmett 1979), and the more recent "Bunte" bedload traps (Bunte et al. 2004, 2007), which are more suitable for coarse-bedded mountain streams (Fig. 12.6). In general, these instruments are much cheaper than permanent stations, but their deployment presents relevant difficulties at high flows and—importantly—they cannot provide continuous bedload data. Helley-Smith samplers (both with 76 and 152 mm intakes) have been widely used in proglacial rivers in the 1980s (e.g., Ashworth and Ferguson 1986; Gurnell et al. 1988; Warburton 1992). Subsequently, at least to our knowledge, two decades passed without any published work on their use in such fluvial systems, until the recent investigations carried out in the Skeldal River (Greenland; Stott 2002), Saint Pierre River (France; Liu et al. 2008; Meunier et al. 2006), Erdalen and Bødalen streams (Norway; Beylich and Laute 2015) and in the Fagge River (Austria; Morche et al. 2012, 2014; Chap. 13). Helley-Smith samplers are relatively easy to deploy (Fig. 12.6), but present certain limitations. Indeed, Vericat et al. (2006) showed that their sampling efficiency is reduced in coarse grained streams due to larger grains blocking the sampler entrance or causing the sampler to rest in a "perched" position on the bed. In order to ameliorate the efficiency of portable samplers for bedload in Mountain Rivers, coarser-meshed (4 mm) traps lying on metal plates, in turn installed flush with the bed, were developed by Bunte et al. (2004). In addition, the fact that these traps are secured to the bed thanks to vertical steel rods facilitate their deployment in fast-flowing (but still wadeable) streams, making them more effective than Helley-Smith samplers in proglacial streams, especially at medium to

Fig. 12.6 Direct bedload sampling using: **a** an Helley-Smith sampler in the Erdalen River, Norway (figure reproduced from Beylich and Laute 2015, with permission from Elsevier); **b** Bunte-type sampler in the Saldur River (Italy); **c** multiple river bedload traps in a cross section of the Scott River, Spitsbergen (figure reproduced from Kociuba 2017, with permission from Elsevier); **d** Bunte-type sampler operated by truck in the Sulden River (Italy)

high (i.e., near-bankfull) flows. "Bunte" traps were deployed in glacierized catchments of Switzerland (Schneider et al. 2016), South Tyrol (Italy, Dell'Agnese et al. 2014), and of the Chilean Andes (Mao et al. 2016). A modification of such traps to operate them by truck-mounted cranes—and thus to sample bedload in streams and at stages where wading is impossible or unsafe at best—was first introduced in Austria (Aigner et al., 2017), and more recently in Italy (Vignoli et al. 2016), in both cases specifically for glacier-fed rivers (Fig. 12.6). A further variation of the "Bunte" trap is the system—operated from the channel banks—developed by Kociuba and Janicki (2015) to monitor bedload transport in proglacial rivers of the Svalbard islands (Fig. 12.6). It is worth noticing that several "Bunte" traps can be employed simultaneously along a cross section, allowing to explore the lateral variability of bedload transport rates and grain sizes in glacier-fed rivers (e.g., Kociuba and Janicki 2014).

Indirect methods for bedload monitoring in mountain rivers rely on different sensors to detect the vibrations generated by the collision of transported clasts on the channel bed or with fixed artificial structures placed in the bed. Such vibrations can be detected and recorded by piezoelectric sensors and geophones (i.e., velocimeters) mounted on plates deployed on the riverbed

(Bogen and Møen 2003; Rickenmann et al. 2014), seismometers placed near the riverbed (Roth et al. 2016), and passive acoustic sensors (microphones) installed either on the bed (Mizuyama et al. 2010) or "listening" to the entire water column ("hydrophones," Geay 2013). Active acoustic methods (i.e., ADCP with bottom tracking capability) can be also utilized (Rennie and Church 2010) but they are not so effective in mountain streams, where shallow water depth and large moving grain sizes inhibit their performance.

The first application of bedload monitoring by vibrational/acoustic sensors (plates or pipes) in proglacial streams dates to the mid-1990s. Rickenmann and McArdell (2008) and Turowski and Rickenmann (2009) reported on the successful use of piezoelectric bedload impact sensors installed on the water intake of a Tyrolean weir in the proglacial area of the Pitzbach stream (Austria; drainage area of 27 km^2 basin, 60% of which covered by glaciers). The piezoelectric sensors could be calibrated using weighting cells in the associated settling basin, allowing continuous monitoring of bedload intensity. More recently, Wyss et al. (2016) reported on the successful calibration of Swiss-type geophones, i.e., 15-mm-thick steel plates (0.36 × 0.5 m) installed in several sites worldwide, including glacierized basins in Switzerland, Austria, and Italy. Flume-based relations could be integrated by field data, thus permitting to derive empirical relationships to estimate bedload transport masses and grain size from continuous records provided by the geophone plates (Wyss et al. 2016). A glacier-fed stream monitored through geophones plates is the Riedbach (Switzerland), where Schneider et al. (2016) calibrated the plate sensors and could explore the effects of channel slope on sediment transport, reporting that traditional bedload equations developed for low-gradient rivers could be successfully applied to mountain streams provided that flow resistance partitioning due to macro-roughness is taken into account. In the glacier-fed Sulden/Solda River (South Tyrol, Italy, see also Fig. 12.1), Swiss-type geophone plates were installed on a check dam (Fig. 12.7) in 2014, and they showed no damages after the passage of an infrequent flood in the same year transporting up to 1-m size boulders. However, the concrete crest of the check dam was severely eroded, possibly not only due to this single flood but as a consequence of sustained high bedload rates, and thus required maintenance.

Continuous measurements of coarse bedload transport in glacierized environments have also

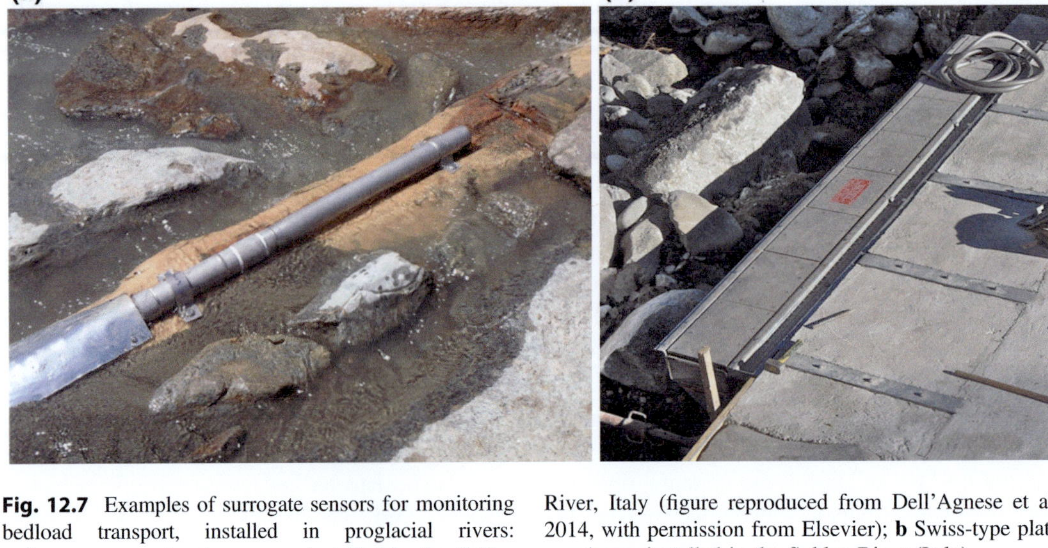

Fig. 12.7 Examples of surrogate sensors for monitoring bedload transport, installed in proglacial rivers: **a** Japanese acoustic pipe sensor installed in the Saldur River, Italy (figure reproduced from Dell'Agnese et al. 2014, with permission from Elsevier); **b** Swiss-type plate geophones installed in the Sulden River (Italy)

been achieved using Japanese-type acoustic pipes (Mizuyama et al. 2010) in two streams of the South Tyrol (Italy) (Saldur, since 2011, Dell'Agnese et al. 2014; and Sulden, in 2014, Vignoli et al. 2016; Fig. 12.7) and in the Estero Morales, a stream of the central Chilean Andes since 2013 (Mao et al. 2016). Compared to geophone plates, these devices are more sensitive as they can detect down to fine gravel size transport (Dell'Agnese et al. 2014). Japanese-type acoustic pipes were calibrated using Bunte traps in both geographical contexts. Unfortunately, the acoustic pipe in the Sulden River got damaged by the 2014 flood mentioned above.

Furthermore, attempts to monitor bedload by seismometers placed on the banks of a proglacial river were performed in France (Burtin et al. 2011). Seismic monitoring of bedload processes is currently increasing due to the relatively low cost and the promising potential of separating effectively bedload from turbulent flow-induced signal (e.g., Gimbert et al. 2014; Burtin et al. 2016). Applications on proglacial rivers are still incipient, but with remarkable recent advances that demonstrated the potential for exploring sediment processes in proglacial rivers and the relationships with sediment production by glacier abrasion or plucking (Gimbert et al. 2016).

Alternative methods for bedload monitoring that have been effectively used in proglacial streams include the assessment of scour-and-fill depths coupled to particle velocities, which nowadays can be obtained by tagging clasts with passive integrated transponders (PIT tags). PIT tags are cheap, durable, and increase considerably the recovery rates after floods (e.g., Lamarre et al. 2005; Bradley and Tucker 2012; Chapuis et al. 2015) if compared with painted particles that were used in the past also in proglacial rivers (e.g., Ashworth and Ferguson 1986). In the Saldur River, Mao et al. (2017) deployed 629 PIT tags and installed stationary antennas to calculate the virtual velocity of clasts, investigating the effects of antecedent flows on incipient motion and displacement distance of coarse sediments.

12.5 Bedload Dynamics in Proglacial Rivers

In proglacial rivers, the positive relationship between bedload transport rate and water discharge (and thus shear stress and stream power) seems to be weaker, with much larger data scatter, than in non-glacierized systems. For the Saldur River, Dell'Agnese et al. (2014) report bedload rates spanning four orders of magnitude (0.00002–0.14 kg m^{-1} s^{-1}) for discharge ranging from 1.3 to 3.7 m^3 s^{-1}, which represent the difference between baseflow and maximum daily discharge during the glacier ablation season. Mean daily bedload rates can vary considerably. Indeed, for the glacierized Scott River, Kociuba (2017) also showed that the mean daily bedload transport rate varied from 0.02 to 43.4 kg m^{-1} day^{-1}. Apart for being very scattered, the relationship between bedload transport rate and liquid discharge in proglacial rivers often shows a change of slope at a certain critical discharge, discriminating different sediment transport regimes, often named phases. Analyzing bedload transport data from the Bas Glacier d'Arolla (Switzerland), Warburton (1992) observed three phases of transport: (1) only fine gravels flushed from pools transported over a stable coarse armor layer; (2) breakup of the armour layer and transport of coarser fractions; (3) breakup of bedforms (i.e., steps) and transport. The same bedload phases have been reported for both glacierized (e.g., Schneider et al. 2016) and non-glacierized environments (e.g., Lenzi et al. 2004; Ma et al. 2014; Recking et al. 2016).

Continuous monitoring of bedload transport allows observing short-scale fluctuations of sediment transport rates, due to daily discharge cycles in the ablation season of proglacial rivers. At the scale of single daily event, the relationship between water discharge and bedload rate is usually nonlinear, as hysteresis loops can be identified (Fig. 12.8). As for suspended sediment transport, the loops can be clockwise or counter-clockwise, and their directions have been

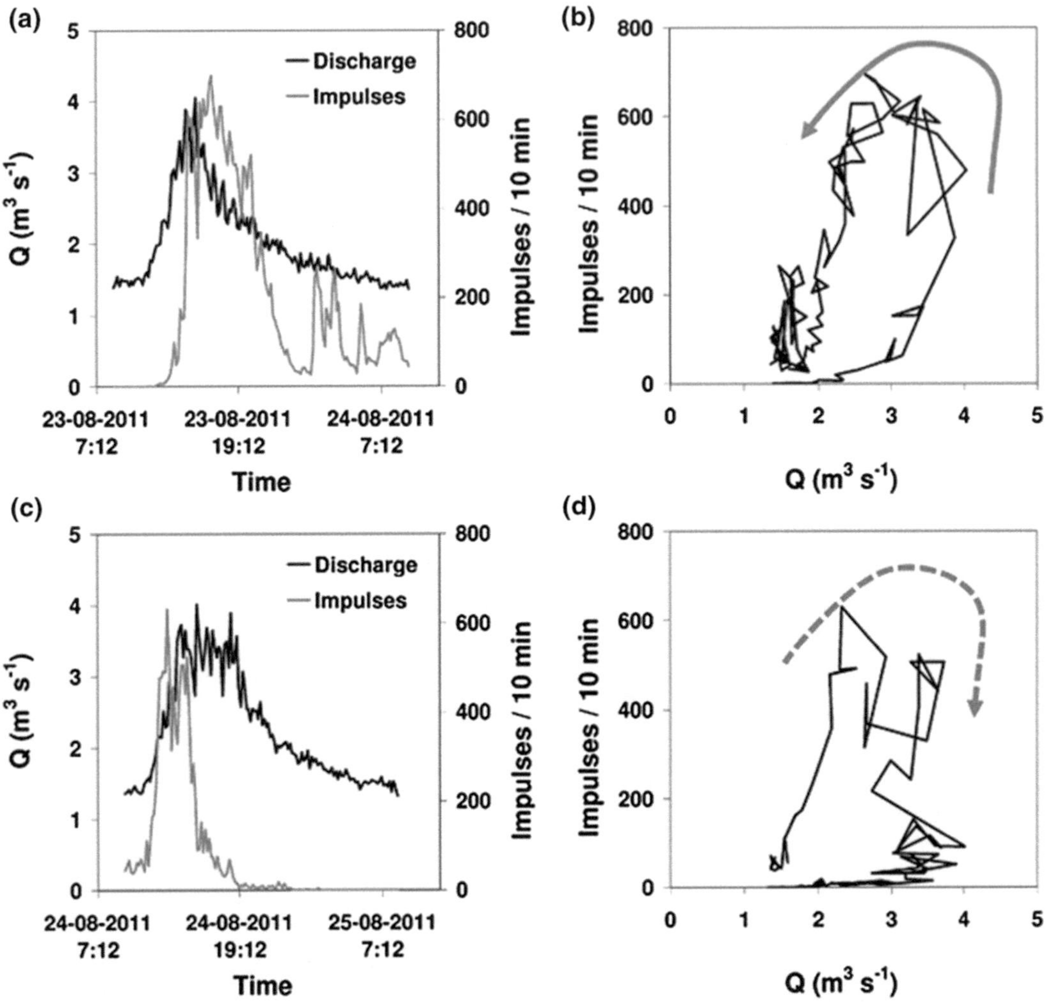

Fig. 12.8 Water discharge and bedload impulses measured by a Japanese acoustic pipe in the Saldur River (Italy). Above is an event that showed a counter-clockwise loop (Q peaking before bedload), whereas below is clockwise event (Q peaking after bedload). Figure reproduced from Mao et al. (2014) with permission from John Wiley and Sons

related to the dynamics of the armor layer or degree of activation of sediment sources (e.g., Hsu et al. 2011; Krein et al. 2016). In proglacial rivers, hysteresis at the scale of daily fluctuations of discharge has increasingly being reported as continuous monitoring of bedload is becoming easier due to the application of surrogate techniques. Using data provided by Swiss geophone plates installed in the Riedbach (a glacier-fed stream in Switzerland), Schneider et al. (2016) reported both seasonal and daily clockwise hysteresis effects. The former was associated to higher transport rates in the early rather than the late glacier melting season (due to sediment depletion of sediment sources from June to September), whereas the daily clockwise hysteresis (higher bedload before the peak of Q) has been explained by the fact that the water surface slope is commonly steeper during rising than during falling limb of hydrographs. Using a Japanese acoustic pipe in the Saldur River, Mao et al. (2014) showed that loops were mostly clockwise during the snowmelt period, and then increasingly more counter-clockwise during the following glacier melting period. They interpreted this change of hysteresis loops in terms of closeness and activity of sediment sources, relating counter-clockwise hysteresis during glacier melting with the time lag needed for sediment provided by the glacial and periglacial areas to be transported to the monitoring section. Lane et al. (1996) also showed that sediment was supplied with a certain time lag from the Haut Glacier d'Arolla (Switzerland), which would result in higher sediment transport during the falling limb of glacier melt hydrographs.

Analyzing seismic signal caused by subglacial water flow at the Mendenhall Glacier in Alaska, Gimbert et al. (2016) recently interpreted different subglacial channel regimes, suggesting that changes in subglacial channel size and pressure gradient can influence sediment transport. They report that increased water pressure gradient is responsible for a higher sediment transport capacity, implying that sediment transport in the proglacial area can be related more with sediment production from the dynamics at the bottom of the glacier than with sediment transport capacity of the glacier melting discharges. Indeed, higher discharges from the Mendenhall Glacier did not necessarily determine higher bedload transport (Gimbert et al. 2016). Similar dynamics were also reported by Mao et al. (2014) for the Saldur River, where bedload rates during glacier melt periods are much higher than during early summer, when the glacier drainage system is not active in terms of sediment transport. However, it must be remembered that hysteretic loops of sediments may depend on the location of the measuring site relative to the glacier, and also on the temporal variation of sediment sources due to glacier migration overtime and changes in sediment connectivity (Micheletti and Lane 2016; Chap. 16).

12.6 Sediment Yield in Proglacial Rivers and Its Partitioning

Hallet et al. (1996) provided a good review of sediment evacuation loads by glaciers (for sediment budgets see Chaps. 15, 17). Sediment yield tends to increase with basin size, and the erosion rates increase from small cirque glaciers to large, fast moving glaciers. Reported rates of erosion range from 0.01 mm $year^{-1}$ for polar glaciers, to 0.1 mm $year^{-1}$ for temperate valley glaciers, to 1.0 mm $year^{-1}$ for small glaciers in the Alps, and to 10–100 mm $year^{-1}$ for large valley glaciers in Alaska. Hallet et al. (1996) also reported higher sediment yields in glacierized than in comparable non-glacierized basins. More recently, analysing a wide dataset of sediment loads of alpine river, Hinderer et al. (2013) confirmed that glacierized basins feature the highest sediment yields of up to 7000 t km^{-2} $year^{-1}$, which are on average 5–10 times higher than in non-glacierized basins. Also, sediment yield and glacial cover were found to be positively correlated (Fig. 12.9).

As to the volumes of sediments transported in suspension by proglacial rivers, Lawler (1994) estimated a mean annual suspended sediment yield of about 9000 t km^{-2} $year^{-1}$ from the Jôkulsâ a Solheimasandi River (Iceland). Riihimaki et al. (2005) reported 1650 t km^{-2} $year^{-1}$ from the Bench Glacier (7.5 km^2 in a 11 km^2 basin). However, much lower values of suspended sediment yield have also been reported in literature. For example, Stott et al. (2014) estimated 31 t km^{-2} $year^{-1}$ from the Sermitsiaq Glacier (Greenland; drainage area of 55 km^2, 26% of which are glacierized), partly due to the presence of proglacial lakes (c.f. Chap. 14). For Himalayan rivers, Wulf et al. (2012) showed that rivers draining glacierized basins of similar sizes can produce quite different suspended sediment fluxes, with the Hunza River

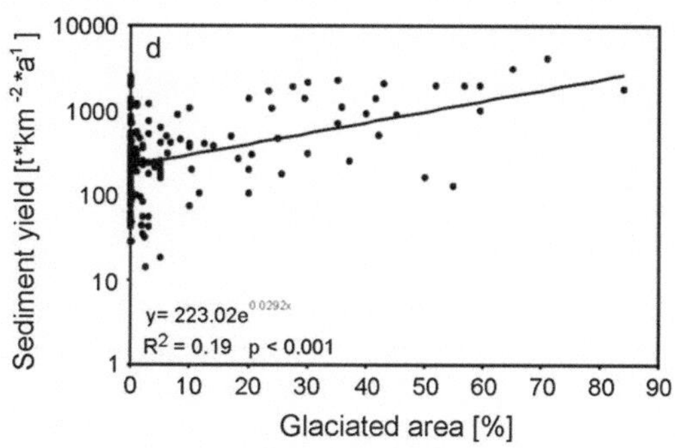

Fig. 12.9 Sediment yield of alpine rivers versus percentage of glacierized area. Figure reproduced from Hinderer et al. (2013) with permission from Elsevier

(13,157 km^2 catchment area, hosting the mighty Baltura glacier, 389 km^2 in size transporting up to 3373 t km^{-2} $year^{-1}$, whereas the Spiti River (12,477, 876 km^2 of which are glacierized) transporting only 499 t km^{-2} $year^{-1}$.

Regarding the relative importance of suspended versus bedload transport in proglacial rivers, based on a wide literature review, Turowski et al. (2010) reported that there is only a weak increase of the relative importance of bedload transport with the increase of the glacierized percentage of the basin. Bogen et al. (2015) showed that the receding Nigardsbreen glacier (Norway) transported to its proglacial lake on average 561 t km^{-2} $year^{-1}$, approximately half as suspended and half as bedload. On another Norwegian glacier (Engabreen), the same authors reported a similar specific sediment yield of 566 t km^{-2} $year^{-1}$, 60% of which transported in suspension. Analysing a long-term series of sediment transport yield from the Nigardsbreen glacier, Hallet et al. (1996) reported that the average yield was 412 t km^{-2} $year^{-1}$, 54% of which transported in suspension. For the proglacial stream draining the Hilda glacier (Canadian Rocky Mountains), Hammer and Smith (1983) reported that bedload represented more than 50% of the total sediment load. A 50% relative contribution of bedload to the total sediment yield (340 t km^{-2} $year^{-1}$) was also reported by Schiefer et al. (2010) for the glacierized Fitzsimmons Creek (western Canada). Recently, Beylich and Laute (2015) showed that in the Erdalen and Bødalen glacierized basins (Norway), bedload accounts for one third of the total sediment transport. They discriminated the sources of sediments, stressing the importance of different transfer processes (e.g., rockfalls, avalanches, bank erosion) and type and location of sinks (e.g., proglacial lakes) in determining sediment load of proglacial streams. In particular, proglacial lakes can play a major role as sediment traps, reducing considerably the sediment delivery to proglacial rivers (see Geilhausen et al. 2013; Carrivick and Tweed 2013; Chap. 14).

12.7 Final Remarks and Perspectives

Sediment transport is operatively challenging to monitor in rivers, especially in proglacial systems, characterized by high flow velocities, high suspended sediment concentrations, and intense bedload rates. As described above, suspended and bedload fractions in the annual sediment transport are apparently of equal relevance in proglacial rivers, so that both of them have to be quantified as precisely as possible in order to estimate sediment yields in such systems. Due to the large temporal variability of both suspended and bedload sediment transport in proglacial rivers, their continuous monitoring is highly advised as it permits to capture the whole spectrum of fluctuations in transport rates and loads, from minutes to years.

As to bedload, both Swiss-type geophone plates and Japanese acoustic pipes have been successfully deployed in proglacial rivers permitting to get valuable insights into daily scale fluctuations in bedload rates, thanks to site-specific calibrations performed by direct methods. It is indeed worth stressing that direct sampling will still be needed in order to properly calibrate indirect sensors. Direct bedload sampling should be carried out using "Bunte-derived" samplers that allow for longer sampling time and have higher catching efficiency compared to the "traditional" Helley-Smith samplers. In perspective, seismic sensors (seismometers and potentially geophones alike) installed on the channel banks offer the great advantage of being less prone to damages during high-magnitude events, as well as of easier and relatively cheaper installation. On the other side, reliable calibration curves for such devices are more challenging to obtain than for impact-type sensors (plates and pipes). With reference to suspended sediment transport, beside the installation of large range turbidimeters (up to 3000 NTU, but if possible reaching 4000 NTU or more), it is advisable to collect numerous water samples during different times of the day and throughout the year (at least from May to October) in order to establish robust NTU-SSC relationships. Such direct sampling is ideally performed using automatic pumping water samplers.

However, in these environments, morphological changes due to bank erosion, aggradation/degradation, and also avulsions (see Baewert and Morche 2014; Chap. 13; Carrivick and Heckmann 2017) may be quite frequent. As a consequence, "light" monitoring stations could be frequently damaged, with potential loss of expensive devices. In order to avoid damages to the sensors due to snow pack accumulations, avalanches or extreme low winter temperatures, it is also suggested to operate the stations only during the ablation seasons. Also, the possibility to install monitoring sites at hydropower weirs would offer the advantages of a stable section and of settling basin/flushing operation data to partly calibrate indirect sensors.

Acknowledgements For the unpublished data coming from the Sulden/Solda River monitoring station ("Aquased" project, www.aquased.net), we thank Gianluca Vignoli (CISMA srl), Silvia Simoni (Mountaineering srl), Walter Bertoldi (University of Trento), Rudolf Pollinger, Roberto Dinale, and Rudi Nadalet (Civil Protection Agency of the Aut. Province of Bozen-Bolzano), and Arnold Richter (Environmental Agency of the Aut. Prov. of Bozen-Bolzano). Michael Engel and Andrea Dell'Agnese (Free University of Bozen-Bolzano) are warmly thanked in particular for the unpublished data from the Saldur River.

References

Aich V, Zimmermann A, Elsenbeer H (2014) Quantification and interpretation of suspended-sediment discharge hysteresis patterns: how much data do we need? CATENA 122:120–129. https://doi.org/10.1016/j.catena.2014.06.020

Aigner J, Kreisler A, Rindler R, Hauer C, Habersack H (2017) Bedload pulses in a hydropower affected alpine gravel bed river. Geomorphology 291:116–127. http://dx.doi.org/10.1016/j.geomorph.2016.05.015

Andermann C, Crave A, Gloaguen R, Davy P, Bonnet S (2012) Connecting source and transport: suspended sediments in the Nepal Himalayas. Earth Planet Sci Lett 351–352(July 2015):158–170. https://doi.org/10.1016/j.epsl.2012.06.059

Arora M, Kumar R, Kumar N, Malhotra J (2014) Assessment of suspended sediment concentration and load from a large Himalayan glacier. https://doi.org/10.2166/nh.2013.129

Ashworth PJ, Ferguson R (1986) Interrelationships of channel processes, changes and sediments in a Proglacial Braided river. Geogr Ann 68(4):361–371. https://doi.org/10.2307/521527

Asselman NEM (1999) Suspended sediment dynamics in a large drainage basin: the River Rhine. Hydrol Process 13:1437–1450

Baewert H, Morche D (2014) Coarse sediment dynamics in a proglacial fluvial system (Fagge River, Tyrol). Geomorphology 218:88–97. https://doi.org/10.1016/j.geomorph.2013.10.021

Beylich AA, Laute K (2015) Sediment sources, spatiotemporal variability and rates of fluvial bedload transport in glacier-connected steep mountain valleys in western Norway (Erdalen and Bødalen drainage basins). Geomorphology 228:552–567. https://doi.org/10.1016/j.geomorph.2014.10.018

Bezinge A, Clark MJ, Gurnell AM, Warburton J (1989) The management of sediment transported by glacial melt-water streams and its significance for the estimation of sediment yield. Ann Glaciol 13:1–5

Bogen J (1980) The hysteresis effect of sediment transport systems. Norsk Geografisk Tidsskrift—Norw J Geogr 34(1):45–54

Bogen J, Bonsnes T (2003) Erosion and sediment transport in High Arctic rivers Svalbard. Polar Res 22(2):175–189

Bogen J, Møen K (2003) Bed load measurements with a new passive acoustic sensor. In: Bogen J, Fergus T, Walling DE (eds) Erosion and sediment transport measurement in rivers: technological and methodological advances, vol 283. IAHS Publication, IAHS, Wallingford, pp 181–192

Bogen J, Xu M, Kennie P (2015) The impact of pro-glacial lakes on downstream sediment delivery in Norway. Earth Surf Proc Land 40(7):942–952. https://doi.org/10.1002/esp.3669

Bradley N, Tucker GE (2012) Measuring gravel transport and dispersion in a mountain river using passive radio tracers. Earth Surf Proc Land 37(10):1034–1045. https://doi.org/10.1002/esp.3223

Bunte K, Abt SR, Potyondy JP, Ryan SE (2004) Measurement of coarse gravel and cobble transport using portable bedload traps. J Hydraul Eng 130(9):879–893. https://doi.org/10.1061/(ASCE)0733-9429(2004)

Bunte K, Swingle KW, Abt SR (2007) Guidelines for using bedload traps in coarse-bedded mountain streams : construction, installation, operation, and sample processing. General Technical Reports RMRS-GTR-191. Fort Collins, CO

Burtin A, Cattin R, Bollinger L, Vergne J, Steer P, Robert A, … Tiberi C (2011) Towards the hydrologic and bed load monitoring from high-frequency seismic noise in a braided river: the "torrent de St Pierre", French Alps. J Hydrol. https://doi.org/10.1016/j.jhydrol.2011.07.014

Burtin A, Hovius N, Turowski JM (2016) Seismic monitoring of torrential and fluvial processes. Earth Surf Dyn. https://doi.org/10.5194/esurf-4-285-2016

Carrivick JL, Tweed FS (2013) Proglacial lakes: character, behaviour and geological importance. Quatern Sci Rev 78:34–52. https://doi.org/10.1016/j.quascirev.2013.07.028

Carrivick JL, Heckmann T (2017) Short-term geomorphological evolution of proglacial systems. Geomorphology. http://dx.doi.org/10.1016/j.geomorph.2017.01.037

Chapuis M, Dufour S, Provansal M, Couvert B, de Linares M (2015) Coupling channel evolution monitoring and RFID tracking in a large, wandering, gravel-bed river: insights into sediment routing on geomorphic continuity through a riffle-pool sequence. Geomorphology. https://doi.org/10.1016/j.geomorph.2014.12.013

Chen F, Cai Q, Sun L, Lei T (2016) Discharge-sediment processes of the Zhadang glacier on the Tibetan Plateau measured with a high frequency data acquisition system. Hydrol Process. https://doi.org/10.1002/hyp.10900

Comiti F, Mao L (2012) Recent advances in the dynamics of steep channels. In: Church M, Biron PM, Roy AG (eds) Gravel-bed rivers: processes, tools, environments. Chichester, UK, Wiley

Dell'Agnese A, Mao L, Comiti F (2014) Calibration of an acoustic pipe sensor through bedload traps in a glacierized basin. Catena 121:222–231. https://doi.org/10.1016/j.catena.2014.05.021

Emmett WW (1979) A field calibration of the sediment-trapping characteristics of the Helley-Smith bedload sampler. U. S. Department of the Interior. Geological Survey, Denver, Colorado

Engel M, Penna D, Bertoldi G, Dell'Agnese A, Soulsby C, Comiti F (2016) Identifying run-off contributions during melt-induced run-off events in a glacierized alpine catchment. Hydrol Process 30:343–364. https://doi.org/10.1002/hyp.10577

Gao P, Josefson M (2012) Event-based suspended sediment dynamics in a central New York watershed. Geomorphology 139–140:425–437. https://doi.org/10.1016/j.geomorph.2011.11.007

Gao P, Pasternack G (2007) Dynamics of suspended sediment transport at field-scale drain channels of irrigation-dominated watersheds in the Sonoran desert, southeastern California. Hydrol Process 21:2081–2092. https://doi.org/10.1002/hyp.6398

Geay T (2013) Mesure acoustique passive du transport par charriage dans les rivières [Passive acoustic measurement of bedload transport in rivers]. Université Joseph Fourier, Grenoble, France

Geilhausen M, Morche D, Otto J, Schrott L (2013) Sediment discharge from the proglacial zone of a retreating Alpine glacier, vol 57(December 2012), pp 29–53

Gimbert F, Tsai VC, Amundson JM, Bartholomaus TC, Walter JI (2016) Subseasonal changes observed in subglacial channel pressure, size, and sediment transport. Geophys Res Lett 3786–3794. https://doi.org/10.1002/2016GL068337

Gimbert F, Tsai VC, Lamb MP, Gimbert F, Tsai VC, Lamb MP (2014) A physical model for seismic noise generation by turbulent flow in rivers. J Geophys Res Earth Surf 119:2209–2238. https://doi.org/10.1002/2014JF003201

Gray JR, Laronne JB, Marr JDG (2010) Bedload-surrogate monitoring technologies, US geological survey scientific investigations report 2010–5091. US Geological Survey, Reston, VA, 37 pp. http://pubs.usgs.gov/sir/2010/5091

Gray JR, Landers MN (2014) Measuring suspended sediment. In: Ahuja S (ed) Comprehensive water quality and purification, vol 1. United States of America: Elsevier, pp 157–204. https://doi.org/10.1016/B978-0-12-382182-9.00012-8

Gurnell AM, Warburton J, Clark MJ (1988) A comparison of the sediment transport and yield characteristics of two adjacent Glacier Basins, Val d' Herens, Switzerland. In: Sediment budgets. IAHS Publication, p 174

Habersack H, Kreisler A, Rindler R, Aigner J, Seitz H, Liedermann M, Laronne JB (2017) Integrated automatic and continuous bedload monitoring in gravel bed rivers. Geomorphology 291:80–93. http://dx.doi.org/10.1016/j.geomorph.2016.10.020

Hallet B, Hunter L, Bogen J (1996) Rates of erosion and sediment evacuation by glaciers: a review of field data and their implications. Global Planet Change 12:213–235

Hammer KM, Smith ND (1983) Sediment production and transport in a proglacial stream: Hilda Glacier, Alberta, Canada. Boreas 12:91–106

Hinderer M, Kastowski M, Kamelger A, Bartolini C, Schlunegger F (2013) River loads and modern denudation of the Alps—a review. Earth Sci Rev. https://doi.org/10.1016/j.earscirev.2013.01.001

Hodgkins R (1996) Seasonal trend in suspended sediment transport from an Arctic glacier, and implications for drainage-system structure. Ann Glaciol 22:147–151

Hsu L, Finnegan NJ, Brodsky EE (2011) A seismic signature of river bedload transport during storm events. Geophys Res Lett 38(13). https://doi.org/10.1029/2011GL047759

Jeje LK, Ogunkoya OO, Oluwatimilehin JM (1991) Variation in suspended sediment concentration during storm discharges in three small streams in upper osun basin, Central Western Nigeria. Hydrol Process. https://doi.org/10.1002/hyp.3360050404

Klein M (1984) Anti clockwise hysteresis in suspended sediment concentration during individual storms. Catena 11:251–257. https://doi.org/10.1016/S0341-8162(84)80024-7

Kociuba W (2017) Determination of the bedload transport rate in a small proglacial high Arctic stream using direct, semi-continuous measurement. Geomorphology 287:101–115. https://doi.org/10.1016/j.geomorph.2016.10.001

Kociuba W, Janicki G (2014) Continuous measurements of bedload transport rates in a small glacial river catchment in the summer season (Spitsbergen). Geomorphology 212:58–71. https://doi.org/10.1016/j.geomorph.2013.05.001

Kociuba W, Janicki G (2015) Changeability of movable bed-surface particles in natural, gravel-bed channels and its relation to bed load grain size distribution (Scott River, Svalbard). Geografiska Ann, Ser A: Phys Geogr 97(3):507–521. https://doi.org/10.1111/geoa.12090

Krein A, Schenkluhn R, Kurtenbach A, Bierl R, Barrière J (2016) Listen to the sound of moving sediment in a small gravel-bed river. Int J Sedim Res 31(3):271–278. https://doi.org/10.1016/j.ijsrc.2016.04.003

Lamarre H, MacVicar B, Roy AG (2005) Using passive integrated transponder (PIT) tags to investigate sediment transport in gravel-bed rivers. J Sedim Res 75 (4):736–741. https://doi.org/10.2110/jsr.2005.059

Lane SN, Richards KS, Chandler JH (1996) Discharge and sediment supply controls on erosion and deposition in a dynamic alluvial channel. Geomorphology. https://doi.org/10.1016/0169-555X(95)00113-J

Lane SN, Bakker M, Gabbud C, Micheletti N, Saugy J-N (2016) Sediment export, transient landscape response and catchment-scale connectivity following rapid climate warming and Alpine glacier recession. Geomorphology. https://doi.org/10.1016/j.geomorph.2016.02.015

Langlois JL, Johnson DW, Mehuys GR (2005) Suspended sediment dynamics associated with snowmelt runoff in a small mountain stream of lake Tahoe (Nevada). Hydrol Process 19(18):3569–3580. https://doi.org/10.1002/hyp.5844

Lawler DM (1994) Variability in Stream Erosion and Sediment Transport (Proceedings of the Canberra Symposium, December 1994). In Recent changes in rates of suspended sediment transport in the Jôkulsâ a Solheimasandi glacial river, southern Iceland. IAHS Publ. no. 224, 1994, 343–350

Lawler DM, Petts GE, Foster IDL, Harper S (2006) Turbidity dynamics during spring storm events in an urban headwater river system: the Upper Tame, West Midlands, UK. Sci Total Environ 360(1–3): 109–126. https://doi.org/10.1016/j.scitotenv.2005.08.032

Leggat MS, Owens PN, Stott TA, Forrester BJ, Déry SJ, Menounos B (2015) Hydro-meteorological drivers and sources of suspended sediment flux in the pro-glacial zone of the retreating Castle Creek Glacier, Cariboo Mountains, British Columbia Canada. Earth Surf Process Land 40(11):1542–1559. https://doi.org/10.1002/esp.3755

Lenzi MA, Marchi L (2000) Suspended sediment load during floods in a small stream of the Dolomites (northeastern Italy). Catena 39:267–282

Lenzi MA, Mao L, Comiti F (2004) Magnitude-frequency analysis of bed load data in an Alpine boulder bed stream. Water Resour Res 40:W0720. https://doi.org/10.1029/2003wr002961

Liébault F, Jantzi H, Klotz S, Laronne JB, Recking A (2016) Bedload monitoring under conditions of ultra-high suspended sediment concentrations. J Hydrol 540:947–958. https://doi.org/doi.org/10.1016/j.jhydrol.2016.07.014

Liu Y, Métivier F, Lajeunesse É, Lancien P, Narteau C, Ye B, Meunier P (2008) Measuring bedload in gravel-bed mountain rivers: averaging methods and sampling strategies. Geodin Acta 211(2):81–92. https://doi.org/doi.org/10.3166/ga.21.81-92

Lucía A, Recking A, Martín-Duque JF, Storz-Peretz Y, Laronne JB (2013) Continuous monitoring of bedload discharge in a small, steep sandy channel. J Hydrol 497:37–50. https://doi.org/doi.org/10.1016/j.jhydrol.2013.05.034

Ma H, Heyman J, Fu X, Mettra F, Ancey C, Parker G (2014) Bed load transport over a broad range of timescales: Determination of three regimes of fluctuations. J Geophys Res, Earth Surf 119(12):2653–2673. https://doi.org/10.1002/2014JF003308

Mao L, Carrillo R (2017) Temporal dynamics of suspended sediment transport in a glacierized Andean basin. Geomorphology 287:116–125. https://doi.org/10.1016/j.geomorph.2016.02.003

Mao L, Carrillo R, Escauriaza C, Iroume A (2016) Flume and field-based calibration of surrogate sensors for monitoring bedload transport. Geomorphology 253: 10–21. https://doi.org/10.1016/j.geomorph.2015.10.002

Mao L, Dell'Agnese A, Comiti F (2017) Sediment motion and velocity in a glacier-fed stream. Geomorphology 291:69–79. https://doi.org/j.geomorph.2016.09.008

Mao L, Dell'Agnese A, Huincache C, Penna D, Engel M, Niedrist G, Comiti F (2014) Bedload hysteresis in a glacier-fed mountain river. Earth Surf Process Land, n/a-n/a. https://doi.org/10.1002/esp.3563

Meunier P, Métivier F, Lajeunesse E, Mériaux AS, Faure J (2006) Flow pattern and sediment transport in a braided river: the "torrent de St Pierre" (French Alps). J Hydrol 330(3–4):96–505. https://doi.org/10.1016/j.jhydrol.2006.04.009

Micheletti N, Lane SN (2016) Water yield and sediment export in small, partially glaciated Alpine watersheds in a warming climate. Water Resour Res 52:1–20. https://doi.org/10.1002/2016WR018774

Mizuyama T, Laronne JB, Nonaka M, Sawada T, Satofuka Y, Matsuoka M, Yamashita S, Sako Y, Tamaki S, Watari M, Yamaguchi S, Tsuruta K (2010) Calibration of a passive acoustic bedload monitoring system in Japanese mountain rivers. U.S. G.S. Scientific Investigations Report 2010–5091

Morche D, Haas F, Baewert H, Heckmann T, Schmidt KH, Becht M (2012) Sediment transport in the proglacial Fagge River (Kaunertal/Austria). In: Erosion and sediment yields in the changing environment, pp 11–15

Morche D, Schuchardt A, Dubberke K, Baewert H (2014) Channel morphodynamics on a small proglacial braid plain (Fagge River, Gepatschferner, Austria). In: IAHS-AISH proceedings and reports, vol 367, pp 109–116. https://doi.org/10.5194/piahs-367–109-2015

O'Farrell CR, Heimsath AM, Lawson DE, Jorgensen LM, Evenson EB, Larson G, Denner J (2009) Quantifying periglacial erosion: Insights on a glacial sediment budget, Matanuska Glacier, Alaska. Earth Surf Proc Land 34:2008–2022. https://doi.org/10.1002/esp.1885

Orwin JF, Smart CC (2004) The evidence for paraglacial sedimentation and its temporal scale in the deglacierizing basin of Small River Glacier, Canada. Geomorphology 58:175–202. https://doi.org/10.1016/j.geomorph.2003.07.005

Penna D, Engel M, Bertoldi G, Comiti F (2017) Towards a tracer-based conceptualization of meltwater dynamics and streamflow response in a glacierized catchment. Hydrol Earth Syst Sci 21:23–41. https://doi.org/10.5194/hess-21-23-2017

Penna D, Engel M, Mao L, Dell'Agnese A, Bertoldi G, Comiti F (2014) Tracer-based analysis of spatial and temporal variation of water sources in a glacierized catchment. Hydrol Earth Syst Sci 18:5271–5288. https://doi.org/10.5194/hess-18-5271-2014

Rainato R, Mao L, García-Rama A, Picco L, Cesca M, Vianello A, Lenzi MA (2017) Three decades of monitoring in the Rio Cordon instrumented basin: sediment budget and temporal trend of sediment yield. Geomorphology 291:45–56. https://doi.org/10.1016/j.geomorph.2016.03.012

Recking A, Piton G, Vazquez-Tarrio D, Parker G (2016) Quantifying the morphological print of bedload transport. Earth Surf Proc Land 41:809–822. https://doi.org/10.1002/esp.3869

Rennie CD, Church M (2010) Mapping spatial distributions and uncertainty of water and sediment flux in a large gravel bed river reach using an acoustic Doppler current profiler. J Geophys Res 115:F03035. https://doi.org/10.1029/2009JF001556

Richards G, Moore RD (2003) Suspended sediment dynamics in a steep, glacier-fed mountain stream, Place Creek, Canada. Hydrol Process. https://doi.org/10.1002/hyp.1208

Rickenmann D, McArdell BW (2008) Calibration of piezoelectric bedload impact sensors in the Pitzbach mountain stream. Geodin Acta 21(1–2):35–52

Rickenmann D, Turowski JM, Fritschi B, Klaiber A, Ludwig A (2012) Bedload transport measurements at the Erlenbach stream with geophones and automated basket samplers. Earth Surf Proc Land 37(9):1000–1011. https://doi.org/10.1002/esp.3225

Rickenmann D, Turowski JM, Fritschi B, Wyss C, Laronne J, Barzilai R, … Habersack H (2014) Bedload transport measurements with impact plate geophones: comparison of sensor calibration in different gravel-bed streams. Earth Surf Process Land 39:928–942. https://doi.org/10.1002/esp.3499

Riihimaki CA, MacGregor KR, Anderson RS, Anderson SP, Loso MG (2005) Sediment evacuation and glacial erosion rates at a small alpine glacier. J Geophys Res: Earth Surf. https://doi.org/10.1029/2004JF000189

Roth DL, Brodsky EE, Finnegan NJ, Rickenmann D, Turowski JM, Badoux A (2016) Bed load sediment transport inferred from seismic signals near a river. J Geophys Res F: Earth Surf. https://doi.org/10.1002/2015JF003782

Rovira A, Batalla RJ (2006) Temporal distribution of suspended sediment transport in a Mediterranean basin: the lower Tordera (NE SPAIN). Geomorphology 79:58–71. https://doi.org/10.1016/j.geomorph.2005.09.016

Sawada M, Johnson PG (2000) Hydrometeorology, suspended sediment and conductivity in a large glacierized basin, Slims River, Yukon Territory, Canada (1993–94). Arctic

Schiefer E, Hassan M, Menounos B, Pelpola CP, Slaymaker O (2010) Interdecadal patterns of total sediment yield from a montane catchment, southern Coast Mountains, British Columbia, Canada. Geomorphology 118(1–2):207–212. https://doi.org/10.1016/j.geomorph.2010.01.001

Schneider JM, Rickenmann D, Turowski JM, Schmid B, Kirchner JW (2016) Bed load transport in a very steep mountain stream (Riedbach, Switzerland): measurement and prediction. Water Resour Res. https://doi.org/10.1002/2016WR019308

Singh P, Haritashya UK, Ramasastri KS, Kumar N (2005) Diurnal variations in discharge and suspended sediment concentration, including runoff-delaying characteristics, of the Gangotri Glacier in the Garhwal Himalayas. Hydrol Process 19:1445–1457. https://doi.org/10.1002/hyp.5583

Srivastava D, Kumar A, Verma A, Swaroop S (2014) Characterization of suspended sediment in Meltwater from Glaciers of Garhwal Himalaya. Hydrol Process 28:969–979. https://doi.org/10.1002/hyp.9631

Stott T (2002) Bedload transport and channel bed changes in the Proglacial Skeldal River, Northeast Greenland. Arct Antarct Alp Res 34(3):334–344

Stott T, Nuttall A-M, Biggs E (2014) Observed run-off and suspended sediment dynamics from a minor glacierized basin in south-west Greenland. Geografisk Tidsskrift-Danish J Geogr 114(2):1–16. https://doi.org/10.1080/00167223.2013.862911

Tena A, Vericat D, Batalla RJ (2014) Suspended sediment dynamics during flushing flows in a large impounded river (the lower River Ebro). J Soils Sediments 14(12):2057–2069. https://doi.org/10.1007/s11368-014-0987-0

Turowski JM, Rickenmann D (2009) Tools and cover effects in bedload transport observations in the Pitzbach Austria. Earth Surf Process Land 34(1):26–37. https://doi.org/10.1002/esp.1686

Turowski JM, Rickenmann D, Dadson SJ (2010) The partitioning of the total sediment load of a river into suspended load and bedload: a review of empirical data. Sedimentology 57(4):1126–1146. https://doi.org/10.1111/j.1365-3091.2009.01140.x

Vericat D, Church M, Batalla RJ (2006) Bed load bias: comparison of measurements obtained using two (76 and 152 mm) Helley-Smith samplers in a gravel bed river. Water Resour Res 42:W01402. https://doi.org/10.1029/2005WR004025

Vignoli G, Simoni S, Comiti F, Dell'Agnese A, Bertoldi W, Dinale R, Nadalet R, Macconi P, Staffler J, Pollinger R (2016) Monitoring sediment fluxes in Alpine rivers: the AQUASED project. Conf Proc Interprevent 2016:426–433

Walling DE, Webb BW (1982) Sediment availability and the prediction of storm-period sediment yields. Recent Dev Explanation Prediction Erosion Sedim Yield (Proc Exeter Symp) 137:327–337

Warburton J (1990) An Alpine Proglacial Fluvial sediment budget. Geografiska Annaler Ser A Phys Geogr 72(3/4):261. https://doi.org/10.2307/521154

Warburton J (1992) Observations of bed load transport and channel bed changes in a Proglacial mountain stream. Artic Alpine Res 24(3):195–203

Williams G (1989) Sediment concentration versus water discharge during single hydrologic events in rivers. J Hydrol 111:89–106

Wood PA (1977) Controls of variation in suspended sediment concentration in the RIver Rother, West Sussex, England. Sedimentology 24:437–445

Wulf H, Bookhagen B, Scherler D (2012) Climatic and geologic controls on suspended sediment flux in the Sutlej River Valley, western Himalaya. Hydrol Earth Syst Sci 16:2193–2217. https://doi.org/10.5194/hess-16-2193-2012

Wyss CR, Rickenmann D, Fritschi B, Turowski JM, Weitbrecht V, Travaglini E, Bardou E, Boes RM (2016) Laboratory flume experiments with the Swiss plate geophone bed load monitoring system: 2. application to field sites with direct bed load samples. Water Res Res. https://doi.org/10.1002/2016WR019283

Yeshaneh E, Eder A, Blöschl G (2013) Temporal variation of suspended sediment transport in the Koga catchment, North Western Ethiopia and environmental implications. Hydrol Process 5984:5972–5984. https://doi.org/10.1002/hyp.10090

Zuecco G, Penna D, Borga M, van Meerveld HJ (2016) A versatile index to characterize hysteresis between hydrological variables at the runoff event timescale. Hydrol Process 30:1449–1466. https://doi.org/10.1002/hyp.10681

13 Fluvial Sediment Transport in the Proglacial Fagge River, Kaunertal, Austria

David Morche, Henning Baewert, Anne Schuchardt, Matthias Faust, Martin Weber and Taimur Khan

Abstract

This study analyzes data from 2012 to 2015 for Fagge River, a glacier-fed river in the Austrian Alps. The observation period covers four consecutive ablation seasons. Suspended sediment and bedload were measured at a gauging station, which was installed directly downvalley of the Gepatschferner glacier snout in 2012. During the following years, the glacier retreated with a high melt rate. This, in turn, caused problems for the operation of the gauging station (due to channel migration and destruction of probes), forcing the station to be moved to the location of the new channel. The majority of sediment was transported downvalley as suspended load. Bedload, in this case, was of minor importance for the whole observational period. However, higher bedload quantities were observed in the Fagge River during single flood events. In turn, the 4-year average of suspended sediment load to bedload ratio was 3.02.

Keywords

PROSA project · Proglacial river · Sediment transport · Sediment yield · Rating curves

13.1 Introduction

Fluvial sediment transport is one of the most important geomorphological processes in glacier forefields (Carrivick et al. 2013; Lane et al. 2017; Chap. 12). Meltout of ice-cored till and topographic forcing can lead to dramatic changes in channel configuration of proglacial areas (Marren and Toomath 2014; Shugar et al. 2017). In addition to dramatic shifts in the proglacial cryosphere, extreme events (e.g., rainstorms, water pocket outbursts, and glacial lake outburst floods) may trigger great morphological surface alterations (Haeberli 1980; Warburton and Fenn 1994; Marren 2005). In contrast to the importance of glacier forefields in terms of alpine sediment fluxes, long-term data for sediment transport dynamics in proglacial rivers is rarely reported in the scientific literature.

D. Morche
University of Halle-Wittenberg, Halle, Germany

D. Morche (✉)
Environmental Authority of Saalekreis District, Merseburg, Germany
e-mail: david.morche@gmx.de

H. Baewert · A. Schuchardt · M. Faust · M. Weber · T. Khan
Institute for Geosciences and Geography, Martin Luther University of Halle-Wittenberg, Von-Seckendorff-Platz 4, 06120 Halle, Germany

T. Heckmann and D. Morche (eds.), *Geomorphology of Proglacial Systems*, Geography of the Physical Environment, https://doi.org/10.1007/978-3-319-94184-4_13

This chapter presents results of a 4-year long study on fluvial sediment transport in an alpine glacier forefield channel.

13.2 Study Area

This study is part of the PROglacial Systems of the Alps (or PROSA; Chap. 1): a project that aims to investigate the consequences of the ongoing glacier melt in terms of landform development through erosion, transport, and deposition of sediment material. The catchment area upstream of the Gepatsch Reservoir (Fig. 13.1) covers an area of about 62 km^2 (35% glacier cover) and is drained by the Fagge River, which originates from the Gepatschferner glacier, one of the largest glaciers in the Austrian Alps. The glacier has receded by more than 2 km since its Little Ice Age maximum (Heckmann et al. 2012). This melting has caused key changes in the sediment dynamics downvalley.

Metamorphic rocks, like paragneiss (biotite-plagioclase gneiss) and granitic orthogneiss, dominate the lithology of the study area (Vehling et al. 2017). Additionally, a hydrometeorological station (operated by TIWAG, Tyrolean Hydropower Company) is located close to Weisssee (2540 m above sea level), a proglacial lake (see Chap. 14 on proglacial lakes) downvalley the Weissseferner glacier.

In order to quantify sediment export from the glacier system by running meltwater, our investigations were concentrated on solid load determination (cf Chap. 12) in the Fagge River proximal to the glacier snout. The catchment draining into the gauging station covers 20 km^2, and about 80% of the catchment area is covered by glacier ice in 2012.

13.3 Methods

Flow velocity was measured with an OTT Nautilus current meter at wadable conditions (water depth <65 cm and flow velocity <3 ms^{-1}; Fig. 13.2). The channel width was determined, and the water level was recorded in 50 cm intervals along the cross section. A two-point measurement (at 20 and 80% water depth) or a one-point measurement (at 40% water depth) of the flow velocity was performed, depending on the stage at each vertical segment of the cross section. Due to technical limits of the current meter, a one-point measurement was performed if the stage was less than 10 cm. This data was then used to estimate the discharge using the velocity-area method.

Furthermore, the salt dilution method was used for discharge measurements during highest discharge in summer months. Under such non-wadable conditions, we applied the dilution method (Moore 2005). Therefore, a certainly known mass of salt was slug injected upstream into the river and the transit time of this tracer was determined by measuring the electrical conductivity ca. 50 m downstream. Following Herschy (2009), the discharge was then determined using the integration method.

These measurements were repeated during different water stages. Overall, 83 discharge measurements were carried out during the 4 years of observation (Table 13.1). For each discharge measurement, the water level was also recorded. Additionally, the water level was measured in a 15-min interval using a datalogger (OTT Orphimedes). The annual hydrograph was then estimated using a rating curve for the stage–discharge relation of each ablation period (see Table 13.2).

High variations of discharge and fast changing channel bed made reconstructions of the water level sensor necessary, since a stable measuring cross section did not exist for the entire period of investigation. This resulted in some inconsistency in data collection over the period of investigation.

A portable Helley–Smith sampler was used to measure the bedload for each discharge measurement (Fig. 13.2). Channel width determined the number of measurements taken per cross section. In general, bedload was measured in 1 m intervals. The duration of the measurement for each 1-m increment was 60 s. This measuring interval was reduced to 30 s for very high bedload transport. The bedload sampler has a square 76.2 mm entrance nozzle and a sample bag constructed from 0.2 mm mesh polyester.

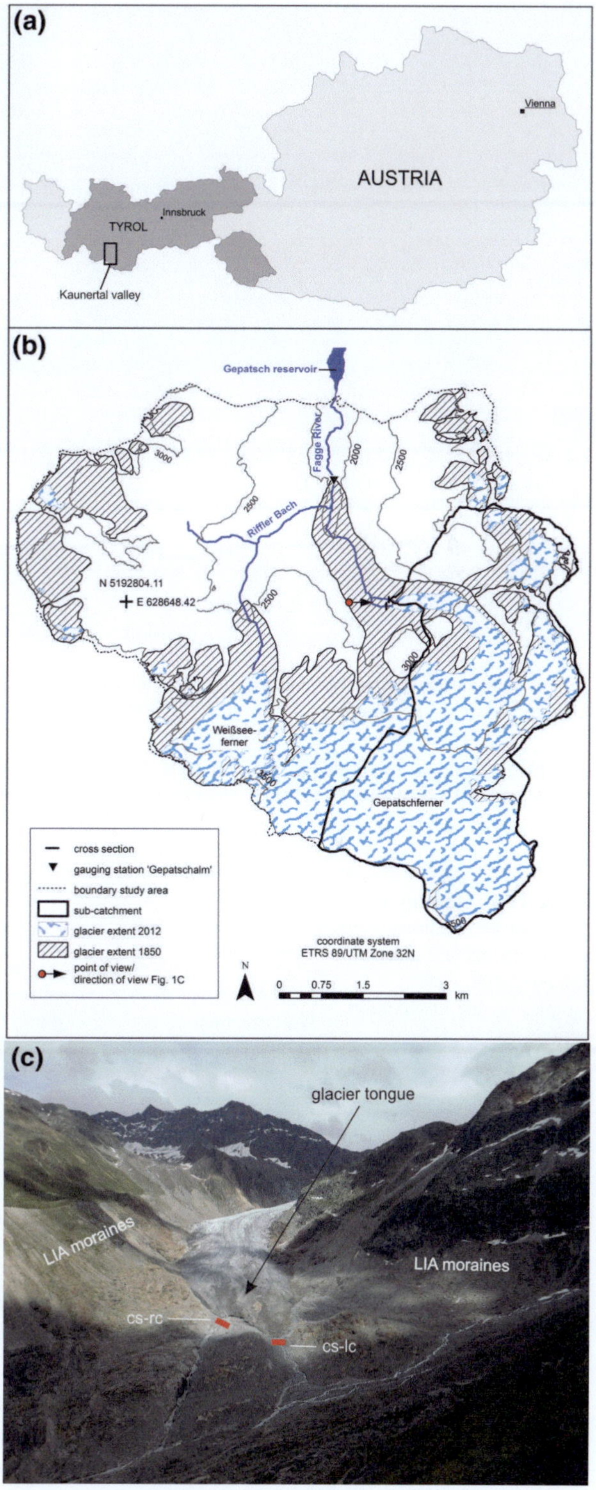

Fig. 13.1 Location of the study area in Kaunertal (**a**), map of the study area with glacier cover at LIA maximum and in 2012 (**b**), the glacier tongue of the Gepatschferner at the beginning of the investigations with its two drainage channels (**c**), the cross section of the right channel (cs-rc) was used in 2012, the cross section of the left channel (cs-lc) since 2013 for the discharge and bedload measurements

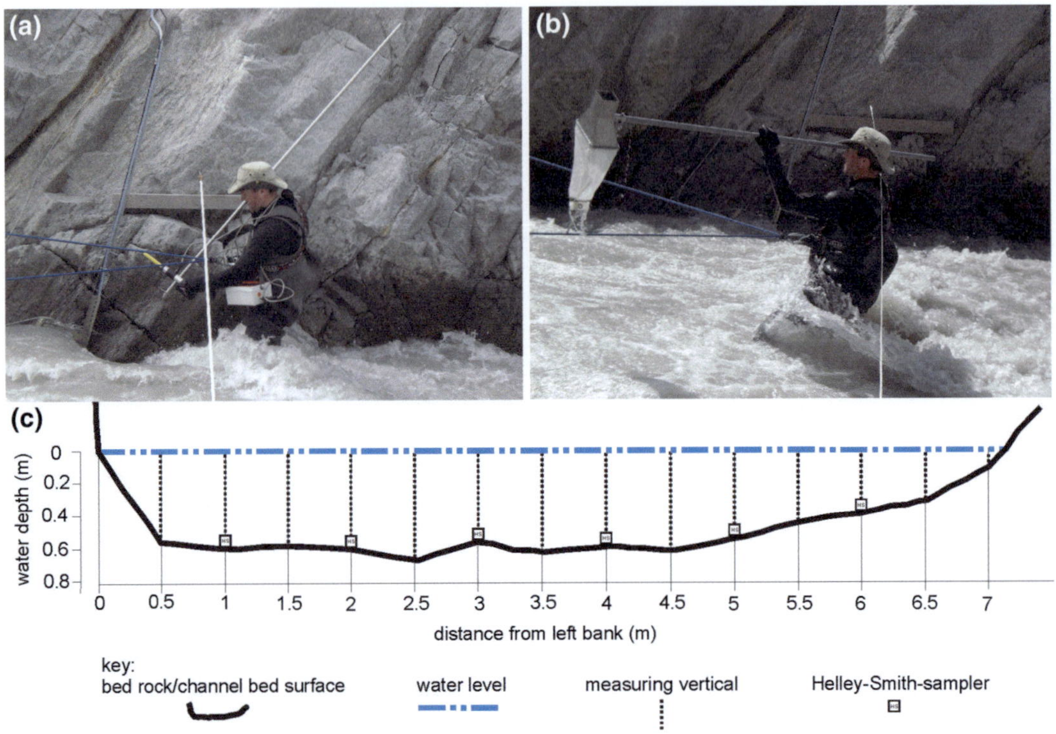

Fig. 13.2 Nautilus flow velocity (**a**) and Helley–Smith bedload (**b**) measurements at the right channel (cs-rc) in 2012 and a sketch (**c**) of the cross section at the gauging station of the Fagge River close to the glacier snout (Photographs by David Morche)

Table 13.1 Quantity of discharge and bedload measurements and manual and automatic water samples for the observation periods 2012–2015

	2012	2013	2014	2015
Discharge	22	17	25	19
Bedload	22	14	29	11
Water/manual	48	25	34	11
Water/automatic	160	79	108	91

Table 13.2 Examples of power-law rating curves for the ablation season 2015

	R^2	n
$Q = 7.57W^{1.4}$	0.745	8
$BL = 51.35Q^{1.3}$	0.26	39
$SSC = 4.82Q^{2.7}$	0.362	85

W Water level (m), *Q* discharge ($m^3\ s^{-1}$), *BL* bedload ($g\ s^{-1}$) for 2013–2015, *SSC* suspended sediment concentration ($mg\ l^{-1}$)

Bedload was measured 76 times during the 4 years of observation (Table 13.1), and it ranged from 3 to 1500 $g\ m^{-1}\ s^{-1}$.

To calculate the transported bedload, the sediment samples from the field were dried and weighed in the laboratory. Additionally, the active channel bed width was determined, which is the channel width where active bedload transport occurs (Ashmore et al. 2011). Bedload transport rate of each measurement was calculated by multiplying the measured total mass of bedload with the active channel bed width. This data was then used to create a rating curve for discharge and bedload for each observed time period of the relevant year (e.g., Table 13.2). Using the annual hydrograph and the rating curve, the bedload yield was determined for the time during which water level measurements were made.

To determine suspended sediment transport in the Fagge River between 2012 and 2015, 118 manually collected water samples and 348 automatically collected water samples were used (Table 13.1). At each point where discharge was measured, a water sample was taken as well. Additionally, an automatic water sampler (AWS) was used to take samples in intervals of 1–24 h during the ablation period (Fig. 13.3). In order to collect suspended sediment samples that were not influenced by bedload, the intake nozzle of the AWS was fixed in a moderately turbulent part of the river that was 20–30 cm above the channel bed.

However, due to changing channel bed conditions, readjustments of the intake nozzle were necessary multiple times, e.g., when it was covered with sediment or when water level was low at the end of the ablation period.

The water samples were filtered through 0.45 μm micropore filters to obtain the suspended sediment concentration (SSC) of each sample. The suspended sediment concentration in the water samples ranged from 26 to 10,570 mg l^{-1}. Similar to the bedload transport, a rating curve was established, and total suspended sediment yield for the measuring period was calculated (Table 13.2).

As a final step, the hydrograph and the rating curves were used to calculate suspended and bedload transport, and thus the total solid load output during the observed period. Because of data gaps in the hydrograph due to changing conditions at the station and malfunction of data

Fig. 13.3 New gauging station installed at the left channel (cs-lc) in 2013, note freshly exposed whalebacks and the glacier tongue in the back (view upvalley, photograph August 18, 2015, David Morche)

loggers, the entire ablation season could not be observed consistently. Therefore, the observed daily mean transport was used to estimate the total season output. The whole season covered the time between early snowmelt (April 1) and September 30 (183 days).

Due to the high variability in suspended and bedload sediment transport, there is an uncertainty in the calculated rates of transport and yields. For estimating the total error in sediment output, the standard error was calculated for suspended sediment as well as bedload for each ablation period.

Grain sizes of the bedload samples were determined by sieving the samples according to a logarithmic Udden–Wentworth scale.

The values of the D_{16}, D_{50}, and D_{84} were used to plot an annual trend in comparison with the associated discharge.

13.4 Results

Sediment load in the Fagge River was measured during four ablation seasons from 2012 to 2015. Because the gauging station installed in June 2012 at the right cross section (cs-rc) was avulsed during an extreme event in late August 2012 (Baewert and Morche 2014), new installations were necessary at the left channel (cs-lc) during the beginning of the measuring season in 2013.

In 2012, the measurements started at the beginning of June when the snow cover decreased and discharge in the Fagge River started to increase (Fig. 13.4). Over the ablation period, a mean discharge of 4.6 $m^3 s^{-1}$ was observed (Table 13.2). Bedload material caught during the June and July measurements was relatively fine. At the beginning of August also coarser particles, D_{84} up to 50 mm, were observed to be mobile (Fig. 13.4). The peak discharge in 2012 was recorded at August 23. Consequently, the value of 3157 t of bedload transport over the observation period or 7503 t over the whole ablation season 2012 does not take into account the bedload that has been transported during and after the extreme event. Baewert and Morche (2014) investigated the geomorphic effects of this particular event and measured the surface changes along the Fagge River using terrestrial laser scanning. The suspended load dominated the total solid load of the Fagge River in 2012, weighing more than 37,500 t (Table 13.1). Again, numbers for the big event were not available.

In 2013, a new gauging station was built at the left channel. Due to the rapidly changing channel bed during the snowmelt, reliable data could only be recorded after mid-July (Fig. 13.5) when a relatively stable left bank had been formed in the bedrock (Fig. 13.3). In 2013, the data collection period lasted between July 22 and September 26. However, discharge data could not be calculated for the entirety of the observational period due to failure of the measuring probes, which restricted the data collection to 43 days during this season. A relatively lower mean discharge of 2.9 $m^3 s^{-1}$ was observed over this ablation period, compared to 2012 (Table 13.3), with the peak discharge of 5.4 $m^3 s^{-1}$ observed on July 27 (Fig. 13.5). Bedload grain size was the coarsest at the beginning of August, with a maximum of 42 mm for D_{84} (Fig. 13.5). A mean bedload of 42 t per day was calculated for 2013, which added up to 1806 t of sediment being transported during the observation period. Over the entirety of the 2013 ablation season, 7851 t of bedload was transported (Fig. 13.5). As for suspended load in the Fagge River in 2013, a mean of 94 t of sediment was transported downvalley daily, accumulating to 4033 t for the entire year (Fig. 13.5). Furthermore, temperature data obtained from the TIWAG Station Weisssee showed maximum temperatures of 16 °C in late July. This time period also experienced the highest precipitation during the 2013 observational phase, with maximum precipitation reaching 5.3 mm (Fig. 13.5). Additionally, temperatures reached −4 °C in late September at the data collection site, with a maximum snow depth of 22 cm (Fig. 13.5).

In 2014, discharge and bedload grain size data were obtained for the time period between June 4 and October 22 (117 days). A mean discharge of 2.4 $m^3 s^{-1}$ was observed over this time period, with the peak discharge observed as 16.4 $m^3 s^{-1}$

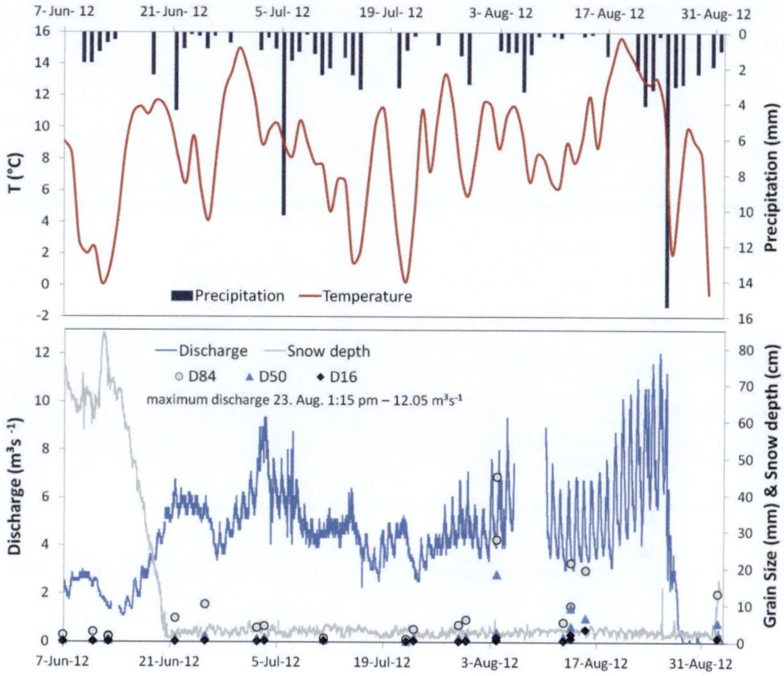

Fig. 13.4 Observation period 2012: graphs of air temperature, precipitation and snow depth (TIWAG Station Weisssee), and the hydrograph of the Fagge River at the glacier Gepatschferner with additional information on grain size of bedload samples. *Note* the bedload measurement at the end of August has been done at the left cross section (cr-cl in Fig. 13.1)

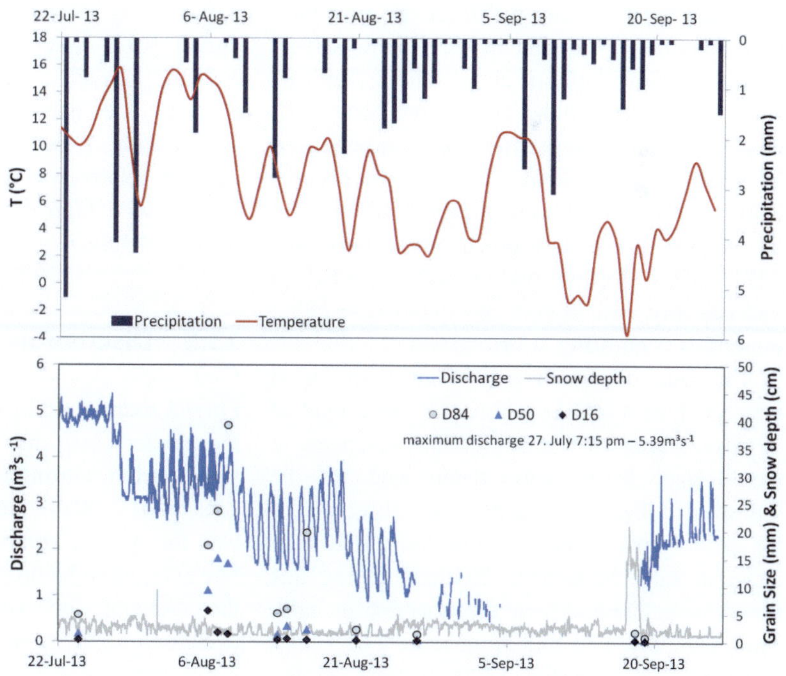

Fig. 13.5 Observation period 2013: graphs of air temperature, precipitation and snow depth (TIWAG Station Weisssee), and the hydrograph of the Fagge River at the glacier Gepatschferner with additional information on grain size of bedload samples

on July 31 (Fig. 13.6). The Fagge River carried coarser bedload in 2014, with the coarsest material being transported end-June to mid-July. During this time period, the D_{84} value exceeded more than 50 mm (Fig. 13.6). A mean bedload mass of around 37 t/day was calculated, adding up to 6771 t for the 2014 ablation season. In contrast, suspended load in 2014 was estimated

Table 13.3 Discharge and solid river load at the Fagge River gauging station close to the Gepatschferner during the observation periods 2012–2015

		2012	2013	2014	2015
Observation period		June 6–August 31	July 22–September 26	June 4–October 10	June 4–September 24
	No. of days with data recordings	76.9	42.9	117	77.8
Discharge	Mean [m^3 s^{-1}]	4.6	2.9	2.4	4.8
	Min [m^3 s^{-1}]	0.7	0.5	0.3	1.3
	Peak [m^3 s^{-1}]	12.1	5.4	16.4	9.9
Bedload	Mean [t/d]	41 ± 16.5	42 ± 22	37 ± 13.5	50 ± 14
	Observation period [t]	3157 ± 1269	1806 ± 944	4329 ± 1580	3890 ± 1089
	183-day ablation season [t]	7503 ± 3020	7851 ± 4026	6771 ± 2471	9150 ± 2562
Suspended sediment load	Mean [t/d]	205 ± 11	94 ± 10	182 ± 23	35 ± 4
	Observation period [t]	15,765 ± 846	4033 ± 429	21,294 ± 2691	2723 ± 311
	183-day ablation season [t]	37,515 ± 2013	17,202 ± 1830	33,306 ± 4209	6405 ± 732

to be 182 t per day, with a total of 33,306 t for the entire ablation season. Moreover, precipitation data shows snowfall in early June, with a maximum snow depth of about 50 cm. Temperature during this time period reached 0 °C; however, similar temperatures were experienced in mid-July without any snow. Temperatures below freezing point were experienced again in late August. The highest precipitation during the 2014 data collection period was recorded in mid-late July, with highest precipitation reaching more than 5 mm (Fig. 13.6).

The data collection period in 2015 lasted between June 4 and September 21 for a total of 78 days. During this time, the mean discharge in the Fagge River was determined to be 4.8 m^3 s^{-1}. Peak discharge was observed as 9.8 m^3 s^{-1} twice, once in late June and then again in early September. Bedload during this observational period was the coarsest in early June, late July, and mid-August, as D_{84} values reached 18–20 mm (Fig. 13.7). In total, 9150 t were transported during the 2015 ablation season. However, a relatively lower suspended load was calculated in the Fagge River during 2015 when compared to the previous year, with 35 t of sediment being transported daily. This accumulates to a total of 6405 t of sediment during 2015. A snow depth of over 60 cm was reported in early June, which gradually melted away by late June (Fig. 13.7). Temperature and precipitation data provided by TIWAG also showed relatively higher temperatures for the season, with high precipitation events recorded mainly in July.

13.5 Discussion

Fluvial sediment transport in the Fagge River at the glacier snout was variable over the 4-year study period. During the first three ablation seasons (2012–2014), more suspended sediment was transported compared to bedload. In 2015, however, more bedload was transported. Overall, the amount of suspended sediment load decreased in 2015. While hydrological data does not suggest a change in the system, higher mean discharge during the recording period could be used to argue that the retreating Gepatschferner glacier left enormous amounts of loose and coarse material ready for bedload transport near

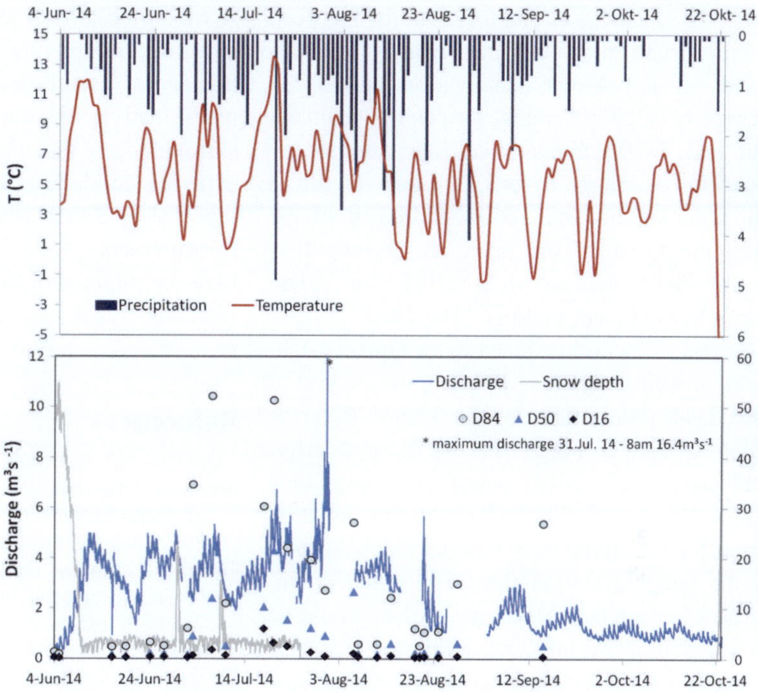

Fig. 13.6 Observation period 2014: graphs of air temperature, precipitation and snow depth (TIWAG Station Weisssee), and the hydrograph of the Fagge River at the glacier Gepatschferner with additional information on grain size of bedload samples

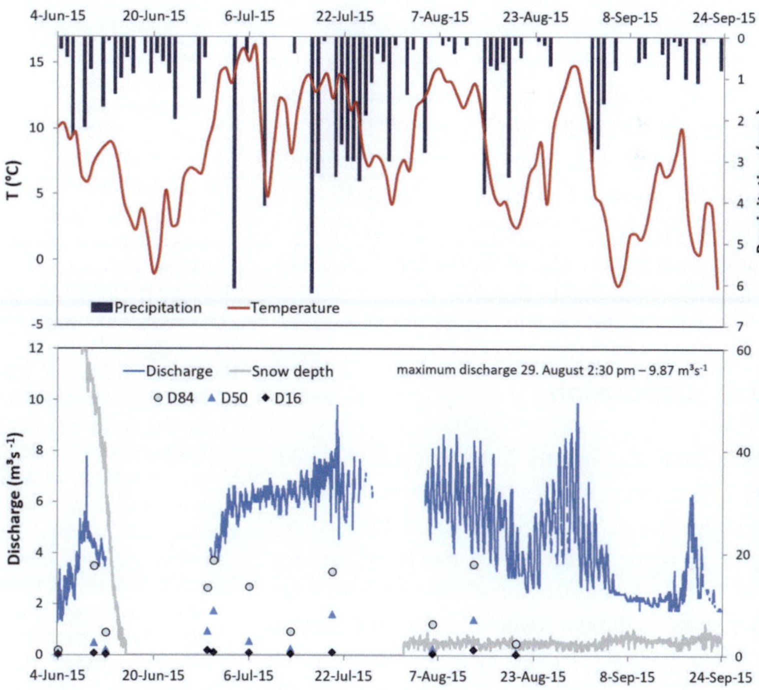

Fig. 13.7 Observation period 2015: graphs of air temperature, precipitation and snow depth (TIWAG Station Weisssee), and the hydrograph of the Fagge River at the glacier Gepatschferner with additional information on grain size of bedload samples

and in the channel and that this material has been mobilized and transported downvalley in 2015. Based on the 4-year observation period, an average of 23,607 t annual suspended sediment load and 7819 t annual bedloads were determined for the Fagge River (see Chap. 17 for an estimation of the total sediment budget of the study area), which can also be expressed as specific yields; bedload yield is 391 t km^{-2}, and suspended sediment yield is 1180 t km^{-2}.

Compared to other glacierized alpine catchments of similar size (ca. 20 km^2), the amount of solid load transported in the Fagge River is within a normal range. In the Eastern Alps, Geilhausen et al. (2013) calculated a suspended sediment load of 7000–10,000 t per ablation period for the alpine river Obersulzbach/Austria. As the gauging station there is located downstream of a proglacial lake, no bedload could be measured. But the lake trap efficiency in Obersulzbach was determined to be >90% for suspended sediment and 100% for bedload, therefore higher solid loads are expected in the rivers draining to the Obersulzbachsee lake.

In the Western Alps, the annual suspended sediment yield of the 25 km^2 Bas Arolla catchment in Switzerland is 1100 t km^{-2} (Gurnell 1995).

Mao and Carillo (2017) recently measured suspended sediment yield in the Andean river Estero Morales/Chile. Their average annual suspended sediment yield of 470 t km^{-2} is lower compared to the Fagge River.

13.6 Conclusion

Glacier retreat is greatly influencing fluvial sediment transport dynamics in the active channel of the Fagge River. Thus, glacier forefields are very active in terms of sediment transfer (Carrivick et al. 2013) as meltwater channels transport large amount of sediment over the ablation period (Geilhausen et al. 2013). The Fagge River in the Gepatschferner forefield is carrying sediment originating from the glacier itself and from proglacial sources like the unstable channel banks built up by morainic and glacio-fluvial deposits. While the range of total sediment load in Fagge River is similar to other high-mountain proglacial streams, interpreting this as a general phenomenon is scrutinized by the high morphodynamic activity in the Kaunertal caused by the shrinking of the Gepatschferner glacier. As experienced during the study period, and also reported by Sanders et al. (2013), the direct field measurement of sediment transport in high-mountain proglacial areas is still a difficult (and sometimes impossible) task.

References

Ashmore P, Bertoldi W, Gardner JT (2011) Active width of gravel-bed braided rivers. Earth Surf Proc Land 36:1510–1521. https://doi.org/10.1002/esp.2182

Baewert H, Morche D (2014) Coarse sediment dynamics in a proglacial fluvial system. Geomorphology 218:88–97. https://doi.org/10.1016/j.geomorph.2013.10.021

Carrivick JL, Geilhausen M, Warburton J, Dickson NE, Carver SJ, Evans AJ, Brown LE (2013) Contemporary geomorphological activity throughout the proglacial area of an alpine catchment. Geomorphology 188:83–95. https://doi.org/10.1016/j.geomorph.2012.03.029

Geilhausen M, Morche D, Otto JC, Schrott L (2013) Sediment discharge from the proglacial zone of a retreating alpine glacier. Z Geomorphol (Supplementary Issue) 57(2):29–53. https://doi.org/10.1127/0372-8854/2012/s-00122

Gurnell AM (1995) Sediment yield from Alpine glacier basins. In: Foster IDL, Gurnell AM, Webb BW (eds) Sediment and water quality in river catchments. Wiley, Chichester, pp 407–435

Haeberli W (1980) Morphodynamische Aspekte aktueller Gletscherhochwasser in den Schweizer Alpen. Regio Basilensis 21:38–78

Heckmann T, Haas F, Morche D, Schmidt K-H, Rohn J, Moser M, Leopold M, Kuhn M, Briese C, Pfeiffer N, Becht M (2012) Investigating an alpine proglacial sediment budget using field measurements, airborne and terrestrial LiDAR data. IAHS Press, Wallingford IAHS Publ 356:438–447

Herschy RW (2009) Streamflow measurement. Routledge, Taylor & Francis, London 507 pp

Lane SN, Bakker M, Gabbud C, Micheletti N, Saugy JN (2017) Sediment export, transient landscape response and catchment-scale connectivity following rapid climate warming and Alpine glacier recession. Geomorphology 277:210–227. https://doi.org/10.1016/j.geomorph.2016.02.015

Mao L, Carrillo R (2017) Temporal dynamics of suspended sediment transport in a glacierized Andean basin. Geomorphology 287:116–125. https://doi.org/10.1016/j.geomorph.2016.02.003

Marren PM (2005) Magnitude and frequency in proglacial rivers: a geomorphological and sedimentological perspective. Earth Sci Rev 70:203–251. https://doi.org/10.1016/j.earscirev.2004.12.002

Marren PM, Toomath SC (2014) Channel pattern of proglacial rivers: topographic forcing due to glacier retreat. Earth Surf Proc Land 39:943–951. https://doi.org/10.1002/esp.3545

Moore RD (2005) Slug injection using salt in solution. Streamline Watershed Manage Bull 8:1–6

Sanders JW, Cuffey KM, MacGregor KR, Collins BD (2013) The sediment budget of an alpine cirque. Geol Soc Am Bull 125:229–248. https://doi.org/10.1130/B30688.1

Shugar DH, Clague JJ, Best JL, Schoof C, Willis MJ, Copland L, Roe GH (2017) River piracy and drainage basin reorganization led by climate-driven glacier retreat. Nat Geosci 10:370–375. https://doi.org/10.1038/NGEO2932

Vehling L, Baewert H, Glira P, Moser M, Rohn J, Morche D (2017) Quantification of sediment transport by rockfall and rockslide processes on a proglacial rock slope (Kaunertal, Austria). Geomorphology 287:46–57. https://doi.org/10.1016/j.geomorph.2016.10.032

Warburton J, Fenn CR (1994) Unusual flood events from an Alpine glacier: observations and deductions on generating mechanisms. J Glaciol 40:176–186

14 Proglacial Lakes in High Mountain Environments

Jan-Christoph Otto

Abstract

Lakes are a characteristic feature of glacial landscapes. They are found in the vicinity of current or past glaciers and ice sheets, in high alpine cirques, inner-alpine valleys as well as lowlands. In recent years, new lakes have emerged in glacier forefields and surface areas and volumes of many proglacial lakes are reported to be growing in many mountain areas due to climate-induced glacier melt. Some proglacial lakes have attracted public and scientific attention due to disastrous events such as lake outburst floods or increasing hazard potential and risk downstream. Proglacial lake formation is the result of glacier retreat exposing a topographic bedrock depression or space behind a sediment dam that inhibits runoff and provokes storage of water and sediment. Proglacial lakes thus are first-order sediment sinks and interrupt the sediment cascade and sediment transfer dynamics from uplands to lowlands. They often are of societal relevance in mountain areas ranging from water supply to energy production, hazard and risk, as well as tourism issues. This review summarises the role of proglacial lakes for geomorphic systems in high mountain environments. We start with a look at the basic terminology and formation principles, followed by an overview of global lake distribution patterns. The geomorphologic significance of proglacial lakes is discussed with respect to the current state of knowledge. Recent developments that allow a modelling of potential future lakes in mountain areas, once the glacier melt, are presented and discussed in the light of natural hazards and risks, as well as socio-economic dimensions of proglacial lake formation.

Keywords

Proglacial lakes distribution · Overdeepening
Lake sedimentation · Sedimentary archive
Outburst flood

14.1 Introduction

Lakes are a characteristic feature of glacial landscapes. They are found in the vicinity of current or past glaciers and ice sheets, in high alpine cirques, inner-alpine valleys as well as lowlands. In recent years, new lakes have emerged in glacier forefields and surface areas and volumes of many proglacial lakes are reported to be growing in many mountain areas due to climate-induced glacier melt (Wang et al.

J.-C. Otto (✉)
Department of Geography and Geology, University of Salzburg, Salzburg, Austria
e-mail: jan-christoph.otto@sbg.ac.at

T. Heckmann and D. Morche (eds.), *Geomorphology of Proglacial Systems*,
Geography of the Physical Environment, https://doi.org/10.1007/978-3-319-94184-4_14

2015; Song et al. 2017; Carrivick and Quincey 2014). Some proglacial lakes have attracted public attention due to disastrous events such as lake outburst floods or increasing hazard potential and risk downstream (Wang and Jiao 2015; Watanabe 1996; Somos-Valenzuela et al. 2016; ICIMOD 2011). Increasing public awareness of climate change and related hazards, as well as scientific interest from landscape evolution studies, has put glacial lakes in the spotlight as documented by recent review articles and a large number of case studies (Arnaud et al. 2016; Bogen et al. 2015; Carrivick and Tweed 2013, 2016; Cook et al. 2016; Emmer et al. 2016; Haeberli et al. 2016a; Maanya et al. 2016; Clague and O'Connor 2015).

Proglacial lake formation is the result of glacier retreat exposing a topographic bedrock depression or space behind a dam that retards runoff and provokes storage of water and sediment. Proglacial lakes thus are first-order sediment sinks and interrupt the sediment cascade and sediment transfer dynamics from uplands to lowlands (Chaps. 15 and 16). The stored sediment represents an important sedimentary archive enabling the reconstruction of glacier conditions in the lake catchment and past climatic conditions of mountain ranges (Larsen et al. 2011; Leemann and Niessen 1994; Blass et al. 2007). Proglacial lakes at glacier margins positively feedback on the glacier system resulting in increased ice velocities, changed mass balances and enhanced melting. They often are of societal relevance in areas ranging from water supply to energy production and tourism issues. Furthermore, a rapid drainage of these lakes can constitute a severe natural hazard for mountain areas by the formation of glacial lake outburst floods (GLOFs). The effects of these catastrophic floods can cause severe damage and produce significant landscape changes. Formation of new lakes systematically increases flood risks in downvalley zones of mountain regions.

This review summarises the importance of proglacial lakes for geomorphic systems in high mountain environments. We start with a discussion of the general terminology and an introduction to lake formation processes and lake types. An overview of the current global distribution highlights the relevance of proglacial lakes for mountain environments. In the main part of this chapter, the geomorphologic significance of proglacial lakes is discussed with respect to the current state of knowledge. Recent developments in modelling of potential future lakes are presented that allow discussing potential changes on mountain environments following temperature-induced glacier melt. Finally, we address the issue of natural hazards resulting from proglacial lakes in mountain regions and discuss the socio-economic dimensions of proglacial lakes.

14.2 Formation and Types of Proglacial Lakes

The formation of proglacial lakes is closely linked to glacier dynamics (Chaps. 2–4) and environmental conditions (Fig. 14.1). The terminology defining a lake with a distinct relationship to glaciers is far from clear-cut. The terms proglacial lake, glacial and glacier-fed lake are often used similarly and commonly associated with glacial meltwater impounded behind a barrier (Ashley 2002). In a general sense, glacial lakes are water bodies directly linked to glacier activity, either by shaping the depression in the land surface, by providing material and landforms for impoundment or simply by delivering meltwater to the lake (U.S. Department of Agriculture 2016). Some definitions for proglacial lakes (in contrast to distal glacial lakes) stress a direct influence by a glacier margin (also termed ice-contact lakes) or subaerial glacial meltwater (Carrivick and Tweed 2013), a circumstance that will only be of limited temporal significance during glacier retreat. Other sources also include supraglacial lakes, i.e. lakes that form on the glacier itself (Zhang et al. 2015). These lakes differ significantly in terms of formation, lifetime and impact on sediment flux and will not be considered in detail here.

Various classification schemes are published which are related to glacial and non-glacial processes responsible for the impoundment

Fig. 14.1 Proglacial lake Obersulzbachtal (Eastern Alps, Austria). *Image* Keuschnig 2015

(Carrivick and Tweed 2013; ICIMOD 2011; Iturrizaga 2014). The storage of water in proglacial areas is either related to a topographic depression or to the existence of a dam or blockage in the landscape that hinders surface runoff. The most commonly used classification includes four main types of glacial lakes: (1) bedrock-dammed lakes, (2) moraine-dammed lakes, (3) ice-dammed lakes and (4) landslide-dammed lakes (Fig. 14.2). While lake types 1 and 2 are restricted to the proglacial zone (Chap. 1), the latter types also occur outside the proglacial area. Since they are typical for mountain regions and exhibit similar impacts on mountain environments, these lake types will be included in the classification here.

Topographic depressions, also termed *overdeepenings*, are the result of glacial erosion by quarrying and abrasion (Benn and Evans 2010; Cook and Swift 2012). Erosion induced by sediment-laden subglacial meltwater may also contribute to their formation (Beaud et al. 2015; Creyts and Clarke 2010). Quarrying is associated with increased stress at the top of bedrock obstacles in the glacier bed coupled with water-pressure fluctuations at the lee side of the bump that results in crack growths (Hooke 2005; Iverson 2002). Abrasion is the mechanical grinding of bedrock by rock fragments frozen to the glacier base. A subglacial overdeepening is a bowl-shaped landform characterised by a steep headwall and a reverse slope at the lower end (Fig. 14.3). A recent study suggests that the overall geometry of surface area, length, width and depth varies substantially and does not reveal any correlations or regularities (Haeberli et al. 2016b). The role of the reverse slope is especially critical with respect to subglacial sediment transport. If a critical slope angle is reached (usually by a factor of 1.7–2 times steeper than the glacier surface slope), two processes happen: (1) supercooling of meltwater and (2) water dissipation from the conduits underneath the ice (Creyts and Clarke 2010; Hooke 1991). Both effects reduce flow velocities and result in the deposition of sediment on the reverse slope at the glacier base.

Bedrock depressions form at various positions below the glacier along its flowline. They are found in cirques, tributary-trunk junctions and in the foreland of ice sheets. Many of the perialpine lakes in Europe (e.g., Lake Geneva, Lake Constance) resulted from ice-scouring and overdeepening by Pleistocene glaciers. Frequently, depressions are related to glacier margins or confluence zones where two ice masses meet (Cook and Swift 2012). Large valleys and fjords often contain multiple bedrock depressions.

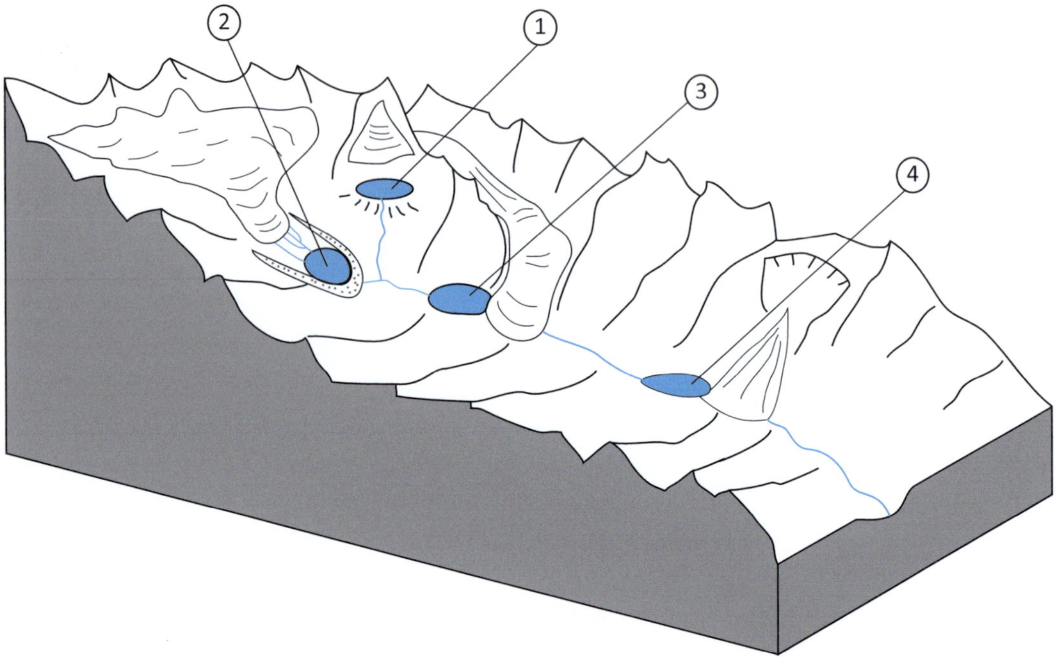

Fig. 14.2 Types of glacial lakes: (1) bedrock-dammed, (2) moraine-dammed, (3) ice-dammed, (4) landslide-dammed

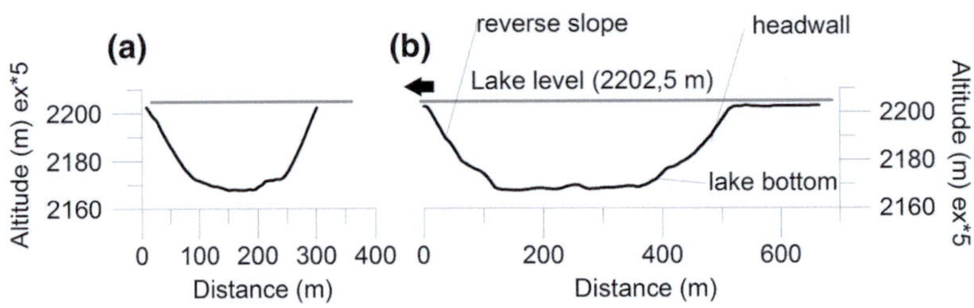

Fig. 14.3 Cross section (**a**) and longitudinal section (**b**) of the proglacial lake at Obersulzbach glacier, Eastern Alps, Austria. Data derived from radio-echo sounding indicating the water–sediment interface (black line). The arrow indicates lake outflow and downstream direction (*Data source* Hydrological Survey of Salzburg 2015). The longitudinal section displays the bowl shape with the steep headwall and reverse slope

Lakes formed in depressions are usually termed *bedrock-dammed lakes*, although bedrock is often covered by glacial till without a discrete damming landform observable.

Moraine-dammed lakes form behind terminal, and lateral moraines (Fig. 14.4) are linked to a glacier advance and subsequent retreat. Lakes also form in basal moraines and formation may be related to a buried bedrock overdeepening (see above). If the depression results from the melting of dead-ice, the hollow is termed kettle-hole. Frontal moraines are formed during ice advance and have a wall-like shape with steep slopes and a width-to-height ratio similar to anthropogenic dams (Clague and Evans 2000). Moraine deposits can act as dams due to their sedimentary composition. Sediment dumped by glaciers of sub-, en- and supraglacial origin is characterised by a mixture of all grain sizes. Coarse and blocky material is commonly embedded into a matrix of fines. Components are often poorly sorted but can also be stratified. Sediments can be more or less compacted and often contain an ice core of either buried glacier

Fig. 14.4 Small moraine-dammed lake in the Eastern Alps (Felbertal, Austria)

ice or due to permafrost conditions (Benn and Evans 2010; Chap. 6). A combination of these characteristics can lead to a reduced infiltration capacity of the moraine deposit and the impoundment of meltwater behind the sediment barrier. *Moraine-dammed lakes* often form in front of large valley and piedmont glaciers but may also develop within cirques. One prerequisite for *moraine-dammed lakes* is the presence of a continuous moraine that blocks the drainage way of meltwater. A moraine dam is continuous when the debris supply at the glacier margin is exceeding the transport capacity of the meltwater stream (Benn and Evans 2010). *Moraine-dammed lakes* are especially susceptible to GLOFs when the internal or external conditions that stabilize the dam (e.g. ice content, weakening by seepage, lake-level rise) change and the dam stability is significantly reduced (Richardson and Reynolds 2000; Clague and Evans 2000; see section below).

Ice-dammed lakes form behind barriers of glacier ice and are not restricted to the proglacial zone (Fig. 14.2). A barrier of ice either results from a glacier advance or from glacier melt (Fig. 14.5). Advancing glaciers tongues can block parts of a valley floor and inhibit drainage, for example from a neighbouring catchment. Variable melt of glaciers in combination with topographic conditions can enable the formation of ice-marginal lakes that form in the space between lateral bedrock or moraines and the glacier. *Ice-dammed lakes* are commonly of more temporary nature compared to other lakes since the barrier made of ice is mobile and less stable compared to the other types of dams. Drainage of *ice-dammed lakes* can be rapid and catastrophic and has led to the largest floods recorded (O'Connor and Costa 2004). However, many ice-dammed lakes drain slowly through subglacial meltwater passages that are affected by glacier movement and sediment load, and open or close according to the thermal and hydrological conditions of the glacier. This results in variable lake levels with repeated filling and drainage on various timescales.

Landslide-dammed lakes are the consequence of catastrophic slope failures that block a meltwater river course by filling an entire valley cross section with debris (Fig. 14.2). Rockfalls or landslides in the vicinity of glaciers are often associated with paraglacial slope adjustment (McColl 2012; Ballantyne 2002; Chaps. 8–11). Glacial erosion produces steep slopes and modifies the stress conditions within slopes. Deglaciation has preparatory and triggering

Fig. 14.5 Lake formed in the degrading glacier tongue of the Turtmann glacier (Central Alps, Switzerland). The lake is ice-dammed and formed due to collapse of an ice cave/tunnel. The lake existed only for a few months before the water drained through loose deposits at the ice margin

effects on slope failures either related to debuttressing, i.e. the removal of the load previously imposed by the glacier mass, or by exposing areas of weakness within the slope by glacial erosion (McColl 2012; Chaps. 8 and 9). Catastrophic failure of preconditioned slopes can be caused by various triggering factors like seismic shaking, extreme rainfall events, melting of ice, snow or permafrost, leading to an excess in cleft water pressures (e.g. Sartori et al. 2003). Landslide dams are often higher and have a smaller width-to-height ratio compared to the other dam types. They can also be located at greater distance from the glaciers when conditioned by local stability factors (Korup and Tweed 2007).

14.3 Global Distribution of Glacial Lakes

A global analyses of lake distribution reveals that the highest lake densities are found within the extent of the last glacial maximum in the northern hemisphere followed by most high mountain zones of the Andes, Himalaya, European Alps and Rockies (Messager et al. 2016). Glacial erosion and deposition are therefore responsible for a vast majority of lakes on earth. Recent studies in high mountains report an increase in both, lake number and lake size (Wang et al. 2015; Zhang et al. 2015; Paul et al. 2007; Veettil et al. 2016; Buckel et al. 2018; Emmer et al. 2016). Glacial lake inventories have been compiled for different mountain areas. Lake mapping has either been carried out in the field, or remotely by manual mapping of aerial or satellite imagery. Automated mapping of larger regions has been based on image analyse of multi-spectral satellite images (e.g. LANDSAT, ASTER, ALOS) (e.g. Gardelle et al. 2011; Allen et al. 2016).

Based on manual satellite image analysis Zhang et al. (2015) report more than 5700 lakes (including supraglacial lakes) within a maximum distance of 10 km from existing glaciers in the Himalaya region and the Tibetan Plateau. Glacial lakes in this region cover an area of 682 ± 110 km^2 whereas glaciers cover 40,800 km^2 (Bolch et al. 2012). Between 1990 and 2010, lake number has increased by 1000 (+21%) and lake area by 23% (Zhang et al. 2015). One hotspot of glacial lake formation is the region of Nepal and Bhutan in the Central and Western Himalaya. More than 3900 lakes (Nepal 1400, Bhutan 2500) with a total area of 170 km^2 make this region especially susceptible to GLOFS (Mool et al. 2001; ICIMOD 2011). Another hotspot of glacial lake formation and related GLOF hazards is the Cordillera Blanca massive in Peru, South America (Reynolds 1992; Iturrizaga 2014; Carey 2005). Vilímek et al. (2016) mapped more than 2300 glacial lakes in the area using Google Earth imagery. Emmer et al. (2016) estimated more than 880 lakes (lake size > 2000 m^2) in the Cordillera Blanca. They report an upward trend in lake distribution indicating a shift of new lake formation towards higher elevations since the 1950s. A glacial lake inventory of the Eastern Alps in Austria, Europe, reveals more than 1400 lakes (lake size > 1000 m^2) at altitudes above 1700 m ASL (Buckel et al. 2018). This lake inventory is based on orthophoto mapping and contains additional lake attributes like lake size, dam type, catchment size, geology of catchment and relative age. Comparison of lake data with an existing glacier inventory for five different time stages (Little Ice Age, 1920, 1969, 1989, 2006; Fischer et al. 2015) indicates a strong increase in the number of new lakes since the turn of the twenty-first century (Fig. 14.6).

14.4 Geomorphic Significance of Proglacial Lakes

Within the alpine sediment cascade, proglacial lakes represent important sediment sinks often located at elevated locations along the cascade path. Formation of proglacial lakes significantly reduces sediment output from glaciated watersheds thus impacting on proglacial sediment budgets (cf. Chaps. 15 and 16) and downstream fluvial sediment availability. The reduction of flow velocity when entering a lake causes sedimentation and formation of delta and lake deposits (cf. quantification of delta aggradation in Chap. 17). Accommodation space, grain size and flow velocity are the dominant parameters determining the amount of material deposited within or transferred through the lake. The ratio between sediment amount delivered to the lake and exported from the lake is defined as trapping efficiency (TE). TE is influenced by the geometry of the lake, the retention time of water in the lake and runoff characteristics (Verstraeten and Poesen 2000). It directly determines the lifetime of the lake before it is filled in with sediments. TE is

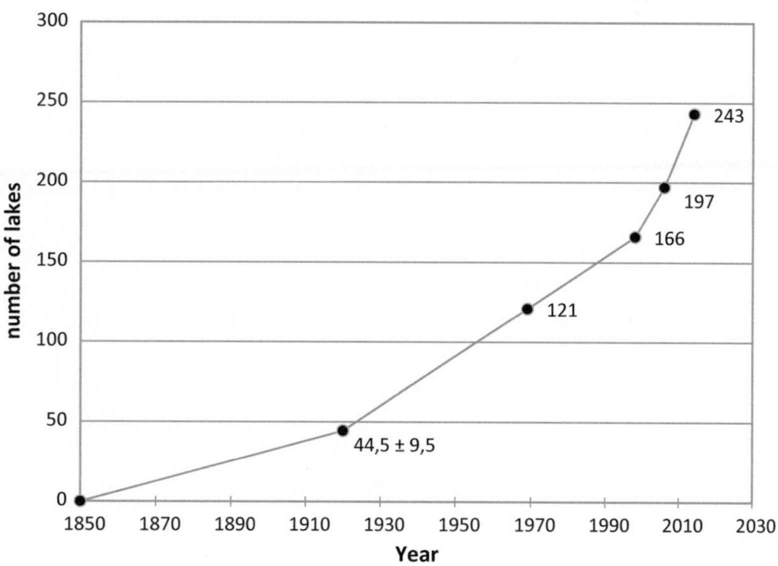

Fig. 14.6 Number of proglacial lakes in Austria at different stages of glacier extent since the Little Ice Age (data from: Buckel et al. 2018)

frequently studied and quantified for reservoirs and non-glacial lakes, but little information on trapping efficiency of proglacial lakes exist (Kummu et al. 2010; Mulu and Dwarakish 2015). Bogen et al. (2015) compare suspended sediment flux of three glacial lakes in Norway with downstream sediment delivery and estimate trapping efficiencies between 16 and 64%. For a small lake in the Canadian Coastal Mountains, Schiefer and Gilbert (2008) calculated a TE between 50 and 70% and based on sediment flux measurements. They conclude that the lake will fill up within less than 100 years. While lake formation is mainly controlled by glacier melt rate (Fig. 14.7), sediment input rate conditions the lifetime of the lake. Lake formation, once initiated can be very rapid. For example in the Obersulzbach valley, Austria Alps, a lake initiated in 1998 increase to its full size of 150,000 m^2 within 13 years (Fig. 14.7). In contrast, lake lifetime of high alpine lakes is highly variable spanning between several years and thousands of years. For example, in Austria, more than 1300 natural lakes exist above 1700 m ASL (see below). Two hundred and forty of these lakes have formed since the Little Ice Age (Buckel et al. in prep.), while the majority of these lakes have formed at the end of Pleistocene and are present since more than 10,000 years. Consequently, sedimentation rates also vary by several orders of magnitude from mm/ to m/a (Fleisher et al. 2003; Koinig et al. 2002; Schiefer 2006).

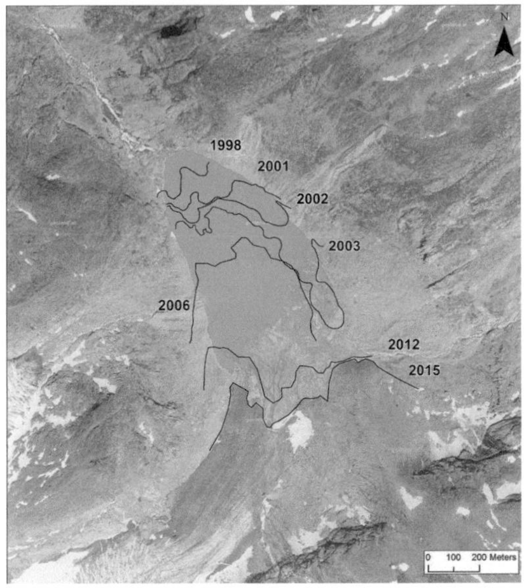

Fig. 14.7 Changing positions of the glacier front and consequent formation of the Sulzsee, Obersulzbachtal, Austria (cf. Figs. 14.1 and 14.3) between 1998 and 2015. The lake reached its maximum size in 2011 when the glacier margin retreated from the lake area. Sediment input from the glacier led to the formation of a delta at the southern lake margin between 2011 and onwards (background image: basemap.at)

14.4.1 Sediment Delivery and Sedimentation

There are several ways of sediment input to proglacial lakes. Sediment delivery differs between ice-contact and distal lakes. Glacial, glaciofluvial (Chaps. 12 and 13), fluvial and gravitational processes operating at or around the lake margins deliver sediments to the lake (Fig. 14.8). In the presence of glacier ice, sediment is discharged from the glacier to the lake by supra- and subglacial meltwater flows, by direct deposition from the ice front (sliding of supraglacial debris) and by dumping of sediment stored in icebergs floating on the lake (Fig. 14.8a). Due to the position of the glacier front and the mobility of floating ice, sediment is delivered to various different locations of the lake. After the retreat of ice from the lake sediment will only be delivered to the lake margins by meltwater streams (Chaps. 12 and 13) or from lateral slopes (Chaps. 10 and 11). Non-glacial processes mostly include reworking of sediments previously deposited by the glacier (e.g. lateral moraines) in the way of paraglacial landform adjustment (Ballantyne 2002; Church and Ryder 1972) (Fig. 14.8b). Vital to the influence of non-glacial sediment delivery is the position of the lake within the catchment, the catchment topography and process connectivity (Rubensdotter and Rosqvist 2009; Brierley et al. 2006; Chap. 16).

Sedimentation patterns within the lake are affected by several factors: (a) lake type, (b) thermal conditions, (c) relative density of lake

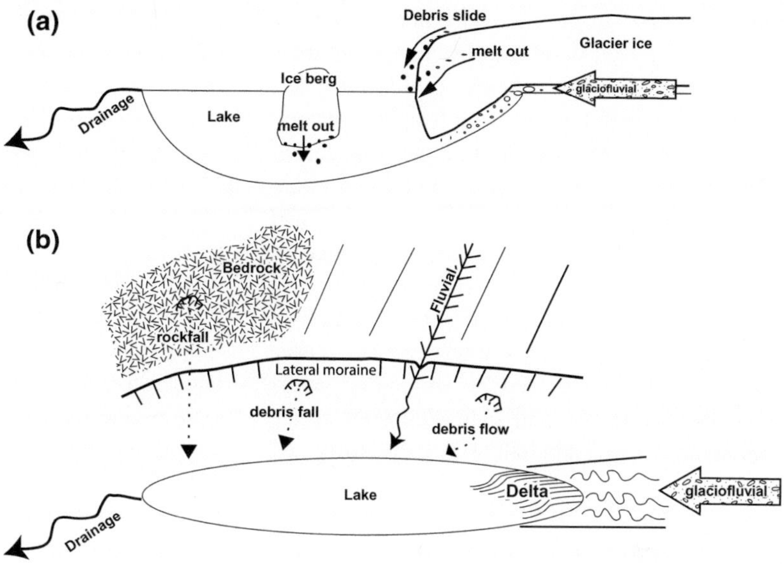

Fig. 14.8 Sources of sediment into a proglacial lake. **a** Sediment input processes related to the presence of glacier ice (ice-contact lakes only). **b** Non-glacial input processes (ice-contact and distal lakes)

and meltwater, (d) position of meltwater inflow and input sources, (e) number of sediment sources, (f) lake basin geometry, (g) slope stability, (h) ice rafting and (i) extent of the ice cover (Ashley 2002). At the stream inlet, physical and thermal conditions of stream and lake water, for example temperature, density and thermal stratification, influence sediment movement and deposition. Thermal stratification of the lake water is created when the upper layers are warmed during summer and less dense water accumulates at the surface while colder, more dense water sinks to the bottom. In case of ice-contact lakes, a thermal convection process can establish. Water temperatures in the contact zone decrease to 0 °C leading to a rise to the surface. These surface waters drift away from the ice front and warm due to solar radiation. Heating of the water increases density and causes sinking. If this water moves back towards the ice front, for example due to an inclined lake bottom, the convention cycle is closed (Kääb and Haeberli 2001). This process is responsible for notch formation at the water–ice interface and also affects calving from the ice front.

In case of thermal stratification, the different layers are separated by the thermocline where the change of temperature with depth is greatest (Ashley 2002). This density distribution affects the mode of sedimentation and sediment reworking (Bennet and Glaser 2009). Depending on the relative density of the meltwater entering the lake, the fine sediment load is either directed to the lower parts (meltwater with sediment denser than lake water), enters the lake at an intermediate depth (similar density of meltwater and lake), or stays at the surface (meltwater less dense than lake water). These conditions determine sedimentation style and depocenter location within the lake. Additionally, grain size and flow velocity determine the transport path of sediment delivered by the stream. Fine sediment is either transported within the upper layer or moves into deeper parts by high-density turbidity flows (downflow). These turbidity flows are often associated with diurnal discharge maxima following increased glacier melt. Sediment introduced by dense meltwater rapidly sinks to the bottom of the lake (Bennet and Glaser 2009). Coarse sediment is rapidly deposited due to reduced flow velocities usually leading to the formation of a delta that progrades into the lake. Fine silt and clays stay in suspension until flow velocity drops below $<0.1\ \mathrm{m\ s^{-1}}$ (Ashley 2002). Besides the thermal balance between inflow and lake water, wind also plays a role in mixing of water and dispersal of sediments within a lake. In winter, ice cover and reduced flow velocities and sediment input from the meltwater stream, as well as eliminated wind impact change the thermal and flow characteristics in proglacial lakes. Surface temperatures are significantly lowered and thermal stratification is removed. In consequence, sediment delivery is reduced and transport within the lakes stops causing suspended particles to sink down. This annual change in depositional conditions between summer and winter can produce a rhythmical composition of lake-bottom sediments termed varves. Varve deposition is characterised by alternating layers of coarser layers (silt and sand) deposited in the summer and fine layers (silt and clay) settling out of suspension during calm conditions in winter. Varve thickness is a function of discharge of suspended sediment and related to runoff, summer temperatures and glacier changes (Leemann and Niessen 1994; Desloges and Gilbert 1994; Menounos et al. 2005). The presence of varved sediment in glacial lakes highlights why glacial lakes are such excellent sedimentary archives for reconstruction of paleo-environments and catchment conditions (Thomas and Briner 2009).

Except fluvial input from non-meltwater streams, sediment input from all other non-glacial processes is less influenced by internal lake conditions (see above) but strongly controlled by the lake surrounding, the distance to sediment source areas and connectivity of erosional processes (Lane et al. 2017). Proglacial environments are characterised by (a) large amounts of loose deposits, (b) free destabilized rock faces and (c) limited vegetation cover (Chap. 19). Erosional processes will thus be highly effective in eroding bedrock and delivering sediment (Chaps. 10, 11, 15 and 17) from slopes, moraines and valley floors. Glacial deposits are poorly sorted and include a great variety of clast sizes dominated by silt, sand and gravel components. These deposits are frequently reworked by debris flows, slides and fluvial processes and delivered to the lake. Gravitational processes on bedrock slopes, like rockfalls and rock avalanches (Chaps. 8 and 9) also contribute sediment to the lake. These processes can operate at high-energy levels and rapidly displace lake water so that large waves develop and initiate GLOFs (see below). Many glacial lakes in mountain terrain are located in proximity to permafrost terrain. Changes in subsurface thermal regime can result in destabilisation of steep slopes or debris deposits, contribute to an high process activity (Deline et al. 2015) and thus increase sediment input.

14.4.2 Geomorphologic Impacts Induced by GLOFs

Major geomorphic work is performed when moraine- and ice-dammed lakes burst generating GLOFs. Water volume, sediment availability and high potential energy due to the elevated topographic position make GLOFs one of the most effective geomorphological processes in mountain environments characterised by huge amounts of deposition, very long transport distances and large boulder sizes in motion. The effectiveness of GLOFs results from the increased shear stress and stream power magnitude related to high gradients and high flow depth that may exceed other floods in steep channels by two orders of magnitude. The large water volume significantly affects river courses and downstream river characteristics for example by modifying floodplains, channels, river terraces and other landforms. Outburst floods from moraine-dammed lakes contain a high amount of sediments entrained from moraine deposits and other sources along the river. High debris loads can turn the flood into debris flows that exert increased destructive forces on downstream terrain, but are restricted to steep channel sections (Clague and Evans 2000). The formation of landforms deposited by GLOFs, their characteristics and clast size is a function of flow type and size, flow attenuation, material characteristics and channel/floodplain morphology (Cenderelli and Wohl 2003; Clague and O'Connor 2015).

14.5 Lake-Glacier Feedbacks

Ice contact lakes interact with the glacier tongue and impact on the glacier margin morphology and glacier dynamics in several ways.

The presence of a body of standing water reduces friction and allows floating of ice with occasional calving. Additionally, basal water pressure rises as subglacial meltwater is slowed down on entering the lake. The increased pressure will help lifting the glacier tongue (Tsutaki et al. 2011). All these factors affect the longitudinal stress balance of the glacier and increase ice flow velocity. This feedback on the glacier mass balance is fully decoupled from climatically induced changes (Carrivick and Tweed 2013). Glacial lakes affect the thermal regime of the glacier. Warm lake water causes increased basal melting of the glacier tongue or internal melt if introduced from lateral ice-dammed lakes into the glacial plumbing system. Within the lake, basal melting reduces ice thickness by forming a notch and initiating the formation of crevasses (Diolaiuti et al. 2006; Sakai et al. 2009). As a consequence, calving occurs. Calving is documented to be the most frequent cause for GLOF initiation and thus extremely relevant from a hazard perspective (see below). Glaciers terminating in lakes therefore experience an increased mass loss compared to glaciers without ice-contact lakes (Song et al. 2017). A more extensive review of impacts of glacial lakes on glacier dynamics is presented by Carrivick and Tweed (2013).

14.6 Future Distribution of Proglacial Lakes

To estimate future proglacial lake formation, several techniques have been designed that focus on identification of bedrock depressions that may trap water as lakes. These studies are based on the assumption that the shape of a glacier bed can derived from the glacier surface and follow the logic of glacier ice thickness models developed in recent years (Binder et al. 2009; Clarke et al. 2009; Farinotti et al. 2008; James and Carrivick 2016; Linsbauer et al. 2012; Frey et al. 2010, 2014; Linsbauer et al. 2009). Simplified glaciological principles (e.g. basal shear stress, mass conservation) are used to calculate ice thickness along central flowlines or at random points and is then extrapolated onto the entire glacier area (Farinotti et al. 2016a). Models based on the perfect plasticity assumption (Paterson 1994) rely entirely on ice surface slope to calculate ice thickness. Flux-based models further need a constrained specific mass balance as input which adds additional uncertainty due to parametrization difficulties. The ability of both model types to predict subglacial patterns of convexities and concavities depends on the general assumption that the slope of today's glacier surface represents subglacial topography. For modelling glacial bedrock depressions, digital elevation models of the glacier surface and of the contemporaneous glacier outline are needed. The quality of the modelling results strongly depends on DEM resolution and the quality of the glacier outline (Haeberli et al. 2016b; Linsbauer et al. 2016).

The GIS tools GlabTOP (Glacier Bed Topography; Linsbauer et al. 2012) and its successor GlabTOP2 (Frey et al. 2014; Linsbauer et al. 2016) have been applied to the Swiss Alps and the Himalaya-Karakorum region. For the Swiss Alps, Linsbauer et al. (2012) identify 394 potential lakes (size > 10^4 m^2) with a total volume of >1.2 km^3 that can develop once the glaciers are disappeared. For the Himalaya-Karakorum region, Linsbauer et al. (2016) predict more than 16,000 overdeepenings as sites with possible future lakes (size > 10^4 m^2) with a total volume of more than 120 km^3.

A validation of the modelled bedrock surfaces is performed using geophysical data on ice thickness (e.g. ground-penetrating radar (GPR) or seismics). Due to the limited availability of ice thickness data for many parts of the world, model validation is often constrained to single glaciers, only. Further issues with validation are related to difficulties of comparing point information of measured ice thickness with interpolated modelled thickness values or the accuracy of validation data like GPR measurements on glaciers that are affected by various parameters (e.g. antenna frequency, sampling frequency, ice thickness) (cf. Robinson et al. 2012). From comparison with GPR soundings, Linsbauer et al. (2012) for example report an average uncertainty of estimated ice thicknesses of ±30% for their model GlabTOP. Another way to validate modelling results is to model glacier bed topography using past glacier extents and compare the results to existing lakes.

14.6.1 Hazards from Proglacial Lakes—GLOFs

Glacial lakes have been identified as a source of natural hazards in high mountain areas due to the potential to generate GLOFs (ICIMOD 2011; Wang and Jiao 2015; Iribarren Anacona et al. 2014; Richardson and Reynolds 2000; Korup and Tweed 2007; Carrivick and Tweed 2016; Clague and Evans 2000; Clague and O'Connor 2015). They often develop below high and steep rock slopes and in catchments with abundant unconsolidated surface deposits. In high mountain terrain, permafrost and glacier occurrence impacts on slope conditions (Haeberli et al. 2017). Steep slopes and loose sedimentary deposits frequently exhibit mass movements, e.g. rockfall, rock avalanche, debris flows or ice avalanches. Consequently, glacial lakes can be prone to rapid, high energy ice, rock and debris inputs that evoke instantaneous displacement of the lake water and the formation of flood waves (Carey et al. 2012; Hubbard et al. 2005). This chain of events then results in GLOFs, the rapid and often catastrophic release of lake water, either if the lake dam is breached or by overtopping (Whitemann 2011; Korup and Tweed 2007). Lake dams composed of moraine deposits or ice represent only metastable blockages that become unstable in time or through sudden impacts such as large waves. However, GLOFs also develop due to dam failure without additional impact by waves. The failure of ice dams is documented to be the most common cause of GLOFs (Carrivick and Tweed 2016). Moraine dam failures have been observed resulting from melting of ice cores incorporated in the moraine dam, seepage and piping and surface erosion due to heavy rainfall. Other failures were due to earthquake impacts (Richardson and Reynolds 2000; Worni et al. 2012). The effects of GLOFs can be catastrophic and far-reaching with impacts on communities tens of kilometres away from the lake (Clague and Evans 2000; Watanabe 1996; ICIMOD 2011). Downstream valleys are affected by inundation, erosion, sedimentation and landslides with large damage potentials to infrastructure and livelihoods. In order to assess the hazard potential of glacial lakes, various modelling approaches have been developed. These range from hydro-topographic flow models (Huggel et al. 2003; Khanal et al. 2015) to the combination of different physical-based modelling approaches that include the entire process chain (Somos-Valenzuela et al. 2016; Worni et al. 2014). Additionally, qualitative models of hazard assessment and lake susceptibility and quantitative risk assessment approaches are applied to classify glacial lakes and manage risk potential in various mountain areas (Emmer and Vilímek 2014; Allen et al. 2009, 2016).

14.6.2 Socio-Economic Dimensions of (New) Proglacial Lakes

Despite significant impacts on geomorphological systems, proglacial lakes also have implications on society (cf. section on hazards). Besides hazards, a couple of other aspects have been addressed especially with respect to the formation of new glacial lakes. By their nature, proglacial lakes have hydrological impacts on mountain ecosystems with implications on mountain and foreland societies. Mountains are considered water towers, and high mountain lakes provide a significant temporal storage capacity that contributes to runoff distribution and peak discharge control, for example by rising lake levels during intense rainfall events. Glacial meltwater significantly contributes to river runoff from mountain areas, especially during the summer months. Looking at the on-going glacier melt the contribution of meltwater to the overall discharge from mountains will decline in the long run after a period of rising meltwater during glacier melt (Seibert et al. 2015). In consequence, fluvial run-off in mountain areas will be significantly reduced in the future especially in combination with changing precipitation regimes. Mountain areas and forelands will most likely be facing problems in water supply in the future (Huss 2011). These effects can be counterbalanced by an increasing number of new lakes, even though total lake volume will be drastically lower compared to current glacier volume (Linsbauer et al. 2012). Addressing the issue of a future discharge deficit, Farinotti et al. (2016b) assessed the impact of artificial dams at locations of potential new lakes on future river discharge from the European Alps. They conclude that by installing storage capacity of 1 km^3, future discharge deficits in the second half of the twenty-first century could be resolved.

Proglacial lakes are also used for hydropower generation. When equipped with artificial dams, new lakes can enhance hydropower production in mountain regions. Since most new lakes will form in elevations above existing reservoirs, future lakes/reservoirs can be used as storage lakes. Water pumped between low-lying and high lakes represents storage of energy that can be immediately accessed in times of peak demand (Haeberli et al. 2016a). This capacity to store energy is a prerequisite for the management of renewable energy production using solar and wind resources.

Additional construction of an artificial dam at new natural lakes can also contribute to water supply and hazard reduction issues, by, for example, increasing the dam freeboard to prevent overtopping of displacement waves and the formation of outburst floods (Haeberli et al. 2016a).

From a touristic point of view, lakes represent important elements of landscape diversity that can compensate for the aesthetic loss of glacier disappearance in high mountain zones (Haeberli et al. 2016a). The examples show, however, that new lakes in mountain areas exhibit potentials and threats for society. Approaches to use these potentials can lead to both synergy effects and conflicts and require a managed and adopted approach by mountain societies.

14.7 Conclusions

Proglacial lakes are important components of high mountain sediment systems as well as characteristic landscape elements. They significantly reduce sediment transfer from uplands to lowlands within the alpine sediment cascade. Proglacial lakes are excellent sedimentary archives for the reconstruction of paleo-environments, paleo-climate and glacier changes. The formation of new lakes significantly influences glacier margins with negative effects on glacier mass balance and glacier melt. Ongoing changes of existing proglacial lakes and the formation of new ones in mountain ranges worldwide are direct consequences of climate and environmental change. They contribute to the understanding of glacier dynamics and patterns of glaciation and provide new insights into future mountain landscapes. In times of rapid glacier melt, proglacial areas are highly dynamic geomorphologic environments. The formation of lakes within these zones matters. Investigating the current developments in proglacial areas contributes to the understanding of relict glacial landforms.

Glacial lakes also have wide implications on mountain environments and mountain societies ranging from water supply to natural hazards and risks. Facing the observed and forecasted increase in number of proglacial lakes and volume of retained water, the increased GLOF hazard of in mountain areas result in a need for enhanced hazard and risk assessment and management of these new lakes. Essentially, a research focus on the rapidly changing proglacial environment is required to assess present and future potentials and threats for environment and society on short and long terms.

References

Allen SK, Schneider D, Owens IF (2009) First approaches towards modelling glacial hazards in the Mount Cook region of New Zealand's Southern Alps. Nat Hazards Earth Syst Sci 9(2):481–499. https://doi.org/10.5194/nhess-9-481-2009

Allen SK, Linsbauer A, Randhawa SS, Huggel C, Rana P, Kumari A (2016) Glacial lake outburst flood risk in Himachal Pradesh, India: an integrative and anticipatory approach considering current and future threats. Nat Hazards, 1–23. https://doi.org/10.1007/s11069-016-2511-x

Arnaud F, Poulenard J, Giguet-Covex C, Wilhelm B, Révillon S, Jenny JP, Revel M, Enters D, Bajard M, Fouinat L, Doyen E, Simonneau A, Pignol C, Chapron E, Vannière B, Sabatier P (2016) Erosion under climate and human pressures: an alpine lake sediment perspective. Quatern Sci Rev 152:1–18. https://doi.org/10.1016/j.quascirev.2016.09.018

Ashley GM (2002) Glaciolacustrine environments. In: Menzies J (ed) Modern and past glacial environments. Butterworth-Heinemann, Oxford, pp 335–359

Ballantyne CK (2002) A general model of paraglacial landscape response. Holocene 12(3):371–376

Beaud F, Flowers G, Venditti J (2015) Efficacy of bedrock erosion by subglacial water flow. Earth Surf Dyn Discuss, 125–145. https://doi.org/10.5194/esurfd-3-849-2015

Benn DI, Evans DJA (2010) Glaciers and glaciation. Routledge, London

Bennet M, Glaser N (2009) Glacial geology—ice sheets and landforms. Wiley, Chichester

Binder D, Brückl E, Roch KH, Behm M, Schöner W, Hynek B (2009) Determination of total ice volume and ice-thickness distribution of two glaciers in the Hohe Tauern region, Eastern Alps, from GPR data. Ann Glaciol 50(51):71–79. https://doi.org/10.3189/172756409789097522

Blass A, Bigler C, Grosjean M, Sturm M (2007) Decadal-scale autumn temperature reconstruction back to AD 1580 inferred from the varved sediments of Lake Silvaplana (southeastern Swiss Alps). Quatern Res 68:184–195. https://doi.org/10.1016/j.yqres.2007.05.004

Bogen J, Xu M, Kennie P (2015) The impact of pro-glacial lakes on downstream sediment delivery in Norway. Earth Surf Proc Land 40:942–952. https://doi.org/10.1002/esp.3669

Bolch T, Kulkarni A, Kääb A, Huggel C, Paul F, Cogley JG, Frey H, Kargel JS, Fujita K, Scheel M, Bajracharya S, Stoffel M (2012) The state and fate of Himalayan Glaciers. Science 336(6079):310–314. https://doi.org/10.1126/science.1215828

Brierley G, Fryirs K, Jain V (2006) Landscape connectivity: the geographic basis of geomorphic applications. Area 38(2):165–174. https://doi.org/10.1111/j.1475-4762.2006.00671.x

Buckel J, Otto JC, Prasicek G, Keuschnig M (2018) Glacial lakes in Austria - distribution and formation since the little ice age. Glob Planet Change 164:39–51. https://doi.org/10.1016/j.gloplacha.2018.03.003

Carey M (2005) Living and dying with glaciers: people's historical vulnerability to avalanches and outburst floods in Peru. Glob Planet Change 47(2–4):122–134. https://doi.org/10.1016/j.gloplacha.2004.10.007

Carey M, Huggel C, Bury J, Portocarrero C, Haeberli W (2012) An integrated socio-environmental framework for glacier hazard management and climate change adaptation: lessons from Lake 513, Cordillera Blanca. Peru. Climatic Change 112(3–4):733–767

Carrivick JL, Quincey DJ (2014) Progressive increase in number and volume of ice-marginal lakes on the western margin of the Greenland Ice Sheet. Glob Planet Change 116:156–163. https://doi.org/10.1016/j.gloplacha.2014.02.009

Carrivick JL, Tweed FS (2013) Proglacial lakes: character, behaviour and geological importance. Quatern Sci Rev 78:34–52. https://doi.org/10.1016/j.quascirev.2013.07.028

Carrivick JL, Tweed FS (2016) A global assessment of the societal impacts of glacier outburst floods. Glob Planet Change 144:1–16. https://doi.org/10.1016/j.gloplacha.2016.07.001

Cenderelli DA, Wohl EE (2003) Flow hydraulics and geomorphic effects of glacial-lake outburst floods in the Mount Everest region, Nepal. Earth Surf Process Land 28(4):385–407. https://doi.org/10.1002/esp.448

Church M, Ryder JM (1972) Paraglacial sedimentation: a consideration of fluvial processes conditioned by glaciation. Geol Soc Am Bull 83:3059–3071

Clague JJ, Evans SG (2000) A review of catastrophic drainage of moraine-dammed lakes in British Columbia. Quatern Sci Rev 19(17–18):1763–1783. https://doi.org/10.1016/S0277-3791(00)00090-1

Clague JJ, O'Connor JE (2015) Chapter 14—Glacier-related outburst floods A2 - Shroder, John F. In: Haeberli W, Whiteman C (eds) Snow and ice-related hazards, risks and disasters. Academic Press, Boston, pp 487–519. http://dx.doi.org/10.1016/B978-0-12-394849-6.00014-7

Clarke GKC, Berthier E, Schoof CG, Jarosch AH (2009) Neural networks applied to estimating subglacial topography and glacier volume. J Clim 22(8):2146–2160. https://doi.org/10.1175/2008jcli2572.1

Cook SJ, Swift DA (2012) Subglacial basins: their origin and importance in glacial systems and landscapes. Earth Sci Rev 115(4):332–372. https://doi.org/10.1016/j.earscirev.2012.09.009

Cook SJ, Kougkoulos I, Edwards LA, Dortch J, Hoffmann D (2016) Glacier change and glacial lake outburst flood risk in the Bolivian Andes. Cryosphere 10(5):2399–2413. https://doi.org/10.5194/tc-10-2399-2016

Creyts TT, Clarke GKC (2010) Hydraulics of subglacial supercooling: theory and simulations for clear water flows. J Geophys Res Earth Surf 115:1–21. https://doi.org/10.1029/2009JF001417

Deline P, Gruber S, Delaloye R, Fischer L, Geertsema M, Giardino M, Hasler A, Kirkbride M, Krautblatter M, Magnin F, McColl S, Ravanel L, Schoeneich P (2015) Chapter 15—Ice loss and slope stability in high-mountain regions A2—Shroder, John F. In: Haeberli W, Whiteman C (eds) Snow and ice-related hazards, risks and disasters. Academic Press, Boston,

pp 521–561. http://dx.doi.org/10.1016/B978-0-12-394849-6.00015-9
Desloges JR, Gilbert R (1994) Sediment source and hydroclimatic inferences from glacial lake sediments: the postglacial sedimentary record of Lillooet Lake, British Columbia. J Hydrol 159(1):375–393. https://doi.org/10.1016/0022-1694(94)90268-2
Diolaiuti G, Citterio M, Carnielli T, D'Agata C, Kirkbride M, Smiraglia C (2006) Rates, processes and morphology of freshwater calving at Miage Glacier (Italian Alps). Hydrol Process 20(10):2233–2244. https://doi.org/10.1002/hyp.6198
Emmer A, Klimeš J, Mergili M, Vilímek V, Cochachin A (2016) 882 lakes of the Cordillera Blanca: an inventory, classification, evolution and assessment of susceptibility to outburst floods. Catena 147:269–279. https://doi.org/10.1016/j.catena.2016.07.032
Emmer A, Vilímek V (2014) New method for assessing the susceptibility of glacial lakes to outburst floods in the Cordillera Blanca, Peru. Hydrol Earth Syst Sci 18 (9):3461–3479. https://doi.org/10.5194/hess-18-3461-2014
Farinotti D, Huss M, Bauder A, Funk M, Truffer M (2008) A method to estimate ice volume and ice thickness distribution of alpine glaciers. 00:1-17. https://doi.org/10.3189/002214309788816759
Farinotti D, Pistocchi A, Huss M (2016a) From dwindling ice to headwater lakes: could dams replace glaciers in the European Alps? Environ Res Lett 11(5):054022
Farinotti D, Brinkerhoff D, Clarke GKC, Fürst JJ, Frey H, Gantayat P, Gillet-Chaulet F, Girard C, Huss M, Leclercq PW, Linsbauer A, Machguth H, Martin C, Maussion F, Morlighem M, Mosbeux C, Pandit A, Portmann A, Rabatel A, Ramsankaran R, Reerink TJ, Sanchez O, Stentoft PA, Singh Kumari S, van Pelt WJJ, Anderson B, Benham T, Binder D, Dowdeswell JA, Fischer A, Helfricht K, Kutuzov S, Lavrentiev I, McNabb R, Gudmundsson GH, Li H, Andreassen LM (2016b) How accurate are estimates of glacier ice thickness? Results from ITMIX, the ice thickness models intercomparison experiment. Cryosphere Discuss 2016:1–34. https://doi.org/10.5194/tc-2016-250
Fischer A, Seiser B, Stocker-Waldhuber M, Mitterer C, Abermann J (2015) The Austrian Glacier inventories GI 1 (1969), GI 2 (1998), GI 3 (2006), and GI LIA in ArcGIS (shapefile) format. https://doi.org/10.1594/pangaea.844988
Fleisher PJ, Bailey PK, Cadwell DH (2003) A decade of sedimentation in ice-contact, proglacial lakes, Bering Glacier, AK. Sediment Geol 160(4):309–324. https://doi.org/10.1016/S0037-0738(03)00089-7
Frey H, Haeberli W, Linsbauer A, Huggel C, Paul F (2010) A multi-level strategy for anticipating future glacier lake formation and associated hazard potentials. Nat Hazards Earth Syst Sci 10:339–352. https://doi.org/10.5194/nhess-10-339-2010
Frey H, Machguth H, Huss M, Huggel C, Bajracharya S, Bolch T, Kulkarni A, Linsbauer A, Salzmann N, Stoffel M (2014) Estimating the volume of glaciers in the Himalayan–Karakoram region using different methods. Cryosphere 8(6):2313–2333. https://doi.org/10.5194/tc-8-2313-2014
Gardelle J, Arnaud Y, Berthier E (2011) Contrasted evolution of glacial lakes along the Hindu Kush Himalaya mountain range between 1990 and 2009. Global Planet Change 75(1–2):47–55. https://doi.org/10.1016/j.gloplacha.2010.10.003
Haeberli W, Buetler M, Huggel C, Friedli TL, Schaub Y, Schleiss AJ (2016a) New lakes in deglaciating high-mountain regions—opportunities and risks. Climatic Change, 1–14. https://doi.org/10.1007/s10584-016-1771-5
Haeberli W, Linsbauer A, Cochachin A, Salazar C, Fischer UH (2016b) On the morphological characteristics of overdeepenings in high-mountain glacier beds. Earth Surf Process Land, n/a–n/a. https://doi.org/10.1002/esp.3966
Haeberli W, Schaub Y, Huggel C (2017) Increasing risks related to landslides from degrading permafrost into new lakes in de-glaciating mountain ranges. Geomorphology 293:405–417. https://doi.org/10.1016/j.geomorph.2016.02.009
Hooke RL (1991) Positive feedbacks associated with erosion of glacial cirques and overdeepenings. Geol Soc Am Bull 103:1104–1108. https://doi.org/10.1130/0016-7606(1991)103%3c1104:PFAWEO%3e2.3.CO;2
Hooke RL (2005) Principles of glacier mechanics. Cambridge University Press, Cambridge
Hubbard B, Heald A, Reynolds JM, Quincey D, Richardson SD, Luyo MZ, Portilla NS, Hambrey MJ (2005) Impact of a rock avalanche on a moraine-dammed proglacial lake: Laguna Safuna Alta, Cordillera Blanca, Peru. Earth Surf Process Land 30(10):1251–1264. https://doi.org/10.1002/esp.1198
Huggel C, Kääb A, Haeberli W, Krummenacher B (2003) Regional-scale GIS-models for assessment of hazards from glacier lake outbursts: evaluation and application in the Swiss Alps. Nat Hazards Earth Syst Sci 3 (6):647–662. https://doi.org/10.5194/nhess-3-647-2003
Huss M (2011) Present and future contribution of glacier storage change to runoff from macroscale drainage basins in Europe. Water Resour Res 47(7):n/a–n/a. https://doi.org/10.1029/2010wr010299
ICIMOD (2011) Glacial lakes and glacial lake outburst floods in Nepal. ICIMOD, Kathmandu
Iribarren Anacona P, Norton KP, Mackintosh A (2014) Moraine-dammed lake failures in Patagonia and assessment of outburst susceptibility in the Baker Basin. Nat Hazards Earth Syst Sci 14(12):3243–3259. https://doi.org/10.5194/nhess-14-3243-2014
Iturrizaga L (2014) Glacial and glacially conditioned lake types in the Cordillera Blanca, Peru: a spatiotemporal conceptual approach. Prog Phys Geogr 38:602–636. https://doi.org/10.1177/0309133314546344
Iverson NR (2002) Processes of glacial erosion. In: Menzies J (ed) Modern and past glacier environments. Butterworth-Heinimann, Oxford, pp 131–145

James WHM, Carrivick JL (2016) Automated modelling of spatially-distributed glacier ice thickness and volume. Comput Geosci 92:90–103. https://doi.org/10.1016/j.cageo.2016.04.007

Kääb A, Haeberli W (2001) Evolution of a high-mountain thermokarst lake in the swiss alps. Arct Antarct Alp Res 33(4):385–390. https://doi.org/10.2307/1552546

Khanal NR, Mool PK, Shrestha AB, Rasul G, Ghimire PK, Shrestha RB, Joshi SP (2015) A comprehensive approach and methods for glacial lake outburst flood risk assessment, with examples from Nepal and the transboundary area. Int J Water Resour Dev 31(2):219–237. https://doi.org/10.1080/07900627.2014.994116

Koinig KA, Kamenik C, Schmidt R, Agustí-Panareda A, Appleby P, Lami A, Prazakova M, Rose N, Schnell ØA, Tessadri R, Thompson R, Psenner R (2002) Environmental changes in an alpine lake (Gossenköllesee, austria) over the last two centuries —the influence of air temperature on biological parameters. J Paleolimnol 28(1):147–160. https://doi.org/10.1023/a:1020332220870

Korup O, Tweed F (2007) Ice, moraine, and landslide dams in mountainous terrain. Quatern Sci Rev 26(25–28):3406–3422. https://doi.org/10.1016/j.quascirev.2007.10.012

Kummu M, Lu XX, Wang JJ, Varis O (2010) Basin-wide sediment trapping efficiency of emerging reservoirs along the Mekong. Geomorphology 119(3–4):181–197. https://doi.org/10.1016/j.geomorph.2010.03.018

Lane SN, Bakker M, Gabbud C, Micheletti N, Saugy JN (2017) Sediment export, transient landscape response and catchment-scale connectivity following rapid climate warming and Alpine glacier recession. Geomorphology 277:210–227. https://doi.org/10.1016/j.geomorph.2016.02.015

Larsen DJ, Miller GH, Geirsdóttir Á, Thordarson T (2011) A 3000-year varved record of glacier activity and climate change from the proglacial lake Hvítárvatn, Iceland. Quater Sci Rev 30(19–20):2715–2731. https://doi.org/10.1016/j.quascirev.2011.05.026

Leemann A, Niessen F (1994) Varve formation and the climatic record in an Alpine proglacial lake: calibrating annually- laminated sediments against hydrological and meteorological data. Holocene 4:1–8. https://doi.org/10.1177/095968369400400101

Linsbauer A, Paul F, Hoelzle M, Frey H, Haeberli W (2009) The Swiss Alps without glaciers—a GIS-based modelling approach for reconstruction of glacier beds. In: Purves R, Gruber S, Straumann R, Hengl T (eds) Geomorphometry. Zürich, pp 243–247

Linsbauer A, Frey H, Haeberli W, Machguth H, Azam MF, Allen S (2016) Modelling glacier-bed overdeepenings and possible future lakes for the glaciers in the Himalaya—Karakoram region. Ann Glaciol 57(71):119–130. https://doi.org/10.3189/2016AoG71A627

Linsbauer A, Paul F, Haeberli W (2012) Modeling glacier thickness distribution and bed topography over entire mountain ranges with GlabTop: application of a fast and robust approach. J Geophys Res Earth Surf 117 (F3):F03007. https://doi.org/10.1029/2011jf002313

Maanya US, Kulkarni AV, Tiwari A, Bhar ED, Srinivasan J (2016) Identification of potential glacial lake sites and mapping maximum extent of existing glacier lakes in Drang Drung and Samudra Tapu glaciers, Indian Himalaya. Curr Sci 111(3):553–560. https://doi.org/10.18520/cs/v111/i3/553-560

McColl ST (2012) Paraglacial rock-slope stability. Geomorphology 153–154:1–16. https://doi.org/10.1016/j.geomorph.2012.02.015

Menounos B, Clague JJ, Gilbert R, Slaymaker O (2005) Environmental reconstruction from a varve network in the southern Coast Mountains, British Columbia, Canada. Holocene 15(8):1163–1171. https://doi.org/10.1191/0959683605hl888rp

Messager ML, Lehner B, Grill G, Nedeva I, Schmitt O (2016) Estimating the volume and age of water stored in global lakes using a geo-statistical approach. Nat Commun 7:13603. https://doi.org/10.1038/ncomms13603, http://www.nature.com/articles/ncomms13603#supplementary-information

Mool PK, Wangda D, Bajracharya SR, Kunzang K, Gurung DR, Joshi SP (2001) Inventory of glaciers, glacial lakes and glacial lake ouburst floods—monitoring and early warning systems in the Hindu Kush-Himalayan region, Bhutan. International Centre for Integrated Mountain Development (ICIMOD), Kathmandu

Mulu A, Dwarakish GS (2015) Different approach for using trap efficiency for estimation of reservoir sedimentation. An Overview. Aquat Procedia 4:847–852. https://doi.org/10.1016/j.aqpro.2015.02.106

O'Connor JE, Costa JE (2004) The world's largest floods, past and present: their causes and magnitudes. U.S. Geological Survey, Reston, Virginia

Paterson W (1994) The physics of glaciers, 480 pp. Pergamon, New York

Paul F, Kääb A, Haeberli W (2007) Recent glacier changes in the Alps observed by satellite: consequences for future monitoring strategies. Glob Planet Change 56:111–122. https://doi.org/10.1016/j.gloplacha.2006.07.007

Reynolds J (1992) The identification and mitigation of glacier-related hazards: examples from the Cordillera Blanca, Peru. In: McCall G, Laming D, Scott S (eds) Geohazards. Chapman and Hall, London, pp 143–157

Richardson SD, Reynolds JM (2000) An overview of glacial hazards in the Himalayas. Quatern Int 65–66:31–47. https://doi.org/10.1016/S1040-6182(99)00035-X

Robinson M, Bristow C, McKinley J, Ruffell A (2012) 1.5.5 Ground penetrating radar. Geomorphological Techniques (Online Edition). British Society for Geomorphology, London

Rubensdotter L, Rosqvist G (2009) Influence of geomorphological setting, fluvial-, glaciofluvial- and mass-movement processes on sedimentation in alpine lakes. Holocene 19:665–678. https://doi.org/10.1177/0959683609104042

Sakai A, Nishimura K, Kadota T, Takeuchi N (2009) Onset of calving at supraglacial lakes on debris-covered glaciers of the Nepal Himalaya. J Glaciol 55(193):909–917. https://doi.org/10.3189/002214309790152555

Sartori M, Baillifard F, Jaboyedoff M, Rouiller JD (2003) Kinematics of the 1991 Randa rockslides (Valais, Switzerland). Nat Hazards Earth Syst Sci 3(5):423–433. https://doi.org/10.5194/nhess-3-423-2003

Schiefer E (2006) Contemporary sedimentation rates and depositional structures in a montane lake basin, southern Coast Mountains, British Columbia, Canada. Earth Surf Process Land 31(10):1311–1324. https://doi.org/10.1002/esp.1332

Schiefer E, Gilbert R (2008) Proglacial sediment trapping in recently formed Silt Lake, upper Lillooet Valley, Coast Mountains, British Columbia. Earth Surf Proc Land 33(10):1542–1556. https://doi.org/10.1002/esp.1625

Seibert J, Jenicek M, Huss M, Ewen T (2015) Chapter 4—Snow and ice in the hydrosphere A2—Shroder, John F. In: Haeberli W, Whiteman C (eds) Snow and ice-related hazards, risks and disasters. Academic Press, Boston, pp 99–137. http://dx.doi.org/10.1016/B978-0-12-394849-6.00004-4

Somos-Valenzuela MA, Chisolm RE, Rivas DS, Portocarrero C, McKinney DC (2016) Modeling glacial lake outburst flood process chain: the case of Lake Palcacocha and Huaraz, Peru. Hydrol Earth Syst Sci Discuss 2016:1–61. https://doi.org/10.5194/hess-2015-512

Song C, Sheng Y, Wang J, Ke L, Madson A, Nie Y (2017) Heterogeneous glacial lake changes and links of lake expansions to the rapid thinning of adjacent glacier termini in the Himalayas. Geomorphology 280:30–38. https://doi.org/10.1016/j.geomorph.2016.12.002

Thomas EK, Briner JP (2009) Climate of the past millennium inferred from varved proglacial lake sediments on northeast Baffin Island, Arctic Canada. J Paleolimnol 41(1):209–224. https://doi.org/10.1007/s10933-008-9258-7

Tsutaki S, Nishimura D, Yoshizawa T, Sugiyama S (2011) Changes in glacier dynamics under the influence of proglacial lake formation in Rhonegletscher, Switzerland. Ann Glaciol 52(58):31–36. https://doi.org/10.3189/172756411797252194

U.S. Department of Agriculture NRCS (2016) National soil survey handbook, title 430-VI—Part 629—Glossary Of Landform And Geologic Terms. U.S. Department of Agriculture. http://www.nrcs.usda.gov/wps/portal/nrcs/detail/soils/ref/?cid=nrcs142p2_054242. Accessed 21 May 2016

Veettil BK, Bianchini N, de Andrade AM, Bremer UF, Simões JC, de Souza Junior E (2016) Glacier changes and related glacial lake expansion in the Bhutan Himalaya, 1990–2010. Reg Environ Change 16(5):1267–1278. https://doi.org/10.1007/s10113-015-0853-7

Verstraeten G, Poesen J (2000) Estimating trap efficiency of small reservoirs and ponds: methods and implications for the assessment of sediment yield. Prog Phys Geogr 24(2):219–251. https://doi.org/10.1177/030913330002400204

Vilímek V, Klimeš J, Červená L (2016) Glacier-related landforms and glacial lakes in Huascarán National Park, Peru. J Maps 12(1):193–202. https://doi.org/10.1080/17445647.2014.1000985

Wang SJ, Jiao S (2015) Evolution and outburst risk analysis of moraine-dammed lakes in the central Chinese Himalaya. J Earth Syst Sci 124(3):567–576

Wang W, Xiang Y, Gao Y, Lu A, Yao T (2015) Rapid expansion of glacial lakes caused by climate and glacier retreat in the Central Himalayas. Hydrol Process 29(6):859–874. https://doi.org/10.1002/hyp.10199

Watanabe TRD (1996) The 1994 Lugge Tsho Glacial Lake Outburst Flood, Bhutan Himalaya. Mountain Research and Development 16:77–81

Whitemann CA (2011) Cold regions hazards and risks. Wiley-Blackwell, Chichester

Worni R, Stoffel M, Huggel C, Volz C, Casteller A, Luckman B (2012) Analysis and dynamic modeling of a moraine failure and glacier lake outburst flood at Ventisquero Negro, Patagonian Andes (Argentina). J Hydrol 444–445:134–145. https://doi.org/10.1016/j.jhydrol.2012.04.013

Worni R, Huggel C, Clague JJ, Schaub Y, Stoffel M (2014) Coupling glacial lake impact, dam breach, and flood processes: a modeling perspective. Geomorphology 224:161–176. https://doi.org/10.1016/j.geomorph.2014.06.031

Zhang G, Yao T, Xie H, Wang W, Yang W (2015) An inventory of glacial lakes in the third pole region and their changes in response to global warming. Glob Planet Change 131:148–157. https://doi.org/10.1016/j.gloplacha.2015.05.013

Part IV

Proglacial Sediment Cascades and Budgets

Sediment Budgets in High-Mountain Areas: Review and Challenges

15

Ludwig Hilger and Achim A. Beylich

Abstract

The changes in the sediment transport regimes of high-mountain areas as a consequence of global warming have received growing attention by geomorphologists, not only because these changes can imply a heightened threat to human infrastructure. While many studies dealing with high-mountain sediment transport processes (e.g., rockfall, debris flows, avalanches, stream transport) have focused on one process only, few studies have tried to establish a holistic view of the sediment transport in high-mountain catchments. This review chapter identifies the need for research in high-mountain sediment budgets, aims at providing an overview of studies that have contributed to this goal, and discusses the methodological state of the art in the different steps necessary for sediment budget construction. In addition, relevant research gaps will be identified, thereby showing potential for future research.

Keywords

Sediment budget · Sediment cascades
Geomorphic processes · Quantification
Morphological budget

15.1 Introduction and Background

The climatic characteristics of high-mountain areas all over the world are currently changing at high rates (e.g., Gobiet et al. 2014; Pepin et al. 2015). This causes follow-up changes in, for example, altitudinal zonation of vegetation (e.g., Körner 1998; Nicolussi et al. 2005), permafrost stability (e.g., Marchenko et al. 2007; Krautblatter 2009; Pavlova et al. 2014) and glacier coverage (e.g., Zemp et al. 2015).

Also, the processes altering the surface of high-mountain areas are dependent on climate (or weather as its manifestation). This influence is exercised not only indirectly (e.g., via the stabilizing effect of vegetation cover), but also directly as factors like temperature and precipitation are drivers of sediment transfer (Beylich et al. 2011).

These developments imply a new impetus to scientists studying geomorphic processes as changing climatic regimes will also lead to different magnitude–frequency relationships of

L. Hilger (✉)
Chair of Physical Geography, Catholic University of Eichstätt-Ingolstadt, Eichstätt, Germany
e-mail: corran@altmuehlnet.de

A. A. Beylich
Geomorphological Field Laboratory (GFL), Trondheim—Selbu, Sandviksgjerde Strandvegen 484 NO-7584, Selbustrand, Norway

T. Heckmann and D. Morche (eds.), *Geomorphology of Proglacial Systems*,
Geography of the Physical Environment, https://doi.org/10.1007/978-3-319-94184-4_15

geomorphic processes. It is very likely that this will have implications for flora and fauna, agriculture, and water resources and might also represent a threat to the human environment (Knight and Harrison 2014).

In general, predictions about the development of geomorphological systems under (present) changing climate are very difficult as even the climatic component alone is hard to model. In addition, feedback mechanisms between climate, geomorphology, and geology are complex and nonlinear, with considerable time lags between changes in forcing and system behavior (Knight and Harrison 2014). As modeling of future developments requires a spatially and temporally well-defined baseline (ideally unaffected by the geomorphic signal of direct human action), opportunities for such predictions are rare. Contemporary high-mountain cold environments often provide the opportunity to study sediment transport in areas where anthropogenic impacts are still less important than the effects of climate change. Accordingly, it is still possible in such environments to develop a library of baseline fluvial yields and sedimentary budgets before the natural environment is completely transformed (Slaymaker 2008; Beylich 2016).

One possible approach to tackle the difficulty of trying to predict future system behavior is to study geomorphic system behavior in the context of past climate change (Knight and Harrison 2014). Therefore, it is worthwhile to build a body of high-quality data on past and present sediment fluxes in high-mountain areas. Ideally, this data should distinguish between different types of processes and facilitate an evaluation of the relative importance and potential interplay of these processes. Studies on sediment budgets meet these criteria by definition (see below), which is why the establishment of recent sediment budgets in high-mountain areas is seen as an important contribution to high mountain and climate change research (Beylich et al. 2011; Beylich et al. 2016).

This chapter aims at providing an overview of studies that have contributed to this goal. Also, relevant research gaps will be identified, thereby showing potential for future research.

15.2 Quantitative Studies on High-Mountain Sediment Budgets

Global warming is most evident in the rapid shrinking of mountain glaciers all over the world. Comparatively, high intensities in geomorphic processes have been observed on recently deglaciated terrain for decades. Reasons for this include, among others, the availability of large stores of unconsolidated sediment and increased river discharge as a consequence of glacier melting.

With these proglacial areas[1] now growing rapidly, an increase of the spatial focus of geomorphological studies on these areas is a self-evident consequence. Within proglacial areas, it has been observed that not only transport intensities are decreasing with time since deglaciation, but that also the relative importance of different transport processes changes (e.g., Ballantyne 1995, 2002a; Matthews et al. 1998; Hinchliffe and Ballantyne 1999). Where mass wasting processes are typically dominating sediment transport in areas close to the current glacier margins, it is often only the fluvial (hillslope) processes transferring masses close to the LIA terminal moraines (Carrivick et al. 2013). It has been hypothesized that the enlargement of proglacial areas also leads to an increased sediment yield from the catchment. While this is certainly the case for many high-mountain areas, this only holds true in catchments of high connectivity along the transport system between the proglacial area and the outlet (Lane et al. 2017; Chap. 16).

While process intensities in proglacial areas have been studied on the scale of a single landform or a hillslope for multiple times (e.g., Curry et al. 2006; Laute and Beylich 2014; Haas et al. 2012), studies investigating multiple processes

[1]There has been a considerable amount of confusion even in the scientific literature about the differentiation of the terms "proglacial," "paraglacial," and "periglacial" (see Slaymaker 2009, 2011 for a discussion). The definition adopted in this chapter subsumes all locations having been deglaciated since the end of the Little Ice Age under the term "proglacial." See also Chap. 1 of this book for terminology.

within a proglacial area on a larger scale are less frequent. As the proglacial areas typically comprise a number of different landforms or hillslopes, the methodological characteristics of such studies are closer to those of catchment-scale studies. This is why this chapter will also discuss studies that investigated sediment transport on the scale of a whole high-mountain catchment.

When it comes to quantitative studies on sediment transfers in proglacial areas and larger high-mountain catchments, there has traditionally been a strong emphasis on the measurement of fluvial catchment export rates (e.g., Walling 1983; Phillips 1986; Warburton 1990; Chaps. 12 and 13; see also Parsons 2012). These export rates are generally determined either by direct measurement of fluvial sediment transport or by quantifying sedimentation rates in lakes at the catchment outlet (e.g., Hicks et al. 1990; Müller 1999; Hasholt et al. 2000; Hinderer 2001; Loso et al. 2004; Schiefer and Gilbert 2008; Bogen et al. 2015).

Sometimes, such values are reported as "sediment budgets." A true sediment budget, however, does not only state how much sediment leaves a catchment. It also quantifies the mobilization of sediment by different processes, enabling statements on their relative importance within the catchment, their proportionate contribution to the overall sediment yield (Dietrich et al. 1982; Warburton 1990; Slaymaker 2003; Beylich et al. 2017), and the effectiveness of different processes in delivering sediment to the fluvial mainstem (Beylich and Laute 2015). As studies occupied with the measurement of sediment transport in the stream draining the catchment alone acquire one value integrating all processes, it does not help in understanding the processes operating within the catchment.

The measurement of all relevant processes in a high-mountain catchment, however, is a challenging task. This is probably the main reason why many sediment budget studies focus on the contribution of sediment to the main channels, neglecting internal sediment production and transfer. Examples for negligence of internal sediment transfer include the omission of sediment transfer in rock glaciers or solifluction lobes or the production of sediment by chemical processes in the formation of soil. These fluxes are often not accounted for because they are often conceptualized as sediment transfer through one single storage element which is most of the time the smallest unit in sediment budget studies (cf. below and Dietrich et al. 1982; Parsons 2012). Although honoring these processes is certainly desirable, the sediment transfer through the soil mantle or other low-intensity processes is often not looked into, especially in large-scale studies. The same holds true for biogenic transport, i.e., transport by processes like tree throw or animal burrowing. In addition, a distinct identification of gravel roads or other anthropogenic features as a potential sediment source is commonly not undertaken in sediment budget studies, although advocated by Dietrich et al. (1982). It has been shown, however, that contributions of unmeasured processes can be substantial. Often, processes not measured are lumped together to explain the error term in the budget or are used for budget closure by assigning imbalances in the budget to these processes. This is problematic, as this might give the impression that the budget is perfectly balanced, although nothing is known about these processes (Parsons 2012).

15.3 Challenges in Quantitative High-Mountain Sediment Budget Studies

High-mountain sediment budgets have been researched for multiple decades, and considerable progress has been made. However, there is still much room for methodological improvement and questions to answer.

In general, sediment budgets can be constructed by spatially and temporally integrating the total amount of mobilized sediment by different sediment moving processes within a defined catchment (Dietrich et al. 1982; Becht et al. 2005). This implies that a thorough measurement campaign is needed as a prerequisite for a sediment budget construction.

Despite Jäckli's (1957; see also Embleton-Hamann and Slaymaker 2006) and Rapp's

(1960) seminal work, it was only until two to three decades ago that the importance of different processes in high-mountain areas was discussed quantitatively (Beylich 2000). Since then, quantitative studies have been on the rise and have become almost standard. Examples include Dietrich and Dunne (1978), Maizels (1979), Hammer and Smith (2008), Fenn (1983), Gurnell and Clark (1987, i.e., contributions therein), Warburton (1990), Becht (1995), Sass and Wollny (2001), Johnson and Warburton (2002), Matsuoka (2008), Gärtner-Roer and Nyenhuis (2009), Eckerstorfer et al. (2012), Bennett et al. (2012), Liermann et al. (2012), Moore et al. (2013), Sanders et al. (2013), Fuller (2014), Laute and Beylich (2016). Please see also Chap. 12 for more information on quantitative work in glacio-fluvial systems.

Quantitative work on high-mountain area sediment transport processes poses some challenges to the researcher: While some processes are more or less constantly active, albeit at different intensities throughout the year, others can be seen as occurring as discrete events (Becht et al. 2005). Many low-magnitude processes (e.g., rockfall <10 m^3) operate at (quasi-) continuous scales, and their occurrence is characterized by a very high spatial and temporal variability (e.g., Dietrich and Dunne 1978; Becht et al. 2005; Carrivick et al. 2013). The measurement of such processes requires the maintenance of a monitoring program honoring the spatial heterogeneity of the study area over a longer time period. Other processes (e.g., debris flows) occur in discrete events. This requires the scientist to be flexible concerning the planning of measurement campaigns.

With many different processes of different activity characteristics, an integration of these processes to a sediment budget is no easy task: It requires (Dietrich and Dunne 1978; Dietrich et al. 1982) a catchment-wide identification, mapping, and quantification of all relevant sediment transport processes (Sect. 15.3.1), a localization and monitoring of the storage elements in the sediment transport system (Sect. 15.3.2), and a localization of their interaction areas (Sect. 15.3.3).

15.3.1 Identification, Mapping, and Quantification of Sediment Transport Processes

The decision of which processes to include in the sediment budget of a certain catchment is often based on observations during exploratory field visits in the study area and initial measurement results.

Each of the processes active in the landscape can be ascribed to a distinctive process domain (or area), that is, the zones within the catchment where it is active and alters the Earth's surface. This area can then either be determined by field mapping (Hubbard and Glasser 2005; Knight 2012; Götz et al. 2013), rule-based classification from remote sensing data (Bartsch et al. 2002; Schneevoigt et al. 2008) or by spatial modeling (Becht et al. 2005; Wichmann 2009).

Although the development of new monitoring techniques (e.g., LiDAR) has facilitated a spatially and temporally better representation of the landscape and its changes during the last two decades (e.g., Kraus and Pfeifer 1998; Favey et al. 2000; Heritage and Hetherington 2007; Götz et al. 2013), a quantification of all identified processes literally everywhere in the catchment is often impossible. This is especially true for low-magnitude processes, such as creep or small-scale rockfall as the resolution provided by airborne laser scanning is not good enough to measure these processes.

This is why measurements are usually undertaken at defined test sites/measurement plots. The empirical measurement data obtained at these sites is then usually used for a regionalization of rates to the whole study area. The basis for this step was outlined by Dietrich et al. (1982: 6): "The only way to generalize from a few measurement sites to a landscape is to develop predictive models of the relation of each transport process to its controls." This step usually implies the identification (e.g., via regression analysis) and study area-wide mapping of these controls. As variable factors (like precipitation or temperature) can hardly be mapped, it is usually the static (on a given timescale) factors (e.g., slope, exposition, permafrost distribution) that are used as a basis for regionalization rules. However, the extrapolation of data collected from (usually) only a few measurement locations to the scale of a whole catchment usually yields suboptimal results (Dietrich and Dunne 1978).

With airborne laser scanning (ALS) having become available during the last decade, the need for regionalization has been decreasing. Although ALS has been available for the generation of digital terrain models for at least 20 years now (Kraus and Pfeifer 1998), the quality and cost of the data acquisition and processing were not good enough for large-scale morphological budgeting. With the quality of the data now being better, sediment movement by some processes (e.g., debris flows, see Hilger 2017) can be monitored study area-wide to a satisfying degree on the scale of whole valleys. As every regionalization/modeling result is associated with an error bigger than the errors of the measurements the model is based on, this can be seen as a step forward. In addition, airborne laser scanning data can be combined with terrestrial laser scanning data to alleviate the shortcomings of the respective measurement geometries.

LiDAR data has already been used in geomorphological studies in alpine catchments, also for sediment budgeting purposes. Carrivick et al. (2013), for example, combined repeat TLS surveys with ALS data of a density of 2 pt m^{-2} to produce erosion rates in the proglacial area of Ödenwinkelkees in the Hohe Tauern mountain range, Austria. They do not strictly differentiate between different processes in their area of interest, but their results (see Fig. 15.1) convincingly show the potential of LiDAR for sediment transfer quantification in proglacial/high-mountain areas.

According to these authors (p. 2), however, the ability of LiDAR to "determine geomorphological changes within alpine catchments is presently limited" because LiDAR data is usually "acquired on a campaign basis, rather than as part of routine monitoring strategies." In addition, it is the volume of such data that impedes the usage of LiDAR data as a main basis for sediment budget constructions in catchments bigger than a few km^2. The evaluation of such data still requires an enormous amount of manpower (see Hilger 2017). Nevertheless, Carrivick et al. (2013) and Orwin et al. (2010) state that a direct differencing of repeated high-frequency topographical surveys is the best way to quantify geomorphological activity to date. Carrivick et al. (2013) assert that LiDAR data has "yet to be exploited for holistically examining multi-scale sediment fluxes within highly dynamic alpine catchments [… and] alpine catchment-wide use of ALS and TLS datasets is still new and developing."

Another example of an evaluation of catchment-wide LiDAR data for sediment transfer measurement is Fuller (2014). He utilized a combination of ground survey, airborne, and terrestrial laser scanning to assess the state of connectivity in a small catchment in the New Zealand's east coast range.

One of the greatest weaknesses in current sediment transport research in high-mountain areas using LiDAR data is that where more than one process sub-system are being looked at, processes are often not discriminated (by mapping, for example), and all surface change is lumped together (e.g., Staines et al. 2015; Rascher and Sass 2016). This makes it impossible to use the obtained information for the construction of a sediment budget. A separation of sediment transferred by different processes using LiDAR data is only possible by interpreting the spatial pattern of DoD values and mapping of correspondent process domains (as in Hilger 2017). In addition, this usually requires short monitoring intervals to prevent coalescence of subsequent events (Bennett et al. 2012).

In addition, detailed identification on the spatial distribution of dead ice (Bosson et al. 2015) or permafrost thaw ("thermoerosion") is missing most of the time. This, however, is necessary to separate surface change caused by actual sediment transport from negative surface changes resulting from ice melt out. The importance of such as distinction has been shown by Irvine-Fynn et al. (2011) who attribute a discrepancy of 12.5–60% in LiDAR-detected

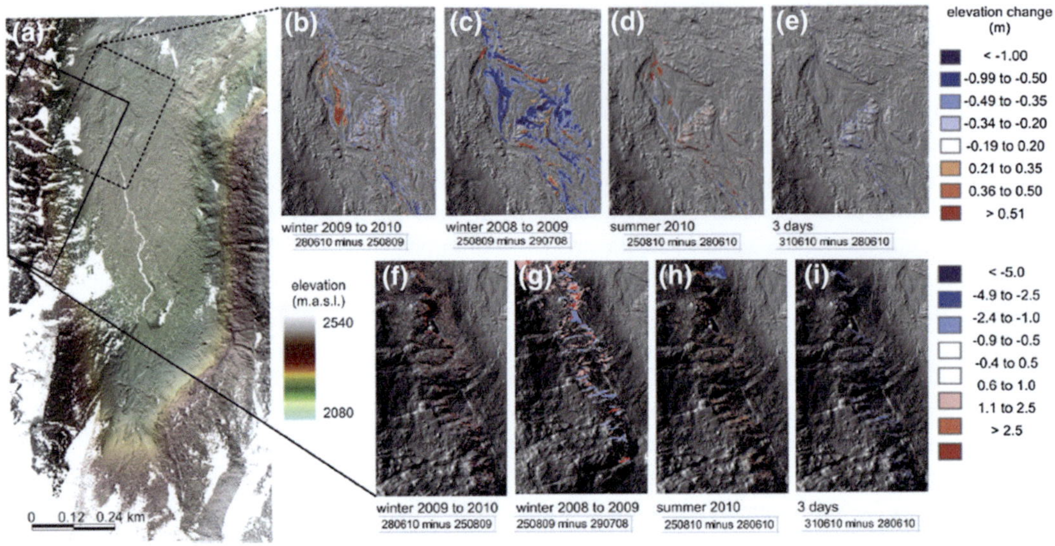

Fig. 15.1 Spatial distribution of surface changes caused by different sediment transfer processes obtained by Carrivick et al. (2013) via differencing of LiDAR data Reprinted from geomorphology, 188, Carrivick et al., contemporary geomorphological activity throughout the proglacial area of an alpine catchment, 83–95, © 2013, with permission from Elsevier

erosion to fluvial sediment output for their study area to ice melt and proceed to apply different models to examine the role of thermoerosion in their budgeting of a lateral moraine of Midtre Lovénbreen, Svalbard. It is quite obvious that studies that do not account for such factors cannot be successful in determining reliable sediment transfer rates. However, DEMs of difference (DoD) are still often being calculated for the whole study areas indiscriminately. A process distinction is not undertaken most of the time, probably because it requires a lot of additional work.

A similar challenge is posed by positive and negative DoD signals caused by the horizontal movement of alternating ridges and depressions (e.g., on the surface of rock glaciers). Just as with the ice melt out areas, these zones need to be masked out from the DoDs before sediment transfer rates can be obtained.

However, if rates are to be calculated for different processes, a mapping of the process domains is necessary and rates would be calculated for only these process domains. This would include an exclusion of the problematic surfaces anyway.

15.3.2 Localization and Monitoring of the Storage Elements in the Sediment Transport System

Typically, most of the sediment mobilized within a catchment is not immediately transported to the mainstem of the river but is deposited along the way, building up a defined storage element of the sediment budget. It is therefore very important for sediment budget studies to identify these storage elements and quantify the sediment transfers both into and out of them. These landscape elements can be seen as nodes in the sediment transfer system (see Sect. 15.3.3 for the conception of the sediment transport system).

Changing conditions of the storage elements can strongly affect the total amount of sediment exported at the outlet of the catchment (by changing the connectivity of the system), and their current configuration can yield important information for an interpretation of the sediment budget in high-mountain areas (Cossart and Fort 2008; Otto et al. 2008, 2009).

If a sediment budget is to be constructed for a time period starting at a point before any systematic data is available, a quantification and interpretation of the structure of the storage elements is often the only way to arrive at a sediment budget of a catchment. Several studies have established sediment budgets for the postglacial time period using this approach: Otto et al. (2008), for example, discuss various approaches on sediment storage quantification in high-mountain areas at different scales, and Otto (2006) and Otto et al. (2009) have shown that sediment storage on the scale of medium sized (c. 10–150 km^2) catchments can be quantified using a combination of geophysical surveys and bedrock surfaces interpolated from DTMs. A similar approach has helped Tunnicliffe et al. (2012) to estimate a postglacial sediment budget of the very large (1230 km^2) Chilliwack Valley, British Columbia. With a focus on the paraglacial characteristics of the reworking of the sediments since the last glacial, they estimate the volume, sediment type, and fractional size distribution of legacy glacial materials. The sedimentation history of the outlet fan is disclosed by the means of ^{14}C dating and paleo-surface interpolation. Newer approaches combine the methodology mentioned above with LiDAR. One example is Götz et al. (2013), who quantify the total postglacial sediment storage in Gradenmoos basin (4.5 km^2, Schober Mountains, Carinthia, Austrian Alps) applying terrestrial laser scanning and geomorphological mapping together with geophysical prospection, drilling, and subsequent bedrock surface reconstruction to estimate total sediment storage for the catchment and also obtain respective sediment transport rates for rockfall.

15.3.3 Localization of Process Interaction Areas

The identification of areas of process interaction is of utmost importance for the establishment of a sediment budget as they constitute the transfer locations of one tributary process to other (transfer) processes of, finally, the fluvial mainstem and the catchment outlet (Dietrich and Dunne 1978). Without this information, it is not possible to distinguish sediment being added to the total discharge from transport of the same material along its sequential path to the basin outlet (i.e., in the main fluvial channels). In other words, spatial identification of process links is the only way to discriminate between contributions by different (slope) processes quantitatively (Dietrich et al. 1982).

The process domains discussed in Sect. 15.3.1 can be further distinguished into zones of erosion, transport, and deposition (Engelen and Venneker 1988). Where a deposition zone of one process overlaps the erosional zone of another, material can be passed from one process domain to another (Becht 1995). Ultimately, whole chains of processes can form, thereby constituting sediment cascades (Warburton 2009; Burt and Allison 2010) that, in turn, form cascading systems when chains of processes form a network (Becht et al. 2005). The conceptualization of sediment transport forming a network or system with landforms/storages constituting the network nodes and processes the network edges has become popular since Chorley and Kennedy (1971) and is the prevalent one in sediment budget studies (e.g., Ballantyne 2002b). It has also been termed "morphosystem analytical approach" (Beylich 2000) or "landsystem approach" (Carrivick et al. 2013).

When all process areas [which are often congruent with certain landform (combinations)] and the interaction areas in the catchment are identified, a representation of the sediment transport system can be established. It is then also possible to summarize the sediment transport network as in Fig. 15.2. Such a summary gives an overview of both sediment storages and processes linking them and is very similar to a sediment budget (being only short of transport rates).

Cascades can be established topologically, i.e., by finding locations of unique process zones overlap [which have been termed "Geomorphological Process Units" (GPUs) by Bartsch et al. (2002)]. Another possibility for cascade identification is the mapping of landforms along toposequences (Otto 2006), combined with an activity assessment of these landforms via

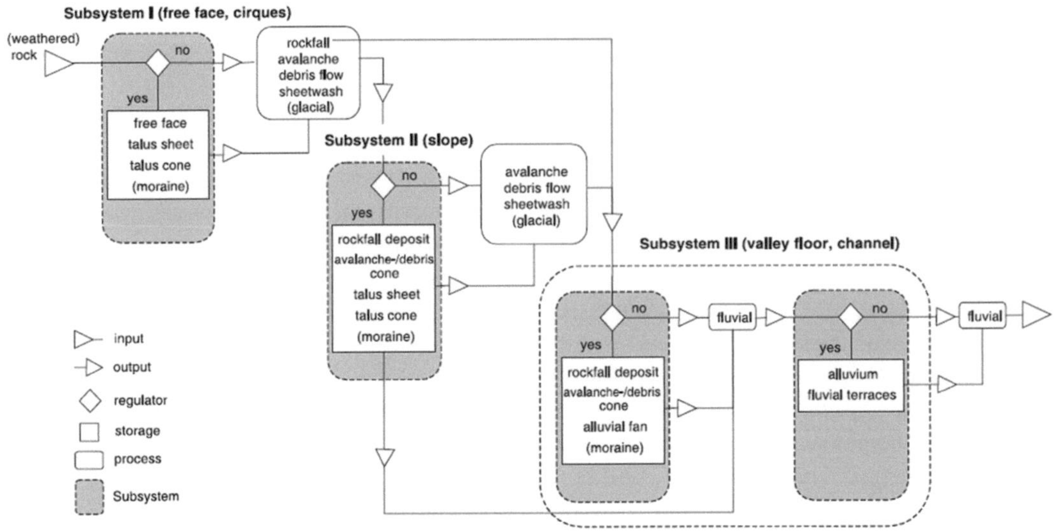

Fig. 15.2 Conceptual model of the sediment cascade network in the Rein Valley as put forth by Schrott et al. (2003) Reprinted from geomorphology, 55, Schrott et al., spatial distribution of sediment storage types and quantification of valley-fill deposits in an alpine basin, Reintal, Bavarian Alps, Germany, 45–63, © 2003, with permission from Elsevier

process measurement (as discussed in Sect. 15.3.1). A statistical summary of all cascades found in a catchment helps to identify important process combinations for sediment transport in a catchment (e.g., Wichmann 2009; Heckmann and Schwanghart 2013a, b).

Depending on the relative location in the sediment cascade/transport network, landscape elements can be classified as sediment sources (e.g., a rockwall), storages (e.g., a talus cone) or sediment sinks (e.g., a lake). Unfortunately, a catchment-wide systematic approach for an identification of sources, links, and sinks is often not undertaken, especially not on the landform scale. As Carrivick et al. (2013) states, the localization of sediment sources, storages, and sinks is one of the most demanding tasks of alpine sediment budgeting, the main reason being that catchment-wide models are generally qualitative/conceptual (Caine 1974; Dietrich and Dunne 1978; Carrivick et al. 2013).

One of the greatest challenges in establishing a representation of the sediment transport network by finding process link locations is the fact that the existence of links between process domains highly depends on the magnitude and frequency of the processes in question at a specific location (Warburton 1990). A debris flow, for example, might reach the process area of the main fluvial system in one year, but not in the following one (e.g., Schrott et al. 2006; Hilger 2017). Especially, the influence of fluvial sediment transport is characterized by a very high temporal variability. Missing high magnitude–low-frequency events in a measurement campaign can lead to significant underestimations of longer-term of sediment delivery (Kirchner et al. 2001; Carrivick et al. 2013). Another problem is constituted by episodically moving sediment bodies as deep-seated slope failures. When the temporal scale of a study is short, the determination of recurrence intervals is especially difficult (Dietrich et al. 1982).

It is not possible to fit a regression model between events (as a debris flow event) and the recurrence interval of their most important driving factors (such as a thunderstorm of very high intensity). This is prevented by too many other, site-specific factors on both sides of the model and lag effects from previous events (such as refilling of sediment bodies available for mobilization) (Dietrich et al. 1982).

This problem could only be tackled by the establishment of magnitude–frequency relationships of the processes in question. To reduce the potential error in process measurements due to short project durations, it is common practice to produce measurement rates from either historical imagery (e.g., Dietrich and Dunne 1978; Schiefer and Gilbert 2007; Staines et al. 2015) or by adopting a sedimentological approach, i.e., by qualitatively and quantitatively looking into the stratigraphy of the sediments in the catchment.

This requires additional expenditure and manpower, and very often, magnitude–frequency relationships are therefore not honored in sediment budget studies. Nevertheless, future research should incorporate such functions as much as possible.

It is a unifying characteristic of many sediment budget studies that they are of limited temporal duration. Although it is very obvious that data on sediment transfers are more reliable the bigger the database is, many studies are set up for only a few years. This is mainly due to the nature of research projects as third-party funded undertakings. When discussing sediment budgets, the findings of Lu et al. (2005) (cited in Parsons 2012) should be kept in mind: "at the plot scale, a few decades of erosion measurements are needed to get statistically meaningful long-term average erosion rates" and "characteristic timescales of sediment yield at the catchment scale can be expected to be much longer" (both p. 5).

15.3.4 Combining the Sediment Transport Network with Measured Rates

As stated by Slaymaker (1991) and Becht et al. (2005), it is desirable to give quantitative data (rates of sediment flux) for the sediment pathways between landforms/storage elements. On the scale of a whole catchment, this remains a challenging task, or as Becht (1995) puts it already 20 years ago: "Coupling the measured sediment transport rates with the modeled process domains, however, remains a challenging problem and is subject to further research."

Nevertheless, there is an increasing trend toward more quantitatively oriented studies which aim at (i) representing the cascading system of whole alpine valleys and, to a lesser degree, (ii) quantifying the element links, i.e., processes quantitatively based on measurements. Some representative work will be reviewed in Sect. 15.3.5.

The ratio of sediment delivered at the catchment outlet to gross erosion within the basin, that is, the sediment delivery ratio (Walling 1983), a proxy for the effectiveness of the sediment transport system, has received growing attention in recent years under the notion of "connectivity." Examples of relevant work include Caine (1989), Borselli et al. (2008), Chiverrell et al. (2009), Thiel (2013), Heckmann and Schwanghart (2013a, b) and Messenzehl et al. (2014). While some confusion between the terms coupling and connectivity exist in the literature, "connectivity" is used here to refer to an attribute of the whole sediment transport system in a catchment or proglacial area. Simply put, a landscape system with many consecutively coupled elements/landforms is a system with a high connectivity and is likely to have a high sediment delivery ratio. A more detailed treatise of connectivity in the context of high-mountain sediment transport can be found in Chap. 16.

15.3.5 The Lack of Truly Comprehensive Quantitative Sediment Budget Studies

Most studies, as stated above, only aim to quantify single processes. This holds true independently of the methodology used, but is especially conspicuous when it comes to repeated topographic surveys as with LiDAR or dGPS. As pointed out by Beylich and Laute (2015), there has been a decidedly strong focus on the sediment transport processes in the main streams of high-mountain catchments (e.g., Milan et al. 2007; Bertoldi et al. 2010; Carrivick et al. 2013). One reason is the fact that sediment transport measurement in the main fluvial system can provide the overall gross sediment output, which is very often substituted as a sediment budget (cf. above). A second reason is certainly the dominance in geomorphological activity of the fluvial system in many high-mountain area valley systems (Wohl 2000; Orwin et al. 2010; Carrivick et al. 2013).

The studies concerned exclusively with the fluvial system in high-mountain environments apply a wide range of methodologies to either quantify sediment fluxes or to frame conceptual models for the geomorphic systems (Wohl 2000). There is also a wide variety on the spatial and temporal scale defined by these studies: Sediment budgeting studies on a relatively large spatial and temporal scale often use sedimentological data such as deposition rates in (proglacial) lakes to arrive at overall export rates (see Chap. 14). An example is Müller (1999), who calculated late- and postglacial sedimentation rates from drillings in the lake Walensee, forming the outlet of a 269 km^2 large catchment in Eastern Switzerland. On a much smaller scale, Hicks et al. (1990) investigated the sedimentation in a proglacial lake of New Zealand's Southern Alps. The authors found that proglacial lakes can be very efficient in trapping sediment delivered to them and go on to compare the relative importance of glacial and non-glacial erosive processes to the overall catchment output. Another example of a study occupied with the sedimentation in evolving proglacial lakes is Bogen et al. (2015). The authors measured fluvial in and output in three Norwegian proglacial lakes, leading to insights into their trapping efficiency and the suspended sediment yield of their catchments. Hicks et al. (1990) and Bogen et al. (2015), being very good examples of modern sediment transport research, are also investigating possible future changes (Sect. 15.1) in high-mountain/proglacial sediment transport systems. Similar studies were put forth by Hasholt et al. (2000), who used sediment cores from two proglacial lakes in Greenland to determine the sediment yields of the respective catchments, and Loso et al. (2004), who use varve counting to determine the sediment yield into Iceberg Lake, Alaska for the post-LIA period. Schiefer and Gilbert (2008) investigated the sedimentation dynamics in a proglacial lake using varve chronology, showing the importance of longitudinal decoupling by proglacial lakes, while Hinderer (2001) quantified the sediment yield from 16 major Alpine catchments by determining the sediment volumes accumulated in overdeepened valleys and lake basins; this allowed a comparison of denudation rates over different timescales.

Marren (2005) focuses on the magnitude–frequency relationships of proglacial stream discharges and their implications for sandur development and not on sediment budgets. However, this study can serve as an example that a parametrization of a magnitude–frequency model is important in interpreting results from budgeting studies conducted over a few years. Sandur sediments are also looked at in Carrivick and Rushmer (2009). Their study uses sedimentological data, cross sections, and mapping to infer a conceptual framework of the proglacial systems of two glaciers in New Zealand. Their focus is predominantly on the sandur sediments but allows statements concerning the relative importance of glacial and non-glacial processes (cf. below). An early example of terrestrial laser scanning use for the monitoring of sediment transport in high-mountain areas in general is Milan et al. (2007). They could demonstrate how LiDAR data can be used to monitor sediment erosion and deposition in high-mountain cold environments at high spatial resolution.

The direct measurement of sediment transport in (proglacial) high-mountain streams (see Chap. 12) is often hampered by the high variability in flow and sediment transfer rates (Orwin et al. 2010, see also the short discussion above) on the one hand and unstable channel layouts on the other. Recommendations on how to measure fluvial sediment fluxes in different settings of high-mountain/cold environments have been given in Orwin et al. (2010).

When it comes to whole proglacial areas, Schiefer and Gilbert (2007) pioneered the use of historical images of multiple volumes for surface reconstruction in proglacial areas. The authors did not only show the feasibility of this method but also elaborate on georeferencing and error control in difficult terrain. A thorough study involving point cloud data stemming from photogrammetry and LiDAR was presented by Staines et al. (2015), who monitored surface change of the Sólheimajökull glacier foreland (Iceland). Due to the configuration and geomorphology of their study site, the focus of their work was mainly on the development of the proglacial fluvial system as well. Their work can serve as an excellent example of how historical aerial imagery can be taken into account to extend the temporal data basis for studying sediment dynamics in proglacial areas.

Only few studies area is available to inform on the sediment production by glaciers. The main reason for this is the difficulty of monitoring glacial sedimentary processes below or within the ice masses (Warburton 1990; Chap. 5). As a result, there is a very limited number of studies informing about the absolute and relative importance of glacial versus non-glacial processes with respect to sediment supply to the main streams of mountain catchments although there has been an ongoing debate for a long time in the scientific literature (Brocklehurst and Whipple 2002; Spotila et al. 2004; O'Farrell et al. 2009; Beylich et al. 2017). Most published rates on erosion of high-mountain glaciers rely on the measurement of sediment yield in proglacial rivers (see Hallet et al. 1996ab, for a review), thereby neglecting the fact that these rates also integrate the fluxes of processes delivering sediment to the glacier from above or the sides. In addition, such rates cannot distinguish between sediment produced by glacial erosion at the glacier bed and material stemming from storages below the ice or next to the glacier margins (Warburton 1990; Harbor and Warburton 1993; O'Farrell et al. 2009). It is therefore of utmost importance to gain knowledge about the sediment input by other processes to the glacial process area. Taking a downstream sandur and its sedimentological characteristics as a data basis, there is evidence that sediment input from the valley walls is much more important than sediment flux directly from the glacier in the catchment studied by Carrivick and Rushmer (2009). Burki et al. (2010) have investigated proglacial stratigraphies and combined geophysical measurements with their results to arrive at glacial denudation rates for the time after AD 1650.

Sediment budgets on various scales but with a strong focus on very large ones (such as whole mountain belts) are reviewed and discussed by Hinderer (2012). The same author and colleagues have also investigated sediment transport on the scale of the entire Alps by analyzing river load and reservoir sedimentation data from 202 drainage basins (between ca. 1 and 10,000 km^2) all over that mountain range. Their results show that glaciated basins have much higher sediment yields than non-glaciated basins and came up with a threefold higher denudation rate of the Western Alps than the Eastern Alps (with a general mean of the Alps of about 0.32 mm year^{-1}, Hinderer et al. 2013).

While Hinderer's study integrates all processes into one final export value, there is also an increasing body of studies incorporating all relevant processes of the area of interest into true sediment budgets enabling the comparison of different processes. Warburton (1990) presented a quantitative sediment budget of the proglacial zone of Bas Glacier d'Arolla, Val d'Hérens, Switzerland. He could show that fluvial sediment transport in the main braidplain was by far the most important process in the geomorphic system of his study area, i.e., that hillslope processes and tributary channels contributed only small amounts of sediment. His final result is depicted in Fig. 15.3. He could also show that short low-frequency, high-magnitude hydrological events had an enormous impact on the overall sediment budget of that catchment. In fact, 95% of the sediment eroded in the proglacial zone was mobilized by a meltwater flood event that lasted only three days. Nevertheless, Warburton's seminal study similarly lacks information on glacial sediment production and is based on a short duration (two months) field campaign only. In general, there are not many studies providing such a comprehensive image of the sediment

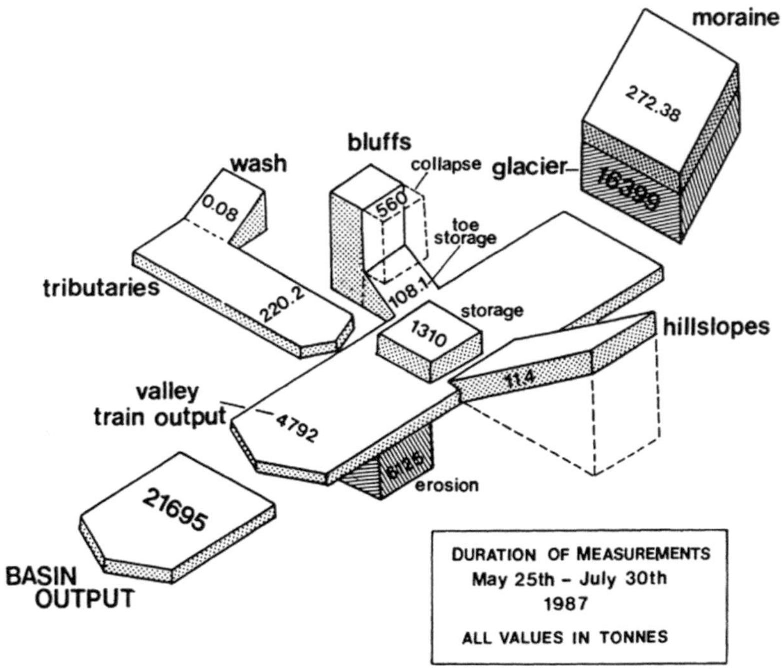

Fig. 15.3 Sediment budget of the proglacial area of Bas Glacier d'Arolla (Warburton 1990) Reprinted from Geografiska Annaler. A 72 (3/4), Warburton, An Alpine Proglacial Fluvial Sediment budget, 261–272, © 1990, with permission from Taylor & Francis

transfer system in high-mountain basins as Warburton (1990). Some studies differentiate between surface change within the fluvial system and the geomorphic work on valley slopes. An example is Irvine-Fynn et al. (2011), who use ALS data from two different dates to quantify surface change on a lateral moraine on the one hand, and the fluvial system on the other. They make no process distinction, however, probably because this is not possible with the low spatial resolution (<1 pt m^{-2}) of their data sets.

An integration of different methods has helped O'Farrell et al. (2009) to compare different process regimes within the proglacial area of Matanuska Glacier, Alaska. Using cosmogenic nuclide erosion rates for bedrock erosion, a combination of field measurements to quantify slope processes and suspended measurements from tributary glacial and non-glacial streams, they were able to juxtapose glacial (approximated by fluvial yield in the proglacial area) and non-glacial processes on the Holocene timescale. They do not differentiate between single processes but use volume measurements for two classes of sediment fans to come up with three different process zones which they quantify separately and extrapolate to the whole study area using morphometric rules.

A sediment budget incorporating different processes including glacial bedrock erosion and rockwall retreat has been presented by Sanders et al. (2013) for a small Alpine cirque. They conclude that rates of rockwall retreat and glacial bedrock erosion are roughly equivalent. They distinguish sources, sediment pathways, and storage elements and proceed to quantify (on a reference period of 1–80 years, depending on the respective process) each of the sediment budget components using different measurement approaches (rockfall monitoring, suspended sediment export measurements, moraine profiles) as shown in Fig. 15.4.

The authors present a very detailed sediment budget for several processes in an almost optimal setting for testing new approaches toward a higher resolved model of alpine sediment budgets and cascades.

One of the most detailed study approaches of an alpine landsystem configuration was presented by Heckmann and Schwanghart (Heckmann and Schwanghart 2013a, b). They explore steps on how the landsystem could be modeled

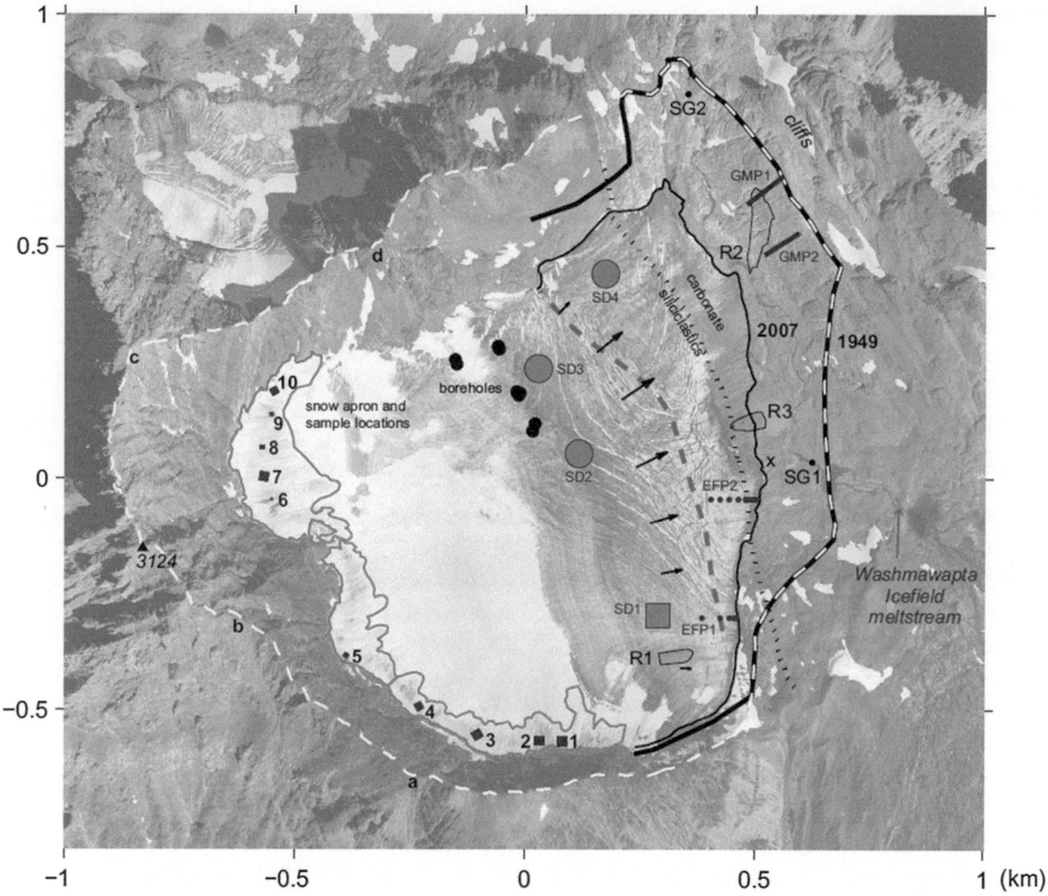

Fig. 15.4 Configuration of the process monitoring network established by Sanders et al. (2013) in Helmet mountain cirque, British Columbia, Canada: SD: Snow apron and supraglacial debris sample boxes, R: large rockfalls, EFP: englacial flux profiles, SG: stream gauge, GMP: ground moraine profile; the principal outlet stream emerges at the x Reprinted with permission from Geological Society of America Bulletin, 125 (1), Sanders et al. (2013), the sediment budget of an alpine cirque, 229–248, © 2013

for a 15 km^2 catchment based on a digital elevation model and numerical spatial process models for different processes. They also present a methodological solution to one of the main problems in both the quantitative and qualitative analysis of landsystems. Numerical models are used to delineate potential sediment pathways based on a DEM. These pathways form the edges of a directed graph, linking the corresponding start and end points. Network analysis then allows for the analysis of areal patterns (i.e., groups of nodes), paths (i.e., edge sequences interpreted as sediment cascades), and the structure of the whole network. Moreover, pathways starting or ending on different spatial units (e.g., landforms) can be aggregated (see Heckmann et al. 2016) for different spatial units (landform, landform type, subcatchment).

15.4 Conclusion and Prospects

Earlier studies on sediment cascades in high-mountain areas successfully determined and visualized sediment pathways (e.g., Wichmann 2009; Heckmann and Schwanghart 2013a, b). These studies, however, developed models without the incorporation of the masses transported and could therefore only give topological information on the sediment transport between spatial units. An example of how measured transport rates can be combined with such information was given by Heckmann et al. (2016): The authors use a numerical model to delineate the process domain for small-magnitude rockfall in the Upper Kaunertal (Austria; c.f. Chap. 9, 17) and proceed to use a representation of the particle trajectories as a graph to analyze the sediment pathways on the scale of DEM pixels, storage elements, and the whole catchment. The trajectories are assigned sediment transport rates that were determined by either (i) production rate classes based on a regression from a rock mass stability index or (ii) rates drawn from a distribution based on values obtained by net measurements and TLS-based transport rates. As a result, the incorporation of measured sediment masses in such or related models would constitute a step forward in sediment budget and cascade research (Chap. 17). Concerning methodological challenges for quantification, Carrivick et al. (2013) state: "High-resolution changes to individual landforms have been measured on an episodic campaign basis, but this technique has not been applied to entire proglacial areas. Quantification of geomorphological work within alpine catchments has been restricted to hydrological gauging of total suspended sediment."

It is clear that more research is needed on full sediment budgets, allowing quantitative statements on the relative importance of all relevant denudational processes in high-mountain basins. This goal can only be achieved by applying different quantification methodologies: While airborne laser scanning can be used for the measurement of high-magnitude processes, it can also form the basis for measurement plot identification and regionalization efforts for low-magnitude processes like creep, fluvial hill slope transport, or avalanches. In the near future, it will still only be possible to quantify these low-magnitude processes using more traditional measurement methods. Future research should also invest effort in establishing better representations of the sediment transport system by identifying sediment process link locations (cf. Sect. 15.3.3). This assessment is also supported by Carrivick et al. (2013): "Future studies should look to (i) utilize repeated ALS to determine geomorphic changes over a whole catchment, (ii) quantify the spatial organization/fragmentation of geomorphological activity."

In utilizing catchment-wide LiDAR data, the problem of separating geomorphic change by sediment transfer from change caused by permafrost and ground ice thaw is probably the most urgent methodological problem. This requires a fast and reliable way to automatically classify such surface changes to produce masks to be used in DoD calculation.

As discussed in Sect. 15.3.3, the incorporation of magnitude–frequency relationships for each process in the sediment budget might be the most challenging but also most rewarding task for future research.

This is closely related with the need for the identification of reliable ways to (i) generate a spatially distributed representation of the fragmented landscape (e.g., a catalog of methods/manual for automated storage element segmentation or process domain mapping) and (ii) locate points of process domain coupling to establish representations of the sediment transport system.

Also, desirable is the integration of different timescales in sediment budgets of the same catchments, that is, by integrating results from sediment storage quantification giving a perspective on the sediment transport regime in the postglacial and recent measurements (Otto and Schrott 2010). The recommendation by Slaymaker (1993, Abstract) who proposed the integration of four timescales—"geological, pleistocene, postglacial, and contemporary" should be honored strongly by the sediment budget research community.

References

Ballantyne CK (1995) Paraglacial debris-cone formation on recently deglaciated terrain, western Norway. Holocene 5:25–33. https://doi.org/10.1177/095968369500500104

Ballantyne CK (2002a) A general model of paraglacial landscape response. Holocene 12:371–376. https://doi.org/10.1191/0959683602hl553fa
Ballantyne CK (2002b) Paraglacial geomorphology. Quatern Sci Rev 21:1935–2017. https://doi.org/10.1016/s0277-3791(02)00005-7
Bartsch A, Gude M, Jonasson C, Scherer D (2002) Identification of geomorphic process units in Karkevagge, northern Sweden, by remote sensing and digital terrain analysis. Geogr Ann Ser A: Phys Geogr 84:171–178. https://doi.org/10.1111/j.0435-3676.2002.00171.x
Becht M (1995) Untersuchungen zur aktuellen Reliefentwicklung in alpinen Einzugsgebieten. GEOBUCH, München
Becht M, Haas F, Heckmann T, Wichmann V (2005) Investigating sediment cascades using field measurements and spatial modelling. IAHS Publ 291:206–213
Bennett GL, Molnar P, Eisenbeiss H, McArdell BW (2012) Erosional power in the Swiss Alps: characterization of slope failure in the Illgraben. Earth Surf Proc Land 37:1627–1640. https://doi.org/10.1002/esp.3263
Bertoldi W, Zanoni L, Tubino M (2010) Assessment of morphological changes induced by flow and flood pulses in a gravel bed braided river: The Tagliamento River (Italy). Geomorphology 114:348–360. https://doi.org/10.1016/j.geomorph.2009.07.017
Beylich A, Lamoureux S, Decaulne A (2011) Developing frameworks for studies on sedimentary fluxes and budgets in changing cold environments. Quaestiones Geographicae. https://doi.org/10.2478/v10117-011-0001-5
Beylich AA (2000) Geomorphology, sediment budget, and relief development in Austdalur, Austfirdir, East Iceland. Arct Antarct Alp Res 32:466. https://doi.org/10.2307/1552396
Beylich AA, Laute K (2015) Sediment sources, spatiotemporal variability and rates of fluvial bedload transport in glacier-connected steep mountain valleys in western Norway (Erdalen and Bødalen drainage basins). Geomorphology 228:552–567. https://doi.org/10.1016/j.geomorph.2014.10.018
Beylich AA (2016) Controls and variability of solute and sedimentary fluxes in alpine/mountain environments. In: Beylich AA, Dixon JC, Zwolinski Z (eds) Source-to-sink fluxes in undisturbed cold environments. Cambridge University Press, Cambridge, pp 378–382
Beylich AA, Dixon JC, Zwoliński Z (eds) (2016) Source-to-sink fluxes in undisturbed cold environments. Cambridge University Press, Cambridge
Beylich AA, Laute K, Storms JEA (2017) Contemporary suspended sediment dynamics within two partly glacierized mountain drainage basins in western Norway (Erdalen and Bødalen, inner Nordfjord). Geomorphology 287:126–143. https://doi.org/10.1016/j.geomorph.2015.12.013
Bogen J, Xu M, Kennie P (2015) The impact of pro-glacial lakes on downstream sediment delivery in Norway. Earth Surf Proc Land 40:942–952. https://doi.org/10.1002/esp.3669
Borselli L, Cassi P, Torri D (2008) Prolegomena to sediment and flow connectivity in the landscape: a GIS and field numerical assessment. Catena 75:268–277. https://doi.org/10.1016/j.catena.2008.07.006
Bosson J-B, Deline P, Bodin X et al (2015) The influence of ground ice distribution on geomorphic dynamics since the Little Ice Age in proglacial areas of two cirque glacier systems. Earth Surf Proc Land 40:666–680. https://doi.org/10.1002/esp.3666
Brocklehurst SH, Whipple KX (2002) Glacial erosion and relief production in the Eastern Sierra Nevada, California. Geomorphology 42:1–24. https://doi.org/10.1016/s0169-555x(01)00069-1
Burki V, Hansen L, Fredin O et al (2010) Little Ice Age advance and retreat sediment budgets for an outlet glacier in western Norway. Boreas. https://doi.org/10.1111/j.1502-3885.2009.00133.x
Burt TP, Allison RJ (eds) (2010) Sediment cascades: an integrated approach. Hoboken, NJ, Wiley
Caine N (1974) The geomorphic processes of the alpine environment. Arctic and alpine environments Methuen, London, pp 721–748
Caine SF (1989) Geomorphic coupling of hillslope and channel systems in two small mountain basins. Z fur Geomorphol 33:189–203
Carrivick JL, Geilhausen M, Warburton J et al (2013) Contemporary geomorphological activity throughout the proglacial area of an alpine catchment. Geomorphology 188:83–95. https://doi.org/10.1016/j.geomorph.2012.03.029
Carrivick JL, Rushmer EL (2009) Inter- and intra-catchment variations in proglacial geomorphology: an example from Franz Josef Glacier and Fox Glacier, New Zealand. Arct Antarct Alp Res 41:18–36. https://doi.org/10.2307/40305853
Chiverrell RC, Foster GC, Marshall P, Harvey AM, Thomas GSP (2009) Coupling relationships: Hillslopefluvial linkages in the Hodder catchment, NW England. Geomorphol 109:222–235
Chorley RJ, Kennedy BA (1971) Physical geography: a systems approach. Prentice-Hall, London
Cossart E, Fort M (2008) Sediment release and storage in early deglaciated areas: towards an application of the exhaustion model from the case of Massif des Écrins (French Alps) since the Little Ice Age. Nor Geogr Tidsskr Norw J Geogr 62:115–131. https://doi.org/10.1080/00291950802095145
Curry AM, Cleasby V, Zukowskyj P (2006) Paraglacial response of steep, sediment-mantled slopes to post-'Little Ice Age' glacier recession in the central Swiss Alps. J Quat Sci 21:211–225. https://doi.org/10.1002/jqs.954
Dietrich WE, Dunne T (1978) Sediment budget for a small catchment in mountainous terrain. Zeitschrift fuer Geomorphol N.F. Suppl 29:191–206
Dietrich WE, Dunne T, Humphrey NF, Reid LM (1982) Construction of sediment budgets for drainage basins. In: Sediment budgets and routing in forested drainage basins: proceedings of the symposium, 31 May–1 June

1982, Corvallis, Oregon Gen Tech Rep PNW-141 Portland, Oregon: Pacific Northwest Forest and Range Experiment Station, Forest Service, US Department of Agriculture, pp 5–23

Eckerstorfer M, Christiansen HH, Rubensdotter L, Vogel S (2012) The role of cornice fall avalanche sedimentation in the valley Longyeardalen, Central Svalbard. Cryosphere Discuss 6:4999–5036. https://doi.org/10.5194/tcd-6-4999-2012

Embleton-Hamann C, Slaymaker O (2006) Classics revisited: Jäckli H 1957: Gegenwartsgeologie des Bündnerischen Rheingebietes. Ein Beitrag zur exogenen Dynamik Alpiner Gebirgslandschaften. Prog Phys Geogr 30:779–783. https://doi.org/10.1177/0309133306071958

Engelen G, Venneker R (1988) ETA (Erosion Transport Accumulation) systems, their classification, mapping and management

Favey E, Pateraki M, Baltsavias EP, Bauder A, Bösch H (2000) Surface modelling for alpine glacier monitoring by Airborne laser scanning and digital photogrammetry. In: Digital photogrammetry." international archives of photogrammetry and remote sensing, vol XXXIII, Part B3

Fenn CR (1983) Proglacial Streamflow series: measurement, analysis and interpretation. Ph.D. thesis, University of Southampton

Fuller IC (2014) Towards an understanding of catchment-scale sediment dynamics: cascades and connectivity in steepland systems. Int J Erosion Control Eng 7:1–8. https://doi.org/10.13101/ijece.7

Gärtner-Roer I, Nyenhuis M (2009) Volume estimation, kinematics and sediment transfer rates of active rock glaciers in the Turtmann Valley, Switzerland. In: Landform—structure, evolution, process control. Springer, Berlin, Heidelberg, pp 185–198

Gobiet A, Kotlarski S, Beniston M et al (2014) 21st century climate change in the European Alps—a review. Sci Total Environ 493:1138–1151. https://doi.org/10.1016/j.scitotenv.2013.07.050

Götz J, Otto J-C, et al (2013) Postglacial sediment storage and rockwall retreat in a semi-closed inner-Alpine sedimentary basin (Gradenmoos, Hohe Tauern, Austria). Geografia Fisica e Dinamica Quaternaria, pp 63–80. https://doi.org/10.4461/gfdq.2013.36.5

Gurnell AM, Clark MJ (1987) Glacio-fluvial sediment transfer. Wiley

Haas F, Heckmann T, Hilger L, Becht M (2012) Quantification and modelling of debris flows in the proglacial area of the Gepatschferner, Austria, using ground-based LiDAR. In: IAHS-AISH publication. International Association of Hydrological Sciences 356, pp 293–302

Haeberli W, Maisch M (2003) Die rezente Erwärmung der Atmosphäre—Folgen für die Schweizer Gletscher. Geographische Rundschau 55:4–12

Hallet B, Hunter L, Bogen J (1996a) Rates of erosion and sediment evacuation by glaciers: a review of field data and their implications. Global Planet Change 12:213–235. https://doi.org/10.1016/0921-8181(95)00021-6

Hallet B, Hunter L, Bogen J (1996b) Rates of erosion and sediment evacuation by glaciers: a review of field data and their implications. Global Planet Change 12:213–235. https://doi.org/10.1016/0921-8181(95)00021-6

Hammer KM, Smith ND (2008) Sediment production and transport in a proglacial stream: Hilda Glacier, Alberta, Canada. Boreas 12:91–106. https://doi.org/10.1111/j.1502-3885.1983.tb00441.x

Harbor J, Warburton J (1993) Relative rates of glacial and non glacial erosion in alpine environments. Arct Alp Res 25:1. https://doi.org/10.2307/1551473

Hasholt B, Walling DE, Owens PN (2000) Sedimentation in arctic proglacial lakes: Mittivakkat Glacier, south–east Greenland. Hydrol Process 14:679–699. https://doi.org/10.1002/(sici)1099-1085(200003)14:4<679:aid-hyp966>3.0.co;2-e

Heckmann T, Hilger L, Vehling L, Becht M (2016) Integrating field measurements, a geomorphological map and stochastic modelling to estimate the spatially distributed rock fall sediment budget of the Upper Kaunertal, Austrian Central Alps. Geomorphology 260:16–31. https://doi.org/10.1016/j.geomorph.2015.07.003

Heckmann T, Schwanghart W (2013a) Geomorphic coupling and sediment connectivity in an alpine catchment—exploring sediment cascades using graph theory. Geomorphology 182:89–103. https://doi.org/10.1016/j.geomorph.2012.10.033

Heckmann T, Schwanghart W (2013b) Geomorphic coupling and sediment connectivity in an alpine catchment—exploring sediment cascades using graph theory. Geomorphology 182:89–103. https://doi.org/10.1016/j.geomorph.2012.10.033

Heritage G, Hetherington D (2007) Towards a protocol for laser scanning in fluvial geomorphology. Earth Surf Proc Land 32:66–74. https://doi.org/10.1002/esp.1375

Hicks DM, McSaveney MJ, Chinn TJH (1990) Sedimentation in Proglacial Ivory Lake, Southern Alps New Zealand. Arctic Alpine Res 22:26. https://doi.org/10.2307/1551718

Hilger L (2017) Quantification and regionalization of geomorphic processes using spatial models and high-resolution topographic data: a sediment budget of the Upper Kauner Valley, Ötztal Alps. PhD thesis. Catholic University of Eichstätt-Ingolstadt, Eichstätt

Hinchliffe S, Ballantyne CK (1999) Talus accumulation and Rockwall retreat, Trotternish, isle of Skye, Scotland. Scott Geogr J 115:53–70. https://doi.org/10.1080/00369229918737057

Hinderer M (2001) Dénudation quaternaire récente dans les Alpes, remplissage des vallées et des lacs, charge solide des rivières actuelles. Geodin Acta 14:231–263. https://doi.org/10.1016/s0985-3111(01)01070-1

Hinderer M (2012) From gullies to mountain belts: a review of sediment budgets at various scales. Sed Geol 280:21–59. https://doi.org/10.1016/j.sedgeo.2012.03.009

Hinderer M, Kastowski M, Kamelger A et al (2013) River loads and modern denudation of the Alps—a review.

Earth Sci Rev 118:11–44. https://doi.org/10.1016/j.earscirev.2013.01.001
Hubbard B, Glasser NF (2005) Field techniques in glaciology and glacial geomorphology. Wiley, Chichester, West Sussex, England ; Hoboken, NJ
Irvine-Fynn TDL, Barrand NE, Porter PR et al (2011) Recent high-arctic glacial sediment redistribution: a process perspective using airborne lidar. Geomorphology 125:27–39. https://doi.org/10.1016/j.geomorph.2010.08.012
Jaeckli H (1957) „Gegenwartsgeologie des Bündnerischen Rheingebietes". Beiträge zu Geologischen Karte der Schweiz, Geotechnische Serie 36. Bern: Kümmerly & Frey
Johnson RM, Warburton J (2002) Annual sediment budget of a UK mountain torrent. Geogr Ann, Ser A: Phys Geogr 84:73–88. https://doi.org/10.1111/1468-0459.00162
Kirchner JW, Finkel RC, Riebe CS et al (2001) Mountain erosion over 10 year, 10 ky, and 10 my time scales. Geology 29:591–594
Knight J (2012) Uses and limitations of field mapping of lowland glaciated landscapes. In: Smith M, Paron P, Griffiths JS (eds) Geomorphological mapping. Methods and applications. Elsevier, Amsterdam [etc.], pp 533–550
Knight J, Harrison S (2014) Mountain glacial and paraglacial environments under global climate change: lessons from the past, future directions and policy implications. Geogr Ann: Ser A, Phys Geogr 96:245–264. https://doi.org/10.1111/geoa.12051
Körner C (1998) World wide positions of alpine treelines and their causes. In: The impacts of climate variability on forests, pp 221–229
Kraus K, Pfeifer N (1998) Determination of terrain models in wooded areas with airborne laser scanner data. ISPRS J Photogrammetry Remote Sens 53:193–203. https://doi.org/10.1016/s0924-2716(98)00009-4
Krautblatter M (2009) Detection and quantification of permafrost change in alpine rock walls and implications for rock instability. PhD thesis, University of Bonn, Bonn
Lane SN, Bakker M, Gabbud C et al (2017) Sediment export, transient landscape response and catchment-scale connectivity following rapid climate warming and Alpine glacier recession. Geomorphology 277:210–227. https://doi.org/10.1016/j.geomorph.2016.02.015
Laute K, Beylich AA (2016) Sediment delivery from headwater slope systems and relief development in steep mountain valleys in western Norway. In: Beylich AA, Dixon JC, Zwolinski Z (eds) Source-to-sink fluxes in undisturbed cold environments. Cambridge University Press, Cambridge, pp 293–312
Laute K, Beylich AA (2014) Environmental controls, rates and mass transfers of contemporary hillslope processes in the headwaters of two glacier-connected drainage basins in western Norway. Geomorphology 216:93–113. https://doi.org/10.1016/j.geomorph.2014.03.021
Liermann S, Beylich AA, van Welden A (2012) Contemporary suspended sediment transfer and accumulation processes in the small proglacial Sætrevatnet sub-catchment, Bødalen, western Norway. Geomorphology 167–168:91–101. https://doi.org/10.1016/j.geomorph.2012.03.035
Loso MG, Anderson RS, Anderson SP (2004) Post-Little Ice Age record of coarse and fine clastic sedimentation in an Alaskan proglacial lake. Geology 32:1065. https://doi.org/10.1130/g20839.1
Lu H, Moran CJ, Sivapalan M (2005) A theoretical exploration of catchment-scale sediment delivery. Water Resour Res. https://doi.org/10.1029/2005wr004018
Maizels JK (1979) Proglacial aggradation and changes in braided channel patterns during a period of glacier advance: an alpine example. Geogr Ann Ser A, Phys Geogr 61:87. https://doi.org/10.2307/520517
Marchenko SS, Gorbunov AP, Romanovsky VE (2007) Permafrost warming in the Tien Shan mountains, Central Asia. Global Planet Change 56:311–327. https://doi.org/10.1016/j.gloplacha.2006.07.023
Marren PM (2005) Magnitude and frequency in proglacial rivers: a geomorphological and sedimentological perspective. Earth Sci Rev 70:203–251. https://doi.org/10.1016/j.earscirev.2004.12.002
Matsuoka N (2008) Frost weathering and rockwall erosion in the southeastern Swiss Alps: long-term (1994–2006) observations. Geomorphology 99:353–368. https://doi.org/10.1016/j.geomorph.2007.11.013
Matthews JA, Shakesby RA, Berrisford MS, McEwen LJ (1998) Periglacial patterned ground on the Styggedalsbreen glacier foreland, Jotunheimen, southern Norway: micro-topographic, paraglacial and geoecological controls. Permafrost Periglac Process 9:147–166. https://doi.org/10.1002/(sici)1099-1530(199804/06)9:2<147:aid-ppp278>3.0.co;2-9
Messenzehl K, Hoffmann T, Dikau R (2014) Sediment connectivity in the high-alpine valley of Val Müschauns, Swiss National Park—linking geomorphic field mapping with geomorphometric modelling. Geomorphology 221:215–229. https://doi.org/10.1016/j.geomorph.2014.05.033
Milan DJ, Heritage GL, Hetherington D (2007) Application of a 3D laser scanner in the assessment of erosion and deposition volumes and channel change in a proglacial river. Earth Surf Proc Land 32:1657–1674. https://doi.org/10.1002/esp.1592
Moore JR, Egloff J, Nagelisen J et al (2013) Sediment transport and bedrock erosion by wet snow avalanches in the Guggigraben, Matter Valley, Switzerland. Arct Antarct Alp Res 45:350–362. https://doi.org/10.1657/1938-4246-45.3.350
Müller B (1999) Paraglacial sedimentation and denudation processes in an Alpine valley of Switzerland. An approach to the quantification of sediment budgets. Geodin Acta 12:291–301. https://doi.org/10.1016/s0985-3111(00)87046-1

Nicolussi K, Kaufmann M, Patzelt G et al (2005) Holocene tree-line variability in the Kauner Valley, Central Eastern Alps, indicated by dendrochronological analysis of living trees and subfossil logs. Veget Hist Archaeobot 14:221–234. https://doi.org/10.1007/s00334-005-0013-y
O'Farrell CR, Heimsath AM, Lawson DE et al (2009) Quantifying periglacial erosion: insights on a glacial sediment budget, Matanuska Glacier, Alaska. Earth Surf Proc Land 34:2008–2022. https://doi.org/10.1002/esp.1885
Orwin JF, Lamoureux SF, Warburton J, Beylich A (2010) A framework for characterizing fluvial sediment fluxes from source to sink in cold environments. Geogr Ann: Ser A, Phys Geogr 92:155–176
Otto J, Goetz J, Schrott L (2008) Sediment storage in Alpine sedimentary systems–quantification and scaling issues. IAHS Publ 325:258
Otto J-C (2006) Paraglacial sediment storage quantification in the Turtmann Valley, Swiss Alps. PhD Thesis, University of Bonn
Otto J-C, Schrott L (2010) Quantifizierung von rezenten und postglazialen Sedimentspeichern und Sedimentflüssen-Konzeptionelle Ansätze und aktuelle Studien aus den Ostalpen. Salzburger Geogr Arbeiten 46:1–13
Otto J-C, Schrott L, Jaboyedoff M, Dikau R (2009) Quantifying sediment storage in a high alpine valley (Turtmanntal, Switzerland). Earth Surf Proc Land 34:1726–1742. https://doi.org/10.1002/esp.1856
Parsons AJ (2012) How useful are catchment sediment budgets? Prog Phys Geogr 36:60–71. https://doi.org/10.1177/0309133311424591
Pavlova I, Jomelli V, Brunstein D, Grancher D, Martin E, Déqué M (2014) Debris flow activity related to recent climate conditions in the French Alps: a regional investigation. Geomorphology 219:248–259. https://doi.org/10.1016/j.geomorph.2014.04.025
Pepin N, Bradley RS, Diaz HF et al (2015) Elevation-dependent warming in mountain regions of the world. Nature Clim Change 5:424–430. https://doi.org/10.1038/nclimate2563
Phillips JD (1986) Sediment storage, sediment yield, and time scales in landscape denudation studies. Geogr Anal 18:161–167. https://doi.org/10.1111/j.1538-4632.1986.tb00089.x
Rapp A (1960) Recent development of mountain slopes in Kärkevagge and surroundings Northern Scandinavia. Geogr Ann 42:65. https://doi.org/10.2307/520126
Rascher E, Sass O (2016) Constructing a sediment budget for the Johnsbach, Styria—adding up numbers and drawing arrows? Geophys Res Abstr 18
Sanders JW, Cuffey KM, MacGregor KR, Collins BD (2013) The sediment budget of an alpine cirque. Geol Soc Am Bull 125:229–248. https://doi.org/10.1130/b30688.1
Sass O, Wollny K (2001) Investigations regarding Alpine talus slopes using ground-penetrating radar (GPR) in the Bavarian Alps, Germany. Earth Surf Proc Land 26:1071–1086. https://doi.org/10.1002/esp.254
Schiefer E, Gilbert R (2007) Reconstructing morphometric change in a proglacial landscape using historical aerial photography and automated DEM generation. Geomorphology 88:167–178. https://doi.org/10.1016/j.geomorph.2006.11.003
Schiefer E, Gilbert R (2008) Proglacial sediment trapping in recently formed Silt Lake, upper Lillooet Valley, Coast Mountains, British Columbia. Earth Surf Proc Land 33:1542–1556. https://doi.org/10.1002/esp.1625
Schneevoigt NJ, van der Linden S, Thamm H-P, Schrott L (2008) Detecting Alpine landforms from remotely sensed imagery. a pilot study in the Bavarian Alps. Geomorphology 93:104–119. https://doi.org/10.1016/j.geomorph.2006.12.034
Schrott L, Götz J, Geilhausen M, Morche D (2006) Spatial and temporal variability of sediment transfer and storage in an Alpine basin (Reintal valley, Bavarian Alps, Gemany). Geogr Helv 61:191–200. https://doi.org/10.5194/gh-61-191-2006
Schrott L, Hufschmidt G, Hankammer M et al (2003) Spatial distribution of sediment storage types and quantification of valley fill deposits in an alpine basin, Reintal, Bavarian Alps, Germany. Geomorphology 55:45–63. https://doi.org/10.1016/s0169-555x(03)00131-4
Slaymaker O (1991) Mountain geomorphology: a theoretical framework for measurement programmes. Catena 18:427–437. https://doi.org/10.1016/0341-8162(91)90047-2
Slaymaker O (1993) The sediment budget of the Lillooet River Basin, British Columbia. Phys Geogr 14:304–320. https://doi.org/10.1080/02723646.1993.10642482
Slaymaker O (2003) The sediment budget as conceptual framework and management tool. Hydrobiologia 494:71–82. https://doi.org/10.1023/a:1025437509525
Slaymaker O (2008) Sediment budget and sediment flux studies under accelerating global change in cold environments. Z Geomorphol Supplementary Issues 52:123–148. https://doi.org/10.1127/0372-8854/2008/0052s1-0123
Slaymaker O (2009) Proglacial, periglacial or paraglacial? In: Knight J, Harrison S (eds) Periglacial and Paraglacial processes and environments. The Geological Society Publishing House, London, pp 71–84
Slaymaker O (2011) Criteria to distinguish between periglacial, proglacial and paraglacial environments. Quaest Geogr 30:85–94
Spotila JA, Buscher JT, Meigs AJ, Reiners PW (2004) Long-term glacial erosion of active mountain belts: example of the Chugach–St. Elias Range, Alaska. Geology 32:501. https://doi.org/10.1130/g20343.1
Staines KE, Carrivick JL, Tweed FS et al (2015) A multi-dimensional analysis of pro-glacial landscape change at Sólheimajökull, southern Iceland. Earth Surf Proc Land 40:809–822
Swanson FJ, Janda RJ, Dunne T, Swanston DN (1982) Workshop on sediment budgets and routing in forested drainage basins: proceedings. U.S. Department of Agriculture, Forest Service, Pacific Northwest Forest and Range Experiment Station, Portland, OR

Thiel M (2013) Quantifizierung der Konnektivität von Sedimentkaskaden in alpinen Geosystemen. PhD thesis, University of Eichstaett-Ingolstadt

Tunnicliffe J, Church M, Clague JJ, Feathers JK (2012) Postglacial sediment budget of Chilliwack Valley, British Columbia. Earth Surf Proc Land 37:1243–1262. https://doi.org/10.1002/esp.3229

Walling DE (1983) The sediment delivery problem. J Hydrol 65:209–237. https://doi.org/10.1016/0022-1694(83)90217-2

Warburton J (1990) An Alpine proglacial fluvial sediment budget. Geogr Ann Ser A, Phys Geogr 72:261–272. https://doi.org/10.1080/04353676.1990.11880322

Warburton J (2009) Mountain environments. In: Perry C, Taylor K (Hrsg) Environmental sedimentology. Wiley, New York, NY

Wichmann V (2009) A new modelling approach to delineate the spatial extent of alpine sediment cascades: GIS and SDA applications in geomorphology. Geomorphology 111:70–78

Wohl E (2000) Mountain rivers. Water resources monograph, vol 14, p 320. American Geophysical Union, Washington DC

Zemp M, Frey H, Gärtner-Roer I et al (2015) Historically unprecedented global glacier decline in the early 21st century. J Glaciol 61:745–762. https://doi.org/10.3189/2015jog15j017

16 Sediment Connectivity in Proglacial Areas

Marco Cavalli, Tobias Heckmann and Lorenzo Marchi

Abstract

Sediment connectivity is an emerging property of geomorphic systems and has become a key issue in research on geomorphic processes and sediment cascades. Sediment connectivity represents coupling relationships between system compartments and elementary units, and thus its understanding has important implications for the behaviour of hydro-geomorphic systems. The investigation and characterization of sediment connectivity and its evolution through time are of particular importance in proglacial areas and high-mountain environments since they are subject to intense morphodynamics and frequent changes in their structure and subsequent variations in sediment connectivity. This chapter aims to review the state of the art of sediment connectivity in proglacial and high-mountain environments studies, provides a synopsis of the most widespread landforms in mountain headwater catchments and describes their role with respect to coarse sediment connectivity. In addition, a section of the chapter is dedicated to the description of a recently developed topography-based sediment connectivity index. An example application to two contrasting alpine glacier forefields shows the effectiveness of this index for investigating and interpreting spatial patterns of connectivity in high-mountain catchments. Finally, we sketch avenues for future research regarding sediment connectivity (not only) in proglacial systems.

Keywords

Sediment connectivity · Geomorphic coupling Sensitivity · Connectivity index · Sediment delivery ratio

16.1 Introduction

In high-altitude alpine environments, geomorphic processes occur with high intensity due to a wide variety of factors ranging from high relief and steep slopes to the thin vegetation cover and climatic conditions that facilitate the triggering of widespread instabilities. Erosional and depositional landforms are shaped and reworked by geomorphic processes on multiple timescales (Stoffel and Marston 2013), and feedbacks exist

M. Cavalli (✉) · L. Marchi
CNR IRPI, Padua, Italy
e-mail: marco.cavalli@irpi.cnr.it

T. Heckmann
Physical Geography, Catholic University of Eichstätt-Ingolstadt, Eichstätt, Germany

T. Heckmann and D. Morche (eds.), *Geomorphology of Proglacial Systems*,
Geography of the Physical Environment, https://doi.org/10.1007/978-3-319-94184-4_16

between landforms and processes. The identification of extent and location of sediment sources and their characterization in a catchment is a fundamental requirement to predict sediment fluxes and estimate sediment supply to downstream areas at the catchment scale (Brardinoni et al. 2015a). In high-mountain catchments, several agents (e.g. gravity, snow, ice, water) contribute to sediment dynamics causing a large variability in the geomorphological activity. The recognition of different transportation agents led Engelen and Venneker (1988) to implement a distributed system approach for sediment dynamics based on various Erosion Transport Accumulation (ETA) systems; glacial, nival, gravitational and fluviatile are the most relevant ETA systems in high-alpine areas.

High-mountain catchments are often very inefficient systems as to the export of sediment, since sediment output at the outlet is much lower than the total sediment eroded within the catchment (Becht 1995; Slaymaker 2006). This so-called sediment delivery problem is described more generally by Walling (1983) and de Vente et al. (2007). In proglacial environments, this imbalance can be explained by sediment deposition in temporary or permanent storage landforms such as talus slopes, debris cones or along the channel (Fig. 16.1). Sediment conveyance is strongly controlled by morphology that in mountain regions is generally rather rugged and complex. This morphological complexity results in a large variability in the efficiency of sediment transfer (Baartman et al. 2013) by geomorphic processes characterizing these regions. As an example, on steep slopes, highly efficient sediment transfer processes, such as debris flows, can favour sediment delivery to the main streams, while some morphological settings and landforms, such as glacial cirques and hanging valleys, can trap the sediment delivered from the upper parts of the catchment preventing an effective downstream transfer of the sediment. Considering mountain catchments as cascading systems (Burt and Allison 2010), a key role is thus attributed to zones of sediment retention and storage in attenuating sediment delivery to the basin outlet and to the geomorphic processes reworking the stored sediment (Schrott et al. 2002, 2006; Straumann and Korup 2009). In this context, geomorphic coupling, i.e. the physical linkage of the sediment sources, stores and the channel network (Harvey 2001) plays a fundamental role in controlling catchment sediment dynamics and yield. In particular, when the sediment produced on the hillslopes directly reaches the channel network, a hillslope-channel coupling relationship is identified (Fig. 16.1). Conversely, a decoupling relationship refers to the case where the supplied material stops on the hillslope or in a buffer zone at the footslope before reaching the channel. Building upon earlier definitions (e.g. Hooke 2003; Brierley et al. 2006; Faulkner 2008; Heckmann and Schwanghart 2013; Cavalli et al. 2013; Bracken et al. 2015), we define sediment connectivity as a state variable of a geomorphic system, representing coupling relationships between elementary units (landforms, slope units, subcatchments) and reflecting the potential of water/sediment to move through the system. Connectivity results from the (dis)continuity of run-off and sediment pathways at a given point in time, and we distinguish structural (governed by the spatial configuration of system components) and functional connectivity (the actual transfer of sediments between these components by geomorphic processes).

Brierley et al. (2006) developed a conceptual framework for sediment connectivity, further developed by Fryirs et al. (2007), and Fryirs (2013), focusing on the limiting factors constraining the efficiency of sediment transfer relationships considering lateral (hillslope-channel), longitudinal (between river reaches) and vertical (surface–subsurface) transfers of water and sediments. Different types of blockages, termed buffers (e.g. alluvial fans, broad floodplains), barriers (e.g. valley constrictions, dams) and blankets (e.g. bed armouring), can disrupt lateral, longitudinal and vertical connectivity in catchments. It is thus fundamental to identify and map the spatial distribution of landforms affecting sediment dynamics and sediment transport pathways in a catchment in order to characterize sediment (dis)connectivity.

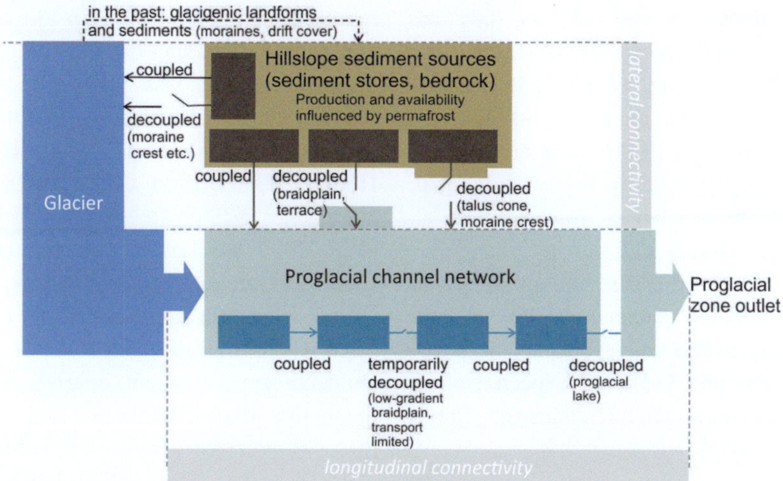

Fig. 16.1 Conceptual model of sediment transfer through a proglacial system, highlighting the role of sediment connectivity between hillslopes, channels and the glacier (lateral connectivity), and along the channel network (longitudinal connectivity). Landforms in brackets are exemplary causes of disconnectivity (buffers and barriers sensu Fryirs et al. 2007). Figure slightly modified from Carrivick and Heckmann (2017) with permission from Elsevier

Bracken et al. (2015) address sediment connectivity as being controlled by different processes that can be ordered along a spectrum between non-hydrological and entirely hydrological control of sediment (i) detachment and (ii) transport. This is of particular importance in alpine environments where mass movements that are not fully hydrologically controlled represent important factors of connectivity (Korup 2005; Heckmann and Schwanghart 2013; Meßenzehl et al. 2014; Heckmann et al. 2016). Since sediment dynamics are not uniform over time, another important issue that has to be considered in connectivity analysis is the temporal scale: the coupling relations within a landscape will change in time as a function of events of a different magnitude and frequency (Harvey 2002; Fryirs 2013), but also through sustained changes in system structure. Accordingly, sediment storage in catchments can be divided into sediment stores, i.e. temporary storage areas (e.g. bars, benches) and sinks, i.e. more permanent areas of storage where sediments can reside for a long time (Fryirs and Brierley 2001).

Besides controlling the efficiency of sediment transfer through the catchment, geomorphic coupling and sediment connectivity have implications for the sensitivity of geomorphic systems to change; in a well-coupled system, changes are supposed to be easily propagated in the up- and downstream direction, while a buffered system hampers the propagation of change (e.g. Brunsden and Thornes 1979; Harvey 2001; Fryirs 2017); this has been highlighted also in the context of paraglacial dynamics in deglaciating mountain systems (Knight and Harrison 2014)

Proglacial areas have been described as systems in equilibrium between glacial and non-glacial processes (Slaymaker 2011); however, the rapid deglaciation occurring in most proglacial areas, specifically since the end of the Little Ice Age (LIA), makes them highly dynamic landscapes in transition from glacial to non-glacial conditions (Knight and Harrison 2014; Heckmann et al. 2016). This transition implies paraglacial morphodynamics (Ballantyne 2002) and frequent changes in system structure (and hence, sediment connectivity) for example:

- Glacigenic sediment sources previously covered by glacier ice are coupled to the proglacial channel network more efficiently (Lane et al. 2017): lateral moraines or drift-mantled slopes turn into sediment

sources (Chaps. 10 and 11) as they are undercut by proglacial channels (Bogen and Bønsnes 2003; Chaps. 12 and 13) and dissected by headward extension of sidewall streams (Curry et al. 2006; Lane et al. 2017; Chaps. 10 and 11) or debris flows (Holm et al. 2004; Haas et al. 2012; Brardinoni et al. 2012; Chaps. 10 and 11).
- Hillslope sources that previously delivered sediments onto the glacier now deliver sediment directly to the active channel network; transport rates and yield are expected to be higher than supraglacial sediment fluxes (Lane et al. 2017).
- The downstream propagation of sediment pulses derived from such sediment sources leads to important off-site implications for the biosphere, the human environment, and for policy and planning (Knight and Harrison 2014); some of these downstream effects constitute natural hazards (e.g. Chiarle et al. 2007).
- High-magnitude events rearrange the spatial pattern of the channel network (Marren 2005; Baewert and Morche 2014) and the associated connectivity.

Hence, the investigation of sediment connectivity and its evolution through time is of particular importance in these environments whose spatial extent is expected to increase even more in future (Knight and Harrison 2014; Lane et al. 2017).

Given the variety of geomorphic agents that drive sediment dynamics and the morphological complexity of the high-mountain catchment, the assessment of sediment connectivity that represents a fundamental step to understand contemporary and predict future sedimentary fluxes is quite a difficult task. While previous studies focused on the generation and interpretation of geomorphological maps (e.g. Harvey 2001; Hooke 2003; Schrott et al. 2002), more recent approaches include spatial modelling (Wichmann et al. 2009; Heckmann and Schwanghart 2013; Chap. 17), indices of connectivity derived from Digital Elevation Models (DEMs) (Borselli et al. 2008; Cavalli et al. 2013; Lane et al. 2017) and connectivity assessment inferred from measurements of surface changes and associated sediment fluxes (Fuller and Marden 2011; Micheletti et al. 2015; Cavalli et al. 2017; Lane et al. 2017; Heckmann and Vericat 2018).

In Sect. 16.2, we provide an overview on proglacial landforms and their role on sediment connectivity. The following Sects. 16.3 and 16.4 are dedicated to the description of a recently developed sediment connectivity index (Cavalli et al. 2013) and an example application to two contrasting alpine glacier forefields.

16.2 Proglacial Landforms and Sediment Connectivity

Natural landforms have an obvious and well-known role in controlling sediment connectivity across different spatial scales. In high-mountain environments, the importance of natural landforms is emphasized by the limited presence of vegetation and the generally scarce influence of anthropogenic changes of topography. Table 16.1 outlines the role of landforms

Table 16.1 Landforms and sediment connectivity in proglacial environments

Landform	Source	Store / Sink	Link	Buffer	Barrier
Rock slope	*				
Rock channel	*		*		
Scree slope/cone	*		*	*	
Moraine	*				*
Rock glacier	*			*	*
Protalus rampart					*
Proglacial plain/valley floor		*		*	
Lake		*			

Fig. 16.2 Scree cone and debris flow at the outlet of a rocky gully, (Velička dolina, High Tatras, Slovakia)

commonly found in alpine proglacial areas towards sediment connectivity. Some of these landforms are not specific to proglacial (or periglacial) environments, but are widespread in alpine headwaters where proglacial processes occur.

Scree slopes and cones play a multiple role in the frame of sediment dynamics in high-mountain areas: they act as sediment stores and buffer zones between upslope rock faces and rock gullies and lower catchment sectors (Morche et al. 2008; Trevisani et al. 2009). Scree slopes and cones, however, are sometimes entrenched by debris flows (Fig. 16.2; Chaps. 10 and 11), usually originating at the interface with overhanging rock gullies (Marchi et al. 2008). In this case, debris flows establish lateral connectivity: when this process occurs, scree slopes become a source of debris, and the debris-flow channels entrenched into the scree provide a link connecting rock slopes to downslope catchment sectors. Moreover, scree slopes and cones can be undercut by the receiving streams, thus becoming coupled sediment sources (Fig. 16.3).

(Termino-) lateral moraines and protalus ramparts (Fig. 16.4) typically act as barriers for sediment, preventing its downslope transfer (Heckmann and Schwanghart 2013; Heckmann et al. 2016). The failure of moraines, however, may give rise to paroxysmal sediment transport

Fig. 16.3 Debris-flow cone undercut by receiving channel (Val Boite, Dolomites, Italy)

Fig. 16.4 Protalus ramparts (Mondeval, Dolomites, Italy)

phenomena, often occurring as debris flows. In such a case, the failure of the moraine, in addition to producing the mobilization of sediment stored upslope, causes the moraine itself to act as a source of sediment (Chaps. 10 and 11). Thus, moraines can be considered both as temporary sediment storages and intermittent sediment sources (Carrivick et al. 2013).

Depending on their location with respect to sources of sediment and to the stream system, rock glaciers may play multiple roles in sediment dynamics in proglacial areas (Brardinoni et al. 2015b). When they are located below rock slopes and steep headwater channels, rock glaciers, because of the remarkable thickness they can reach, may act as effective barriers to downslope sediment transfer. Moreover, because of their rugged surface, rock glaciers may also reduce the run-out of rockfall and large debris flows originating from upslope areas, thus acting as stores and buffers for debris delivered by these processes. The role of rock glaciers as sources of sediment is associated with debris production occurring especially at their front. This process is enhanced by the increase of the movement rate of the rock glacier under increasing temperature and can be associated with increased debris-flow activity (Lugon and Stoffel 2010; Kummert and Delaloye 2015; Kummert et al. 2017).

Low-slope alluvial plains and valley floors (Fig. 16.5) are effective sediment storage areas in alpine valleys (Otto et al. 2009), represent preferential deposition sites for sediment transported by proglacial streams and may also act as buffers separating sources of sediment, such as rock slopes and debris-flow tracks, from the channel network. However, where a braidplain is directly adjacent to sediment storage landforms such as alluvial or debris cones, the latter can be undercut during floods (e.g. Baewert and Morche 2014; Chaps. 11, 12 and 13).

Lakes (Chap. 14), which are common in glacial cirques and glacial valleys and often occupy depressions formed after glacial retreat, represent the main and more effective sink for sediment in a number of high-mountain environments. Regarding the persistency and time evolution of lakes as traps for sediment, it is necessary to distinguish lakes whose downstream boundary corresponds to an erosion-resistant rock threshold from lakes and ponds dammed by moraines or other potentially failure-prone barriers, such as landslide dams or debris-flow cones. Lakes in glacial cirques and hanging valleys (Fig. 16.6) bounded by rock thresholds are long-lasting water bodies able to decouple the upper part of the catchment from downstream sediment transfer (while hydrological connectivity is preserved through surface or subsurface pathways), until sediment fills the lake completely. Even after lake filling by inflowing sediment, low-slope alluvial deposits hamper effective linkage of sediment sources in the upper sector of the catchment with downstream channels. Outbursts from lakes bounded by failure-prone barriers, including glacier dams (Clague and Evans 2000; O'Connor et al. 2001; Hewitt and Liu 2010; Westoby et al. 2014), combine the effect of sudden release of large water volume with large availability of loose sediment from the failed barrier, as mentioned before regarding the role of moraines as sources of sediment. If combined

Fig. 16.5 Proglacial alluvial plain downstream of Tierberggletscher, Switzerland

Fig. 16.6 An alpine lake acting as sediment trap in a hanging valley of the Dolomites (Antermoia Lake, Italy)

with steep slope of the downstream flow path, which is quite common in alpine proglacial areas, debris flows (Chiarle et al. 2007), hyperconcentrated flows of processes intermediate between hyperconcentrated flows and stream flow or debris floods (Cenderelli and Wohl 1998) may result, delivering huge sediment volumes in time intervals commonly varying from minutes to few hours.

16.3 An Index of Sediment Connectivity

The methods most commonly used for the analysis of sediment connectivity include geomorphological and sedimentological field observations, and monitoring of sediment fluxes by means of field instrumentation. In recent years, an increasing effort has been devoted to developing new approaches in order to qualitatively address sediment connectivity.

IC is a distributed geomorphometric index developed by Borselli et al. (2008) and applied by the same authors in agricultural and forest environments in the context of soil erosion studies. IC is mainly focused on the influence of topography on sediment connectivity and can be easily derived from a Digital Terrain Model (DTM) with some required information on land use cover in its original version. Cavalli et al. (2013) proposed a new version of IC with refinements to deal with debris flows and sediment transport processes dominating sediment dynamics in mountain catchments, and to use the index with high-resolution DTMs. This version of IC aims at representing the linkage between different parts of the catchment by evaluating the potential connection between hillslopes and features of interest such as catchment outlet, main channel network or a given cross section along the channel. The IC is defined as:

$$\mathrm{IC} = \log_{10}\left(\frac{D_{\mathrm{up}}}{D_{\mathrm{dn}}}\right) \tag{16.1}$$

where D_{up} and D_{dn} are the upslope and downslope components of connectivity (Fig. 16.7), respectively. IC is defined in the range of [$-\infty$, $+\infty$], with connectivity increasing for larger IC values.

The upslope component D_{up} is the potential for downward routing of the sediment produced upslope and is estimated as follows:

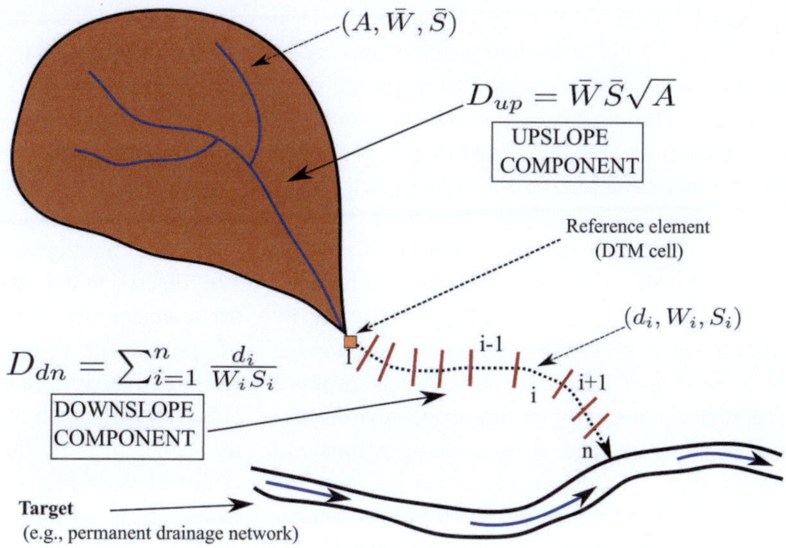

Fig. 16.7 Schematic view of IC computation (after Borselli et al. 2008)

$$D_{\rm up} = \overline{WS}\sqrt{A} \tag{16.2}$$

where $\overline{W}$ is the average weighting factor of the upslope contributing area, $\bar{S}$ is the average slope gradient of the upslope contributing area (m/m), and A is the upslope contributing area (m^2).

The downslope component $D_{\rm dn}$ takes into account the flow path length that a particle has to travel to arrive at the nearest target or sink. Therefore, $D_{\rm dn}$ can be expressed as:

$$D_{\rm dn} = \sum_i \frac{d_i}{W_i S_i} \tag{16.3}$$

where d_i is the length of the flow path along the ith cell according to the steepest downslope direction (m), and W_i and S_i are the weighting factor and the slope gradient of the ith cell, respectively. It is worth noting that d_i can assume two values: cell size (l) in the case of cardinal direction and $l\sqrt{2}$ in the case of diagonal direction.

Main refinements made to the original model by Borselli et al. (2008) are related to slope, contributing area and weighting factor computation. According to the approach by Cavalli et al. (2013), slope is computed along the direction of flow and takes into account the effect of very steep slopes that are quite common in mountain environments, contributing area is calculated using the multiple flow D-infinity approach (Tarboton 1997) to capture flow paths on hillslopes where divergent flow occurs, and weighting factor is derived on the basis of surface roughness characteristic (Cavalli and Marchi 2008; Trevisani and Cavalli 2016). Further details on the methodology can be found in Cavalli et al. (2013).

IC was firstly applied in two small adjacent headwater catchments of the Eastern Alps (Gadria and Strimm, north-eastern Italy), characterized by contrasting morphology and by a distinct efficiency of sediment routing. Results showed that the upper part of the Strimm catchment, where a hanging valley was shaped by past glacial activity and the former glacial and the present periglacial processes imparted generally gentle slopes, is characterized by a low sediment connectivity. In contrast, the downstream-most part of the catchment, a more confined valley featuring steep, straight and rather short paths on the hillslopes, showed higher values of sediment connectivity. Overall high sediment connectivity characterizes the Gadria catchment where the sediment supply from thc hillslopes is decidedly high and debris flow is the main observed process.

Meßenzehl et al. (2014) applied IC in combination with geomorphic field mapping to investigate sediment dynamics of a formerly glaciated alpine valley in Switzerland (Val Müschauns). This study demonstrates once again the key role of the overall valley morphology shaped by glacial activity in controlling sediment conveyance in mountain catchments: V-shaped valley cross sections in the lower part of the catchment were characterized by a high sediment connectivity, whereas a large portion of the valley resulted decoupled from the channel network due to glacial cirque morphology. The work by Meßenzehl et al. (2014) highlights the potential of IC as a tool for semi-quantitatively estimate sediment connectivity in mountain environment but, at the same time, stresses the importance of geomorphic field mapping to better interpret and evaluate index results and for a holistic understanding of mountain cascading systems. IC has also proved useful to analyse the evolution of connectivity patterns due to the retreat of the Zinal Glacier in Navizence catchment (Southern Swiss Alps; Goldin et al. 2016).

16.4 Case Study: Application of the IC Connectivity Index to Two Contrasting Study Areas

As an example, we apply the index described in the previous section to two proglacial areas in the Austrian Central Alps (Fig. 16.8), the Kaunertal (*K*) and the Zwieselbachtal (*Z*). The two areas are characterized by similar lithology (gneisses), climate (mean annual precipitation below 1000 mm) and relief (present-day forefields: $K = 2000\text{–}2200$ m; $Z = 2500$ m), but contrasting morphology: while the Gepatschferner glacier has been retreating through a narrow valley further confined by large lateral moraines (Kaunertal; Heckmann et al. 2012; Chap. 1), the forefield of the Zwieselbachferner glacier (Zwieselbachtal; see e.g. Becht 1995; Heckmann and Schwanghart 2013) is c. 600-m wide. A high-resolution (1 m cell size) raster DEM generated from airborne LiDAR data is available in both study areas and was used to create IC maps; thanks to the high resolution of the DEM data, we followed the approach by Cavalli et al. (2013) and derived the surface roughness from the DEM in order to use it as the weighting factor of the IC. To some extent, high-resolution DEMs are capable of representing, via surface roughness, different grain sizes that are known to influence connectivity (Lane et al. 2017). In both cases, we used the contemporary channel network (mapped from aerial photographs/in the field) as the target. We do so under the assumption that lateral (dis)connectivity represents the major component of connectivity in proglacial areas and that sediment delivered to the proglacial channels can be effectively transported further. As a consequence, the values of IC in the two study areas are more easily comparable, despite their different areal extent (it is worth remembering that both the upslope contributing area and the downslope flow length depend on the size of the catchment).

The Kaunertal map (Fig. 16.9) highlights the rapid recession of the Gepatschferner by more than 2 km since the year 1850. The proglacial area is confined between steep drift-mantled slopes on the left and a distinct lateral moraine on the right valley wall. The area between the glacier extents of 1850 and 1953 is characterized by comparatively low lateral connectivity, where the proglacial river flows on a braidplain; two tributaries, however, are characterized by high IC values indicating sediment delivery from these tributary catchments. Near the 1953 glacier extent, the steep lateral moraine on the right side of the channel is intensely coupled to the channel network. This moraine is heavily dissected by gully erosion and debris flows (Haas et al. 2012; Baewert and Morche 2014; Chap. 11), forming debris cones that are undercut by the proglacial river. The high IC values on the left-hand side originate from steep slopes and low roughness—however, as the river bypasses a bedrock landform (a whaleback), sediment delivery is expected to be much lower than from the moraines and debris cones on the opposite side. Finally, the area between the 1953 and 2006 glacier extent has high IC values, which is caused by the steep lateral moraine and drift-mantled slopes directly adjacent to the proglacial river; this includes two tributary catchments S and SW of the present-day glacier that are well coupled to the

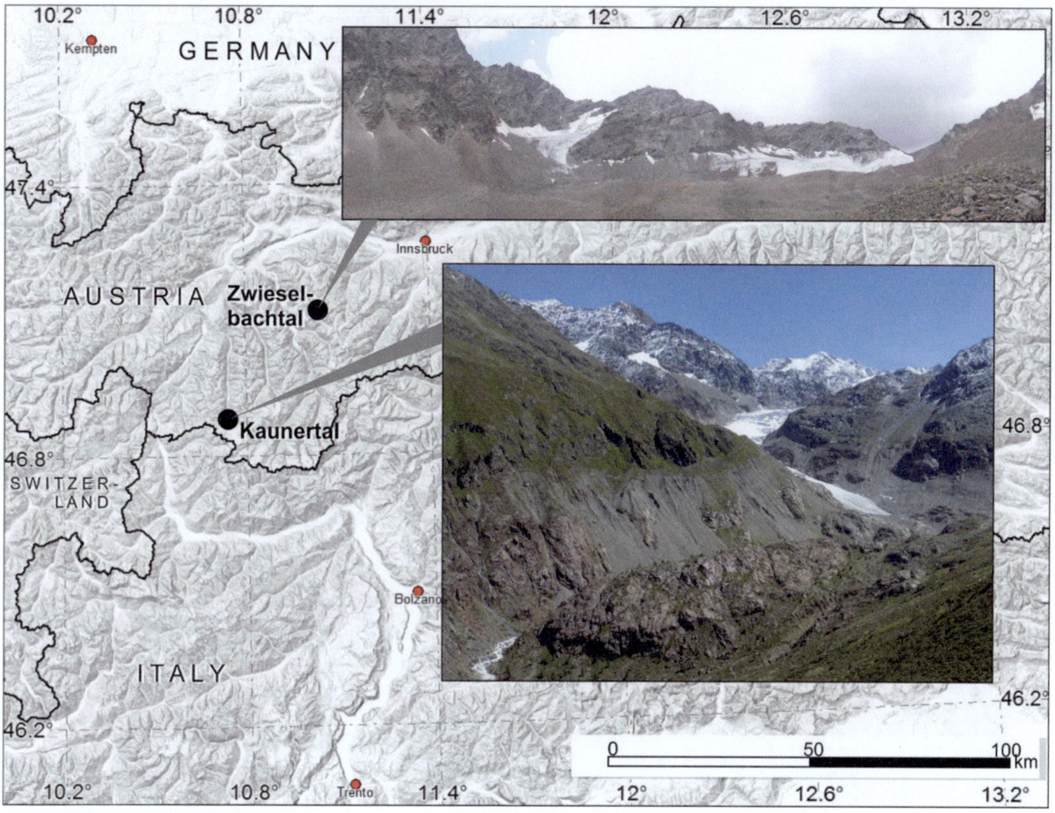

Fig. 16.8 Location of the two study areas where we applied the IC connectivity index to a high-resolution DEM. Inset photographs (Kaunertal: photograph by M. Becht) show the contrasting characteristics of the present-day proglacial area: wide and smooth in the Zwieselbachtal area, and confined by steep valley walls and a lateral LIA moraine in the Kaunertal area

respective channels. The very high IC values on the right-hand lateral moraine extend upslope near the 1953 extent; this area of high connectivity is bounded by an older, postglacial moraine crest. The 1850 lateral moraine, however, also features a marked crest, forming the boundary between high- and medium-connected slopes near the river and poorly connected slopes further upslope, until the postglacial moraine forms another barrier to sediment delivery from the valley side. The tributary catchments north of the present-day glacier tongues appear virtually unconnected; although channels exist in these catchments, sediments are delivered to the glacier rather than to its proglacial channel.

The proglacial area of Zwieselbachferner glacier (Fig. 16.10) is similar in width (ca. 500 m), but comparatively flat, and the distance of glacier recession since 1850 is only half of the Gepatschferner area, which is a consequence of the different glacier characteristics: the Zwieselbachferner originates on an *N*-facing headwall, while the Gepatschferner is fed by a large plateau; therefore, the latter reaches lower elevations. Unlike the Kaunertal, extensive areas of high connectivity are limited to the catchment section downstream of the 1850 termino-lateral moraine. This is despite the fact that the latter has been breached by present-day meltwater channels; these channels are accompanied by zone of higher connectivity. Due to the undulating ground moraine terrain, this zone does not extend far to the east, so that the largest part of the proglacial area is virtually unconnected to the main channel network. The 1850 lateral moraine forms a clear boundary on the right-hand side;

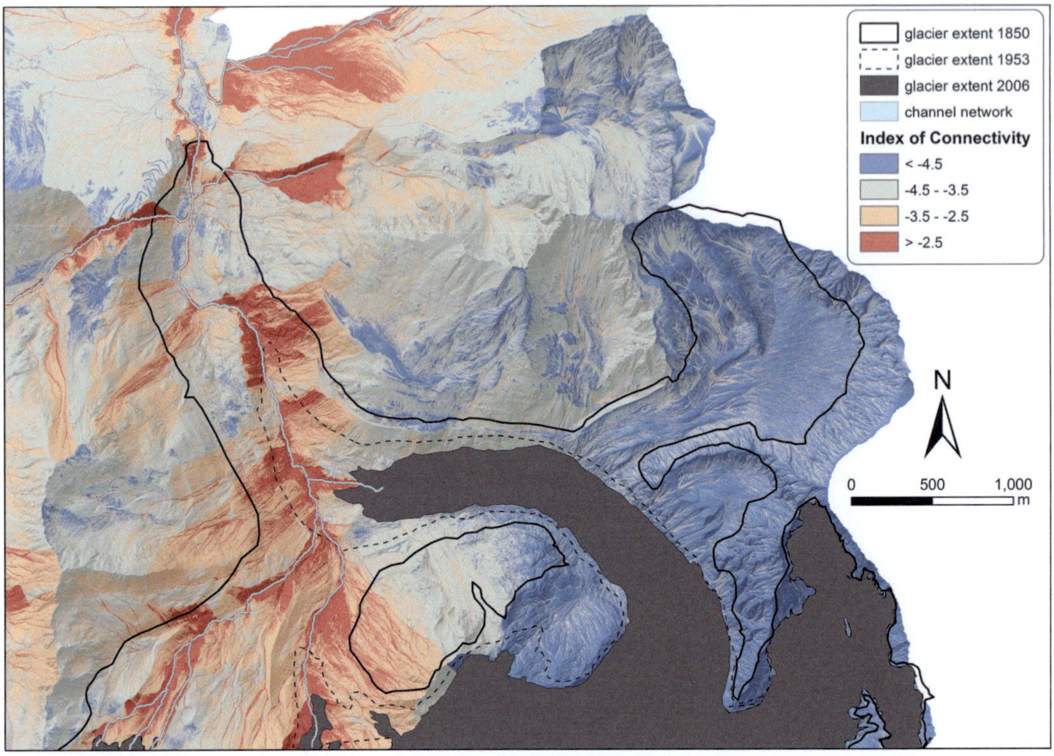

Fig. 16.9 IC map of the Kaunertal proglacial area, including the present-day, 1953 and 1850 extent of the Gepatschferner glacier (glacier extent 1850 taken from Hartl 2010)

sediments from the tributary rockwalls and young talus cones (that are already being dissected by debris flows) have to bypass it to reach the main channel. Downstream of the 1850 moraine, the tributary hillslopes appear well connected to the trunk stream (although the latter is discontinuous as it infiltrates in coarse sediments some 500 m before the outlet of the valley section). This is not so much true for the hanging valleys on the left side of the valley that are characterized by low (or at least lower than the steeper downslope sections) IC values.

The main difference between the two case studies appears to be the topography within the proglacial area. The sidewalls of the Kaunertal are very steep, featuring drift-mantled slopes and a lateral moraine plus paraglacial storage landforms originating from its dissection. On the contrary, the largest part of the proglacial area of the Zwieselbachtal is characterized by an undulating ground moraine; only a narrow section of the area appears to be connected to the channel network, which is mainly due to the short distance to the present-day meltwater channel. Tributary channels (Kaunertal) or talus and debris cones (both sites) directly adjacent to the channel network are the main factors of high connectivity in both valleys. Other studies have already named the topography of the proglacial zone (narrow/confined vs. wide/unconfined) as a factor of a different sediment delivery and morphodynamics (e.g. Laute and Beylich 2014; Temme and Lange 2014; Beylich and Laute 2015).

16.5 Discussion and Outlook

In this chapter, we explained the role of sediment connectivity in high-mountain and proglacial areas. Connectivity that reflects coupling relationships between landforms and the potential of

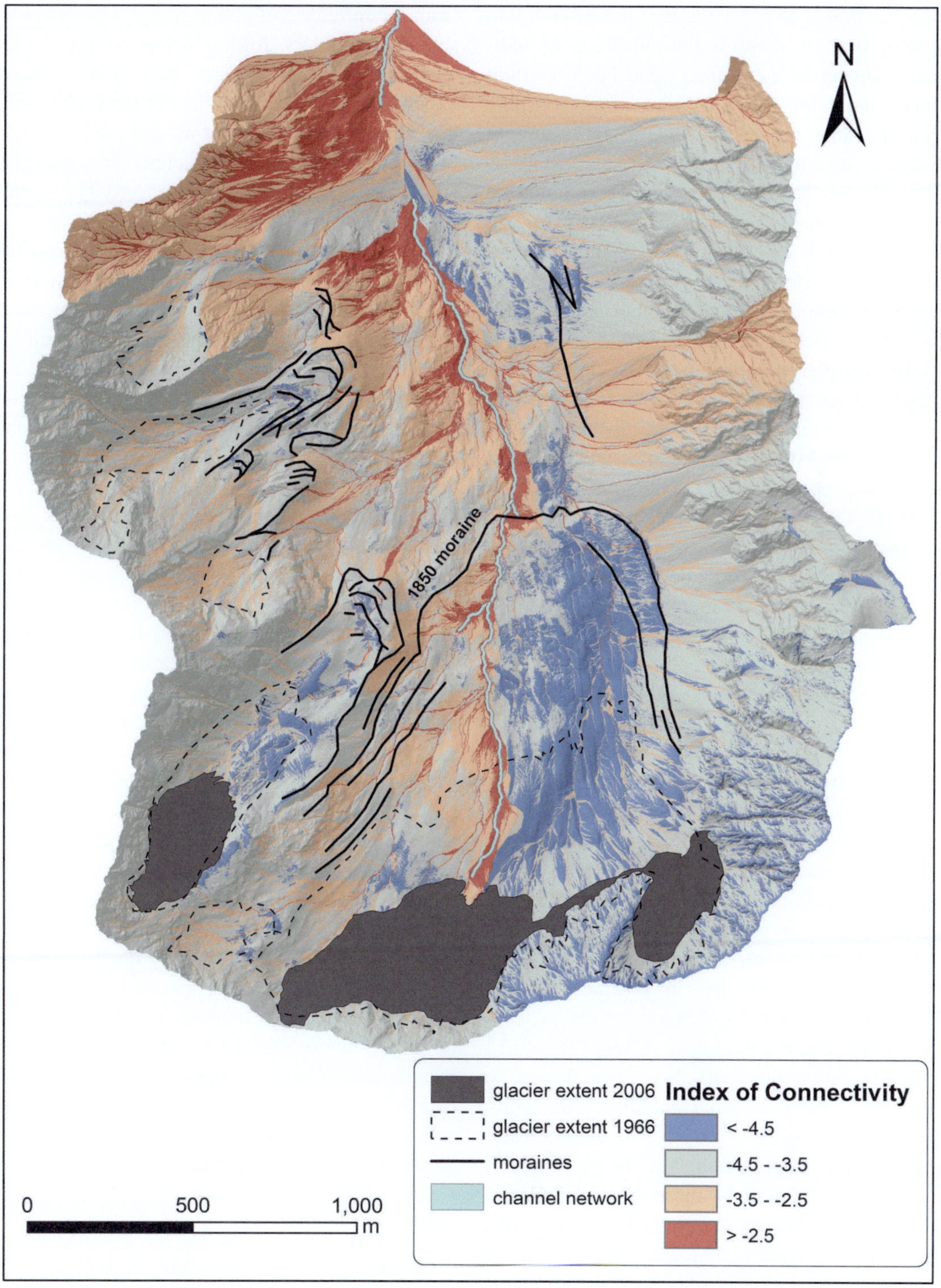

Fig. 16.10 IC map of the upper Zwieselbachtal valley, including the proglacial area of the Zwieselbachferner glacier and its extent today and in 1966, and the LIA moraines (1966 extent and moraines taken from Heuberger 1966)

water/sediment to move through the geomorphic system strongly controls the efficiency of sediment transfer in a catchment and has important implications for the behaviour of geomorphic systems and the establishment of sediment budgets. In addition, sediment connectivity is also relevant for the sensitivity of geomorphic systems to change. This is particularly true in highly dynamic environments as proglacial areas where the transition from glacial to non-glacial conditions implies paraglacial morphodynamics and leads to frequent changes in system structure.

The chapter provided also an overview on proglacial landforms and their role in controlling sediment connectivity highlighting that most of them can act both as storage sites and sediment source depending on local conditions.

If on the one hand connectivity influences the propagation of hillslope and channel changes that occur during deglaciation (glacier retreat, permafrost warming and degradation), on the other hand it is also subject to change itself, especially in high mountain systems. Permafrost warming transiently accelerates the movement of rock glaciers, with a potential establishment or enhancement of sediment connectivity (e.g. Kummert et al. 2017). Permafrost degradation leads to enhanced sediment mobilization from sediment stores and enhanced sediment production from rockwalls; mass movements can couple hillslopes to the channel network, but can also produce barriers (Korup 2005). Furthermore, depressions formed by melt-out of ground ice in sediment bodies induce disconnectivity. Glacier retreat exposes glacigenic material to erosion (Chaps. 10 and 11); enhanced connectivity is expected due to enhanced lateral channel mobility and gully formation (Lane et al. 2017). On the contrary, the formation of proglacial lakes in the course of deglaciation effectively disconnects proglacial systems (Geilhausen et al. 2013; Chap. 14). Another development that decreases connectivity is the accumulation of sediment too coarse for further transport (Lane et al. 2017). Apart from the influence of connectivity, a decrease in sediment delivery from glaciers to proglacial channels (Chaps. 5 and 12) and storage exhaustion also have the effect of declining rates of sediment yield (Lane et al. 2017).

Due to the different influences and multiple feedbacks regarding connectivity, the catchment-scale response to deglaciation is highly complex. We argue that the investigation of sediment connectivity is a key component of proglacial geomorphology. Connectivity indices (e.g. Cavalli et al. 2013; Lane et al. 2017) can be used to investigate and interpret spatial patterns of connectivity. However, the explanatory or predictive capacity of indices regarding system behaviour, specifically sediment delivery under different forcing magnitudes, is a major research gap. We propose three general approaches that enable the comparison of DEM-based indices and field- or model-based observations and measurements: first, high-resolution DEMs of difference may serve to derive sediment delivery ratios (as proxies of functional connectivity; Heckmann and Vericat 2018; Chap. 11), above all at the interfaces of hillslopes and channels. This is facilitated by the increased availability of high-resolution multitemporal surveys (e.g. Carrivick et al. 2013; Micheletti and Lane 2016; Cavalli et al. 2017). Second, particle tagging techniques can be used to investigate within-channel connectivity (e.g. Mao et al. 2017; Chap. 12). Third, landscape evolution models (e.g. Coulthard and van de Wiel 2017) can be used (i) to compare static indices to modelled functional connectivity under different forcing scenarios and (ii) to estimate trajectories of how connectivity changes with time, starting from a "fresh" proglacial landscape and evolving as paraglacial morphodynamics adjust the landscape to non-glacial conditions.

Acknowledgements Work on this chapter was supported by the COST action ES1306 CONNECTEUR (http://connecteur.info) that provided funding for meetings and a training school on sediment connectivity, which is gratefully acknowledged.

References

Baartman JEM, Masselink R, Keesstra SD, Temme AJAM (2013) Linking landscape morphological complexity and sediment connectivity. Earth Surf Process Landf 38:1457–1471. https://doi.org/10.1002/esp.3434

Baewert H, Morche D (2014) Coarse sediment dynamics in a proglacial fluvial system (Fagge River, Tyrol). Geomorphology 218:88–97. https://doi.org/10.1016/j.geomorph.2013.10.021

Ballantyne CK (2002) Paraglacial geomorphology. Quat Sci Rev 21:1935–2017. https://doi.org/10.1016/S0277-3791(02)00005-7

Becht M (1995) Untersuchungen zur aktuellen Reliefentwicklung in alpinen Einzugsgebieten. Münchener Geographische Abhandlungen, A(47):187

Beylich AA, Laute K (2015) Sediment sources, spatiotemporal variability and rates of fluvial bedload transport in glacier-connected steep mountain valleys in western Norway (Erdalen and Bødalen drainage basins). Geomorphology 228:552–567. https://doi.org/10.1016/j.geomorph.2014.10.018

Bogen J, Bønsnes TE (2003) Erosion and sediment transport in High Arctic rivers, Svalbard. Polar Res 22:175–189. https://doi.org/10.3402/polar.v22i2.6454

Borselli L, Cassi P, Torri D (2008) Prolegomena to sediment and flow connectivity in the landscape: a GIS and field numerical assessment. CATENA 75: 268–277. https://doi.org/10.1016/j.catena.2008.07.006

Bracken LJ, Turnbull L, Wainwright J, Bogaart P (2015) Sediment connectivity: a framework for understanding sediment transfer at multiple scales. Earth Surf Process Landforms 40:177–188. https://doi.org/10.1002/esp.3635

Brardinoni F, Church M, Simoni A, Macconi P (2012) Lithologic and glacially conditioned controls on regional debris-flow sediment dynamics. Geology 40:455–458. https://doi.org/10.1130/G33106.1

Brardinoni F, Cavalli M, Heckmann T, Liébault F, Rimböck A (2015a) Guidelines for assessing sediment dynamics in alpine basins and channel reaches. Final Report of the SedAlp Project, Work Package 4. http://www.sedalp.eu/download/dwd/reports/WP4_Report.pdf

Brardinoni F, Sosio R, Scotti R, Marchi L, Crema S, Cavalli M (2015b) Colluvial sediment sources, glacial and periglacial depositional landforms, and geomorphometry-based sediment connectivity in the Saldur River basin, Italy Guidelines for assessing sediment dynamics in alpine basins and channel reaches. Final Report of the SedAlp Project, Work Package 4, Vienna, S 16–23

Brierley G, Fryirs K, Jain V (2006) Landscape connectivity. The geographic basis of geomorphic applications. Area 38:165–174

Brunsden D, Thornes JB (1979) Landscape sensitivity and change. Trans Inst Br Geogr 4:463–484. https://doi.org/10.2307/622210

Burt TP, Allison RJ (2010) Sediment cascades in the environment: an Integrated Approach. In: Burt TP, Allison RJ (eds) Sediment cascades. John Wiley & Sons Ltd, pp 1–15. https://doi.org/10.1002/9780470682876.ch1

Carrivick JL, Heckmann T (2017) Short-term geomorphological evolution of proglacial systems. Geomorphology, Sediment cascades in cold climate geosystems 287:3–28. https://doi.org/10.1016/j.geomorph.2017.01.037

Carrivick JL, Geilhausen M, Warburton J, Dickson NE, Carver SJ, Evans AJ, Brown LE (2013) Contemporary geomorphological activity throughout the proglacial area of an alpine catchment. Geomorphology 188:83–95. https://doi.org/10.1016/j.geomorph.2012.03.029

Cavalli M, Marchi L (2008) Characterisation of the surface morphology of an alpine alluvial fan using airborne LiDAR. Nat Hazards Earth Syst Sci 8:323–333

Cavalli M, Trevisani S, Comiti F, Marchi L (2013) Geomorphometric assessment of spatial sediment connectivity in small Alpine catchments. Geomorphology 188:31–41. https://doi.org/10.1016/j.geomorph.2012.05.007

Cavalli M, Goldin B, Comiti F, Brardinoni F, Marchi L (2017) Assessment of erosion and deposition in steep mountain basins by differencing sequential digital terrain models. Geomorphology 291:4–16. https://doi.org/10.1016/j.geomorph.2016.04.009

Cenderelli DA, Wohl EE (1998) Sedimentology and clast orientation of deposits produced by glacial-lake outburst floods in the Mount Everest Region, Nepal. In: Kalvoda J, Rosenfeld CL (eds) Geomorphological hazards in high mountain areas. Kluwer Academic Publishers, The Netherlands, pp 1–26

Chiarle M, Iannotti S, Mortara G, Deline P (2007) Recent debris flow occurrences associated with glaciers in the Alps. Global Planet Change 56:123–136

Clague JJ, Evans SG (2000) A review of catastrophic drainage of moraine-dammed lakes in British Columbia. Quatern Sci Rev 19:1763–1783

Coulthard TJ, van de Wiel MJ (2017) Modelling long term basin scale sediment connectivity, driven by spatial land use changes. Geomorphology 277:265–281. https://doi.org/10.1016/j.geomorph.2016.05.027

Curry AM, Cleasby V, Zukowskyj P (2006) Paraglacial response of steep, sediment-mantled slopes to post-Little Ice Age glacier recession in the central Swiss Alps. J Quat Sci 21:211–225. https://doi.org/10.1002/jqs.954

de Vente J, Poesen J, Arabkhedri M, Verstraeten G (2007) The sediment delivery problem revisited. Prog Phys Geogr 31:155–178. https://doi.org/10.1177/0309133307076485

Engelen GB, Venneker GW (1988) ETA (Erosion Transport Accumulation) systems, their classification, mapping and management. In Bordas MP, Walling DE (eds) Sediment Budgets, IAHS Publication no. 174, Wallingford (UK), pp 397–412

Faulkner H (2008) Connectivity as a crucial determinant of badland morphology and evolution. Geomorphology, Fluvial systems: Dynamics, Morphology and the Sedimentary Record, Special Issue in honour of Adrian Harvey 100:91–103. https://doi.org/10.1016/j.geomorph.2007.04.039

Fryirs K (2013) (Dis)Connectivity in catchment sediment cascades; A fresh look at the sediment delivery problem. Earth Surf Process Landforms 38:30–46. https://doi.org/10.1002/esp.3242

Fryirs KA (2017) River sensitivity; A lost foundation concept in fluvial geomorphology. Earth Surf Process Landforms 42:55–70. https://doi.org/10.1002/esp.3940

Fryirs K, Brierley GJ (2001) Variability in sediment delivery and storage along river courses in Bega catchment, NSW, Australia: implications for geomorphic river recovery. Geomorphology 38:237–265

Fryirs KA, Brierley GJ, Preston NJ, Kasai M (2007) Buffers, barriers and blankets: the (dis)connectivity of catchment-scale sediment cascades. CATENA 70:49–67. https://doi.org/10.1016/j.catena.2006.07.007

Fuller IC, Marden M (2011) Slope–channel coupling in steepland terrain: a field-based conceptual model from the Tarndale gully and fan, Waipaoa catchment, New Zealand. Geomorphology 128:105–115. https://doi.org/10.1016/j.geomorph.2010.12.018

Geilhausen M, Morche D, Otto J-C, Schrott L (2013) Sediment discharge from the proglacial zone of a retreating Alpine glacier. Zeit fur Geo Supp 57:29–53. https://doi.org/10.1127/0372-8854/2012/S-00122

Goldin B, Rudaz B, Bardou E (2016) Application of a sediment connectivity GIS-based index in a basin undergoing glacier retreat: the case study of the Navizence catchment. ROL 39:35–38. https://doi.org/10.3301/ROL.2016.41

Hartl L (2010) The Gepatschferner from 1850–2006—changes in length, area and volume in relation to climate. Diploma thesis, University of Innsbruck, Innsbruck

Harvey AM (2001) Coupling between hillslopes and channels in upland fluvial systems: implications for landscape sensitivity, illustrated from the Howgill Fells, northwest England. CATENA 42:225–250. https://doi.org/10.1016/S0341-8162(00)00139-9

Harvey AM (2002) Effective timescales of coupling within fluvial systems; geomorphology on Large Rivers. Geomorphology 44:175–201. https://doi.org/10.1016/S0169-555X(01)00174-X

Haas F, Heckmann T, Hilger L, Becht M. (2012) Quantification of debris flows in the proglacial area of the gepatschferner/Austrian Alps using Ground-based and Airborne LIDAR Data. In: Collins AL, Golosov V, Horowitz AJ, Lu X, Stone M, Walling Des E, Zhang X (eds) Erosion and sediment yields in the changing environment. Proceedings of an IAHS International Commission on Continental Erosion Symposium, held at the Institute of Mountain Hazards and Environment, CAS-Chengdu, China, 11–15 October 2012-Wallingford, 2012-S, pp 293–302 (IAHS publication 356). ISBN 978-1-907161-33-9

Heckmann T, Schwanghart W (2013) Geomorphic coupling and sediment connectivity in an alpine catchment—exploring sediment cascades using graph theory. Geomorphology 182:89–103. https://doi.org/10.1016/j.geomorph.2012.10.033

Heckmann T, Vericat D (2018) Computing spatially distributed sediment delivery ratios: inferring functional sediment connectivity from repeat high-resolution digital elevation models. Earth Surf Process Landforms 218:88. https://doi.org/10.1002/esp.4334

Heckmann T, McColl S, Morche D (2016) Retreating ice; research in pro-glacial areas matters. Earth Surf Process Landforms 41:271–276. https://doi.org/10.1002/esp.3858

Heckmann T, Haas F, Morche D, Schmidt K-H, Rohn J, Moser M, Leopold M, Kuhn M, Briese C, Pfeifer N, Becht (2012) Investigating an alpine proglacial sediment budget using field measurements, airborne and terrestrial LiDAR data. In: Collins AL, Golosov V, Horowitz AJ, Lu X, Stone M, Walling Des E, Zhang X (eds) Erosion and sediment yields in the changing environment. Proceedings of an IAHS International Commission on Continental Erosion Symposium, held at the Institute of Mountain Hazards and Environment, CAS-Chengdu, China, 11–15 October 2012-Wallingford, 2012-S, pp 438–447 (IAHS publication 356), ISBN 978-1-907161-33-9

Heuberger H (1966) Gletschergeschichtliche Untersuchungen in den Zentralalpen zwischen Sellrain- und Ötztal, Innsbruck (=Wissenschaftliche Alpenvereinshefte 20)

Hewitt K, Liu J (2010) Ice-dammed lakes and outburst floods, Karakoram Himalaya: historical perspectives on emerging threats. Phys Geogr 31:528–551

Holm K, Bovis M, Jakob M (2004) The landslide response of alpine basins to post-Little Ice Age glacial thinning and retreat in southwestern British Columbia. Geomorphology 57:201–216

Hooke J (2003) Coarse sediment connectivity in river channel systems: a conceptual framework and methodology. Geomorphology 56:79–94. https://doi.org/10.1016/S0169-555X(03)00047-3

Knight J, Harrison S (2014) Mountain glacial and paraglacial environments under global climate change: lessons from the past, future directions and policy implications. Geogr Ann Ser A Phys Geogr 96:245–264. https://doi.org/10.1111/geoa.12051

Korup O (2005) Geomorphic imprint of landslides on alpine river systems, southwest New Zealand. Earth Surf Process Landforms 30:783–800. https://doi.org/10.1002/esp.1171

Kummert M, Delaloye R (2015) Quantifying sediment transfer between the front of an active alpine rock glacier and a torrential gully. In: Jasiewicz J, Zwoliński Z, Mitasova H, Hengl T (eds) Geomorphometry for Geosciences. Adam Mickiewicz University in Poznań—Institute of Geoecology and Geoinformation, Poznań, Poland, Bogucki Wydawnictwo Naukowe, pp 193–196

Kummert M, Delaloye R, Braillard L (2017) Erosion and sediment transfer processes at the front of rapidly moving rock glaciers; systematic observations with automatic cameras in the western Swiss Alps. Permafrost and Periglac Process 40:721. https://doi.org/10.1002/ppp.1960

Lane SN, Bakker M, Gabbud C, Micheletti N, Saugy J-N (2017) Sediment export, transient landscape response and catchment-scale connectivity following rapid climate warming and Alpine glacier recession. Geomorphology 277:210–227. https://doi.org/10.1016/j.geomorph.2016.02.015

Laute K, Beylich AA (2014) Environmental controls, rates and mass transfers of contemporary hillslope processes in the headwaters of two glacier-connected drainage basins in western Norway. Geomorphology

Lugon R, Stoffel M (2010) Rock-glacier dynamics and magnitude–frequency relations of debris flows in a high-elevation watershed: Ritigraben, Swiss Alps. Global Planet Change 73:202–210

Mao L, Dell'Agnese A, Comiti F (2017) Sediment motion and velocity in a glacier-fed stream. Geomorphology 291:69–79. https://doi.org/10.1016/j.geomorph.2016.09.008

Marchi L, Dalla Fontana G, Cavalli M, Tagliavini F (2008) Rocky headwaters in the Dolomites, Italy: field observations and topographic analysis. Arct Antarct Alp Res 40(4):685–694

Marren PM (2005) Magnitude and frequency in proglacial rivers: a geomorphological and sedimentological perspective. Earth Sci Rev 70:203–251

Meßenzehl K, Hoffmann T, Dikau R (2014) Sediment connectivity in the high-alpine valley of Val Müschauns, Swiss National Park—linking geomorphic field mapping with geomorphometric modelling. Geomorphology 221:215–229. https://doi.org/10.1016/j.geomorph.2014.05.033

Micheletti N, Lane SN (2016) Water yield and sediment export in small, partially glaciated Alpine watersheds in a warming climate. Water Resour Res 52:4924–4943. https://doi.org/10.1002/2016WR018774

Micheletti N, Lambiel C, Lane SN (2015) Investigating decadal-scale geomorphic dynamics in an alpine mountainsetting. J Geophys Res Earth Surf 120:2155–2175. https://doi.org/10.1002/2015JF003656

Morche D, Schmidt K-H, Sahling I, Herkommer M, Kutschera J (2008) Volume changes of Alpine sediment stores in a state of post-event disequilibrium and the implications for downstream hydrology and bed load transport. Nor Geogr Tidsskr-Norw J Geogr 62:89–101

O'Connor JE, Hardison, JH 3rd, Costa, JE (2001) Debris flows from failures of neoglacial-age moraine dams in the three sisters and Mount Jefferson wilderness Areas, Oregon, 2001. U.S. Geological Survey Professional Paper 1606, 105 p

Otto J-C, Schrott L, Jaboyedoff M, Dikau R (2009) Quantifying sediment storage in a high alpine valley (Turtmanntal, Switzerland). Earth Surf Proc Land 34 (13):1726–1742

Schrott L, Niederheide A, Hankammer M, Hufschmidt G, Dikau R (2002) Sediment storage in a mountain catchment: geomorphic coupling and temporal variability (Reintal, Bavarian Alps, Germany). Z. Geomorph. N.F. Suppl. 127:175–196

Schrott L, Götz J, Geilhausen M, Morche D (2006) Spatial and temporal variability of sediment transfer and storage in an Alpine basin (Reintal valley, Bavarian Alps, Germany). Geogr Hel 61:191–200

Slaymaker O (2006) Towards the identification of scaling relations in drainage basin sediment budgets. Geomorphology 80:8–19. https://doi.org/10.1016/j.geomorph.2005.09.004

Slaymaker O (2011) Criteria to distinguish between periglacial, proglacial and paraglacial environments. Quaest. Geogr 30:85–94

Stoffel M, Marston RA (2013) 7.1 mountain and hillslope geomorphology: an introduction A2—Shroder. In: John F (ed) Treatise on geomorphology. Academic Press, San Diego, pp 1–3

Straumann RK, Korup O (2009) Quantifying postglacial sediment storage at the mountain-belt scale. Geol 37:1079–1082. https://doi.org/10.1130/G30113A.1

Tarboton DG (1997) A new method for the determination of flow directions and upslope areas in grid digital elevation models. Water Resour Res 33:309–319

Temme AJAM, Lange K (2014) Pro-glacial soil variability and geomorphic activity—the case of three Swiss valleys. Earth Surf Process Landforms 39:1492–1499. https://doi.org/10.1002/esp.3553

Trevisani S, Cavalli M (2016) Topography-based flow-directional roughness; potential and challenges. Earth Surf Dynam 4:343–358. https://doi.org/10.5194/esurf-4-343-2016

Trevisani S, Cavalli M, Marchi L (2009) Variogram maps from LiDAR data as fingerprints of surface morphology on scree slopes. Nat Hazards Earth Syst Sci 9:129–133

Walling DE (1983) The sediment delivery problem. J Hydrol, Scale Problems in Hydrology 65:209–237. https://doi.org/10.1016/0022-1694(83)90217-2

Westoby MJ, Glasser NF, Brasington J, Hambrey MJ, Quincey DJ, Reynolds JM (2014) Modelling outburst floods from moraine-dammed glacial lakes. Earth-Sci Rev 134:137–159

Wichmann V, Heckmann T, Haas F, Becht M (2009) A new modelling approach to delineate the spatial extent of alpine sediment cascades. Geomorphology, GIS and SDA applications in geomorphology 111:70–78. https://doi.org/10.1016/j.geomorph.2008.04.028

A Sediment Budget of the Upper Kaunertal

17

Ludwig Hilger, Jana-Marie Dusik, Tobias Heckmann, Florian Haas, Philipp Glira, Norbert Pfeifer, Lucas Vehling, Joachim Rohn, David Morche, Henning Baewert, Martin Stocker-Waldhuber, Michael Kuhn and Michael Becht

Abstract

This chapter presents the sediment budget of the Upper Kaunertal (Kauner valley, Ötztal Alps, Austria) for the years 2012–2014 as obtained in the framework of the PROSA (high-resolution measurements of morphodynamics in rapidly changing PROglacial Systems of the Alps) research project. An important methodological basis of this high-mountain sediment budget is the usage of study area-wide LiDAR data (TLS and ALS) of comparatively high temporal and spatial resolution to measure rates of erosion and deposition, and to regionalize/upscale rates at the local scale. After several billion measurement points and data from fieldwork, mapping, and modeling efforts had been processed and evaluated, it was possible to identify and quantify sediment transfer by all relevant processes at the scale of the 62 km^2 study area. These processes include rockfall of three different magnitude classes, debris flows, avalanches, creep on talus, fluvial processes (hillslopes and main fluvial system), rock glaciers, and glaciers. After a short presentation of the process-specific methods to obtain catchment-wide rates, we discuss process-specific results and the budget. The sediment budget does not only show the relative importance of the mentioned processes and spatial subunits (proglacial vs. non-proglacial) in the Upper Kaunertal. It also gives insight into the importance of high-magnitude events and the configuration of the sediment transport system.

L. Hilger (✉) · T. Heckmann · F. Haas · M. Becht
Chair of Physical Geography, Catholic University of Eichstätt-Ingolstadt, Eichstätt, Germany
e-mail: corran@altmuehlnet.de

J.-M. Dusik
Bavarian State Agency for Environment (LfU), Geological Survey, Hof/Saale, Germany

P. Glira · N. Pfeifer
TU Vienna, Vienna, Austria

L. Vehling · J. Rohn
University of Erlangen-Nuremberg, Erlangen, Germany

D. Morche
University of Halle-Wittenberg, Halle, Germany

D. Morche
Environmental Authority of Saalekreis District, Merseburg, Germany

H. Baewert
University of Halle-Wittenberg, Halle, Germany

M. Stocker-Waldhuber
Institute for Interdisciplinary Mountain Research, Austrian Academy of Sciences, Innsbruck, Austria

M. Stocker-Waldhuber
Department of Geography, Physical Geography, Catholic University of Eichstätt-Ingolstadt, Eichstätt-Ingolstadt, Germany

M. Kuhn
Institute of Atmospheric and Cryospheric Sciences, University of Innsbruck, Innsbruck, Austria

T. Heckmann and D. Morche (eds.), *Geomorphology of Proglacial Systems*, Geography of the Physical Environment, https://doi.org/10.1007/978-3-319-94184-4_17

Keywords
PROSA project · Sediment budget · Measurement · Geomorphological map Regionalisation · Spatial modelling

17.1 Introduction

Most (pro)glacial sediment budgets established so far have mainly focused on suspended and bedload transport in the main channel(s) of proglacial areas (Chap. 15). Hillslope processes, such as slope-aquatic processes or mass movements (i.e., debris flows, slides, and rockfalls), have received comparatively little attention, causing research need in complete high-mountain sediment budgets, especially under global warming conditions. As high-mountain areas have been comparatively weakly impacted by human action, these areas can be used to establish baselines for the investigation and modeling of geomorphic processes under future climate conditions. This is why the construction of a present-day sediment budget of the Upper Kaunertal is seen as a valuable contribution to both sediment budget and high-mountain geomorphology research. The main objectives in construction of the sediment budget of the Upper Kaunertal are (i) to assess the relative importance of different geomorphic processes in the study area and (ii) to obtain an estimate of the proportion of mobilized sediment that finally reaches the outlet of the catchment, thereby enabling inferences on the state of connectivity in the Upper Kaunertal. [1]

The catchment (c. 62 km^2) under study is located in the Ötztal Alps, a mountain range in the Central Alps of Austria (see Chap. 1). Crystalline rocks make up the rockwalls and form peaks up to 3583 m high. The defined catchment outlet is constituted by the mouth of the main river ("Fagge") into the Gepatsch reservoir at 1810 m. A large share of the catchment is covered by glacier ice (c. 31% as of 2012). At Weißsee weather station (2500 m), the average annual precipitation is 983 mm and the average air temperature is −0.6 °C. Coniferous forest can be found below c. 2150 m. A map of the study area showing some place names used in this chapter is shown in Fig. 17.1.

Processes identified as being of relevance for the Upper Kaunertal sediment budget (based on field observations made during the first field visits and insights gained during the preparation of the geomorphological–geological map) are

- Mass wasting processes, including rockfall of all magnitudes (Sects. 17.4.1 and 17.4.2; see also Chap. 9; Vehling 2016; Vehling et al. 2016, 2017; Heckmann et al. 2016) and rock slides (Sect. 17.4.4; Chap. 9)
- debris flows (Sect. 17.4.3; see also Chap. 11; Haas et al. 2012)
- full-depth avalanches (Sect. 17.4.5)
- hillslope processes such as slope wash and linear erosion (Sect. 17.4.7; see also Chap. 11)
- periglacial creep (rock glaciers; Sect. 17.4.6; Chap. 7; Berger et al. 2004; Hausmann et al. 2007; Krainer et al. 2007; Dusik et al. 2015)
- glacial processes (Chap. 5; Giese 1963)
- fluvial processes in the main channels (Chap. 13; Baewert and Morche 2014).

The dissimilarity in mechanics and controlling factors of these processes requires a wide range of measurement, data processing, mapping, and modeling techniques to arrive at sediment transfer rates valid for the whole catchment. As the focus of this contribution is on results rather than measurement, the discussion of methodological aspects was kept to a minimum. Interested readers are referred to the Ph.D. thesis of the first author (Hilger 2017) and the publications mentioned above. The final sediment budget of the Upper Kaunertal as a meso-scale catchment spanning multiple altitudinal zones is the result of a collaboration of all sub-projects of the PROSA project (see Chap. 1).

[1]This chapter contains results also discussed in the dissertation of the main author (Hilger 2017). While almost no content was directly copied unchanged, it is likely that the way of argumentation and some phrases can also be found in the dissertation.

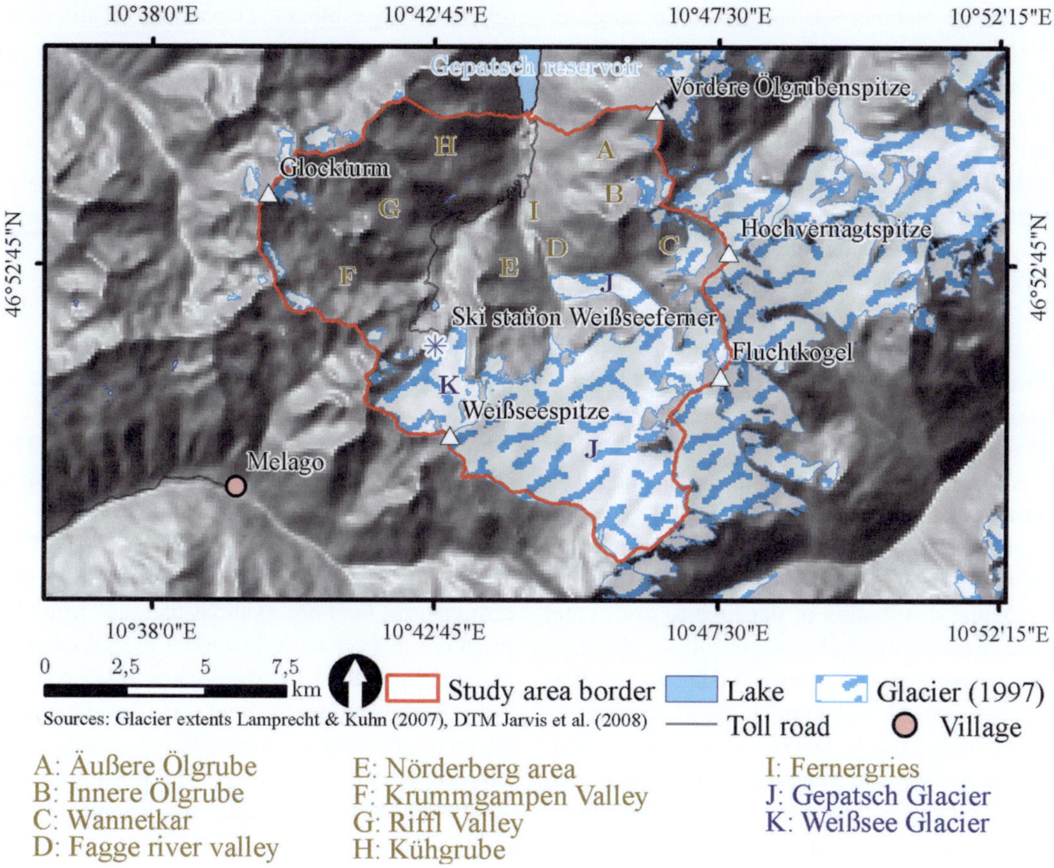

Fig. 17.1 Place names within the PROSA project study area used in this chapter. Figure taken from Hilger (2017)

17.2 Data and Methods

The data used for the construction of the sediment budget were acquired, processed, and evaluated by five different working groups over the span of about three years; in general, the data can be assigned to one of four distinct categories:

- Measurements using "classical" techniques, such as discharge measurements (Chap. 13), rockfall collectors (Chap. 9), hill slope transport measurements with troughs, extensometer measurements in deep-seated gravitational slope deformations (Chap. 9) or geophysical investigations (Chaps. 5 and 7)
- measurements on the hillslope or plot scale using airborne and terrestrial LiDAR (e.g., Chaps. 9 and 11)
- historical aerial imagery, and
- mapping results produced from the former two.

A unifying characteristic of almost all work on the construction of the sediment budget and methodological backbone of PROSA project is the high relevance of light detection and ranging (LiDAR) data. Several billion LiDAR measurement points with a precision of less than 10 cm using both terrestrial (more than 1.5 bn points) and airborne laser scanning (ALS, more than 4.5 bn points) have been acquired, processed, and evaluated for the construction of the sediment budget.

TLS data have been acquired using two different Riegl TLS devices one to four times a year at different monitoring stations. The survey frequency depended on the elevation and accessibility of the monitoring station. In total, 40 monitoring stations spanning a wide range of elevations, aspects, lithologies, and dominant process areas have been maintained, yielding over 200 TLS topographic surveys that were used for the investigation of surface changes and related sediment transfer by different processes (Fig. 17.2). Airborne laser scanning data used for the results presented in this chapter were acquired by Milan Geoservice GmBH in July 2012, October 2012, and July 2014.

After having undergone an elaborate point cloud classification work flow, all LiDAR measurement points were used to construct digital terrain models (DTMs). The relatively high point density of the LiDAR data allowed spatial raster resolutions higher than one meter. Multitemporal digital elevation models (DEMs) were used to compute DEMs of difference (DoD) to interpret the spatial distribution of negative (erosion) and positive (deposition) changes in surface elevation and to compute volumetric sediment budgets ("morphological budgeting"). DEM uncertainty was propagated into DoDs and volumetric budgets using standard Gaussian error propagation and significance tests (Lane et al. 2003; Glira et al. 2014). The resulting surface change patterns allowed for interpretation and process identification. DTMs were also used for the derivation of a wide range of land surface parameters (e.g., slope, roughness). These, in turn, served as input for spatial numerical models used for regionalization and (geomorphological) mapping.

Historical aerial photographs and orthophotos were acquired from the Austrian federal office of geodesy (BEV, Vienna) and Milan Geoservice GmbH for a total of 14 different dates ranging from June 1953 to July 2014. While some of these volumes were orthophotos, imagery from other dates was orthorectified using Trimble's Inpho software package. The followed work flow corresponds to the classical work flow of digital stereo photogrammetry (cf. Schulz and Dornblut 2002 for more information). The procedure resulted in a global accuracy of between 0.4 and 0.5 pixels (between 10 and 25 cm) for the different image blocks. Depending on the scale and resolution of the input images, the ground resolution of the resulting orthophotos was fixed between 20 and 55 cm.

17.3 A Geomorphological Map of the Upper Kaunertal

Given historical imagery of 14 different dates and multiple ALS survey results and their derivatives, it was possible to produce detailed maps of vegetation classes, sediment storages, and geomorphic activity for the whole catchment. The sediment storage map, as one of the most important foundations for an identification or process links, was prepared at a scale of c. 1:6000. In addition to the mentioned DTM and orthophotos, field observations and data from the literature (e.g., data on historical glacial extents, as in Kerschner 1979) were taken into account. From the beginning, the map was conceptualized to serve the aim of model-based process-link analysis. As using standard point- and line-based symbology set forth, e.g., by Kneisel et al. (1998), does not lend itself to GIS-based analysis (such as calculating areal proportion of landform types, analysis of landform topology), the map was constructed in two versions. Non-overlapping polygon objects representing sediment storages and process areas serve as the basis for GIS analysis, while the more traditional version (with storage landforms represented by point symbols) can serve as guidance in the field and the purpose of visualization. The map was generated with contributions from different sub-projects of the PROSA project (see Chap. 1). About 6000 units were mapped by hand and morphologically undifferentiated bedrock class polygons were further subdivided into aspect-based rockwall sections using a moving-window-based approach. The final geomorphological map analyzes the study area into roughly 22,000 different landscape units. Of these, c. 16,000 are bedrock sections. Qualitatively, 41 classes were distinguished, such as

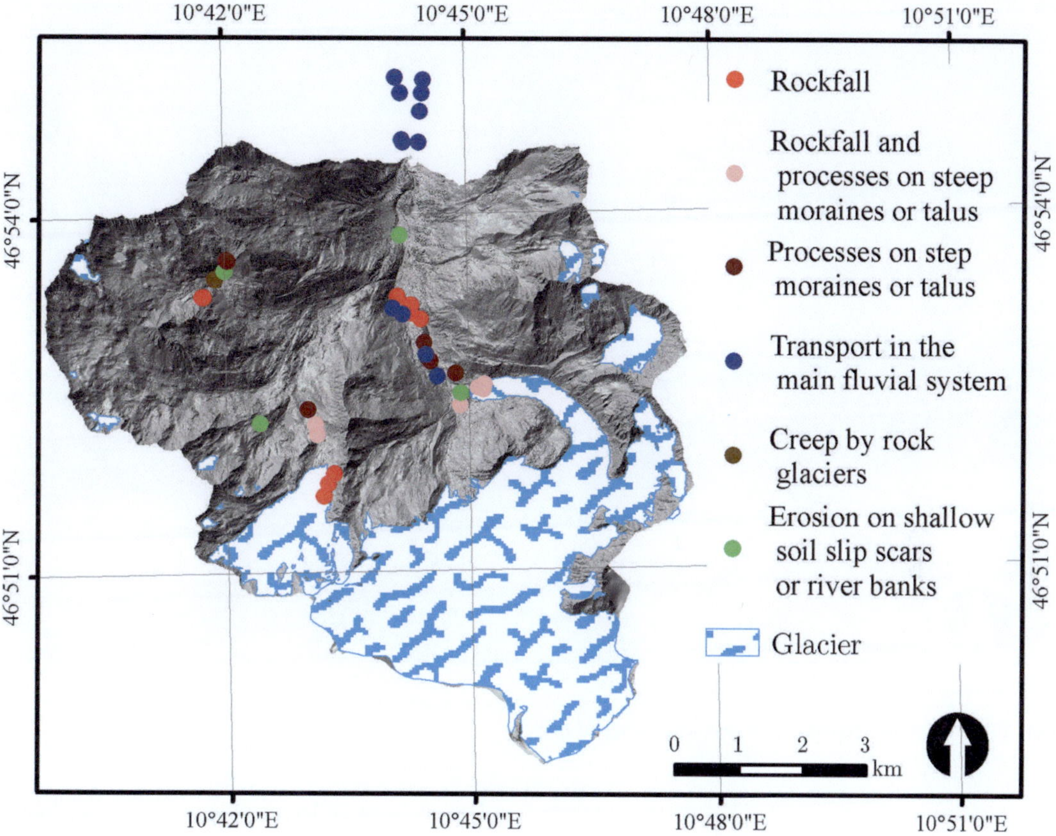

Fig. 17.2 Location of TLS monitoring stations within the study area

units in the active fluvial system (e.g., gravel bars, active channels), glacial storage landforms of two different age-groups (e.g., lateral moraine older than A.D. 1855, ground moraine younger than A.D. 1855), cryogenic landforms (such as rock glaciers of different activity states), bedrock differentiated according to geology, process areas (erosion, transport, and deposition zones of single debris flows), etc. In designing the landform classes, geomorphological maps prepared by others (e.g., Otto 2006) served as a starting point, but classes were further differentiated where it was needed (e.g., separation of moraine ages). The area affected by processes (e.g., debris flows on lateral moraines) was mapped on a multitemporal basis, using historical aerial imagery (see Sect. 17.4.3).

An example section of the resulting map is depicted in Fig. 17.3. For the sake of clarity, the legend (Fig. 17.4) only contains the class units visible in the depicted area and only landforms. Process areas are not visualized.

Table 17.1 gives the surface percentages of different superordinate landform classes in the Upper Kaunertal. The scale and resulting detail of the geomorphological map implies that sediment balances could be established for landscape units much smaller than the whole catchment (as undertaken for the rockfall part by Heckmann et al. 2016). For the catchment-scale budget presented here, areas related to the single processes were collapsed; thus we obtained the mass of sediment mobilized versus transferred to the main fluvial system at the catchment scale.

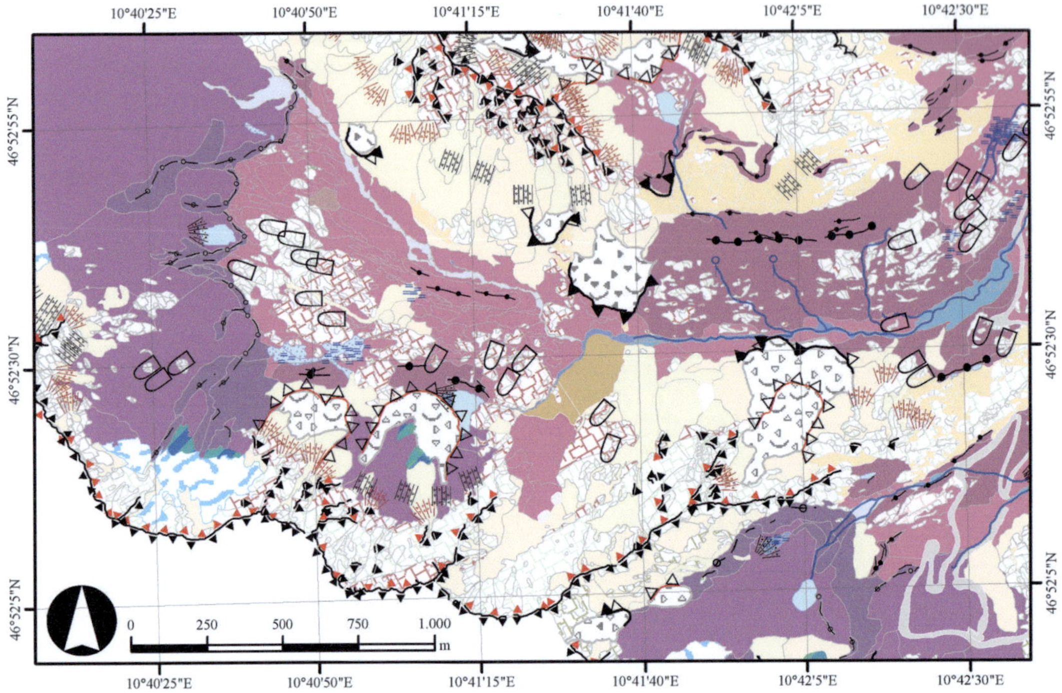

Fig. 17.3 Example section of the geomorphological map

17.4 Process Quantification at the Catchment Scale

With the help of the geomorphological map, a generalized graphical representation of the cascading system (with respect to the relevant processes and storage landforms) was set up (Fig. 17.5).

The process-specific databases, processing and analysis steps taken to obtain sediment transfer estimates will be referred to in the following in dedicated subsections. The treatise will start with mass wasting processes, followed by other hillslope processes and concluding with the construction of the budget using data on glacial processes and fluvial sediment transport in the main channels.

17.4.1 Rockfall

Small-scale rockfall (debris fall) was quantified using both direct measurements of rockfall mass collected from collector nets and volume change quantification from TLS data (see Chap. 9 for details).

With a total of 14 rockfall rates from the PROSA study area, rockfall production rates in kg yr^{-1} m^{-2} were calculated. As the data were acquired using different methods, approaches were slightly different for net-derived and TLS-derived rates. Net-derived production rates significantly depend on the rockwall reference area contributing to the respective nets. This is the main reason why Heckmann et al. (2016) tested three different approaches to obtain the net collector sediment contributing areas (SCA). Here, we used the starting points (and their real surface area computed from the DEM) of modeled rockfall trajectories that ended on the corresponding collector net. To identify potential source cells, a slope threshold of 47.5° was determined using the methodology described in Loye et al. (2009). Rockfall production rates for the TLS-derived mass data were obtained by dividing the collected mass by the real surface

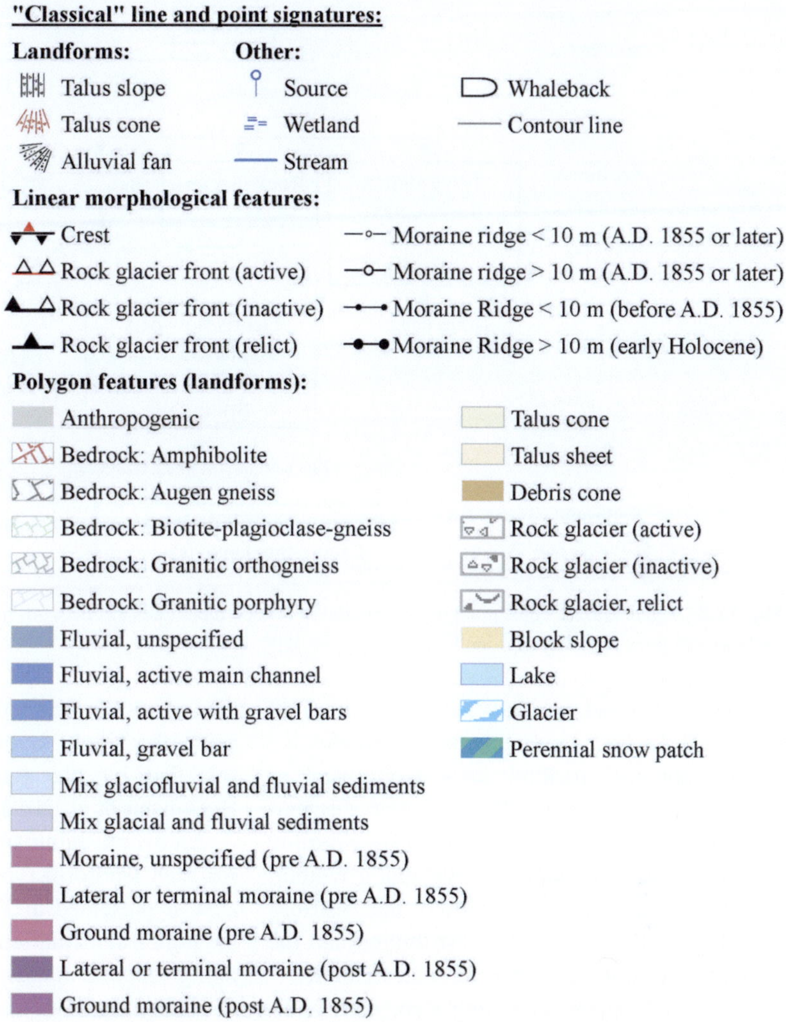

Fig. 17.4 Legend for the geomorphological map (Fig. 17.3)

Table 17.1 Landform coverage

Landform category	Area (km^2)	Percentage
Bedrock	c. 12.92	c. 20.8
Gravitational	c. 6.47	c. 10.5
Periglacial	c. 4.89	c. 7.9
Glacier	c. 19.47	c. 31.4
Glacial	c. 14.02	c. 22.6
Fluvial	c. 1.4	c. 2.2
Anthropogenic	c. 0.84	c. 1.4
Other, mix forms	c. 1.99	c. 3.2

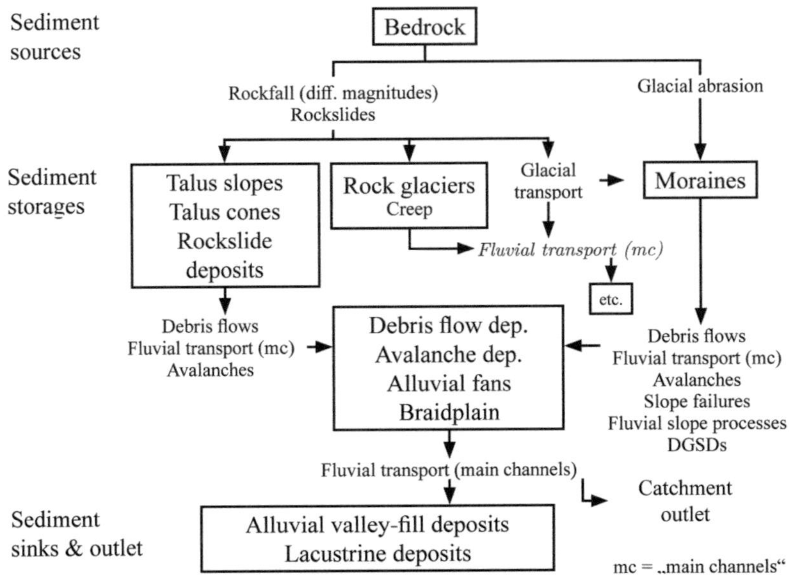

Fig. 17.5 A generalized, conceptual representation of the sediment transport network in the Upper Kaunertal. Relevant processes in italic and (storage) landforms in regular type

area of the bedrock surfaces surveyed with TLS scans and the time interval between samples.

The obtained rockfall rates were used to estimate the parameters of a statistical distribution. A chi-square test, a Kolmogorov–Smirnov test, and a Weibull plot confirmed that a Weibull distribution is suitable to represent this distribution. Figure 17.6 shows this distribution and the rockfall rates obtained with different methods.

In the subsequent modeling step, each identified rockfall source cell in the study area was assigned a rockfall production rate that was drawn from this Weibull distribution. The rockfall produced at each start cell is subsequently distributed downslope via a numerical rockfall model. The model applied has been described in detail by Wichmann and Becht (2005), Wichmann (2006). Since its development, it has been extended to, for example, log modeled trajectories from start to stop cells (Heckmann and Schwanghart 2013). In combination with the geomorphological map (Sect. 17.3), the trajectories were used to set up a graph model of rockfall sediment transfer through the catchment: Raster cells form the nodes that are connected by edges representing potential rockfall trajectories (and their rates of sediment transfer). Nodes (and the corresponding edges) can be aggregated using the single landforms, or landform types, specified by the geomorphological map (c.f. Heckmann et al. 2016) (Fig. 17.7).

An overall summary of the results, aggregated for each landform type represented in the study area, is presented in Table 17.2. The data reflect the high fragmentation of the complex landscape in high-mountain areas. For example, about 26% of small-magnitude rockfall is deposited on glacier surfaces from where it might be transferred to the proglacial areas and via fluvial transport to the basin outlet. Only 2.6% of debris fall is transferred directly into competent channels of the main fluvial system, and about 18% is blocked by glacial landforms like lateral or ground moraines. Finally, the total debris production sums up to about 25,803 t yr^{-1}.

While the approach discussed above honors all empirical measurement data, a crucial weakness is the lack of a spatial distribution of rockfall rates. This is why a second approach was developed to obtain a study area-wide rockfall sediment budget. To facilitate this approach, Vehling (2016) developed a rock mass stability index and mapped six distinct classes study area wide. As the index had proven to be strongly

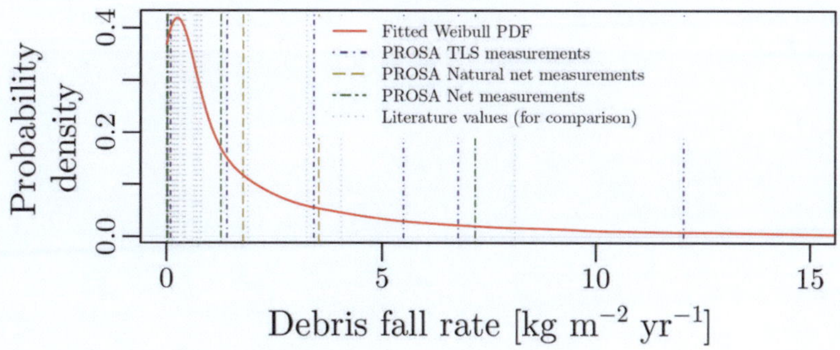

Fig. 17.6 Rockfall production rates and fitted Weibull distribution. Slightly changed, after a figure in Hilger (2017). Values found in other alpine studies are shown for comparison

correlating with rockfall production rates (see Chap. 9), the resulting map could be used in rockfall rate regionalization. However, only rockfall collector net measurements were used to assign rates to the different rock mass stability index classes. While this approach of debris production rate was very different to the one described above, the means of trajectory modeling and analysis was the same.

Using spatially distributed rockfall production rates, a rate of 28,738 t yr^{-1} (c. 463 t yr^{-1} km^{-2}) was estimated for the whole catchment. This corresponds to a total of geomorphic work performed of circa 1565 W (circa 25.3 W km^{-2}). As the use of spatially distributed rates is able to model the true sediment transfer by rockfall more realistically, it is this value that will be used for the construction of the sediment budget in Sect. 17.5.

Vehling (2016) reports a related approach to compute the rockfall sediment budget that better honors the spatial factors controlling debris production on rockwalls. He used two different scenarios (one using a simple regionalization of mapped debris fall production rates based on a mapped rock mass stability index, and one using a nonlinear regression function between the index and debris fall production rates) to obtain a sediment transfer rate for debris fall on the whole catchment scale (i.e., not distributed by landforms or landform types). His first scenario gives an estimate of circa 36,400 t yr^{-1}, while the second one yields an estimate of about 46,934 t yr^{-1}. It is evident that Vehling's estimates are much higher than the one reported above (28,738 t yr^{-1}). The main reason probably is Vehling's use of a lower slope threshold (40° instead of 47.5°) to identify source areas. If we had also used 40° as a threshold (note that we opted for a data-based rather than an arbitrary threshold), a similar total mass as estimated by Vehling would have resulted.

A very detailed documentation of the general work flow presented here, in combination with a discussion of the methodology can be found in Heckmann et al. (2016). Since this publication, several improvements have been made to produce what we consider a more reliable result, for example, the availability of a bigger database (TLS data, longer measurement period at the nets), a further differentiation of the geomorphological map, and the usage of a Weibull instead of a log-normal distribution. Our results are much higher (by a factor of c. 3.5) than the ones published by Heckmann et al. (2016) who used a much smaller database regarding PROSA results and did not apply spatially distributed rockfall rates. The proportions of rockfall sediment delivery to different geomorphic units, however, are fairly consistent, which is due to the use of the same trajectory model.

17.4.2 Rockfall of Higher Magnitude

Sediment transfer rates for rockfalls larger than 100 m^3 for the whole study area were presented by Vehling (2016; see also Chap. 9). Vehling

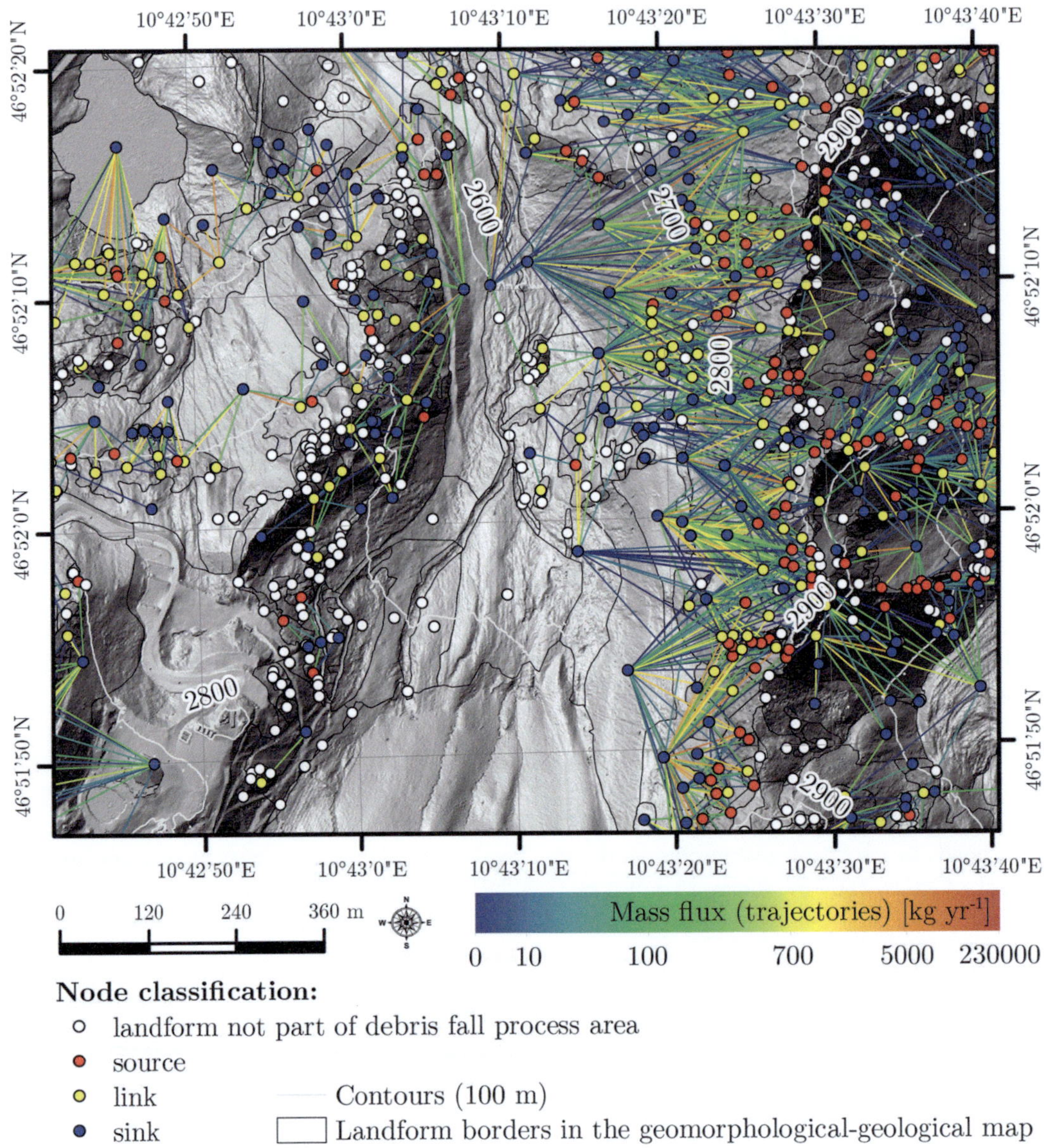

Fig. 17.7 Detail of the debris fall sediment transport network showing the coupling of rockwall sections with individual landforms. Nodes are placed at the centroid coordinates of each polygon and colored according to their classification as either sediment source, link, or sink

mapped and quantified over 100 rockfalls of this magnitude using ALS data from 2006 to 2012 to arrive at a total sediment transfer rate of c. 111,680 t yr^{-1}, while c. 66,000 t yr^{-1} must be attributed to the highly active rock slide/rockfall complex at the "Schwarze Wand" (Vehling 2016) above the glacier tongue. It is conspicuous that also a large proportion of rockfalls over 100 m^3 fall onto the glacier. Of the remaining c. 45,680 t yr^{-1}, only about 1000 t yr^{-1} is probably deposited in the main fluvial system (pers. comm. Lucas Vehling 2016), 7640 t yr^{-1} does not leave the hill slopes, and 37,040 t yr^{-1} falls onto glacier ice at locations other than the "Schwarze Wand." The same author has also identified a Bergsturz event (c. 1,000,000 m^3) in

Table 17.2 Contribution of low-magnitude rockfall to the sediment budget, unraveled to show sediment delivery to different target landforms or process domains present in the study area in scenario G

Landform target class	Percentage study area	Mass (t yr^{-1})	Percentage of total rockfall received	Category	Sum mass for category (t yr^{-1})	Percentage total rock received
Bedrock	20.8	1569.37	6.08	Bedrock	1569.37	6.08
Talus sheets	6.4	4397.87	17.04	Gravitational	9662.48	37.45
Talus cones	3.3	4052.83	15.70			
Debris cones	0.7	1211.78	4.70			
Block slope	4.9	610.88	2.37	Periglacial	1724.25	6.68
Rock glacier, not spec	0.1	87.12	0.33			
Rock glacier, active	1.6	569.46	2.21			
Rock glacier, inactive	0.5	161.53	0.63			
Rock glacier, relictic	0.8	295.26	1.14			
Glacier	31.4	6779.22	26.27	Glacier	6779.22	26.27
Gravitational glacial	0.7	601.07	2.33	Glacial	4669.00	18.10
Moraine, not spec., pre-1855	2.7	258.29	1.01			
Terminal/lateral moraine, pre-1855	5.0	331.03	1.29			
Ground moraine, pre-1855	3.3	218.85	0.85			
Terminal/lateral moraine, post-1855	3.7	1078.48	4.18			
Ground moraine, post-1855	7.9	2181.28	8.45			
Active, competent channel	0.3	396.46	1.54	Fluvial	669.34	2.60
Gravel bars, shallow braided channels and fluvial reworked sediments	0.5	57.13	0.22			
Old fluvial storage	1.4	215.75	0.84			
Anthropogenic	1.3	178.84	0.69	Anthropogenic	178.84	0.69
Lakes, not spec., etc	2.7	550.55	2.13	Other	550.55	2.13
Total	100	25803.05	100	Total	25803.05	100

the Riffle Valley, a sub-basin of the Upper Kaunertal. With an estimated age of c. 11,000 yr and a material density of 2.7 t m^3, this would imply a transport rate of c. 245 t yr^{-1}. Virtually all of the fine particle size content of the deposit has been transported away since the bergsturz event at the end of the Egesen stadium (Vehling 2016). Due to this fact and the immense difference in the duration of the time reference period, it is impossible to use the bergsturz sediment transport in the construction of a recent sediment budget of the study area.

17.4.3 Debris Flows

With ALS-based DEMs covering the whole catchment and TLS surveys conducted at a high temporal resolution at the main debris flow-prone locations, debris flow volumes were measured directly using LiDAR data by mapping depositional bodies on DoD. In order to increase the temporal scale, an area–volume relationship was fitted to 62 debris flow deposits, and this relationship was applied to debris flows whose deposits were mapped on historical aerial photos (for details, see Hilger 2017). In summary, the deposits of 199 debris flows that occurred within the time period 1954–2014 were quantified.

The fact that about 75% of all debris flows detected and quantified for the time period from 2006 on occurred within the proglacial areas of either Gepatsch or Weißsee glaciers underlines the relevance of this process in areas of recent deglaciation (Damm and Felderer 2008; Legg et al. 2014). Depending on the time period chosen as a reference, mass transfer by debris flows in the Upper Kaunertal is between 704 t yr^{-1} (1954–2014, using both measured and estimated debris flow volumes) and 1790 t yr^{-1} (2006–2014, including only those debris flow volumes that were measured using DoD).

The total mass determined is highly dependent on the area–volume relationship used. Ordinary bootstrapping (Canty and Ripley 2016) revealed a standard error of roughly 3600 t (58 t yr^{-1}), or circa 8% of the total sediment transfer estimated for the time period AD 1954–2014.

The measurement of debris flow deposits alone makes it difficult to analyze the proportion of debris flow sediment that has entered the fluvial system. Therefore, we computed budgets (i.e., erosion and deposition volumes) for those debris flows that had reached the fluvial system (as indicated by the spatial pattern of the DoD) in order to obtain the volume of the debris flow that has left the hillslope subsystem. This revealed that, during the years 2006–2014, hillslope-type debris flows transferred a minimum of 455 t of sediment to the fluvial system. The corresponding delivery rate of 62 t yr^{-1} amounts to only 3% of the debris flow volume measured within this period on the catchment scale. On the hillslope scale, however, the proportion of sediment delivered to the channel network can be extremely variable. Haas et al. (2012; see also Chap. 11), for example, report 9 and 59% for two different sections of a lateral moraine in the Kaunertal.

17.4.4 Deep-Seated Gravitational Slope Deformations and Rock Slides

While debris flows are comparatively effective in transferring sediment to the main fluvial system, rock slides and deep-seated gravitational slope deformations (DGSDs) only have an indirect impact on the contribution of hillslope processes to the main fluvial system. Vehling (2016) has identified several of such moving sediment complexes in the Upper Kaunertal, although only some of them show measurable movement rates. Notable DGSDs are located at the Nörderberg mountain and the Kühgrube Valley (Fig. 17.1). While the latter one is rather small (circa 160,000 m^3) and shows very little activity, the former is of interest concerning sediment transport in the study area as shallow soil slips have developed on its surface. These, in turn, provide material to the large Fernergries avalanche(s) that are capable of transferring sediment to the main fluvial system (Sect. 17.4.5). This is an example of how DGSDs are relevant in preparing material for follow-up processes. The currently moving body has a volume of about 300,000 m^3 and moves at velocities of about 5 mm yr^{-1} (Vehling 2016). The sediment transfer by the other rock-slides and DGSDs located in the study area could not be quantified.

17.4.5 Avalanches

It has been repeatedly shown that some avalanches not only transport snow, but also other materials like vegetation and sediment (e.g., Jomelli 1999; Heckmann et al. 2002; Freppaz et al. 2010; Ceaglio et al. 2010; Korup and Rixen 2014). Especially full-depth avalanches and slush avalanches occurring late in winter or early in spring are relevant here (Saemundsson et al. 2008). In the Upper Kaunertal, observations indicate that also surface layer avalanches transport sediment. They do so by flowing through bedrock gullies (Moore et al. 2013) and across hill slope areas already snow-free in early spring (Luckman 1977), or by removing debris that fell onto the snow surface during winter (Jomelli and Bertran 2001). For simplicity, we use the term "sediment transporting avalanche" in this study to refer to all types of avalanches relevant for sediment transport.

Fieldwork for avalanche sediment transport quantification was conducted during the snow melt period (late May and early June of 2013 and 2014) as this is the main time during the year when relevant avalanches are occurring (Gardner 1983; Baggi and Schweizer 2009; Näher 2013). Due to safety concerns, not all parts of the catchment could be mapped for avalanches each spring. As a result, values were extrapolated for the whole catchment based on the mapped area and the mass on this area. Mapping and sampling were undertaken as soon as sediment was found concentrated on top of pure snow on avalanche accumulation bodies, but before signs of significant lateral melting were present.

Mapping was accomplished in a scale of about 1:2500. Following André (2007), Jomelli and Bertran (2001), and Sass et al. (2010), six sediment cover classes were chosen for sediment mapping on avalanche bodies. Except for a self-contained class representing bare snow without any sediment cover (class 0), five equidistant classes representing a sediment cover from >0–20% (class 1) to 80–100% (class 5) were defined for mapping and sampling. Examples of each of the different classes are depicted in Fig. 17.8. As an example, Fig. 17.9 shows a map of two large sediment transporting avalanches recurring every year on the western moraine above the Ferngries.

Sampling plots for the different coverage classes were 0.25 m^2 most of the time (cf. areas delineated by metering rules in Fig. 17.8). Sampling plots were placed randomly within the sediment covered part of an avalanche deposit by blindfolded throwing the metering rule from uphill onto the surface (c.f. Korup and Rixen 2014). As it was not possible to sample the whole snow column below any sampling plot, it cannot be ruled out that some sediment was lost to quantification this way. Investigations by Bell et al. (1990), Heckmann et al. (2002), and Heckmann (2006) have shown that, on average, about 6.6% (s = 6.2%, n = 16) of the mass found on top of such sampling plots is found within the snow body below the surface.

The thickness of the sediment layer varied between 0 and 70 mm (Näher 2013). This is an additional potential source of error, as the percentage of surface cover mapped might be very high and the sediment thickness is comparatively low. It could be observed multiple times that surfaces appearing very dark from afar had only a very thin cover of small-size particles (Näher 2013). All sediment found within a sampling area was collected and taken to the laboratory for drying and weighting. Boulders and cobbles (>64 mm) were weighted in the field using a digital scale to reduce transport effort from sampling sites difficult to access. In total, 20 sampling sites were evaluated in 2013 and 40 sampling sites in 2014. The distribution of the sampling plots over the cover classes and the average sediment mass for each class is shown in Table 17.3.

The very broad confidence intervals reflect the issue of both different sediment thicknesses and the limited number of sampling plots available. The lower limits of zero for classes three and four are certainly not realistic as these classes appear rather dark and definitely contain quite some sediment. However, further samples are necessary to narrow down the interval.

Most avalanches do not transport sediment to localities where material can be re-mobilized by other processes (cf. below for one exception). The total mass moved by sediment transporting avalanches in the Upper Kaunertal is substantial. Taking the results from the winters of 2012/2013 (Näher 2013) and 2013/2014 (Rumohr 2015), an average of 12.6 t km^{-2} was moved by avalanches. This corresponds to a total mass of

Fig. 17.8 Examples of sampling plots representing the different sediment cover classes

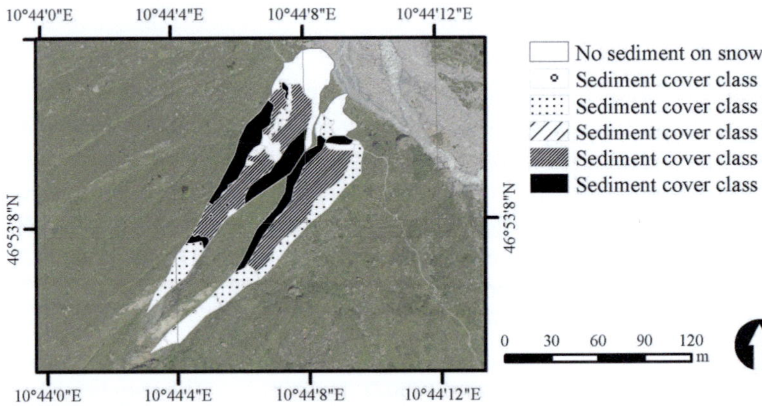

Fig. 17.9 Sediment cover classes mapped on the two large Fernergries avalanches in spring 2014. Figure taken from Hilger (2017)

Table 17.3 Average weight contained on 1 m^2 of each sediment cover class in winter 2013/2014

Sediment cover class	Average weight (kg m^{-2})	Number of plots	Confidence interval (kg m^{-2})[a]
One (0–20%)	0.54	11	[0.11,0.97]
Two (20–40%)	1.09	12	[0.56,1.6]
Three (40–60%)	7.06	5	[0,15.8]
Four (60–80%)	5.5	5	[0,12.1]
Five (80–100%)	18.8	6	[6.33,31.2]

[a]95% confidence level

sediment transported by avalanches in one year of c. 781 t. Only a very small portion of this amount entered the fluvial system as most avalanches act on the hill slopes of tributary valleys with broad valley bottoms (such as the Riffle Valley or Krummgampen Valley). Only

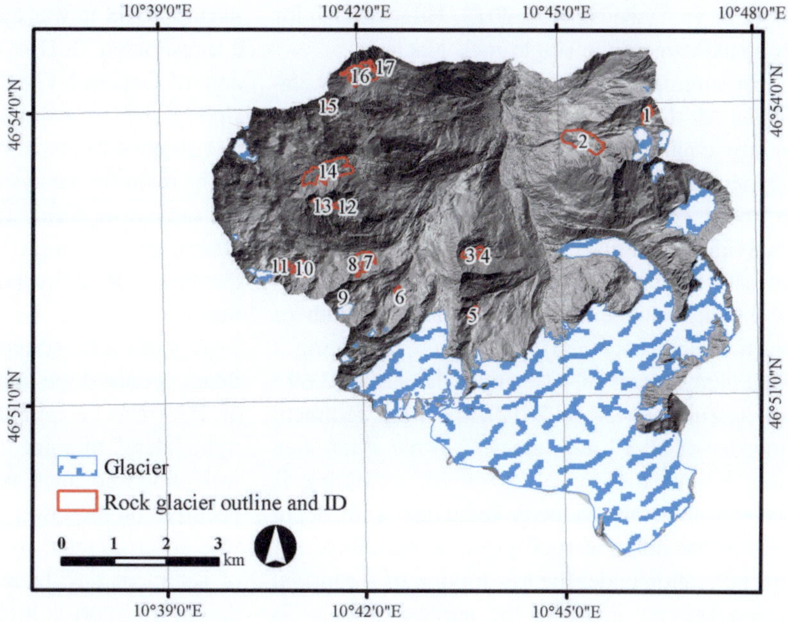

Fig. 17.10 Location of the 17 active rock glaciers in the Upper Kaunertal

one avalanche reached the fluvial system directly; this avalanche, however, has moved quite a significant amount of sediment from the 1855 lateral moraine (and parts of the hillslope above) into the fluvial system: Visual inspection of the outline and sediment cover classes over the gap created by the Faggenbach through the avalanche deposit yields a mass of 13.46 t delivered by the avalanche to the fluvial system in the winter of 2012/2013. Only the results from two winters were used in the calculation of the result presented above. Considering the very high temporal variability of avalanches in general and in the study area (Luckman 1977; Näher 2013), this is a very short time to assess the contribution of avalanches reliably and adds to the uncertainty of the estimate together with sampling strategy weaknesses and inaccessibility of some parts of the study area (see above). Based on an interview of a local avalanche expert, Näher (2013) reports that the winter of 2012/2013 was a "normal" avalanching winter, i.e., "normal" in comparison to historic winters. This interannual variability, however, has a severe impact on the number of avalanches in the Upper Kaunertal and the two winters before 2012/2013 were seen as under average in terms of avalanche activity (Näher 2013).

17.4.6 Creeping Permafrost by Rock Glaciers

Before measurements began, it was already clear that sediment transport by rock glaciers directly to the fluvial system is very low in the Upper Kaunertal. Sediment transfer by rock glaciers is therefore mainly important for internal sediment turnover.

For an estimation of sediment transfer by rock glaciers, both sediment mass and velocity had to be determined. Sediment mass was estimated based on rock glacier extent mapping, thickness estimation via frontal height measurement and sediment percentage estimates (Neugirg 2013). Mean annual displacement rates for the time from 2006 to 2012 were estimated using feature tracking on DTM-derived hill shade grids for each rock glacier (cf. Scambos et al. 1992; Kaufmann and Ladstädter 2003; Dusik et al. 2015; Chap. 7). Seventeen rock glaciers were identified for sediment transfer analysis in the Upper Kaunertal. Figure 17.10 gives an overview of their locations.

The estimation of rock glacier mass yielded values between 0.25 and 22.63 Mt, assuming a sediment content of 40% (Gärtner-Roer 2012). Displacement rates ranged from 0.09 to

0.44 m yr^{-1} (see Neugirg 2013; Hilger 2017) for a breakdown of the single rock glaciers.

In summary, a total mass of 52.92 Mt (0.25–22.63 Mt for individual rock glaciers) is currently contained in rock glacier bodies in the Upper Kaunertal. These move, on average, with a speed of 0.17 m yr^{-1} (0.09–0.44 m yr^{-1} for individual rock glaciers) at the surface. Using the measured annual average horizontal displacement rates at the surface, height and width of each rock glacier body and, again, assuming a sediment content of 40% and a density of 2.69 t m^3 (Hausmann et al. 2012), the total sediment transfer of all 17 rock glaciers in the study area was estimated at c. 21,497 t yr^{-1}. Nothing is known about the velocity reduction with depth within the different rock glaciers; therefore, our quantification under the assumption of a constant creep velocity equal to the surface velocity is very likely an overestimation.

All rock glaciers except one are located too far from the active fluvial system as to directly deliver sediment. In the Krummgampen Valley, a rock glacier is adjacent to the channel, and historical imagery from 1997 shows sediment from the rock glacier having advanced into the channel, with the river flowing through and under the blocks of the rock glacier. It is impossible, however, to estimate the amount of sediment mobilized. Due to the small channel gradient at this location, it probably is very low, and the total amount is probably negligible in comparison to other processes.

17.4.7 Slope Wash and Linear Erosion on Hillslopes

Slope erosion by surface runoff on hill slopes occurs as sheetwash, linear (rill or gully) erosion, or piping (in the subsurface). In comparison to other processes, they have the ability to connect sediment sources relatively far above the main channels to the main channel network.

The net sediment balance of these processes was measured using DEMs (July 4th, 2012; July 18th, 2014; resolution 0.7 m) on the steep lateral moraines. A spatially distributed LoD with a mean of 0.08 m was applied. Figure 17.11 shows a thresholded DoD of the true right lateral moraine of Gepatsch Glacier. The net balance of the slope is 1837 t (using a density of 2.2 t m^{-3}). As just above 400 t have been moved from the slope to the main fluvial system via debris flows during the same time period (c.f. Sect. 17.4.3), it can be concluded that circa 1400 t were transferred to the Fagge river by processes other than debris flows.

A ratio of sediment transported by debris flows versus slope-aquatic and fluvial processes of 1:3.7 can be estimated for the AD 1855 true right lateral moraine of Gepatsch Glacier. It is well possible that some deposition that has occurred on the lower parts of the lateral moraine was not accounted for, and that the true amount of sediment having reached the Fagge river by fluvial transport is higher, because accumulation bodies by fluvial transport are typically spread out and of low height and therefore difficult to detect using DoD. Estimating the contribution of fluvial hill slope erosion to the main fluvial system at only one location in the study area is probably underestimating the total value. It is possible that the true right 1855 lateral moraine of the Weißsee Glacier, for example, is coupled to the upper reaches of the Riffle River during high intensity rainstorm events. As this river passes significant depositional areas downstream, however, it is improbable that much of this sediment reaches the Fagge river. In addition, an unvegetated moraine slope at the true left side of the Upper Fagge river valley (just below the Gepatsch Glacier snout) is close enough to the main fluvial system to contribute sediment. The slope, however, is very small. Nevertheless, the value reported above should therefore be regarded as a minimum value.

Sediment transport also occurs in small channels on vegetated hill slopes. Transport rates measured in sediment traps, however, were relatively low (mean: 5.82 t km^{-2} yr^{-1}) in relation to the sediment transport on the lateral moraines bare of vegetation (5830 t km^{-2} yr^{-1}). These results support the results by Becht (1995), who showed that fluvial transport on non-vegetated moraine material is about 1000 times higher than

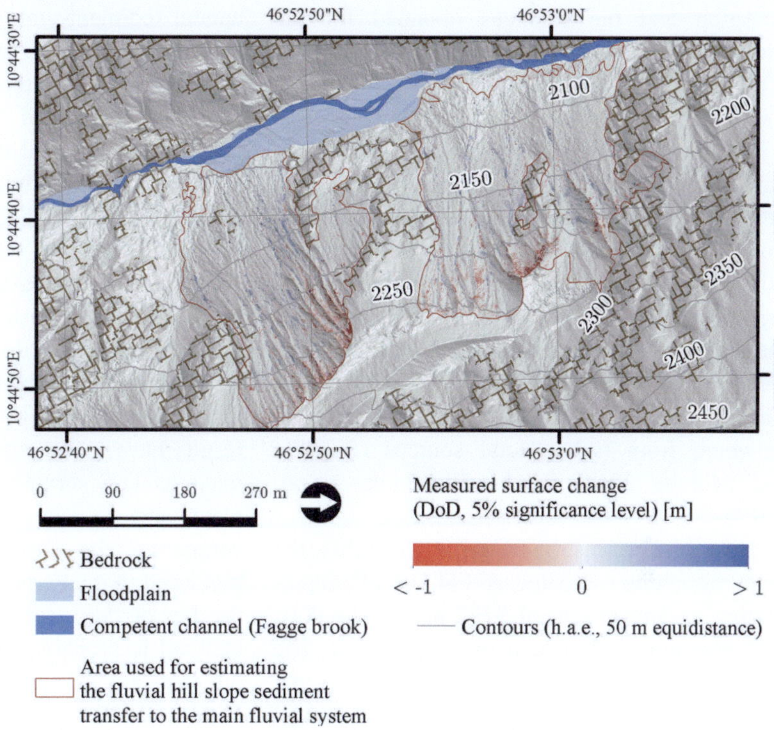

Fig. 17.11 Erosion and accumulation on the true right lateral moraine of Gepatsch Glacier. Figure taken from Hilger (2017)

on other surfaces. The same author reported that fluvial erosion is about 90% lower in vegetated moraines than on unvegetated ones in the Pitz Valley.

It is for the above-mentioned reasons that the value of circa 700 t yr^{-1} obtained via DEM differencing on an area of 0.12 km^2 on the AD 1855 Gepatsch Glacier lateral moraine can be regarded as representative for the sediment delivery to the main fluvial system in the study area.

17.4.8 Sediment Transport in the Channel Network

For the construction of the sediment budget, fluvial sediment transfer measurements at two locations are important: (i) at the glacier snouts and (ii) at the basin outlet. More information on the methodology of these measurements can be found in Chap. 13. Measurements of both bed and suspended load were undertaken at the Gepatsch Glacier snout during the ablation periods of 2012, 2013, and 2014 on an average of c. 96 days per year. The obtained measurements were used to calculate an average daily transport rate of c. 52 t day^{-1} for bed load and c. 173 t day^{-1} for suspended load. Taking the months May to October (184 days) as the period of significant fluvial transport through the glacier snout, a total rate of glacifluvial sediment of 41,401 t yr^{-1} was estimated.

Long-term sediment yield (=export from the catchment) was computed by morphological budgeting using pre-reservoir and recent DEMs of the delta building up in the Gepatsch reservoir at the catchment outlet. The pre-reservoir DEM was reconstructed from a photogrammetric analysis of aerial photos taken in 1953. Terrestrial LiDAR surveys were possible as the lake level is very low every May, exposing a delta at the mouth of the Fagge river into the reservoir (see scan positions north of the study area in Fig. 17.2). In addition, data from a TLS survey conducted in November 2015 by TiWAG (Tiroler Wasserkraft AG) were available for balancing. The obtained overall levels of detection and the balances obtained for the investigated time periods 1954–2015, 2012–2013, and 2013–2015 can be found in Table 17.4.

On August 25/26, 2012, heavy rain presumably triggered the rupture of a water pocket in or below the glacier tongue. The ensuing flood (max. discharge 47.3 m^3 s^{-1}) deposited c. 90,000 m^3 of sediment in the lower proglacial area (e.g., the Fernergries braidplain), while our measurements indicated that only c. 70,000 m^3 were removed from the upper proglacial area. The difference, about 20,000 m^3, must have been delivered from (sub-)glacial sources (Baewert and Morche 2014; Stocker-Waldhuber et al. 2017).

Using the pre-reservoir and 2015 DEMs, the delta volume is estimated at 194,701 m^3, representing an annual rate of 3192 m^3 yr^{-1} (6384 t yr^{-1} assuming a bulk density of 2 t m^{-3}). The annual rate during 2012–2013 is in the same order of magnitude as the long-term average. During three ablation seasons (2013–2015), however, 73,254 m^3 (on average 24,418 m^3 yr^{-1} or 48,836 t yr^{-1}) of sediment were deposited in the delta. Two observations follow from these numbers: First, the sediment volume transferred to the delta since the 2012 flood corresponds to almost 80% of the material deposited on the floodplain during the event; together with the fact that the annual sediment output from the glacier 2012–2014 (41,401 t yr^{-1}) is similar to the annual yield of 48,836 t yr^{-1} at the catchment outlet (years 2013–2015), this highlights the fact that the longitudinal sediment connectivity in the Fagge river system has been comparatively high after the flood. Second, almost 40% of the delta volume can be attributed to the (reworked) sediment mobilized by the 2012 event, emphasizing the exceptional character of this event, and supporting Warburton's (1990) conclusion that large magnitude, low frequency flood events are the dominant control of sediment transfer in the proglacial zone.

Tschada and Hofer (1990) applied specific sediment transport rates measured at the Taschach and Pitz rivers between the years 1964 and 1989 to the Fagge river, estimating the total annual sediment yield to the delta at 58,340 m^3 yr^{-1} (45,930 m^3 yr^{-1} suspended load, 12,410 m^3 yr^{-1} bedload). As these authors used the total catchment area of the Gepatsch reservoir (107 km^2) instead of that at the inflow (62 km^2) for their estimation, we corrected their results by a factor of 62/107 and get a rate of 33,804 m^3 yr^{-1} (26,613 m^3 yr^{-1} suspended load, 7192 m^3 yr^{-1} bedload). Using a density of 1.8 t m^{-3}, the suspended load is estimated at c. 47,900 t yr^{-1}, which is more than double the value of 22,267 t yr^{-1} that we determined for suspended load in the period 2012–2014. On the one hand, our results have to be regarded as a minimum estimate, as TLS surveys could not cover the complete delta. As we estimated bedload transport by comparing the suspended sediment transport measured in the Fernergries to the TLS survey in the fan of the reservoir, the true total mass of bedload is probably higher. On the other hand, our data are measurements, while Tschada and Hofer (1990) report an estimation based on transferring measured data from neighboring catchments to the Fagge river. Furthermore, while delta deposition from tributaries of the reservoir located further north can be considered negligible, it is possible that an unknown quantity of sediment from the Pitz valley water diversion (bedload, however, is fully retained by a Tyrol weir at the water intake) is included in our balance, as water from the Pitz diversion enters the reservoir just north of the area that could be surveyed by TLS.

17.5 Results and Discussion: The Sediment Budget of the Upper Kaunertal

For the establishment and closure of the Upper Kaunertal sediment budget, only data on glacial processes and mobilization of sediment by the main channels are missing.

The sediment balance of the glaciers in the study area is not known as it is impossible to measure the erosional work at the glacier basis. The sediment mass flux and its changes through the glacier tongue of Gepatsch glacier were

Table 17.4 Results of the sediment yield quantification efforts using point cloud data of the Fagge river delta in the Gepatsch reservoir

Time period	LoD (1) (m)	Volume (m^3 yr^{-1})	Mass (t yr^{-1})
1954–2015	0.44	c. 3191	c. 6383
2012–2013	0.12	c. 2404	c. 4808
2013–2015	0.12	c. 24,418	c. 48,836

The period 2013–2015 included the summers of 2013, 2014, and 2015 and was therefore regarded as representing three ablation seasons
Table taken from Hilger (2017)

measured using a variety of methods. Using data from Stocker-Waldhuber et al. (Chap. 5), the sediment production of Gepatsch Glacier is estimated at 1700–170,000 t yr^{-1}, which is consistent with the measured glacifluvial sediment transfer of 41,401 t yr^{-1}. No information is available on the masses deposited as moraines in the forefield to this date. Also, no data are currently available for glaciers other than the Gepatsch glacier, which is why the sediment budget of the Upper Kaunertal remains incomplete concerning the sediment transfer from glaciers. Bringing together the results for the different processes treated in the previous sections, the Upper Kaunertal sediment budget is presented in Fig. 17.12.

The main fluvial system is represented as a thick arrow. The contributions of other processes are indicated by black lines of varying thicknesses indicating their relative importance. Processes are labeled in small caps, with time periods the values refer to added in brackets. Values next to a circular arrow represent internal sediment transfer, that is the amount of sediment redistribution on storage landforms without delivery to the main fluvial system. It is typical of sediment budget studies that their results are associated with high uncertainties, especially in catchments displaying a high relief and topographic heterogeneity (Sanders et al. 2013). This holds true even more when the studied catchment is comparatively large and processes are studied only for a few years (as in this study). This is why the sediment budget requires discussion.

The first fact that is conspicuous from looking at the sediment budget is that hill slope processes contribute only a fraction of the total mass of sediment exported from the catchment. As already mentioned, the extreme event in August 2012 left its mark in the high amount of sediment eroded from the river banks. An estimation via ALS DEM differencing yielded a mass of circa 8250 t yr^{-1} having been mobilized between the summers of 2012 and 2014 in this way. It has been observed that much of this sediment has been deposited just downstream of the locations it has been mobilized from. However, it is unknown which portion of the mobilized mass has been transported through to the outlet in relation to material of other origins. The remaining discrepancy of the sum of processes versus the total export is indicative of a high fluvial sediment delivery from below the glacier.

One source of error independent from the uncertainties of single process estimates is the anthropogenic modification having taken place during the measurement period in the Upper Kaunertal. Sediment has been transported by human action on a regular basis at various locations: Sediment has been excavated from the floodplain just below the Fernergries bridge, from the delta apex at the reservoir and from the proglacial area of the Weißsee Glacier. It is unknown where the sediment has been transported to in all cases.

Rockfalls of different magnitudes are dominating the sediment transfer in the study area, followed by rock glacier transport, fluvial transport in the main channels and debris flows. The importance of glacial processes is very high, although a precise value cannot be given. Avalanches transport a substantial amount of sediment but cannot compete with other processes, while shallow soil slips seem to have ceased activity in recent years. Overall, sediment transfer in the main fluvial system is of greater

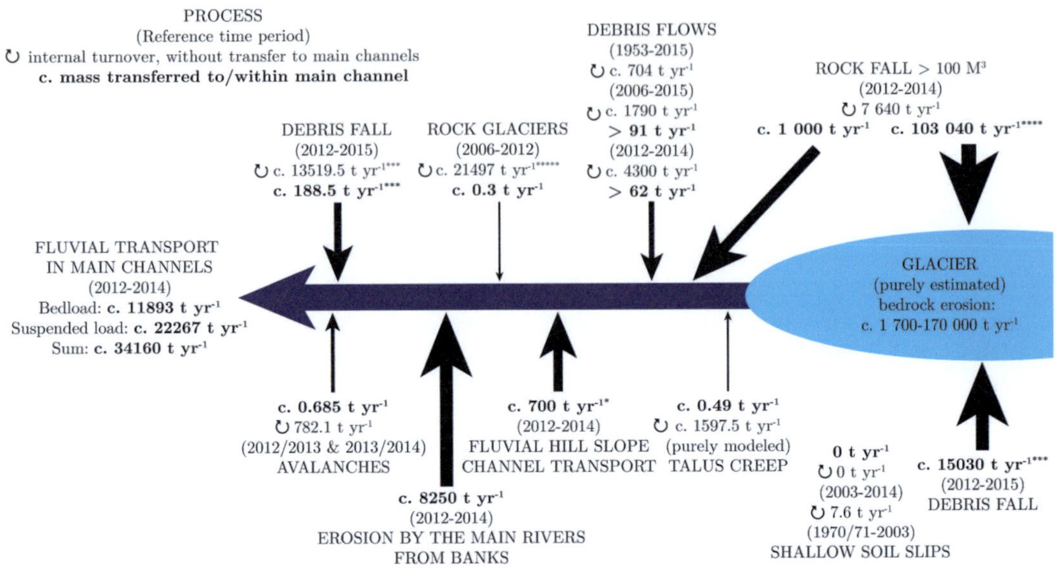

Fig. 17.12 Representation of the Upper Kaunertal sediment budget. Figure adopted from Hilger (2017)

importance to the total amount of sediment exported from the catchment than all hill slope processes combined, which is mainly the result of transport of sediment flushed from below the snout and mobilization from the banks (especially during the extreme event).

Because of their long recurrence intervals, bergsturz events are of minor importance among rockfall processes in the study area. This result is different to what both Krautblatter et al. (2012) and Caine (1986, cited in Vehling 2016) reported from their study areas. Krautblatter et al. (2012) stated that bergsturz events, followed by rockfall <10 m^3, are of greatest importance in the Rein Valley, Northern Calcareous Alps. The importance of bergsturz events in the Rein Valley, however, is caused by large distances between joints in otherwise compact rock making up the high and very steep rockwalls. For the Upper Kaunertal, Vehling (2016) concludes that although extreme magnitudes are less effective than smaller ones, it is "extreme locations" that are very important: The "Schwarze Wand" rockslide complex is a source of small-magnitude rockfall releasing about two times the total mass of what is mobilized at all other rockwalls combined. A dominance of rockfall of small magnitudes is also reported by Sanders et al. (2013). As Vehling (2016) correctly states, however, this could be the result of the short measurement/reference period their results refer to.

Studies that have compared the relative importance of different geomorphic processes in high-mountain areas are rare. Warburton (1990) determined the relative importance of different processes in the proglacial area of Bas Glacier d'Arolla. As his measurements were conducted only for a short time during the 1987 ablation season, they are not directly comparable to the result obtained in this study, based on a monitoring program spanning several years (no data on avalanche sediment transport are available from Warburton 1990, for example). In general, however, the results seem to complement one another: Warburton has found transport in the channels dominating the sediment transfer in his study area, followed by hillslope processes (including rockfall) and slope wash being of minor importance. Beylich (2000) gives a ranking of many geomorphic processes following their relative importance in terms of total mass transfer (i.e., independent to the respective contribution to the main fluvial system) in Austdalur, East

Iceland, ranking aquatic slope denudation first, followed by geochemical denudation, avalanches, debris and boulder falls, talus creep, debris slides, and flows and deflation. This is very different to what has been found in the Upper Kaunertal. Reasons include, among others, the relatively low relief in Beylich's study area (which relates to the occurrence of comparatively few rockwalls), the absence of a glaciers (and corresponding land forms), and rock glaciers. In addition, a much larger percentage of the Austdalur study area is vegetated.

In general, hillslope-channel coupling in the study area is poor; this implies that only a small amount of material is transported directly from sediment storages on the slope to the main channels. A look at the sediment budget (which also gives internal sediment transfer/mobilization rates) reveals that while about 34,160 t yr^{-1} have left the catchment between 2012 and 2014, about 50,000 t yr^{-1} were mobilized but did not reach either the glacier or the main fluvial system. As sediment export 2012–2014 is much higher than the long-term average, and as an additional 120,000 t yr^{-1} of rockfall were deposited on glacier ice, often far from the main fluvial system, it becomes evident how disconnected the sediment transport system in the Upper Kauner Valley really is unless an extreme event occurs. Comparing the total rate of sediment mobilization (almost 180,000 t yr^{-1}) and an assumed export rate of about 7000 t yr^{-1} (educated guess based on the long-term average, excluding the 2012 extreme event), it is clear that only about 4% of all sediment moved in the study area reaches the outlet. This is not untypical for glacial valleys, as their broad valley bottoms often lead to a decoupling of the main fluvial and hill slope subsystems. Similar results are reported by Beylich et al. (2011) from Erdalen, Norway or Austdalur, Iceland. Not many studies have tried to establish a sediment budget for meso-scale catchments or have even quantified more than one sediment transferring process in such a catchment. In addition to the ones discussed in Chap. 8.1, Vorndran (1979) (cited in Becht 1995) has established a mass balance for the Sextner river catchment in Southern Tyrol but has not differentiated between different processes. Becht (1995) quantified different geomorphic processes in the Eastern Alps.

The authors believe that the presented results are a good step forward in quantitatively investigating sediment transport of all relevant processes and can serve as a reference for future research. Maybe, the Kaunertal can become a type locality for sediment budget studies of high-mountain meso-scale catchments.

Acknowledgements The measurement of different processes in a high-mountain area over a time period of several years is a demanding task. In addition to the authors, several student assistances provided very valuable help and support in the field, performed analysis in the laboratory, helped in mapping efforts, and even produced intermediate results. Especially mentionable is the field and laboratory work as well as mass calculations for avalanches by Martin Näher and Phillip Rumohr. Important results on rock glacier movement have been provided by Philip Neugirg, while Florian Riehl worked with data from the sediment traps in hillslope channels. Other student assistants the authors would like to thank are Sarah Betz, Stefan Löser, Sebastian Wiggenhauser, Hendrik Hövel, Kerstin Schlobies, Arnt Luthart, Anne Schuchardt, Matthias Faust, Martin Weber, Eric Rascher and Karolin Dubberke, Alexander Bryk and Shannon Hibbard.

References

André MF (2007) Geomorphic impact of spring avalanches in Northwest Spitsbergen (79° N). Permafrost Periglac Process 1:97–110. https://doi.org/10.1002/ppp.3430010203

Baewert H, Morche D (2014) Coarse sediment dynamics in a proglacial fluvial system (Fagge River, Tyrol). Geomorphology 218:88–97. https://doi.org/10.1016/j.geomorph.2013.10.021

Baggi S, Schweizer J (2009) Characteristics of wet-snow avalanche activity: 20 years of observations from a high alpine valley (Dischma, Switzerland). Nat Hazards 50:97–108. https://doi.org/10.1007/s11069-008-9322-7

Becht M (1995) Untersuchungen zur aktuellen Reliefentwicklung in alpinen Einzugsgebieten. Münchener Geographische Abhandlungen A47. GEOBUCH, München

Bell I, Gardner J, de Scally F (1990) An estimate of snow avalanche debris transport, Kaghan Valley, Himalaya, Pakistan. Arct Alp Res 22:317. https://doi.org/10.2307/1551594

Berger J, Krainer K, Mostler W (2004) Dynamics of an active rock glacier (Ötztal Alps, Austria). Quatern Res

62:233–242. https://doi.org/10.1016/j.yqres.2004.07.002
Beylich AA (2000) Geomorphology, sediment budget, and relief development in Austdalur, Austfirðir, East Iceland. Arct Antarct Alp Res 32:466–477
Beylich A, Lamoureux S, Decaulne A (2011) Developing frameworks for studies on sedimentary fluxes and budgets in changing cold environments. Quaestiones Geographicae. https://doi.org/10.2478/v10117-011-0001-5
Caine T (1986) Sediment movement and storage on alpine slopes in the Colorado Rocky Mountains. Allen and Unwin, Boston
Caine SF (1989) Geomorphic coupling of hillslope and channel systems in two small mountain basins. Zeitschrift fur Geomorphologie 33:189–203
Canty A, Ripley BD (2016) R package boot
Ceaglio E, Freppaz M, Maggioni M et al (2010) Full-depth avalanches and soil erosion: an experimental site in NW Italy, p 15565
Damm B, Felderer A (2008) Identifikation und Abschätzung von Murprozessen als Folge von Gletscherrückgang und Permafrostdegradation im Naturpark Rieserferner-Ahrn (Südtirol); (Identification and assessment of debris flows as a consequence of glacier retreat and permafrost degradation in the area of Rieserferner-Ahrn, South Tyrol (in German). Abhandlungen der Geologischen Bundesanstalt, 29–32
Dusik JM (2013) Vergleichende Untersuchungen zur rezenten Dynamik von Blockgletschern im Kaunertal dargestellt an Beispielen aus dem Riffeltal und der Inneren Ölgrube: M.sc. thesis Cath. University Eichstaett-Ingolstadt
Dusik JM, Leopold M, Heckmann T et al (2015) Influence of glacier advance on the development of the multipart Riffeltal rock glacier, Central Austrian Alps. Earth Surf Proc Land 40:965–980. https://doi.org/10.1002/esp.3695
Freppaz M, Godone D, Filippa G, Maggioni M, Lunardi S, Williams MW, Zanini E (2010) Soil erosion caused by snow avalanches: a case study in the Aosta Valley (NW Italy). Arct Antarct Alp Res 42:412–421. https://doi.org/10.1657/1938-4246-42.4.412
Gardner JS (1983) Observations on erosion by wet snow avalanches, Mount Rae Area, Alberta, Canada. Arct Alp Res 15:271. https://doi.org/10.2307/1550929
Gärtner-Roer I (2012) Sediment transfer rates of two active rockglaciers in the Swiss Alps. Geomorphology 167–168:45–50. https://doi.org/10.1016/j.geomorph.2012.04.013
Giese P (1963) Some results of seismic refraction work at the Gepatsch glacier in the Oetztal Alps. IAHS Publication 61:154–161
Glira P, Briese C, Pfeifer N, Dusik JM, Hilger L, Neugirg F, Baewert H (2014) Accuracy analysis of height difference models derived from terrestrial laser scanning point clouds. European Geosciences Union. Geophys Res Abstracts
Haas F (2008) Fluviale Hangprozesse in Alpinen Einzugsgebieten der Nördlichen Kalkalpen; Quantifizierung und Modellierungsansätze. (=Eichstaetter Geographische Arbeiten 17) Profil-Verl., München/Wien
Haas F, Heckmann T, Becht M, Cyffka B (2011) Ground-based laserscanning—a new method for measuring fluvial erosion on steep slopes. In: Hafeez MM (ed) GRACE, remote sensing and ground-based methods in multi-scale hydrology. Proceedings of the symposium JHS01 held during the IUGG GA in Melbourne (28 June–7 Jullet 2011). IAHS Publication, Wallingford, S 163–168
Haas F, Heckmann T, Hilger L, Becht M (2012) Quantification and modelling of debris flows in the proglacial area of the Gepatschferner/Austria using ground-based LIDAR. In: Collins AL, Golosov V, Horowitz AJ, Lu X, Stone M, Walling DE, Zhang X (eds) Erosion and sediment yields in the changing environment: proceedings of an IAHS international commission on continental erosion symposium, held at the institute of mountain hazards and environment, CAS-Chengdu, China, 11–15 Oct 2012. Wallingford, pp 293–302
Hausmann H, Krainer K, Brückl E, Mostler W (2007) Creep of two alpine rock glaciers: observation and modelling (Ötztal- and Stubai Alps, Austria). Grazer Schriften der Geographie und Raumforschung 43:145–150
Hausmann H, Krainer K, Brückl E, Ullrich C (2012) Internal structure, ice content and dynamics of Ölgrube and Kaiserberg rock glaciers (Ötztal Alps, Austria) determined from geophysical surveys. Austrian J Earth Sci 105:12–31
Heckmann T (2006) Untersuchungen zum Sedimenttransport durch Grundlawinen in zwei Einzugsgebieten der Nördlichen Kalkalpen: Quantifizierung, Analyse und Ansätze zur Modellierung der geomorphologischen Aktivität. (=Eichstaetter Geographische Arbeiten 14) Profil-Verl. München/Wien
Heckmann T, Schwanghart W (2013) Geomorphic coupling and sediment connectivity in an alpine catchment—exploring sediment cascades using graph theory. Geomorphology 182:89–103. https://doi.org/10.1016/j.geomorph.2012.10.033
Heckmann T, Wichmann V, Becht M (2002) Quantifying sediment transport by avalanches in the bavarian alps—first results. Z Geomorph N.F Suppl 127:137–152
Heckmann T, Hilger L, Vehling L, Becht M (2016) Integrating field measurements, a geomorphological map and stochastic modelling to estimate the spatially distributed rockfall sediment budget of the Upper Kaunertal, Austrian Central Alps. Geomorphology 260:16–31. https://doi.org/10.1016/j.geomorph.2015.07.003
Hilger L (2017) Quantification and regionalization of geomorphic processes using spatial models and high-resolution topographic data: a sediment budget of the Upper Kauner Valley, Ötztal Alps (Doctoral

Dissertation Cath. University of Eichstaett-Ingolstadt). https://opus4.kobv.de/opus4-ku-eichstaett/files/381/fertig_pdf_a-1b.pdf
Jomelli V (1999) Les effets de la fonte sur la sédimentation de dépôts d'avalanche de neige chargée dans le massif des Ecrins (Alpes françaises)/The effects of the snow melt on the sedimentation of dirty snow avalanche deposits in the Ecrins Massif (French Alps). Géomorphologie: relief, processus, environnement 5:39–57. https://doi.org/10.3406/morfo.1999.974
Jomelli V, Bertran P (2001) Wet snow avalanche deposits in the French alps: structure and sedimentology. Geogr Ann Ser A Phys Geogr 83:15–28. https://doi.org/10.1111/j.0435-3676.2001.00141.x
Kaufmann V, Ladstädter R (2003) Quantitative analysis of rock glacier creep by means of digital photogrammetry using multi-temporal aerial photographs: two case studies in the Austrian Alps. In: Proceedings of the 8th international conference on permafrost, pp 21–25
Kerschner H (1979) Spätglaziale Getscherstände im inneren Kaunertal (Ötzta-ler Alpen). In: Keller W (ed) Studien zur Landeskunde Tirols und angrenzender Gebiete. Vol. 6. Leidlmair-Festschrift; 2 of Innsbrucker geographische Studien. Inst. für Geographie der Univ. Innsbruck, Innsbruck, pp 235–248
Kneisel C, Lehmkuhl F, Winkler S, Tressel E, Schröder H (1998) Legende für geomorphologische Kartierungen in Hochgebirgen (GMK Hochgebirge). Trierer Geographische Studien 18. Trier
Korup O, Rixen C (2014) Soil erosion and organic carbon export by wet snow avalanches. Cryosphere 8:651–658. https://doi.org/10.5194/tc-8-651-2014
Krainer K, Mostler W, Spötl C (2007) Discharge from active rock glaciers, Austrian Alps: a stable isotope approach. Austrian J Earth Sci 100:102–112
Krautblatter M, Moser M, Schrott L, Wolf J, Morche D (2012) Significance of rockfall magnitude and carbonate dissolution for rock slope erosion and geomorphic work on Alpine limestone cliffs (Reintal, German Alps). Geomorphology 167:21–34
Lane SN, Westaway RM, Murray Hicks D (2003) Estimation of erosion and deposition volumes in a large, gravel-bed, braided river using synoptic remote sensing. Earth Surf Proc Land 28:249–271
Legg NT, Meigs AJ, Grant GE, Kennard P (2014) Debris flow initiation in proglacial gullies on Mount Rainier, Washington. Geomorphology 226:249–260. https://doi.org/10.1016/j.geomorph.2014.08.003
Loye A, Jaboyedoff M, Pedrazzini A (2009) Identification of potential rockfall source areas at a regional scale using a DEM-based geomorphometric analysis. Nat Hazards Earth Syst Sci 9:1643–1653. https://doi.org/10.5194/nhess-9-1643-2009
Luckman BH (1977) The geomorphic activity of snow avalanches. Geogr Ann Ser A, Phys Geogr 59:31. https://doi.org/10.2307/520580
Marquínez J, Duarte RM, Farias P, Sánchez MJ (2003) Predictive GIS-based model of rockfall activity in mountain cliffs. Nat Hazards 30:341–360. https://doi.org/10.1023/B:NHAZ.0000007170.21649.e1
Moore JR, Egloff J, Nagelisen J, Hunziker M, Aerne U, Christen M (2013) Sediment transport and bedrock erosion by wet snow avalanches in the Guggigraben, Matter Valley, Switzerland. Arct Antarct Alp Res 45:350–362. https://doi.org/10.1657/1938-4246-45.3.350
Morche D, Baewert H, Bryk A (2014) Bed load transport in a proglacial river (Fagge, Gepatschferner, Tyrol). EGU. Geophys Res Abstracts
Näher M (2013) Analyse des Sedimenttransports durch Grundlawinen im Kaunertal zur Quantifizierung des Sedimentbudgets mittels Verfahren aus der terrestrischen Photogrammetrie. M.Sc. thesis Cath. University of Eichstaett-Ingolstadt
Neugirg P (2013) Beurteilung der Dynamik und Quantifizierung des Sediment-haushalts von ausgewählten Blockgletschern im Gletschervorfeld des Gepatsch-ferners (hinteres Kaunertal) auf Grundlage von multitemporalen LiDAR-Daten und Luftbildern: B.Sc. thesis Cath. University Eichstaett-Ingolstadt
Otto JC (2006) Paraglacial sediment storage quantification in the Turtmann Valley, Swiss Alps. Doctoral Dissertation, University of Bonn
Rumohr P (2015) Quantifizierung des Sedimenttransports durch Grundlawi-nen im oberen Kaunertal mittels gravimetrischer und terrestrischphotogrammetri-scher Verfahren: M.Sc. thesis Cath. University of Eichstaett-Ingolstadt
Saemundsson Þ, Decaulne A, Jónsson HP (2008) Sediment transport associated with snow avalanche activity and its implication for natural hazard management in Iceland. In: International symposium on mitigative measures against snow avalanches, pp 137–142
Sanders JW, Cuffey KM, MacGregor KR, Collins BD (2013) The sediment budget of an alpine cirque. Geol Soc Am Bull 125:229–248. https://doi.org/10.1130/B30688.1
Sass O (2005) Temporal variability of rockfall in the Bavarian Alps, Germany. Arct Antarct Alp Res 37:564–573
Sass O, Heel M, Hoinkis R, Wetzel K-F (2010) A six-year record of debris transport by avalanches on a wildfire slope (Arnspitze, Tyrol). Zeitschrift für Geomorphologie 54:181–193
Scambos TA, Dutkiewicz MJ, Wilson JC, Bindschadler RA (1992) Application of image cross-correlation to the measurement of glacier velocity using satellite image data. Remote Sens Environ 42:177–186
Schulz E, Dornblut S (2002) Herstellung von Geländemodellen und Ortho-photos im Wettersteingebirge. Diploma thesis, Technische Fachhochschule, Berlin
Söderman G (2013) Slope processes in cold environments of Northern Finland. Fennia-Int J Geogr 158:83–152
Stocker-Waldhuber M, Fischer A, Keller L et al (2017) Funnel-shaped surface depressions—indicator or accelerant of rapid glacier disintegration? A case

study in the Tyrolean Alps. Geomorphology 287:58–72. https://doi.org/10.1016/j.geomorph.2016.11.006

Taylor JR (1997) An introduction to error analysis: the study of uncertainties in physical measurements. University Science Books, Sausalito

Tschada H, Hofer B (1990) Total solids load from the catchment area of the Kaunertal hydroelectric power station: the results of 25 years of operation. IAHS Publication 194:121–128

Vehling L (2016) Gravitative Massenbewegungen an alpinen Felshängen-Quantitative Bedeutung in der Sedimentkaskade proglazialer Geosysteme (Kaunertal, Tirol). Ph.D. thesis, Friedrich-Alexander-Universität Erlangen-Nürnberg, Erlangen. https://opus4.kobv.de/opus4-fau/files/6955/Dissertation_Lucas_Vehling.pdf

Vehling L, Rohn J, Moser M (2016) Quantification of small magnitude rockfall processes at a proglacial high mountain site, Gepatsch glacier (Tyrol, Austria). Zeitschrift für Geomorphologie, Supplementary Issues 60:93–108

Vorndran G (1979) Geomorphodynamische Massenbilanzen (=Augsburger Geographische Hefte 1) University of Augsburg, Ausgburg

Warburton J (1990) An alpine proglacial fluvial sediment budget. Geogr Ann Ser A, Phys Geogr 72:261–272. https://doi.org/10.1080/04353676.1990.11880322

Warburton J (1992) Observations of bed load transport and channel bed changes in a proglacial mountain stream. Arct Alp Res 195–203

Wichmann V (2006) Modellierung geomorphologischer Prozesse in einem alpinen Einzugsgebiet: Abgrenzung und Klassifizierung der Wirkungsräume von Sturzprozessen und Muren mit einem GIS (=Eichstätter Geographische Arbeiten 13). Profil-Verlag München/Wien

Wichmann V, Becht M (2005) Modeling of geomorphic processes in an alpine catchment. In: Atkinson PM, Foody GM, Darby SE, Wu F (eds) GeoDynamics. CRC Press, Boca Raton, S 151–167

Part V

The Role of Soil, Vegetation and Morphodynamics in the Evolution of Proglacial Systems

The Uncalm Development of Proglacial Soils in the European Alps Since 1850

18

Arnaud J. A. M. Temme

Abstract

Glaciers in the Alps and other mountain regions are widely retreating. This contribution focusses on the soils that are forming in the proglacial areas. These soils are important because of the hydrological and ecological effect they will have in future glacierless valleys. A geographical approach is taken that attempts to explain differences in rates of soil formation between proglacial valleys. Through a comparison of published soil chronosequences of European proglacial areas, it is found that age is the most important factor determining rates of soil development —even where morphodynamics are strong. Nonetheless, the effect of geomorphic activity and the effect of vegetation succession have been clearly observed in several studies. The combination of all factors forces us to acknowledge a complex model of soil formation in alpine proglacial valley that among others highlights the heterogenous and dynamic nature of morphodynamics. This model invites us to fill in some blanks in our understanding and suggests that with a larger number of proglacial soil development studies, we may be able to provide pan-alpine information on soils in proglacial areas. This is of importance, for instance, when establishing pan-alpine carbon budgets.

Keywords

Soil development · Proglacial area · Soil types
Chronosequence · Morphodynamics

18.1 Introduction

As poster children of climate change, images of retreating glaciers are ubiquitous in the public domain. Repeat visitors to alpine valleys experience these changes themselves when looking out to glacial remains over usually large expanses of glacial till and side moraines that are coarse, badly sorted, sometimes unstable or unsettled, and above all recent. These are images of lost functionality. Less glacial mass means less melting in summer and hence less water in large continental rivers (Singh and Bengtsson 2005), and less glacial mass can also mean less

A. J. A. M.Temme (✉)
Department of Geography, Kansas State University, Manhattan, KS, USA
e-mail: arnaudtemme@ksu.edu

A. J. A. M.Temme
Soil Geography and Landscape, Wageningen University, Wageningen, The Netherlands

A. J. A. M.Temme
Institute for Arctic and Alpine Research, University of Colorado, Boulder, CO, USA

T. Heckmann and D. Morche (eds.), *Geomorphology of Proglacial Systems*,
Geography of the Physical Environment, https://doi.org/10.1007/978-3-319-94184-4_18

mountain tourism, not only by mountaineers who like to traverse glaciers, but also by tourists who like to look at glaciers from safe ground (Beniston 2003; Carey 2010). The melting process itself can cause the formation and possibly catastrophic failure of glacial dams which is an important natural hazard in mountain regions (Huggel et al. 2002; Richardson and Reynolds 2000). One is justified in asking whether there is any functional gain to offset these losses. The development of soils in the 'new land' in front of retreating glaciers is one potential functional gain and the topic of this chapter.

Soils in proglacial environments are important. First, the properties of soils formed in the usually thin deposits covering bedrock play a large role in the hydrological behaviour of proglacial valleys. Where soils are better developed—with more weathered rocks, a larger clay fraction, better soil structure and more organic matter—they can better retain water during the melting season and after rainstorms (e.g. Wösten et al. 1999). This means that they are better able to dampen peak outflows, and that they are better able to continue to provide water to downstream areas during dry periods. The latter is particularly important when contemplating a future in which most alpine valleys will no longer contain glaciers (e.g. Haeberli et al. 2007), and hence can no longer benefit from a steady flow of glacial meltwater in summers (Xu et al. 2009).

A second important functionality of proglacial soils is the fact that their (pedo-)diversity is the template for biodiversity. If developing soils in proglacial valleys substantially increase their variation over time, they will be better able to support proglacial floral and (micro-)faunal diversity (Khaziev 2011). Over a timescale of decades, such an increase in proglacial biodiversity could become more important than the expected loss in downstream riverine biodiversity due to a larger proportion of cold glacial meltwater (Jacobsen et al. 2012).

In the context of this importance of proglacial soils, the objective of this chapter is to review current knowledge about soil development in proglacial areas, with particular emphasis on the relation between soils and morphodynamics. Geographical focus is on the European Alps, and temporal focus is on the development of soils since the end of the Little Ice Age (LIA), around 1850. In the European Alps, there is usually good control on the retreat history of glaciers since this time (Chaps. 2 and 3), which is a prerequisite for chronosequence studies.

18.2 Proglacial Soil Development Studies in the European Alps

Most studies of soil development and ecological succession in proglacial valleys have focussed on these valleys, because they present beautiful natural experiments where the effect of time on soil development can be well studied. Site selection has therefore prioritised valleys where glacial retreat has been uninterrupted and gradual, and where effects of morphodynamics and other disturbing factors are perceived to be of minimal importance. From these studies, important conclusions have been drawn about the rates of soil development in alpine climates. However, the prioritised valleys represent end-members of the complete set of proglacial valleys. This is because in many proglacial valleys, a wide range of hillslope processes affect the valley (cf. Chaps. 8, 9, 10 and 11), glaciers have not retreated in an uninterrupted manner, parent material is not uniform glacial till, and developing vegetation (Chap. 19) exerts a control on soil development that modulates the influence of time. Recent studies have started to elucidate these additional effects, so that ultimately a more complete picture of soil development in proglacial valleys is emerging.

The end point of soil development in most alpine valleys of intrusive or strongly metamorphic lithology is a (usually skeletic) podzol—a soil type with acidic topsoil characterised by the e- and illuviation of organic matter associated with Fe- and Al-sesquioxides (WRB 2015, Fig. 18.1c). Exceptions have mainly been reported from less common lithologies and drier intra-alpine climates (D'Amico et al. 2014a, b, 2015). However, the podzol as a developmental endpoint is often only reached after about

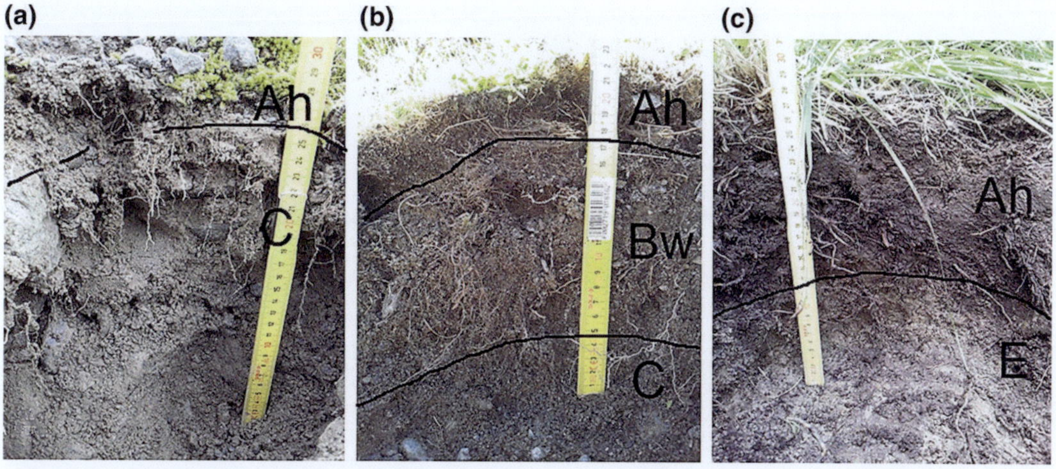

Fig. 18.1 Examples of proglacial soils adapted from Temme et al. 2016. **a** Skeletic regosol. **b** Skeletic cambisol. **c** Haplic podzol. The first two soils are from within a proglacial area, the third is from outside a proglacial area and is older than the LIA

1000 years (Egli et al. 2001; Zech and Wilcke 1977), although some of the oldest post-LIA soils in proglacial valleys in Italy (between 190 and 250 years old) are also podzols (D'Amico et al. 2014a, b).

Typically, the oldest post-LIA proglacial soils in the European Alps are less than 180 years old and hence less developed than that. The most common soils are skeletic, hyperskeletic or humiskeletic leptosols and regosols. Below, I will review the pedological work done in several proglacial valleys in the Alps (Fig. 18.2), starting with more traditional chronosequence work before broadening the field to include studies that focus on a wider range of soil forming factors, including morphodynamics. Inevitably, I am not doing justice to the full breadth of work in alpine proglacial valleys. Given the soil-geographical approach of this chapter, studies that focus purely on soil-related processes outside of the geographical context of soil forming factors are not discussed in detail. This includes much work focussing on the development of microbial soil fauna with time (e.g. Duc et al. 2009; Nicol et al. 2005; Philippot et al. 2011; Schmalenberger and Noll 2009; Schurig et al. 2013; Sigler and Zeyer 2002; Tscherko et al. 2003). In addition, I consider it likely that my literature search has missed some of soil chronosequences explored in the context of ecological research questions.

18.2.1 Morteratsch Glacier

The proglacial valley of the Morteratsch glacier (Fig. 18.2) is possibly the best-studied proglacial valley in the Alps, mainly due to many studies by scientists from the University of Zürich. Building on early ecological work focussed on the time-dependent development of vegetation (Burga 1999; Stöcklin and Bäumler 1996), Egli et al. (2006b) used a detailed soil map of the Morteratsch proglacial area to derive chronosequences for soil types occurring in the valley. Soil types that were observed in this study were mainly lepto- and regosols, with occasionally cambisols in the older parts of the proglacial valley. In addition to soil age, there were indications that soil types varied between landforms and between classes of slope steepness (Egli et al. 2006b). However, this was not tested statistically at the time, and the method to map soils based on 47 soil pits might have partially introduced some of these relations. In an accompanying publication, Egli et al. (2006a) used the relations between ages, landforms and relative occupancy by

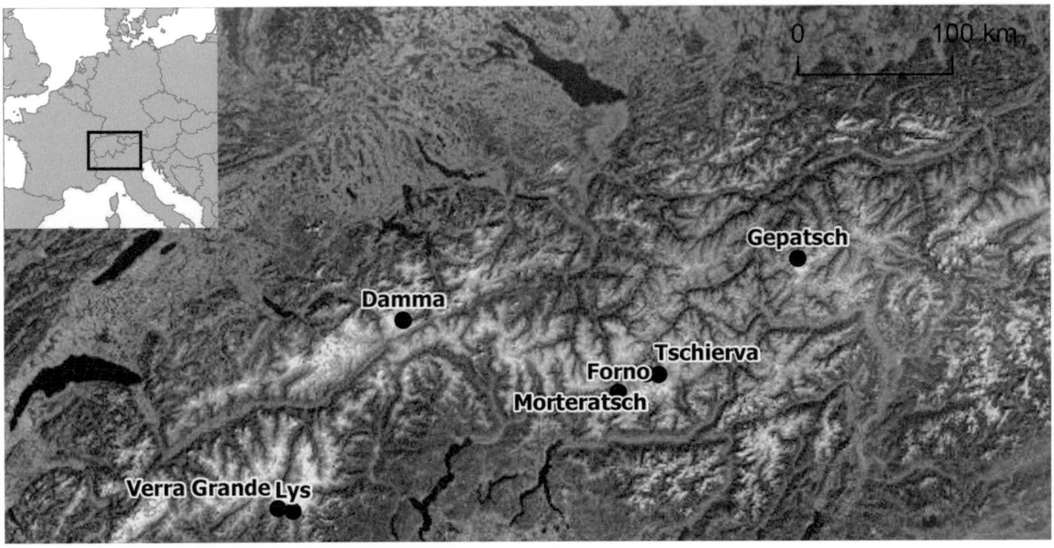

Fig. 18.2 Locations of some proglacial soil chronosequence studies, and location in Europe (inset)

different soil types to simulate a future soil map of the Morteratsch valley after continued retreat of the Morteratsch glacier. The topography of parts of the valley that were still covered by the glacier at the time of writing was simulated based on splines (Jürgens and Biegger 2006).

Later investigations of the Zürich group on soils in the Morteratsch forefield focussed on the weathering and soil-forming processes in more detail (Mavris et al. 2010). Despite the only weak topographic variation in the proglacial valley, it was found that soils on (slightly) north- or polar-facing slopes were better weathered and had higher rates of smectite formation than soils on south- or equator-facing slopes (Egli et al. 2011). It was hypothesised that the availability of water for weathering explained the differences, even in these cold environments where weathering was expected to be temperature-limited. In the alpine proglacial environment, water availability is higher on north-facing surfaces because there is less evaporation. This water-availability hypothesis was later strengthened and detailed by results from the Altai Mountains in Siberia (Egli et al. 2015) and from the Front Range of the Rocky Mountains in Colorado (Langston et al. 2015). Langston et al. (2015) suggested for their study area that the increase in water availability on polar-facing slopes could also be explained by the fact that snow there often accumulates throughout winter and melts only at the end, providing the soil with a stronger pulse of meltwater than on equator-facing slopes, where melting can occur more frequently, but then results in smaller pulses of water in the soil. This mechanism may be valid for the European Alps as well.

A more geographical approach to soil variation in the Morteratsch valley was recently taken by Temme and Lange (2014), who studied soils at an additional 50 point locations in the valley. The focus of this study was on the quantification of relations between soil properties on the one hand, and soil age and terrain parameters on the other hand. In this sense, it mirrored to some extent the first studies by Egli et al. (2006b), who studied this relation for soil type rather than soil properties. Soil age was found to explain about 60% of the variation in pH, and 40% of the variation in the depth of soil formation (A and B horizons) and the vegetation cover. The effect of morphodynamics (based on terrain parameters measured in the field and derived from a 30-m resolution Digital Elevation Model) was significant but less important than the effect of time. Linear regression models of the soil properties based on age, slope steepness, curvature and insolation explained between 80% (pH) and 30% (stoniness) of variation. Based on these results and qualitative field observations, it was hypothesised that the effect of morphodynamics might be best quantified at smaller spatial resolution than 30 m. Many of the small landscape features that appeared to drive soil variation were smaller than this scale.

18.2.2 Forno and Tschierva Glaciers

The Forno and Tschierva glaciers are geographically close to the Morteratsch glacier (Fig. 18.2). This allowed Temme and Lange (2014) and Temme et al. (2015) to consider them climatologically and lithologically relatively similar to that valley. The three valleys differed in terms of glacial retreat history, which was regular in the Forno and Morteratsch valleys, but irregular in the Tschierva valley—and in terms of hillslope effects on the proglacial plain, these were considered absent in the Morteratsch, weak in the Tschierva and strong in the Forno valley. Hence, soil development in the Forno and Tschierva proglacial valleys was expected to be a less strong function of age than in the Morteratsch proglacial valley. This was indeed the case: correlations between soil properties and soil age were weaker in these two valleys (Temme and Lange 2014). However, landscape properties (again resolved at 30 m) did not have a stronger effect than in the Morteratsch valley, resulting in overall less variation in soil properties explained by regression models in the Forno and Tschierva valleys.

The diversity of soil properties, expressed as the coefficient of variation, was found to increase with soil age in the Morteratsch valley, but decrease with age in the Forno and Tschierva valleys (Temme et al. 2015). It was hypothesised that the increase in diversity resulted from positive feedbacks on soil development that amplify small differences in parent material or topographic position, and that the decrease in diversity resulted from hillslope processes that have affected older soils somewhat equally, and more than younger soils.

18.2.3 Damma Glacier

Investigations of soils in the Damma Glacier forefield (Fig. 18.2) have a long history, particularly because of work in the context of the Biglink project (Bernasconi et al. 2008). Much of the work has focussed on establishing the soil microbial and nutrient development with increasing soil age. This microbial development of proglacial soil is strongly interdependent with the development of other soil properties. Overall, soil pH (decreasing), cation exchange capacity, biomass and bacterial and fungal communities (all increasing) changed with soil age (Bernasconi et al. 2011).

The increase in soil carbon is accompanied by a stronger increase in C throughput (Guelland et al. 2013), changing the soil system from one dominated by accumulation to one with high throughput within the approximately 100 years of chronosequence studied. Much of the accumulation of soil carbon was found to be in the form of microbial cell-envelopes (Schurig et al. 2013), highlighting the importance of microbial communities in soil C-dynamics. Overall C accumulation was not only a function of soil age, and a clear negative effect of colder decades on accumulation rates (and on Net Primary Production) was observed (Smittenberg et al. 2012). Also rates of litter mineralisation were not a function of soil age—the main driver appeared to be soil wetness, with drier locations experiencing slower litter breakdown (Guelland et al. 2013).

Vegetation on the young developing soils in the Damma proglacial area was both N- and P-limited. Nitrogen was mainly delivered through atmospheric deposition and rock funnelling, not through N_2 fixing (Göransson et al. 2016). Already in young soils, stocks of S are mainly in the form of organic complexes as opposed to inorganic S (Prietzel et al. 2013). This too highlights the importance of microbial communities: these are needed to degrade organic S complexes to provide S to growing plants. Microbial communities along the chronosequence contained more mycorrhizal fungi at younger age and were dominated by bacteria and other fungi at older age (Welc et al. 2012).

An expected shift in older, more acidic and organic rich soils from bacteria to fungi was not observed within the chronosequence, and must therefore occur later in the development of soils into Podzols.

18.2.4 Gepatsch Glacier

Studies in the forefield of the Gepatschferner in the west of Austria (Fig. 18.2) have focussed on the morphodynamics related to glacial retreat (Heckmann et al. 2012; Chaps. 1, 5, 7, 9, 11, 13, 17). The fact that the area is suited for this purpose already suggests that the Gepatsch proglacial area is an area where soil development cannot be expected to be a clear function of time.

First, the Gepatschferner has not retreated uninterruptedly. It has re-advanced slightly between 1970 and 1997 (Hartl 2010). Second, in contrast to the previously discussed valleys where bedrock was granite, the bedrock in the Gepatsch valley is gneiss with a small fraction of pyrite (Vavtar 1981). Thirdly and most importantly, the proglacial area is affected by fluvial redistribution, undercutting and degrading moraines, and debris flows from surrounding slopes (cf. Chap. 11). The activity of the proglacial stream, the Fagge, is strongly variable in time and affects a substantial fraction of the proglacial area (Morche et al. 2012; Morche et al. 2014; Baewert and Morche 2014; Chap. 13). Moraine degradation and debris flows were quantified using repeated Lidar data and also affect a substantial fraction of the proglacial area (Haas et al. 2012; Chaps. 11 and 17). These morphodynamic processes have a different spatial signature: the effect of fluvial redistribution runs largely parallel to the valley axis, whereas the effect of collapses and debris flows is roughly parallel to the slopes (i.e. perpendicular to the valley axis) and concentrated in the most narrow reach. As a result, their influence on soil–age relations is not easily predicted.

Soils in the Gepatsch proglacial area were studied by Temme et al. (2016), with the objective of leveraging the detailed morphodynamic information to clarify the joint effects of age and morphodynamics on young, developing soils. The sampling design (for 97 locations) ensured that the effects of age and morphodynamics could both be reliably quantified. The main outcome was that despite the morphodynamic activity, age remained a main control on soil properties at the level of the entire proglacial valley. However, the effect of morphodynamics could be much better studied with the detailed digital elevation models (DEMs) available from the morphodynamic studies, than had been the case for earlier chronosequence studies.

In fact, the correlation between a range of soil properties and some morphometric indicators exhibited a clear scale-dependence. This allowed an assessment of the morphodynamic processes that affect soil properties. It was shown that correlations were usually low at the previously used 30-m resolution (e.g. Temme and Lange 2014), proving that morphodynamic processes typically affect soils and soil patterns at smaller scale than 30 m. Profile curvature, which indicates whether a position is likely to experience erosion or deposition, was strongly correlated with pH and the combined silt and clay fraction at resolutions ranging from 1 to about 5 m. Eroding soils had higher pH values and lower amounts of silt and clay. The stream power index, which indicates to which extent a position is exposed to fluvial activity, was strongly correlated with soil properties at resolution ranging up to 25 m. Soils in floodplains had higher pH values and rooting depths, but lower amounts of silt and clay. The main point taken from the study for our purposes here should be that soil development in places where morphodynamics play a substantial role is far from calm and undisturbed—but that morphodynamic effects can be well quantified with detailed altitude data.

18.2.5 Lys and Verra Grande Glaciers

Building on larger-scale soil surveys in the Aosta Valley (D'Amico et al. 2014a), scientists from the University of Turin studied two proglacial valleys that are of interest because of their widely different lithologies: the proglacial valleys of the

Lys and Verra Grande glaciers (Fig. 18.2; D'Amico et al. 2014a, b, 2015). The Lys glacier overlies granitic gneiss and paragneiss bedrock with minor mafic and ultramafic intrusions, whereas the Verra Grande glacier mostly overlies serpentinite bedrock. Retreat histories are similar in the sense that both glaciers retreated from a position closely upslope of the LIA maximum moraine around 1860 (Carnielli 2005; Citterio et al. 2007)—however, the LIA maximum extent itself (and soils developed on it) of the Verra Grande glacier is probably a few hundred years older than 1850 (D'Amico et al. 2015). Compared to the other proglacial valleys reviewed here, the Verra Grande and Lys valleys are also the only two south of the main alpine arc, and the only two that are generally facing south.

Soil development with time is faster in the gneiss-based Lys proglacial valley than in the serpentinite-based Verra Grande proglacial valley (D'Amico et al. 2015). Serpentinite-based soils are generally seen to reduce plant productivity (Alexander et al. 2007). Perhaps because of this limitation of plant productivity, the effect of ultimate colonisation by grasses and larch shrubs was clearly seen in soil properties. Grass and later shrub colonisation of soils in the Verra Grande proglacial area clearly accelerated soil development rates, although these remained lower on serpentinite till than those in the Lys proglacial valley and in parts of the Verra Grande proglacial valley where the parent material was less pure serpentinite. Also in the Lys valley, the effect of vegetation colonisation was clear. In this case, subalpine larch and rhodondendron vegetation accelerated soil formation to the extent that 60-year-old soils displayed bleached E-horizons (D'Amico et al. 2014a). The authors argued that this vegetation type, rather than climatic factors, explains the exceptionally fast soil formation relative to other study areas.

The limitations placed on vegetation succession on serpentinite-dominated till were argued to enhance the vulnerability of these areas to environmental change (D'Amico et al. 2016). In this context, it is good to realise that although serpentinite is common in orogenic belts, it is much less so than granitic or gneiss bedrock, on which vegetation establishment and soil development are much faster. In the European Alps, mafic and ultramafic rocks are mainly limited to the Swiss and western Italian Alps.

18.3 Soil Development as a Function of Time

When comparing the effect of time on soil properties between study sites (Table 18.1), it is necessary to choose a functional form for the soil–time relation in each site. In most cases, rates of change have been found to increase after a few years (or at least from when vegetation is established, D'Amico et al. 2015) and reach a maximum for ages between 100 and 1000 years. Rates of change then drop to low long-term steady-state values (e.g. Egli et al. 2001). This type of development is functionally akin to the humped weathering model that explains how bedrock weathers into soil (Burke et al. 2007). Exceptions exist, such as the slow start of development on serpentinite parent material (D'Amico et al. 2016). During the slow start, accelerating soil development would be well described with an exponential model. Where the maximum value of a development indicator is reached, such as for vegetation cover, and soil development slows down, a logistic growth function would be appropriate. Here, to allow objective comparison and for ease of reporting, the observations provided with the various studies discussed above have all been linearly regressed to age.

As mentioned before, the Verra Grande glacier proglacial area, where serpentinite parent material slows down vegetation establishment, displays some of the lowest rates of development. However, there appears to be also significant variation in rates of development between proglacial areas of the same lithology. Although this may to some extent be due to uncertainty in the estimation of the rates, it is also likely that climatic differences between proglacial areas (for instance, between Lys and Gepatsch) and different morphodynamic settings (for instance, between Morteratsch, Forno and Tschierva)

Table 18.1 Comparison of annual rates of change in various soil properties between the proglacial valleys

Annual rate of change in soil properties					
Location	Lithology	pH (-)	Topsoil OM (M%)	Vegetation cover (%)	Depth of soil formation (cm)
Morteratsch (Temme and Lange 2014)	Granite	−0.017	–	0.53	0.021
Forno (Temme and Lange 2014)	Granite	−0.008	–	0.5	0.061
Tschierva (Temme and Lange 2014)	Granite	−0.014	–	no relation	0.014
Damma (Dümig et al. 2011)	Granite	−0.007	0.034[b]	–	0.027
Lys (D'Amico et al. 2014)	Gneiss	−0.0078	0.025[b]	>0.59	0.07
Gepatsch (Temme et al. 2016)	Gneiss	−0.0035[a]	0.105	–	–
Verra Grande (D'Amico et al. 2015)	Serpentinite	−0.0075	0.009[b]	0.31	~0.03

Rates are calculated using linear regression where no intercept was imposed
[a]For soils older than 20 years. Younger soils experienced a pH rise that was assumed related to pyrite weathering in fresh sediment
[b]Obtained from SOC values using the Van Bemmelen factor (SOC = 0.58 SOM)

cause these differences. It is attractive to contemplate a meta-analysis of development rates at the point-scale rather than the scale of the proglacial area. This could allow a separation of the effect of morphodynamics from the effect of climate, where variation within areas would be mainly caused by morphodynamics and microtopographical differences, and variation between areas by climate. However, a much larger data set of proglacial chronosequences will be needed for this purpose.

18.4 A Complex Model of Soil Development in Proglacial Areas

Beyond the effect of age, which clearly appears to differ between sites, a complex model of proglacial soil development emerges from the various studies reviewed above (Fig. 18.3). In this model, proglacial soil development is neither linear, nor monotonous or calm (cf. Bardgett et al. 2005; Sommer et al. 2008). The calmest case is a proglacial area where moraines protect the valley from hillslope influences and geomorphic activity plays a negligible role, but even here, minor or major differences in the glacial till parent material (Fig. 18.3A) will get amplified over time as initial soil formation, dominated by positive feedbacks, speeds up (D'Amico et al. 2016; Temme et al. 2015).

Moreover, since vegetation and associated soil (micro-)fauna develops strongly nonlinearly over time through various successional stages, vegetation influence on soil development is also nonlinear (either speeding up with time, or reaching a maximum and then decreasing). This can cause further divergence in properties (Fig. 18.3H; D'Amico et al. 2014a, b).

Additionally, in most proglacial areas, there are one or more morphodynamic and geomorphic complicating factors. The most important one is where erosion rates in steep proglacial valleys allow neither soil development nor vegetation establishment. In such sections, the proglacial valley appears headed for a pure bedrock valley floor. The eventual or original absence of till where bedrock crops out (Fig. 18.3D), and the effect of eroding and collapsing moraines (Fig. 18.3B) and debris flow from hillslopes overlying the proglacial area (Fig. 18.3C), all also

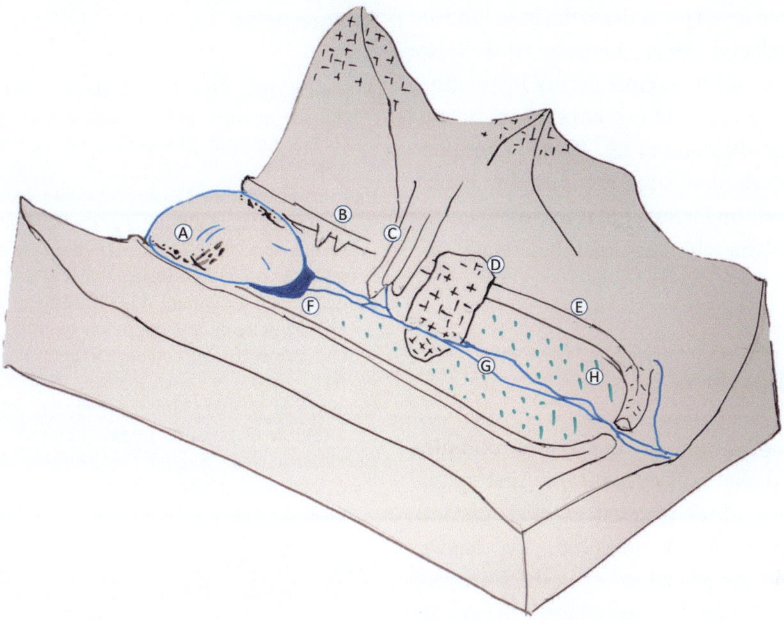

Fig. 18.3 Sketch of a proglacial valley. Factors that can affect the time-dependent development of soils in proglacial areas are indicated with characters. A indicates the provision of heterogenous sub- and supraglacial till to the proglacial area. B indicates the degradation of lateral moraines. C indicates hillslope processes, including debris flows, from surrounding hillslopes. D indicates the absence of till or moraine material where bedrock surfaces. E indicates undisturbed moraine, decoupling the proglacial area from external sediment fluxes. F indicates the effect of glacial lakes, and G indicates fluvial redistribution. H indicates the effect of vegetation transitions and patterning

affect soil formation. The mechanism by which this effect occurs is often the burial of soils under material provided by moraines or hillslopes. This has a negative effect on the level of soil development if the material is provided from fresh, unweathered sources (such as steep-side moraines), but can have a positive effect on development if the material is provided from hillslopes which have often been exposed for much longer than the proglacial area. Both of these examples have been observed in the European Alps (Temme et al. 2015). Reworking of material during consecutive episodes of erosion and deposition complicates this process both in the source zones and in the proglacial area. The imperfect sedimentary connectivity of the morphodynamics of hillslope and fluvial processes causes varying residence times of paraglacial sediment on its way to the proglacial area and ultimately further downstream (Heckmann et al. 2012, Knight and Harrison 2014; Chaps. 15, 16 and 17).

Buried palaeosols are sometimes found in proglacial areas. Their persistence indicates that burial has been fast relative to soil formation speeds. Examples come from the Austrian Alps (Mahaney et al. 2011) and from British Columbia (Karlstrom and Osborn 1992) and Spitsbergen (Kabala and Zapart 2009). In Austria, the buried soils were found under LIA moraines and hence older than the LIA. These palaeosols had a convoluted character, indicating either cryoturbation or pushing by moraines during the LIA (Mahaney et al. 2011).

Soil development is also affected by fluvial morphodynamics (Fig. 18.3G; Chaps. 12 and 13). Incision, lateral erosion and deposition all affect soil properties. The fluvial effect is more substantial where proglacial streams are multichannel and affect a relatively wide floodplain in the proglacial valley. This is often the case in Little Ice Age proglacial areas because fluxes are strongly seasonal with maxima during the melting season, and there is ample presence of unconsolidated sediment (Morche et al. 2014). The alternative, a single-channel stream with stable banks, would affect soils much less, but is more typical for mature catchments with vegetated banks and less seasonal flux.

18.5 Implications

The main implication of this more complex model of proglacial soil evolution is that spatial upscaling from chronosequences and chronofunctions is not easily possible. To answer questions about the global, or even the European alpine carbon budget of proglacial areas, for instance, it is necessary to assess how important each complicating factor is in each proglacial area, or in groups of proglacial areas, and then account for the effect of these factors on the budget. At the same time, it appears that the process knowledge is available that would allow for such an assessment. From this, it appears that data sets that are required and sufficient for a carbon budget of proglacial areas of the European Alps would include the extent of proglacial areas, the extent of bedrock in proglacial areas, retreat histories and retreat predictions for glaciers, geological maps, a digital elevation model, current and predicted vegetation maps and chronofunctions for the rate of below-ground organic matter accumulation as a function of location, vegetation and parent material. Although this is a long list, most of these data are currently available and others can be relatively easily produced. Probably the most difficult aspect remains accounting for the complex effect of active morphodynamics on soil development. Elucidating this morphodynamic effect at appropriate spatial scales should therefore be part of the future research agenda, along with an increased understanding of how different parent materials and climatic settings determine the variation in chronofunctions.

References

Alexander EB, Ellis CC, Burke R (2007) A chronosequence of soils and vegetation on serpentine terraces in the Klamath Mountains, USA. Soil Sci 172:565–576

Baewert H, Morche D (2014) Coarse sediment dynamics in a proglacial fluvial system (Fagge River, Tyrol). Geomorphology 218:88–97. https://doi.org/10.1016/j.geomorph.2013.10.021

Bardgett RD, Bowman WD, Kaufmann R, Schmidt SK (2005) A temporal approach to linking aboveground and belowground ecology. Trends Ecol Evol 20:634–641

Beniston M (2003) Climatic change in mountain regions: a review of possible impacts. Clim Change 59:5–31

Bernasconi SM, Bauder A, Bourdon B et al (2011) Chemical and biological gradients along the Damma Glacier soil chronosequence, Switzerland. Vadose Zone J 10:867. https://doi.org/10.2136/vzj2010.0129

Bernasconi SM, BigLink Project Members (2008) Weathering, soil formation and initial ecosystem evolution on a glacier forefield: a case study from the Damma Glacier, Switzerland. Mineral Mag 72:19–22

Burga CA (1999) Vegetation development on the glacier forefield Morteratsch (Switzerland). Appl Veg Sci 2:17–24

Burke BC, Heimsath AM, White AF (2007) Coupling chemical weathering with soil production across soil-mantled landscapes. Earth Surf Proc Land 32:853–873

Carey M (2010) In the shadow of melting glaciers: climate change and Andean society. Oxford University Press, Oxford

Carnielli T (2005) Snout and area recent variations of Grande di Verra Glacier (Monte Rosa, Alps). Geografia Fisica e Dinamica Quaternaria VII: 79–87

Citterio M, Diolaiuti G, Smiraglia C et al (2007) The fluctuations of Italian glaciers during the last century: a contribution to knowledge about alpine glacier changes. Geogr Ann: Ser A, Phys Geogr 89:167–184

D'Amico ME, Bonifacio E, Zanini E (2014a) Relationships between serpentine soils and vegetation in a xeric inner-alpine environment. Plant Soil 376:111–128

D'Amico ME, Freppaz M, Filippa G, Zanini E (2014b) Vegetation influence on soil formation rate in a proglacial chronosequence (Lys Glacier, NW Italian Alps). CATENA 113:122–137. https://doi.org/10.1016/j.catena.2013.10.001

D'Amico ME, Freppaz M, Leonelli G et al (2015) Early stages of soil development on serpentinite: the proglacial area of the Verra Grande Glacier, Western Italian Alps. J Soils Sediments 15:1292–1310

D'Amico ME, Catoni M, Terribile F, Zanini E, Bonifacio E (2016) Contrasting environmental memories in relict soils on different parent rocks in the south-western Italian Alps. Quat Int 418:61–74

Duc L, Noll M, Meier BE et al (2009) High diversity of diazotrophs in the forefield of a receding alpine glacier. Microb Ecol 57:179–190

Dümig A, Smittenberg R, Kögel-Knabner I (2011) Concurrent evolution of organic and mineral components during initial soil development after retreat of the Damma glacier, Switzerland. Geoderma 163:83–94

Egli M, Fitze P, Mirabella A (2001) Weathering and evolution of soils formed on granitic, glacial deposits: results from chronosequences of Swiss alpine environments. Catena 45:19–47

Egli M, Lessovaia SN, Chistyakov K et al (2015) Microclimate affects soil chemical and mineralogical properties of cold alpine soils of the Altai Mountains (Russia). J Soils Sediments 15:1420–1436

Egli M, Wernli M, Burga C et al (2011) Fast but spatially scattered smectite-formation in the proglacial area Morteratsch: an evaluation using GIS. Geoderma 164:11–21

Egli M, Wernli M, Kneisel C et al (2006a) Melting glaciers and soil development in the proglacial area Morteratsch (Swiss Alps): II. Modeling the present and future soil state. Arct Antarct, Alp Res 38:510–521

Egli M, Wernli M, Kneisel C, Haeberli W (2006b) Melting glaciers and soil development in the proglacial area Morteratsch (Swiss Alps): I. Soil Type Chronosequence. Arct Antarct, Alp Res 38:499–509. https://doi.org/10.1657/1523-0430(2006)38%5b499:MGASDI%5d2.0.CO;2

Göransson H, Welc M, Bünemann EK et al (2016) Nitrogen and phosphorus availability at early stages of soil development in the Damma glacier forefield, Switzerland; implications for establishment of N2-fixing plants. Plant Soil 404:251–261. https://doi.org/10.1007/s11104-016-2821-5

Guelland K, Hagedorn F, Smittenberg RH, Göransson H, Bernasconi SM, Hajdas I, Kretzschmar R (2013) Evolution of carbon fluxes during initial soil formation along the forefield of Damma glacier, Switzerland. Biogeochemistry 113:545–561. https://doi.org/10.1007/s10533-012-9785-1

Haas F, Heckmann T, Hilger L, Becht M (2012) Quantification and modelling of debris flows in the proglacial area of the Gepatschferner, Austria, using ground-based LiDAR. IAHS-AISH publication. International Association of Hydrological Sciences, pp 293–302

Haeberli W, Hoelzle M, Paul F, Zemp M (2007) Integrated monitoring of mountain glaciers as key indicators of global climate change: the European Alps. Ann Glaciol 46:150–160

Hartl L (2010) The Gepatschferner from 1850–2006-changes in length, area and volume in relation to climate. Diploma thesis (unpublished), University of Innsbruck

Heckmann T, Haas F, Morche D, Schmidt K-H, Rohn J, Moser M, Leopold M, Kuhn M, Briese C, Pfeifer N, Becht M (2012) Investigating an alpine proglacial sediment budget using field measurements, airborne and terrestrial LiDAR data, vol 356. IAHS-AISH publication, pp 438–447

Huggel C, Kääb A, Haeberli W et al (2002) Remote sensing based assessment of hazards from glacier lake outbursts: a case study in the Swiss Alps. Can Geotech J 39:316–330

Jacobsen D, Milner AM, Brown LE, Dangles O (2012) Biodiversity under threat in glacier-fed river systems. Nat Clim Change 2:361–364

Jürgens C (2006) A visual system for the interactive study and experimental simulation of climate-induced 3D Mountain Glacier Fluctuations. JSTOR

Kabala C, Zapart J et al (2009) Recent, relic and buried soils in the forefield of Werenskiold Glacier, SW Spitsbergen. Pol Polar Res 30:161–178

Karlstrom ET, Osborn G (1992) Genesis of buried paleosols and soils in Holocene and late Pleistocene tills, Bugaboo Glacier area, British Columbia, Canada. Arct Alp Res 108–123

Khaziev FK (2011) Soil and biodiversity. Russ J Ecol 42:199–204

Knight J, Harrison S (2014) Mountain Glacial and paraglacial environments under global climate change: lessons from the past, future directions and policy implications. Geogr Ann: Ser A, Phys Geogr 96:245–264. https://doi.org/10.1111/geoa.12051

Langston AL, Tucker GE, Anderson RS, Anderson SP (2015) Evidence for climatic and hillslope-aspect controls on vadose zone hydrology and implications for saprolite weathering. Earth Surf Process Land 40:1254–1269

Mahaney WC, Hancock RG, Melville H (2011) Late glacial retreat and Neoglacial advance sequences in the Zillertal Alps, Austria. Geomorphology 130:312–326

Mavris C, Egli M, Plötze M, Blum JD, Mirabella A, Giaccai D, Haeberli W (2010) Initial stages of weathering and soil formation in the Morteratsch proglacial area (Upper Engadine, Switzerland). Geoderma 155:359–371

Morche D, Haas F, Baewert H, Heckmann T, Schmidt K-H, Becht M (2012) Sediment transport in the proglacial Fagge River (Kaunertal/Austria), vol 356. IAHS-AISH Publication, pp 72–80

Morche D, Schuchardt A, Dubberke K, Baewert H (2014) Channel morphodynamics on a small proglacial braid plain (Fagge River, Gepatschferner, Austria), vol 367. IAHS-AISH Publication, pp 109–116. https://doi.org/10.5194/piahs-367-109-2015

Nicol GW, Tscherko D, Embley TM, Prosser JI (2005) Primary succession of soil Crenarchaeota across a receding glacier foreland. Environ Microbiol 7:337–347

Philippot L, Tscherko D, Bru D, Kandeler E (2011) Distribution of high bacterial taxa across the chronosequence of two alpine glacier forelands. Microbial Ecol 61:303–312

Prietzel J, Wu Y, Dümig A, Zhou J, Klysubun W (2013) Soil sulphur speciation in two glacier forefield soil chronosequences assessed by S K-edge XANES spectroscopy: S speciation in glacier forefield soils by XANES. Eur J Soil Sci 64:260–272. https://doi.org/10.1111/ejss.12032

Richardson SD, Reynolds JM (2000) An overview of glacial hazards in the Himalayas. Quat Int 65:31–47

Schmalenberger A, Noll M (2009) Shifts in desulfating bacterial communities along a soil chronosequence in the forefield of a receding glacier. FEMS Microbiol Ecol 71:208–217

Schurig C, Smittenberg RH, Berger J, Kraft F, Woche SK, Goebel MO, Heipieper HJ, Miltner A, Kaestner M (2013) Microbial cell-envelope fragments and the formation of soil organic matter: a case study from a glacier forefield. Biogeochemistry 113:595–612

Sigler W, Zeyer J (2002) Microbial diversity and activity along the forefields of two receding glaciers. Microb Ecol 43:397–407

Singh P, Bengtsson L (2005) Impact of warmer climate on melt and evaporation for the rainfed, snowfed and glacierfed basins in the Himalayan region. J Hydrol 300:140–154

Smittenberg RH, Gierga M, Göransson H, Christl I, Farinotti D, Bernasconi SM (2012) Climate-sensitive ecosystem carbon dynamics along the soil chronosequence of the Damma glacier forefield, Switzerland. Global Change Biol 18:1941–1955. https://doi.org/10.1111/j.1365-2486.2012.02654.x

Sommer M, Gerke H, Deumlich D (2008) Modelling soil landscape genesis—a "time split" approach for hummocky agricultural landscapes. Geoderma 145:480–493

Stöcklin J, Bäumler E (1996) Seed rain, seedling establishment and clonal growth strategies on a glacier foreland. J Veg Sci 7:45–56

Temme AJAM, Heckmann T, Harlaar P (2016) Silent play in a loud theatre—dominantly time-dependent soil development in the geomorphically active proglacial area of the Gepatsch glacier, Austria. Catena 147:40–50

Temme AJAM, Lange K, Schwering MF (2015) Time development of soils in mountain landscapes—divergence and convergence of properties with age. J Soils Sed 15:1373–1382

Temme AJAM, Lange K (2014) Pro-glacial soil variability and geomorphic activity—the case of three Swiss valleys. Earth Surf Process Land 39:1492–1499. https://doi.org/10.1002/esp.3553

Tscherko D, Rustemeier J, Richter A, Wanek W, Kandeler E (2003) Functional diversity of the soil microflora in primary succession across two glacier forelands in the Central Alps. Eur J Soil Sci 54:685–696

Vavtar F (1981) Syngenetische metamorphe Kiesanreicherungen in Paragneisen des Ötztal-Kristallins (Kaunertal, Tirol). Veröff Museum Ferdinandeum Innsbruck, S 151–169

Welc M, Bünemann EK, Fließbach A, Frossard E, Jansa J (2012) Soil bacterial and fungal communities along a soil chronosequence assessed by fatty acid profiling. Soil Biol Biochem 49:184–192. https://doi.org/10.1016/j.soilbio.2012.01.032

Wösten J, Lilly A, Nemes A, Le Bas C (1999) Development and use of a database of hydraulic properties of European soils. Geoderma 90:169–185

WRB (2015) World reference base for soil resources 2014, update 2015 International soil classification system for naming soils and creating legends for soil maps. FAO, Rome

Xu J, Grumbine RE, Shrestha A, Eriksson M, Yang X, Wang Y, Wilkes A (2009) The melting Himalayas: cascading effects of climate change on water, biodiversity, and livelihoods. Conserv Biol 23:520–530

Zech W, Wilke B (1977) Vorläufige Ergebnisse einer Bodenchronosequenzstudie im Zillertal. Mitteilungen der Deutschen Bodenkundlichen Gesellschaft 25: 571–586

Vegetation Succession and Biogeomorphic Interactions in Glacier Forelands

19

Jana Eichel

Abstract

Proglacial areas are not only the stage for glacial processes and paraglacial dynamics, which shape the landscape following glacier retreat. At the same time, the new terrain is colonized quickly by plants and animals. Different plant species follow each other in a sequence of successional stages. This sequence is controlled by both abiotic and biotic processes and depends on local-, landscape- and regional-scale environmental factors, such as soil properties, topography and elevation. Yet, successional sequences are often disrupted or changed by disturbances. For example, geomorphic processes delay vegetation succession, limit its development to pioneer stages or change its pathways. However, vegetation succession is not only changed by disturbances, plants can also actively influence geomorphic processes. These biogeomorphic interactions control patterned ground, glaciofluvial floodplain and moraine slope development. Once geomorphic activity decreases to a certain degree, ecosystem engineer species can establish, e.g. the dwarf shrub *Dryas octopetala* on lateral moraine slopes. When plant biomass reaches a certain volume, it starts to affect geomorphic processes; e.g., interactions change the dominant process on moraine slopes from slope wash and slide to bound solifluction. These biogeomorphic feedbacks stabilize the glacial sediments and facilitate establishment for later successional species, such as trees.

Keywords

Vegetation colonisation · Succession Chronosequence · Disturbance · Biogeomorphic interactions · Ecosystem engineers

19.1 Introduction

As stated by Jochimsen (1962), a glacier foreland is not a desert (‘*Das Gletschervorfeld – keine Wüste*’). Instead, once this new terrain is exposed by the retreating glacier ice, it is quickly colonized by plants and animals. Vegetation succession and geomorphic processes interact in the course of paraglacial landscape adjustment. These ‘biogeomorphic interactions’ shape the glacier foreland and are therefore a very relevant control for proglacial landscape development.

J. Eichel (✉)
Institute of Geography and Geoecology, Karlsruhe Institute of Technology (KIT), Karlsruhe, Germany
e-mail: jana.eichel@kit.edu

T. Heckmann and D. Morche (eds.), *Geomorphology of Proglacial Systems*,
Geography of the Physical Environment, https://doi.org/10.1007/978-3-319-94184-4_19

In future decades, glacier foreland areas will further expand with ongoing glacier retreat due to climate change (Paul et al. 2004; Chap. 2). If we envisage our present-day glacier forelands in 50 years, we do not only expect them to be larger, but also expect the now grey, bare moraine slopes and valley floors to have been colonized by plants. This colonization is important, as it stabilizes glacial sediments (Raymond Pralong et al. 2011), which otherwise frequently fill up reservoirs in glacier forelands (Bogen 1988; Hauenstein 2005; Raymond Pralong et al. 2011). But how quick can plants colonize? It has been shown that high sediment reworking limits plant colonization (Richter 1994; Curry et al. 2006; Moreau et al. 2008), so how long does it take until sediment reworking decreases sufficiently for vegetation to establish? In addition, plant colonization and development depend on species traits, which determine their ability to survive in active geomorphic environments (response traits) and to actively influence the occurring geomorphic processes (effect traits) (Violle et al. 2007; Corenblit et al. 2015). Based on their traits, some species could be more effective in stabilizing sediment in glacier foreland than others. But which species are these 'ecosystem engineers'? And are vegetated sediments permanently stable? This is important to know not only for sediment management, as debris flow occurring on lateral moraine slopes can also represent a natural hazard. It has been shown that high-magnitude geomorphic processes can remove the stabilizing vegetation cover also in later stages of the vegetation succession. So, under which conditions are vegetation covered sediments stable?

The aim of this chapter is to provide some first answers from recent ecological and biogeomorphic research by illustrating when and how plants and geomorphic processes interact in glacier forelands. In the following, the chapter considers these questions:

1. Primary vegetation succession in glacier forelands: How long does it take for vegetation to establish after glacier retreat?
2. Disturbances: When and how do they influence vegetation establishment?
3. Biogeomorphic interactions: When and how are glacial sediments stabilized?

19.2 Vegetation Establishment and Succession

Vegetation succession is a gradual, directional change in the species composition and structure of ecosystems over a longer period of time (Matthews 1992; Dierschke 1994). In this process, a sequence of plant communities with different species compositions follows each other in a certain space (Tansley 1920). Vegetation succession is initiated by processes disturbing the ecosystem (White 1979), for example thunderstorms, volcanic eruptions, fire (Pickett and White 1987). Based on the degree of 'biological inheritance' left after the disturbance, such as seeds and plant fragments, succession is classified into primary or secondary succession (Walker and del Moral 2003). During primary succession, plants and animals colonize areas which were previously not colonized, while during secondary succession, species composition and other biotic properties change in an already existing ecosystem following the disturbance (MacDonald 2002). As the retreating glacier usually exposes bare terrain with no biological material (but see Sharp 1958; Fickert et al. 2007 and others for counterexamples), vegetation succession in glacier forelands is a primary succession (Matthews 1992).

19.3 Vegetation Succession Models

To understand and describe vegetation succession in time, diverse successional models were developed. Relevant models to understand vegetation succession and associated biogeomorphic interactions in glacier forelands are the interaction-based models by Clements (1928) and Connell and Slatyer (1977), the strategy-based models by MacArthur and Wilson (1967) and Grime (1977) and the geoecological succession model by Matthews (1992) (see Table 19.1).

Table 19.1 Overview on important succession models in the context of vegetation succession and biogeomorphic interactions in glacier forelands. Key statements are summarized

Model type	Model	Nature of model
Interaction-based models	Facilitation-based succession (Clements 1928)	Dominant species modify their environment ('reaction'), new species arrive and can establish in modified environment ('facilitation'), replacement of previously dominant species. Result: Sequence of dominant species modifying their environment and subsequently replacing each other
	Facilitation, tolerance, inhibition (Connell and Slatyer 1977)	In addition to facilitation, early-successional species can occur together with later successional species (tolerance) or permanently inhibit other species (inhibition). Result: Sequential, simultaneous or singular occurrence of species
Strategy-based models	r- and K-strategists (MacArthur and Wilson 1967)	Species strategies depend mainly on growth rates and resource requirements. This determines sequential occurrence of species. Result: Quickly growing r-strategists are succeeded by resource-intensive, competitive K-strategists
	CSR primary plant strategies (Grime 1977)	Stress and disturbance intensity determine plant life history strategy types. When these properties change in time, plant strategists can follow after each other. Result: With changing environmental conditions in glacier forelands, increasing disturbance-adapted ruderal strategists are followed by competitive strategists with low disturbance and high resource levels, followed by stress strategists in a low-resource environment
Geoecological models	Geoecological succession model (Matthews 1992)	Importance of physical and biotic processes changes during the successional sequence. Result: In the first years to decades following glacier retreat, abiotic processes mainly control primary vegetation succession; later, biotic processes become the most important controls

Interaction-based succession models assume that species interactions control vegetation succession. The *facilitation-based succession model* by Clements (1928) highlights the role of individual organisms for sequential vegetation development. It assumes that species can change their environment (reaction) and thereby improve environmental conditions. New species arrive and are able to establish due to the positive environment modification (facilitation). They replace the previously dominant species and in turn start to modify their environment. Connell and Slatyer (1977) added two more types of interactions to Clement's facilitation in their *'facilitation, tolerance, inhibition'-model*. In this model, interactions between early and later successional species control the successional change. In comparison with Clements, this interaction cannot only be positive (facilitation), but also neutral (tolerance) or negative (inhibition). This results in different successional pathways. In the tolerance successional pathway, early-successional species do not affect later successional species, and these only appear later because of their life history traits, such as lower growth rates. In the inhibition successional pathway, early-successional species can inhibit or completely exclude the colonization by later successional species. They are only replaced once they die, e.g. through a high-magnitude disturbance. The different successional pathways are not exclusive but a continuum of possibilities, as early-successional species can have more than one effect on later successional species.

Strategy-based models assume a relationship between species traits, especially life history traits (properties of species related to their growth, fertility and survival, e.g. diaspore quantity; Lavorel et al. 2007; Violle et al. 2007), and their environment. The r- and K-strategist model by MacArthur and Wilson (1967) is based on the Verhulst model of population dynamics, which relates population growth rate and carrying capacity of the environment.

Colonization starts with r-strategists, which possess high growth rates and can thereby quickly colonize a new environment. In time, they are replaced by K-strategists, which grow more slowly and need more resources, but can compete better for these. As an extension of the r- and K-strategist model, Grime (1977) classified plant life history strategy types based on species' adaption to stress (factors that limit photosynthesis) and disturbances (destruction of biomass). In a successional sequence, different strategists follow after each other due as stress and disturbance levels change in the environment in time. Once disturbance intensity decreases and resource availability increases in glacier forelands, the disturbance-adapted ruderal (R) strategist is succeeded by the competitor (C). In time, resource availability and disturbance intensity decrease and the stress-tolerator (S) dominates. However, all strategists can also persist in certain ecological contexts.

As the described models mainly emphasize biological processes, they have been criticized by Matthews (1992) for their limited power to explain primary succession in glacier forelands. Here, vegetation succession is also controlled by physical processes. In his *geoecological succession model*, Matthews (1992) combines physical and biological processes and describes their changing importance in the successional sequence. In the beginning, abiotic processes determine primary vegetation succession, while biotic processes control succession later (Raab et al. 2012). Matthew's geoecological model is used to describe an exemplary primary glacier foreland succession in the following section, including its successional stages and trends.

19.4 Primary Vegetation Succession in Glacier Forelands

For primary successions in glacier forelands, the decreasing importance of abiotic processes with increasing terrain age is linked with a set *of abiotic successional trends*, describing changes in the abiotic environment in time, while the increasing importance of biotic processes is related to certain *biotic successional trends*, describing changes in vegetation properties in time (Fig. 19.1; Matthews 1992; Milner et al. 2007).

Abiotic successional trends include improving environmental conditions with increasing terrain age, such as a proceeding soil development and increasing temperatures (Matthews 1992; Andreis et al. 2001; Raffl et al. 2006). Indicators for proceeding soil development are increasing depths of soil horizons, increasing soil organic matter content and decreasing pH values (Matthews 1992; Rydgren et al. 2014). How nutrient availability develops in time is unclear as it can both increase or decrease (Chapin et al. 1994; Rydgren et al. 2014). Temperatures increase with increasing time since deglaciation, as glacier winds become less strong with increasing distance to the retreating glacier and the soil heats more easily due to more organic matter (Kaufmann 2002; Egli et al. 2011).

Biotic successional trends are linked with the abiotic trends. In time, vegetation cover and biomass increase and vegetation stratifies into different layers with a physiognomic vegetation development from lichen and herbaceous plants to trees (Lüdi 1945; Birks 1980; Matthews 1992; Garibotti et al. 2011). With increasing terrain age, more species occur until 40–80 years since deglaciation; afterwards, species numbers decrease (Matthews 1999). Similarly, succession rates are relatively quick until a terrain age of about 50 years and decrease afterwards, as later successional species grow more slowly (Matthews 1992; Rydgren et al. 2014). The

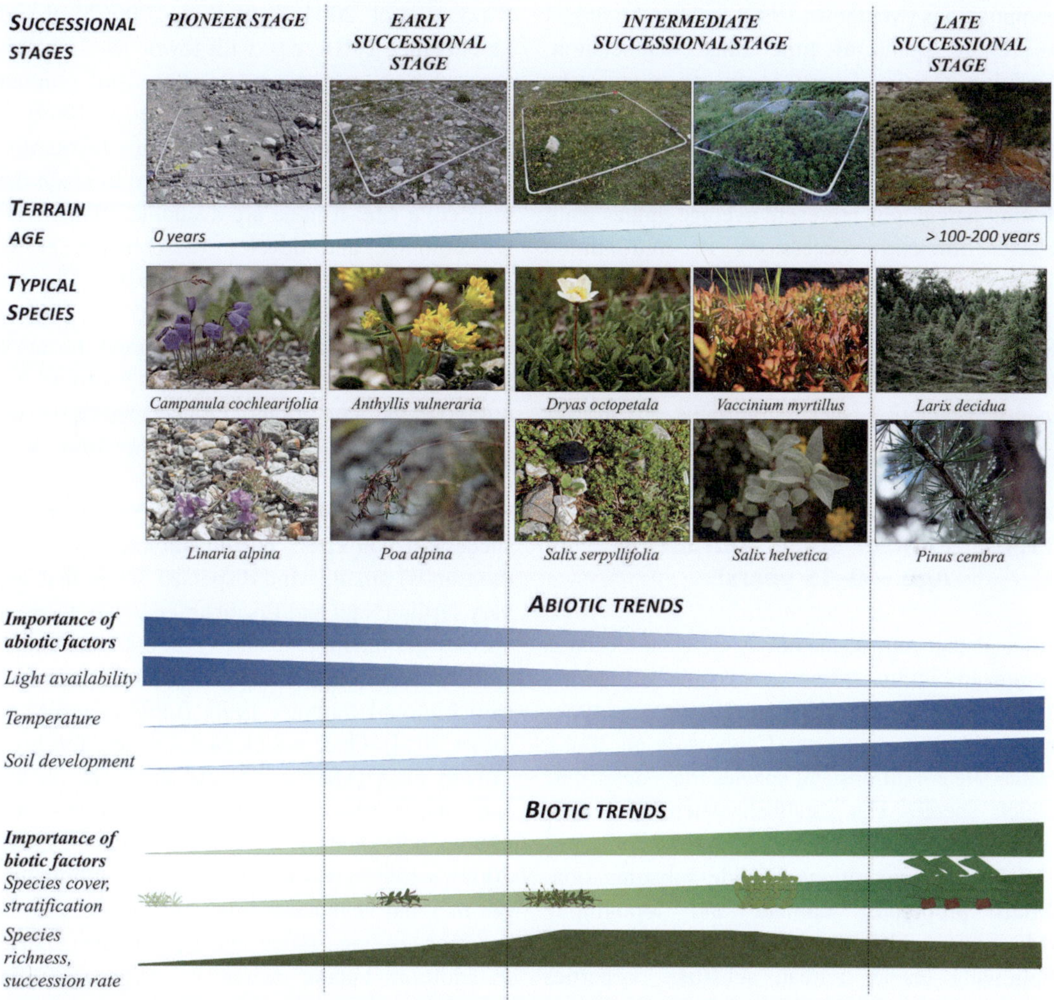

Fig. 19.1 Successional stages and linked abiotic and biotic successional trends for primary succession in glacier forelands. For abiotic and biotic factors in general, their changing importance during vegetation succession is illustrated, while changing values/intensities are illustrated for individual abiotic (light availability, temperature, soil development) and biotic variables (species cover, stratification; species richness, succession rate). Exemplary successional stages and typically occurring species are taken from Turtmann glacier foreland, Switzerland (Eichel et al. 2013, 2016)

linked trends in terrain age, soil development, species numbers and vegetation stratification have been termed 'terrain age factor complex' (Matthews and Whittaker 1987; Matthews 1992). As species composition changes, the dominant life forms, dispersal strategies and other traits also change (Wilmanns 1993; Chapin et al. 1994). Caccianiga et al. (2006) found that ruderal species are replaced by stress-tolerant species with increasing terrain age. This supports Grime's (1977) CSR model. Thus, life history traits associated with species strategies strongly control the colonization process and succession rates (Erschbamer and Mayer 2011; Schwienbacher et al. 2012).

How abiotic and biotic factors are linked to occurring species and their typical traits during succession is described in the following sections using *successional stages*. Successional stages are classified based on specific species compositions (Matthews 1992) and can be linked to certain spans of time since deglaciation. Thereby, they are very useful to anchor the sequence of occurring species compositions in time, despite their limitations (cf. Matthews 1992). The terrain age values used in the following paragraphs represent average values from glacier foreland successions in mid-latitude alpine environments and mesic habitats (Ellenberg 1996; Nagl and Erschbamer 2010; Robbins and Matthews 2010; Eichel et al. 2013). In higher latitudes and more severe environments, vegetation succession rates are slower (Matthews 1992; Robbins and Matthews 2010).

19.4.1 Pioneer Stage (Terrain Age ~ 0–15 years)

After glacier retreat, the newly exposed terrain is colonized within a few years time (Matthews 1992; Nagl and Erschbamer 2010), sometimes even within the first year (Cannone et al. 2008). Initial site conditions and species traits determine which species can establish where and after which time.

Initial site conditions include substrate (material) properties, climate and topography (Matthews and Whittaker 1987; Matthews 1992). Following glacier retreat, substrate properties change quickly through active processes such as deflation, pervection (downwashing of fine material from the surface) and frost sorting (Matthews and Vater 2015). Low temperatures and glacier winds affect species growing close to the glacier (Matthews 1992; Kaufmann 2002). In addition, elevation and aspect control site-specific climatic properties, such as temperature, wind, solar radiation (Körner 2003), which in turn control species compositions and successional pathways in glacier forelands (Raffl et al. 2006; Robbins and Matthews 2014). On a small spatial scale (cm–m extent), seedling germination and establishment often depends on microtopographic safe sites, such as depressions or larger rocks. In these protected areas, conditions are more favourable than elsewhere (Cooper et al. 2004; Raffl et al. 2006; Nagl and Erschbamer 2010), e.g. with lower wind velocities, slower surface water flow and higher moisture availability (Jumpponen et al. 1999).

However, even if conditions are favourable for colonization, diaspores first have to reach the respective site: if none are available, plants cannot colonize and develop (Erschbamer et al. 2008; Nagl and Erschbamer 2010). Pioneer diaspores are often transported by wind (anemochory). Adapted species traits promote seedling germination and establishment under unfavourable conditions. Pioneer species (e.g. *Linaria alpina, Campanula cochlearifolia, Saxifraga aizoides*; Fig. 19.1) are often ruderal strategists (Caccianiga et al. 2006), which are successful colonizers due to numerous (several thousands) small, wind-dispersed seeds that are very fertile (Nagl and Erschbamer 2010). Pioneer species germinate and survive at cold and moist conditions and can tolerate high solar radiation (Stöcklin and Bäumler 1996; Robbins and Mathews 2014). Once established, they exhibit high growth rates (Marcante et al. 2009). Nevertheless, the mortality of seedlings, plantlets and young individuals is still high (Marcante et al. 2009), while individual density (3–4 individuals per m^2) and vegetation cover are low (<10%) in the pioneer stage (Nagl and Erschbamer 2010). In addition, lichen, mosses and algae are also important colonizers at the pioneer stage (Burga et al. 2010; Garibotti et al. 2011).

19.4.2 Early-Successional Stage (Terrain Age ~ 15–40 years)

Within one to two decades, short-lived pioneers can start to die back due to age constraints, pioneers specifically adapted to moist and cold conditions can be excluded from the sites by changing environmental conditions or competition (Stöcklin and Bäumler 1996; Erschbamer et al. 2008). Other species from the pioneer stage can persist in later stages as they are able to survive despite changing environmental conditions, e.g. through clonal growth and a high phenotyphic plasticity (Stöcklin and Bäumler 1996).

Early-successional species (e.g. *Anthyllis vulneraria, Poa alpina, Trifolium pallescens*; Fig. 19.1) start to dominate under improved environmental conditions with higher temperatures, advanced soil development and a stabilized substrate (Matthews and Vater 2015). They are often ruderal or intermediate strategists and possess roots that are associated with bacteria that can fix nitrogen in the developing soil (Kuen and Erschbamer 2002; Caccianiga et al. 2006; Raffl et al. 2006; Schweingruber et al. 2007). At this stage, cryptogamic crusts also commonly develop and stabilize the ground (Matthews and Vater 2015). Due to this facilitation, further species can establish and vegetation cover (30–50%) and species numbers increase (Matthews 1992; Raffl et al. 2006).

19.4.3 Intermediate Successional Stage (Terrain Age ~ 40–80 Years)

Most species occur 40–50 years after glacier retreat and create a high vegetation cover (around 60–70%). The larger and less numerous diaspores of the later successional species are often transported into the foreland by geomorphic processes originating at the steep slopes outside the glacier foreland, such as avalanches and debris flows (Raffl et al. 2006; Nagl and Erschbamer 2010). In the intermediate successional stage, vegetation stratification starts with the occurrence of shrub species (e.g. *Salix* spp.; Fig. 19.1; Andreis et al. 2001; Raffl et al. 2006; Nagl and Erschbamer 2010). Spatially variable snow and moisture conditions lead to the co-occurrence of different plant communities in the intermediate successional stage (Ellenberg 1996). Common communities are initial alpine grasslands (e.g. *Elynetum*), snowbed communities (e.g. *Salix herbaceae* and *Salix retusae-reticulatae*) in areas with late lying snow or other grassland communities dominated by dwarf shrubs (*Dryadeto-Firmetum*; Eichel et al. 2013; Ellenberg 1996; Robbins and Matthews 2014). Intermediate successional species often actively contribute to soil development, for example by producing an organic matter layer (*Elyna myosuroides*) or further nitrogen fixation (*Dryas octopetala*) (Reisigl and Keller 1994).

19.4.4 Late-Successional Stage (Terrain Age > 80–100 Years)

In the late-successional stage, stratification is pronounced with the occurrence of tree species (e.g. *Larix decidua*; Fig. 19.1). In their shadows, shade plants (e.g. *Pyrola minor*) replace less-competitive light-adapted species (Raffl et al. 2006; Schweingruber et al. 2007; Robbins and Matthews 2014). In alpine environments, late-successional species are often stress strategists (e.g. *Carex curvula*) that are less able to disperse and less fertile than pioneer species, but can grow more quickly and survive despite competition (Stöcklin and Bäumler 1996; Caccianiga et al. 2006; Marcante et al. 2009). The mature vegetation in the stage has often been called 'climax community' (mature plant community determined by climatic conditions; Whittaker 1974). However, this view has been strongly challenged, as a variety of alternative mature communities (grassland, dwarf shrub, shrub or forest communities) can occur in one area (Matthews and Whittaker 1987). On a longer timescale, vegetation change is a constant process; thus, a final, permanent 'climax' may never reached as it is replaced other communities (Matthews 1992).

19.5 What Disrupts the Simple Successional Sequence?

In reality, vegetation succession in glacier forelands is often not a linear, straightforward sequence of successional stages. Instead, vegetation succession rates and successional pathways can strongly differ within and between different glacier forelands (Matthews and Whittaker 1987; Matthews 1992; Raffl et al. 2006). On a local scale, spatially variable environmental conditions affect succession pathways and rates, while aspect and topography create different successional dynamics on a landscape scale (Raffl et al. 2006; Garibotti et al. 2011; Rydgren et al. 2014). On a regional scale, they are controlled by bioclimatic gradients such as altitude and continentality (Robbins and Matthews 2010, 2014).

Local-scale spatial variability of soil texture and moisture produces fine-scale vegetation patterns, as fine material enhances plant establishment by providing a steady water supply due to its a high water retention capacity (Jochimsen 1962; Nagl and Erschbamer 2010). In contrast, blocky material is colonized very slow (Lüdi 1958; Ellenberg 1996). Differential successional rates and pathways can result from the differential colonization (Burga et al. 2010). In the course of succession, local-scale differences become less important and successional pathways often converge (Garibotti et al. 2011; Prach and Rachlewicz 2012). However, successional pathways can also diverge in later stages when strong, local-scale environmental differences persist, e.g., in severe physical environments (Matthews and Whittaker 1987; Matthews 1992).

At the landscape scale, Raffl et al. (2006) showed the importance of aspect for vegetation succession in the Rotmoosferner foreland (Austria). They found different successional pathways on the north-east and south-west exposed valley sides, which relate to differences in solar radiation, lithology and the occurrence of geomorphic processes. Furthermore, topographic parameters control vegetation succession dynamics, such as the slope toposequence of lateral moraine slopes (Minami et al. 1997; Vetaas 1997). Species composition is variable along the toposequence gradient due to different successional trajectories at moraine slope base, mid-slope and crest (Andreis et al. 2001; Garibotti et al. 2011). This gradient was attributed to the 'microtopographic factor complex' by Matthews and Whittaker (1987), which relates to snow melt, exposure based on topographic position and wind exposition, and moisture conditions. It is more effective on lateral moraine slopes than at the valley bottom (Andreis et al. 2001; Garibotti et al. 2011). Regionally, altitude and climate characteristics (e.g. continentality) influence succession rates and pathways. At higher altitudes, successional rates are slower (Körner 2003; Robbins and Matthews 2010, 2014) and mature stages often do not contain tree species, as these do not occur above certain altitudinal limits (Ellenberg 1996; Körner 2003; Robbins and Matthews 2010; 2014).

A further important factor influencing succession dynamics at the landscape scale is disturbance (Richter 1994; Matthews 1999; Raffl et al. 2006). How disturbances influence vegetation succession will be described in the following section.

19.6 The Role of Disturbances for Vegetation Succession in Glacier Forelands

Disturbances are destructive events that disrupt ecosystem, community or population structure, irrespective of if they are normal for the system or not (Pickett and White 1987). This disruption also includes changes in resources, in substrate availability or in the physical environment in general. How disturbances modify both the physical environment and vegetation succession was often ignored (Matthews 1999) by preferably selecting stable sites to study undisturbed glacier foreland succession (Prach and Rachlewicz 2012; Robbins and Matthews 2014). This bias in data collection may be a main reason for the fact that terrain age has long been considered more important than other environmental factors such as disturbances (Rydgren et al. 2014). By disrupting time-dependent vegetation development, secondary disturbances limit the applicability of frequently used chronosequence (space for time) approaches (Matthews 1999). These assume that a temporal sequence of development can be represented by space; thus, sites with an increasing distance from the glacier, and therefore an increasing age, are seen as representing sequential successional stages (Matthews 1992; Walker et al. 2010). However, this assumption is not valid when secondary disturbances occur (Matthews 1999).

In glacier forelands, geomorphic processes are most important in disturbing vegetation succession, besides wind-related, animal and human disturbances. Secondary disturbances affect large areas (up to 50%) of the glacier foreland (Oliver et al. 1985) and play a fundamental role in primary succession by modifying the substrate (erosion, deposition, mixing; see also Chap. 18 on soil development), damaging plants and influencing nutrient availability (Matthews 1999; Caccianiga et al. 2006). Thereby, they can delay, slow down or revert the successional process (retrogressive changes; Birks 1980; Matthews 1992; Matthews and Whittaker 1987). Due to disturbances, species composition and successional pathways can differ at terrain of the same age (Oliver et al. 1985; Matthews 1999; Moreau et al. 2008; Eichel et al. 2013). However, disturbances can also positively affect succession by transporting diaspores (Matthews 1999; Raffl et al. 2006).

The glacier and its dynamics of advance and retreat strongly control disturbances in glacier forelands. Based on this relationship, Matthews (1999) classified disturbances in glacier forelands into glacial, glacier-dependent, glacier-conditioned (paraglacial) and glacier-independent.

19.6.1 Glacial Disturbances

Glacial disturbances are disturbances due to direct contact between glacier ice and ecosystems, such as overriding during glacier advances, material deposition directly from the glacier onto plants, especially by debris flows transporting supraglacial moraine material, or slumping of moraines through dead-ice melt. Thus, glacial disturbances are limited to the area close to the glacier (Matthews 1999). Evidences for overriding of mature vegetation by glacial disturbances were found in several glacier forelands. Here, retreating glaciers exposed pieces of wood, organic lake sediments and peat, which were incorporated into the glacier during glacier advances in the Holocene (Nicolussi and Patzelt 2000; Hormes et al. 2001).

19.6.2 Glacier-Dependent Disturbances: Glaciofluvial Processes

Though ecosystems are not in a direct contact with the glacier ice, they can still be influenced by it and its melt. These glacier-dependent disturbances include glacioclimatic disturbances, which are often limited to the area close to the glacier, such as glacier winds; glaciohydrologic disturbances, such as an increased soil moisture close to the glacier due to ice melt and glaciofluvial disturbances by meltwater streams (see Chaps. 12 and 13 on proglacial rivers). In the foreland of the Midtre Lovénbreen glacier on Svalbard (Norway), Moreau et al. (2004, 2005, 2008) showed that glaciofluvial processes delay and disrupt vegetation succession and can more strongly determine species composition than terrain age. Pioneer stages continuously occur next to later successional stages at older terrain due to permanent disturbance by active glaciofluvial run-off (Moreau et al. 2004). In addition, newly developing channels on previously stable, older terrain can also trigger a regressive vegetation development: pioneer stages reoccur and replace later successional (Moreau et al. 2008). Thus, glaciofluvial processes can produce different successional pathways and mosaic of vegetation stages that can be independent from time since deglaciation (Moreau et al. 2008; Cowie et al. 2014).

19.6.3 Glacier-Conditioned (Paraglacial) Disturbances: Periglacial and Hillslope Processes

Glacier-conditioned disturbances are disturbances conditioned by glacier retreat and its effect on material properties, thus paraglacial disturbances (Matthews 1999; Ballantyne 2002).

They include aeolian deflation, loess-like redeposition, pervection, periglacial and hillslope processes (Matthews 1999; Chaps. 8–11, 17) and affect vegetation succession in a similar way as glaciofluvial processes:

Due to periglacial processes (Chaps. 6 and 7) such as cryoturbation and patterned ground development, colonization is delayed, pioneer stages persist on older terrain, and local-scale vegetation patterns occur in glacier forelands (Matthews and Whittaker 1987; Haugland and Beatty 2005; Haugland 2006). Close to the glacier, patterned ground develops quickly within 10 years of deglaciation (Haugland and Beatty 2005). Colonization of the patterned ground landforms by vascular plants is however delayed as frost heave damages their roots and dries up the soil (Haugland and Beatty 2005; Matthews 1999). Lichen and other bryophytes are important colonizers at this stage, as they can withstand low-intensity frost disturbances (Matthews et al. 1998). Pioneer vascular plant species (mainly grasses and sedges) can only colonize once frost activity decreases. This happens first at the patterned ground borders, where frost heave is less active due to reduced moisture in the quickly draining coarse material (Thorn 1976; Haugland and Beatty 2005; D'Amico et al. 2014). Patterned ground centres remain active for a longer time due to their frost-susceptible fine material and are colonized several decades later (Haugland and Beatty 2005). When frost activity declines, shrubs start to replace pioneer species first at the borders and later in the centre. After 70–80 years, the colonization of the centres draws level with the borders; however, a late-successional stage was not reached (Haugland and Beatty 2005; Haugland 2006). Similar to vegetation succession, soil development is delayed by active frost processes. Once activity decreases, horizons and organic matter develop together with plant colonization at the patterned ground borders (Haugland 2006). Thus, periglacial processes result in different successional pathways and a local-scale mosaic of different vegetation and soil development stages in the glacier foreland.

On moraine slopes and in their surroundings, hillslope processes are the main disturbance agent (Moreau et al. 2004; Raffl et al. 2006; Garibotti et al. 2011; Eichel et al. 2016; Chaps. 10, 11 and 17). In the Turtmann glacier foreland (Switzerland), Eichel et al. (2013) found that the sequence of successional stages is disrupted in areas with high slope gradient and elevation at terrain age 81–100 years: the lateral moraine slopes (Fig. 19.2a). While a later successional stage would be expected at this terrain age (Matthews 1992), pioneer, early and late-successional stages occur next to each other on the lateral moraine slopes (Fig. 19.2d). This can be attributed to active paraglacial slope processes: despite their age, the lateral moraine slopes are still reworked by highly or moderately active processes, such as slope wash (Fig. 19.2b; Chaps. 10 and 11), shallow landslides (Fig. 19.2c), debris flows (Fig. 19.2e; Chap. 11) or snow avalanches (Fig. 19.2e; Chap. 17). They produce pioneer stages and early-successional stages, delay or set back vegetation succession, and different successional pathways occur (Moreau et al. 2004; Eichel et al. 2013).

According to the paraglacial concept, the impact of glacier-conditioned disturbances should decrease in time as sediment is consolidated or exhausted and slopes stabilize due to a

Fig. 19.2 Paraglacial impact by slope processes on vegetation succession in the Turtmann glacier foreland. **a** Paraglacial impact showing in the NMDS result of vegetation plot data, overlay with terrain age classes. NMDS1 axis is related to time since deglaciation, ecosystem development (vegetation) cover and a successional sequence. At areas with high slope gradient, elevation (NMDS2 axis) and terrain age (lateral moraine slopes), three different successional stages were found, which can be attributed to a paraglacial impact by slope processes (Reproduced from Eichel et al. 2013 with permission from Elsevier). **b** Active slope wash inhibiting vegetation colonization at the lateral moraine crest. **c** A small shallow landslide removing vegetation at the moraine crest in 2015. **d** Lateral moraine slope in the Turtmann glacier foreland disturbed by different slope processes. **e** *Salix* spp. shrubs at the lateral moraine foot buried by a debris flow. **f** *Larix decidua* tree uprooted at the moraine slope by a snow avalanche ▶

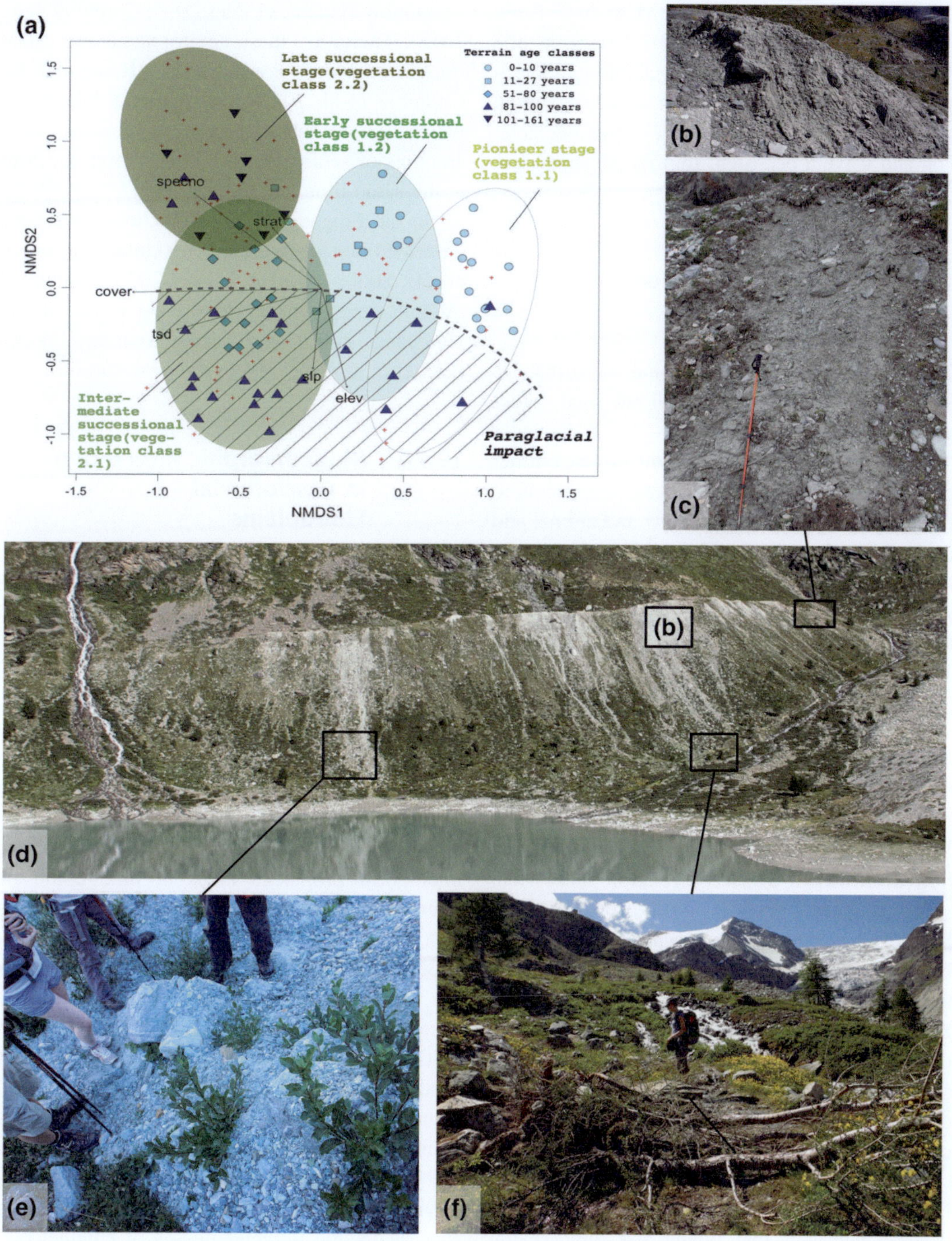
(a)
Late successional stage(vegetation class 2.2)
Early successional stage(vegetation class 1.2)
Pionieer stage (vegetation class 1.1)
Terrain age classes
0-10 years
11-27 years
51-80 years
81-100 years
101-161 years
specno
strat
cover
tsd
slp
elev
Inter-mediate successional stage(vege-tation class 2.1)
Paraglacial impact
NMDS2
NMDS1
1.5
1.0
0.5
0.0
-0.5
-1.0
-1.5
(b)
(c)
(d)
(e)
(f)

protective vegetation cover (Matthews 1999; Robbins and Matthews 2010). However, the described studies show that glaciofluvial, periglacial and slope processes can affect vegetation succession even several decades after deglaciation and leave a persistent imprint on vegetation patterns for more than a century.

19.6.4 Glacier-Independent Disturbances

Glacier-independent geomorphic disturbances include avalanches, debris flows and other gravitative processes (Chaps. 8–11, 17) that enter the glacier foreland from adjacent slopes. Further glacier-independent disturbances are nivation from late lying snow, windthrow and animal and human disturbances. In comparison with glacier related disturbances, these processes can be viewed as exogenous to the glacier foreland and are not controlled by terrain age (Matthews 1999).

19.7 Biogeomorphic Interactions in Glacier Forelands

While it was shown for several glacier forelands how disturbances impact vegetation succession, very few studies investigated the reverse effect: how plants influence the occurrence of disturbances, and how vegetation succession and disturbances interact (Matthews and Vater 2015; Eichel et al. 2016). Biogeomorphology studies these reciprocal interactions between geomorphic processes, landforms, plants and vegetation development. Biogeomorphic research, especially in fluvial, coastal and arid environments, showed not only how geomorphic processes influence plant colonization and vegetation development, but that plants and their spatial patterns can actively control the their occurrence (Corenblit et al. 2007, 2015; Gurnell et al. 2012; Balke et al. 2014; Gurnell 2014). It is believed that biogeomorphic interactions are strongly controlling landscape development in environments with highly active geomorphic and vegetation dynamics (Viles 2004; Corenblit et al. 2015). Therefore, these environments are called '*biogeomorphic ecosystems*' (Balke 2013; Corenblit et al. 2015), and glacier forelands, in which geomorphic processes rework high amounts of sediments while vegetation develops at the same time, are one of them.

So far, biogeomorphic interactions have been described for glaciofluvial floodplains (Gurnell et al. 2000; Moreau et al. 2008), patterned ground development (Matthews et al. 1998; Haugland 2006) and lateral moraine slopes (Eichel et al. 2013, 2016; see next section).

19.7.1 Biogeomorphic Interactions in Glaciofluvial Floodplains

In glaciofluvial floodplains, the channel pattern development is controlled by interactions between riparian vegetation development and glaciofluvial dynamics. Gurnell et al. (2000) describe these interactions in a conceptual model, which links the development of channel patterns with increasing distance to the glacier (downstream gradient) to riparian vegetation succession (Fig. 19.3).

How glaciofluvial and vegetation dynamics interact in proglacial areas in time could be interpreted as a '*fluvial biogeomorphic succession*' sensu Corenblit et al. (2007). This biogeomorphic succession concept describes how the dominant process regime changes within a decadal successional timescale from physical (geomorphic) to biotic (vegetation) after a destructive flood. Four biogeomorphic successional phases show how interactions between plants and fluvial change in time. As for glacier forelands the retreating glacier initiates glaciofluvial floodplain development and vegetation succession, the different phases of channel pattern development could be interpreted as biogeomorphic successional phases (Fig. 19.3). (1) *Geomorphic phase*: In the bar-braided section close to the glacier, sediment input is high, and geomorphic processes dominate and inhibit plant colonization. Channel bank strength is low, as

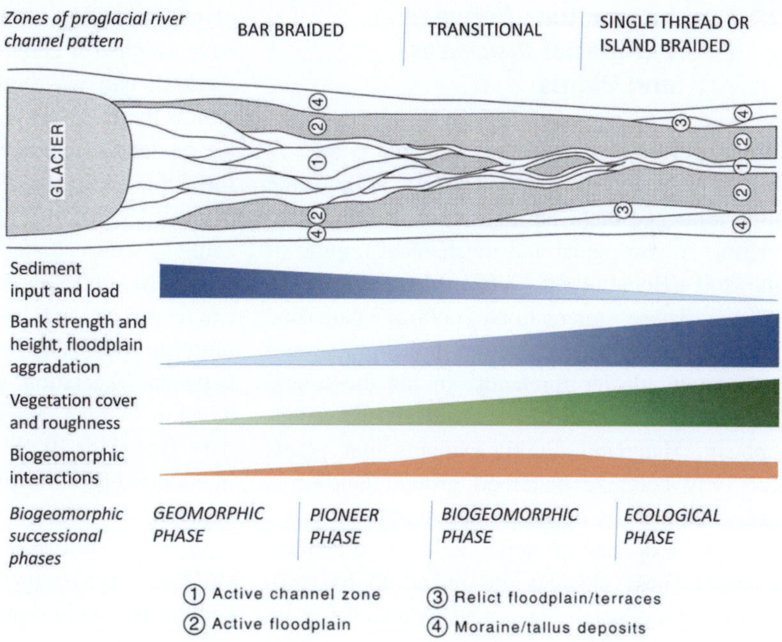

Fig. 19.3 Interactions between riparian vegetation and channel pattern development in a glaciofluvial floodplain. Channel pattern changes and associated changes in sediment and vegetation properties are related to biogeomorphic successional phases sensu Corenblit et al. (2007). Reproduced from Gurnell et al. (2000) with permission from Elsevier

stabilizing vegetation cover is missing. Therefore, the multiple channels constantly shift (Gurnell et al. 2000; Moreau et al. 2008). (2) *Pioneer phase*: With increasing distance to the glacier and terrain age, pioneer species start to colonize the channel banks and bars of the braided channel pattern. With developing vegetation cover, plants' above- and below-ground biomass increases. In the (3) *biogeomorphic phase*, glaciofluvial processes and vegetation start to interact. Below-ground biomass, such as roots, increases bank strength, which in turn limits lateral erosion and thereby reduces channel width. Above-ground vegetation biomass increases surface roughness, reduces flow velocities, protects from scour and enhances rate of sediment deposition. This promotes floodplain aggradation, which in turn increases bank height. Less sediment is supplied to the downstream sections of the river. The channel pattern type changes to a transitional type with a decreasing number of channels (Gurnell et al. 2000; Moreau et al. 2008). In the Midtre Lovénbreen glacier foreland, Moreau et al. (2008) observed that this takes about 30 years. Woody plant species, such as *Salix* spp. shrubs, play a special role for the transition. Through their adapted traits, such as robust root networks, they act as ecosystem engineers in fluvial systems (Gurnell 2014). At this stage, vegetation succession proceeds in channels with only intermittent snow or ice-melt run-off or in inactive channels (Moreau et al. 2008). With increasing vegetation cover and channel stabilization, the channel pattern develops to a single thread or island braided with vegetated islands. In this (4) *ecological phase*, vegetation is made up by comparably stable communities (Moreau et al. 2008) and exerts a major control on the glaciofluvial system (Gurnell et al. 2000). The importance of biogeomorphic interactions decreases, and biotic processes, such as competition, become the dominant drivers of vegetation change.

Downstream disturbances such as lateral inputs of water or sediment from surrounding slopes or the presence of rock steps can disrupt the sequence of biogeomorphic succession (Gurnell et al. 2000). Furthermore, whether interactions occur and how quick the biogeomorphic succession proceeds also depends on factors controlling vegetation succession (see previous sections), such as altitude, successional pathways and late-successional species composition.

19.7.2 Interactions Between Periglacial Processes and Plants

While periglacial processes (Chaps. 6 and 7) disturb vegetation succession, plants at the same time influence them through their effects on the thermal, hydrological and mechanical regime of the soil (Benninghoff 1952; Matthews et al. 1998). These interactions control patterned ground activity and vegetation development and patterns in glacier forelands. In his theoretical geoecological model of patterned ground development, Haugland (2006) explains that plants can only colonize patterned ground landforms once their activity decreases due to reduced water contents, a decline of slope angle and a textural change. These changes are linked to the proceeding paraglacial adjustment (Matthews et al. 1998). Once plants establish, multidirectional interactions between frost activity, plant colonization and soil development start, first at the patterned ground borders and later in the landform centre (Haugland 2006). Vegetation influences frost activity by changing the thermal and moisture regime of the soil, e.g. by shading, retaining moisture and intercepting rain (Benninghoff 1952; Ghestem et al. 2011). Plant roots decrease the soil movement by increasing soil shear strength, related to increased angle of internal friction and soil cohesion (Fattet et al. 2011; Veylon et al. 2015), and reinforcing the soil matrix (Ghestem et al. 2011, 2014).

19.8 A Detailed Case Study: Biogeomorphic Dynamics on Lateral Moraine Slopes in the Turtmann Glacier Foreland, Switzerland

So far, it is not well understood if and how vegetation stabilizes lateral moraine slopes and which role paraglacial processes play for vegetation succession. Curry et al. (2006) and Mercier et al. (2009) challenged the view that vegetation colonization is a major cause for slope stabilization (Ballantyne 2002). They found that plants only colonize moraine slopes once these have reached a certain degree of stability. This research gap was addressed by recent research within the BIMODAL ('Biogeomorphic dynamics on lateral moraines in the Turtmann glacier forefield, Switzerland') project (Fig. 19.4; Eichel et al. 2013, 2016). This project investigates (i) if some species can stabilize the slopes better than others, (ii) under which conditions these species can establish and interact with occurring geomorphic processes and (iii) how interactions between vegetation and paraglacial dynamics develop in time. Mainly permanent plot survey data (see Fig. 19.4b for plot location), which included vegetation properties (species and total vegetation cover), biotic properties (species number, stratification), geomorphic processes and landforms occurring in the plots, geomorphic activity and geomorphometry, material properties (grain sizes), soil properties (horizons) and terrain age, was used to answer these questions (see Eichel et al. 2016 for details). A summary of results at the current state of the project is provided in this section.

19.8.1 Ecosystem Engineering by *Dryas octopetala* L.

Biogeomorphic research showed that some species can more effectively affect their environment than others (Gurnell 2014; Corenblit et al. 2015). These *ecosystem engineers* are species that change their physical environment through their adapted traits and thereby create habitats for other species (Jones et al. 1994; Jones 2012). On lateral moraine slopes, *D. octopetala* was identified as an ecosystem engineer (Eichel et al. 2016). Permanent plot data showed that geomorphic activity permanently decreases once *D. octopetala* reaches a cover value of 35%. Above this cover value, the dominant geomorphic processes change from slope wash and slide to bound solifluction. *D. octopetala* induces these changes through its adapted morphological and root traits. Through its strong and deeply rooted woody root stock (Fig. 19.5b), it is firmly anchored even on moving slopes. Its branching

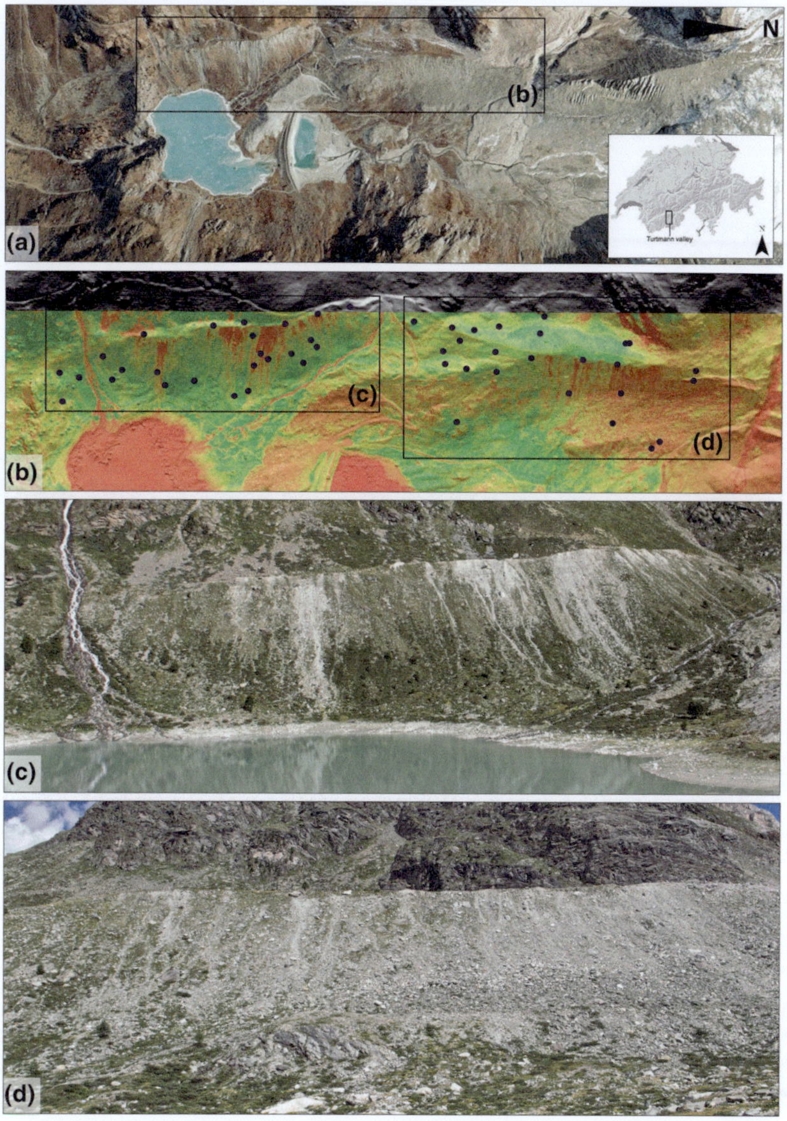

Fig. 19.4 Lateral moraines in the Turtmann glacier foreland, Switzerland. **a** Location of Turtmann glacier foreland and location of lateral moraines within the foreland. **b** Location of permanent plots (blue dots) on the lateral moraine slopes. **c** Lateral moraines with a terrain age of 80–100 years, showing complex patterns of geomorphic activity and vegetation, and **d** Lateral moraine slopes with a terrain age of 60–80 years

fine lateral and coralloid root systems are associated with ectomycorrhiza (Reisigl and Keller 1994; Ellenberg 1996; Kutschera et al. 1997; Harrington and Mitchell 2002). Together, they increase soil aggregate stability and the angle of internal friction and thereby enhance slope stability (Graf et al. 2009; Bast et al. 2014). By growing in a mat with low-lying stems, branches and numerous leaves (Fig. 19.5a), *D. octopetala* accumulates fine sediments, biomass, humus and diaspores (Fig. 19.5c, d), stores moisture and limits soil erosion at the same time (Reisigl and Keller 1994; Welker et al. 1997; Körner 2003). As fine material and soil moisture condition intensify solifluction process (Harris et al. 2008), solifluction is probably favoured below the *Dryas* mat. These interactions between *D. octopetala* and solifluction could promote the development of turf-banked solifluction lobes, making these 'biogeomorphic structures' (Corenblit et al. 2010; Eichel et al. 2016, 2017). By stabilizing the slope, accumulating material and nitrogen fixation through its associated bacteria, *D. octopetala* acts as a nurse plant for other species (Fig. 19.5e,

Fig. 19.5 Engineering traits of *Dryas octopetala* L. **a** Mat growth of *D. octopetala*, covering moving debris. **b** Main root stock, branching after some cm into finer lateral roots. **c** Fine material accumulation in the bottom side of the *Dryas* mat. **d** Humus accumulation and fine roots within the mat, and **e**, **f** facilitation for other species (*Leontopodium alpinum* CASS. and *Campanula cochlearifolia* LAM). Reproduced from Eichel et al. (2016) with permission by Wiley and Sons

f) and promotes further ecosystem development to grassland associations, e.g. the intermediate or later successional *Elynetum* (Elkington 1971; Reisigl and Keller 1994; Ellenberg 1996; Klanderud and Totland 2004). Thus, ecosystem engineering by *D. octopetala* influences both paraglacial adjustment and vegetation succession on lateral moraine slopes (Eichel et al. 2016).

Probably, species with similar traits, such as the *Salix* dwarf shrubs *Salix herbacea*, *Salix reticulata* and *Salix serpyllifolia*, also affect lateral moraine slope development, especially if they grow in mats similar to *D. octopetala* (cf *Salix herbacea*, Wijk 1986; Beerling 1998).

19.8.2 Conditions for Biogeomorphic Feedbacks: A '*Biogeomorphic Feedback Window*'

As ecosystem engineers influence geomorphic processes, they have to be able to establish and reach a certain cover or biomass for biogeomorphic interactions to occur (Balke et al. 2014). Permanent plot data from the Turtmann glacier foreland showed that whether *D. octopetala* occurs and develops depends on the magnitude and frequency of the geomorphic processes, the *geomorphic disturbance regime* (Pickett and White 1987; Turner et al. 2001). Eichel et al. (2016) found that disturbance regime and species composition are interrelated on their lateral moraine slopes. They determined how resistant the dominantly occurring species are towards disturbances (magnitude of disturbance that they can absorb; Holling and Gunderson 2002) and how quickly they can recover after a disturbance (resilience; Gunderson 2000) based on their traits. By linking species resilience and resistance to the geomorphic disturbance regime, Eichel et al. (2016) identified the conditions under which engineer species can establish and biogeomorphic interactions can occur (*biogeomorphic feedback window*) (Fig. 19.6).

When processes with a high magnitude (gullying, debris flows) or frequency (interrill erosion, debris sliding) occur, plant colonization is limited. Only species adapted to these conditions can establish and grow, for example the

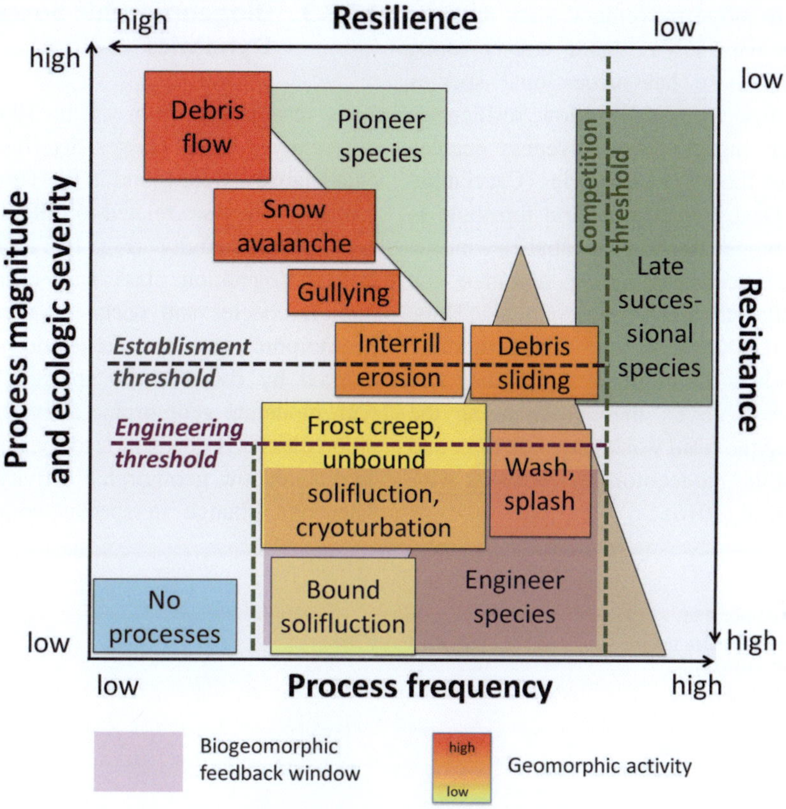

Fig. 19.6 Biogeomorphic feedback window, showing how the relationship between geomorphic disturbance regime (process magnitude, ecological severity, process frequency) and plant species traits (resilience and resistance) controls the occurrence of biogeomorphic feedbacks once an establishment and engineering threshold are passed. Geomorphic (process magnitude and ecological severity, process frequency) and ecological (resilience, resistance) axes are displayed together in this figure to illustrate the conditions for feedbacks, but are not functionally related. Reproduced from Eichel et al. (2016) with permission by Wiley and Sons

pioneer species *Linaria alpina* (Fig. 19.6). Its root system and stems are adapted to debris movements, and it can quickly re-establish after being destroyed by disturbance (Schröter et al. 1926; Rauh 1939; Ellenberg 1996). Therefore, *Linaria alpina* possesses an intermediate resistance, but high resilience. As engineer species, e.g. *D. octopetala,* grow more slowly, they need a long time to re-establish after a disturbance (low resilience) (Hendry and Grime 1993) and do not grow when disturbances occur frequently. When slopes stabilize in the course of paraglacial adjustment, either the process frequency or the magnitude of the occurring processes decreases (Ballantyne 2002; Mercier et al. 2009). This gives the engineer species sufficient time or more suitable conditions (less sediment movements) to successfully establish in '*windows of opportunity*' sensu Balke et al. (2011, 2014). *D. octopetala*, for example, can firmly anchor in the slope due to its high root strength. Engineer cover and biomass increase and can cross an engineering threshold: geomorphic activity permanently decreased when *D. octopetala* covered around 35% of the plot. The biogeomorphic feedback window is reached in which *Dryas* or other engineer species limit sediment transport and change soil texture and moisture. Bound solifluction becomes dominant, and the geomorphic disturbance regime changes. Species that are less adapted to slope movement, such as *Salix* shrub species with low-resistant woody roots, can establish. These late-successional species take a long time to establish (low resilience). However, their higher competitiveness enables them to replace the engineer species (Caccianiga et al. 2006). Once this competition threshold is passed, the biogeomorphic feedback window ends and later successional species dominate in a geomorphically stabilized environment. Thus, biogeomorphic feedbacks on lateral moraine slopes start when the engineer establishes and develops a large cover, thereby changing the disturbance regime from wash and slide to bound solifluction in the biogeomorphic feedback window (Eichel et al. 2016).

19.8.3 Biogeomorphic Succession Dynamics

At a terrain age from 81 to 100 years, three different vegetation classes, determined from plot data analysis, were found in the Turtmann glacier foreland. They are related to different degrees of geomorphic activity (Fig. 19.7a; Eichel et al. 2013). Vegetation class 1 is characterized by pioneer species and occurs in areas with high geomorphic activity. Vegetation class 2 is marked by dwarf shrub species, which occur with moderate geomorphic activity, while vegetation class 3 is characterized by shrub species in areas with low geomorphic activity (Fig. 19.7a) The joint change in species composition and

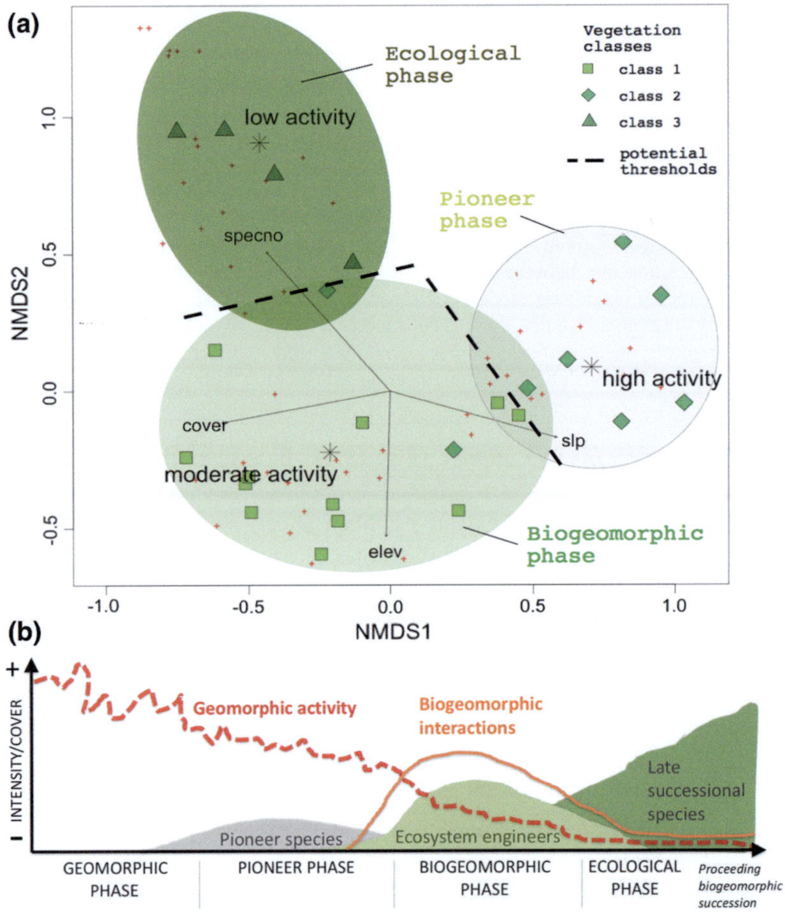

Fig. 19.7 Biogeomorphic succession on lateral moraine slopes. **a** NMDS result from Eichel et al. (2013) showing biogeomorphic succession phases at constant terrain age, based on vegetation and activity classes. Environmental parameters: *specno* = species number, *cover* = vegetation cover, *slp* = slope, *elev* = elevation (Reproduced from Eichel et al. 2013 with permission by Elsevier), and **b** conceptual model of biogeomorphic succession at lateral moraine slopes. Decreasing geomorphic activity and increasing biogeomorphic interactions control the changing species composition and cover and interact with each other

geomorphic activity was interpreted as a biogeomorphic succession with three biogeomorphic successional phases (Fig. 19.7a, b). In the (1) *pioneer phase*, abiotic processes control vegetation establishment; therefore, only vegetation class 1 dominated by pioneer species occurs in this phase in areas with high geomorphic activity. As scree-wanderer and as scree-creeper, the pioneer species *Achillea nana* and *Linaria alpina* are adapted to the moving debris (Schröter et al. 1926). In the (2) *biogeomorphic phase*, abiotic and biotic processes start to interact. As described in the biogeomorphic feedback window, engineer species (dwarf shrubs) in vegetation class 2 actively influence the occurring geomorphic processes. In this phase, biogeomorphic interactions are most intense. Under moderate geomorphic activity, intermediate successional stages, such as grassland associations (*Dryadeto-Firmetum, Elynetum*), are developed. In the (3) *ecological phase,* geomorphic activity is low and later successional species (vegetation class 3) dominate and the ecosystem engineers are excluded as biotic processes such as competition and inhibition become dominant.

In fluvial environments, biogeomorphic successions start after high-magnitude disturbances (large floods). However, Eichel et al. (2016) found that biogeomorphic succession is not related to the terrain age of lateral moraine slopes, which would be expected with the retreating glacier as last major disturbance. Instead, changes in species composition are related to decreasing geomorphic activity (Fig. 19.7b). This decrease in geomorphic activity, related to paraglacial adjustment, controls the biogeomorphic succession until biogeomorphic feedbacks are strong enough to amplify the decrease and finally stop geomorphic activity.

Thus, both vegetation succession and paraglacial adjustment are not necessarily time-dependent on lateral moraine slopes, but driven by interactions between plants and geomorphic processes. These biomorphic interactions control the landscape development following glacier retreat (Eichel et al. 2013, 2016).

19.9 Conclusion

Interactions between vegetation succession and paraglacial adjustment play a key role for the development of proglacial landscapes following glacier retreat. Based on the ecological and biogeomorphic knowledge presented in the previous sections, the questions posed in the introduction can be answered:

(1) ***How long does it take for vegetation to establish after glacier retreat?***
Colonization by pioneer species can start within years after glacier retreat. With increasing terrain age, successional stages follow each other with species adapted to the changing environmental conditions. A species composition similar to that outside of the glacier foreland can be reached after decades to centuries in mid-latitude alpine environments, depending on regional (altitude, continentality), landscape (aspect, slope toposequence) and local (soil texture and moisture) conditions.

(2) ***When and how do disturbances influence vegetation establishment and succession?***
Vegetation establishment and succession can be inhibited, limited, altered or delayed by disturbances for several decades to centuries. At any time after glacier retreat, active geomorphic processes can produce local- and landscape-scale vegetation patterns that disrupt the time-dependent vegetation succession gradient.

(3) ***When and how are glacial sediments stabilized?***
Ecosystem engineer species can establish on glacial sediments when the magnitude or frequency of paraglacial processes decreases to a certain degree, usually after a few decades. Once these species reach a certain cover or amount of below-ground biomass years to decades following establishment, they start to actively influence geomorphic processes by increasing slope stability, limiting soil and lateral erosion, enhancing sediment deposition and changing the

hydrological and thermal regimes. Geomorphic activity strongly decreases in this biogeomorphic feedback window, and later successional species, such as trees, establish in the stabilized environment.

However, in all successional phases, vegetation can get partly or completely destroyed by high-magnitude processes and sediment reworking can recommence with no or pioneer vegetation. Thus, glacial sediments colonized by plants are not necessarily permanently stable. More research is needed to understand biogeomorphic dynamics and their role for sediment stabilization in glacier forelands, which also needs to integrate the changing climate.

Acknowledgements The research presented in some sections of this chapter is part of the BIMODAL (Biogeomorphic dynamics on lateral moraines in the Turtmann glacier forefield, Switzerland) project, funded by the German Research Foundation DFG (DI 414/22-1). Thanks go to two anonymous reviewers, the editor Tobias Heckmann and Daniel Draebing for helpful and constructive comments on the manuscript.

References

Andreis C, Caccianiga M, Cerabolini B (2001) Vegetation and environmental factors during primary succession on glacier forelands: some outlines from the Italian Alps. Plant Biosyst 135:295–310. https://doi.org/10.1080/11263500112331350930

Balke T (2013) Establishment of biogeomorphic ecosystems : a study on mangrove and salt marsh pioneer vegetation

Balke T, Bouma TJ, Horstman EM et al (2011) Windows of opportunity: thresholds to mangrove seedling establishment on tidal flats

Balke T, Herman PM, Bouma TJ (2014) Critical transitions in disturbance-driven ecosystems: identifying Windows of Opportunity for recovery. J Ecol 102: 700–708. https://doi.org/10.1111/1365-2745.12241

Ballantyne CK (2002) Paraglacial geomorphology. Quat Sci Rev 21:1935–2017

Bast A, Wilcke W, Graf F et al (2014) The use of mycorrhiza for eco-engineering measures in steep alpine environments: effects on soil aggregate formation and fine-root development. Earth Surf Process Landf 39:1753–1763. https://doi.org/10.1002/esp.3557

Beerling DJ (1998) Salix herbacea L. J Ecol 86:872–895. https://doi.org/10.1046/j.1365-2745.1998.8650872.x

Benninghoff WS (1952) Interaction of vegetation and soil frost phenomena. Arctic 5:34–44. https://doi.org/10.14430/arctic3898

Birks HJB (1980) The present flora and vegetation of the moraines of the Klutlan Glacier, Yukon Territory, Canada: a study in plant succession. Quat Res 14:60–86. https://doi.org/10.1016/0033-5894(80)90007-1

Bogen J (1988) Glacial sediment production and development of hydro-electric power in glacierized areas. Ann Glaciol 13:6–11

Burga CA, Krusi B, Egli M et al (2010) Plant succession and soil development on the foreland of the Morteratsch glacier (Pontresina, Switzerland): Straight forward or chaotic? Flora-Morphol Distrib Funct Ecol Plants 205:561–576

Caccianiga M, Luzzaro A, Pierce S et al (2006) The functional basis of a primary succession resolved by CSR classification. Oikos 112:10–20

Cannone N, Diolaiuti G, Guglielmin M, Smiraglia C (2008) Accelerating climate change impacts on alpine glacier forefield ecosystems in the European Alps. Ecol Appl 18:637–648. https://doi.org/10.1890/07-1188.1

Chapin FS, Walker LR, Fastie CL, Sharman LC (1994) Mechanisms of primary succession following deglaciation at Glacier Bay, Alaska. Ecol Monogr 64:149–175. https://doi.org/10.2307/2937039

Clements FE (1928) Plant succession and indicators: a definitive edition of plant succession and plant indicators. The Wilson Company, New York

Connell JH, Slatyer RO (1977) Mechanisms of succession in natural communities and their role in community stability and organization. Am Nat 111:1119–1144

Cooper EJ, Alsos IG, Hagen D et al (2004) Plant recruitment in the High Arctic: seed bank and seedling emergence on Svalbard. J Veg Sci 15:115–124. https://doi.org/10.1658/1100-9233(2004)015%5b0115:PRITHA%5d2.0.CO;2

Corenblit D, Baas A, Balke T et al (2015) Engineer pioneer plants respond to and affect geomorphic constraints similarly along water–terrestrial interfaces world-wide. Glob Ecol Biogeogr 24:1363–1376. https://doi.org/10.1111/geb.12373

Corenblit D, Tabacchi E, Steiger J, Gurnell AM (2007) Reciprocal interactions and adjustments between fluvial landforms and vegetation dynamics in river corridors: a review of complementary approaches. Earth Sci Rev 84:56–86

Corenblit D, Steiger J, Delmotte S (2010) Abiotic, residual and functional components of landforms. Earth Surf Process Landf 35:1744–1750. https://doi.org/10.1002/esp.2064

Cowie NM, Moore RD, Hassan MA (2014) Effects of glacial retreat on proglacial streams and riparian zones in the Coast and North Cascade Mountains. Earth Surf Process Landf 39:351–365. https://doi.org/10.1002/esp.3453

Curry AM, Cleasby V, Zukowskyj P (2006) Paraglacial response of steep, sediment-mantled slopes to post-"Little Ice Age" glacier recession in the central Swiss Alps. J Quat Sci 21:211–225. https://doi.org/10.1002/jqs.954

D'Amico ME, Freppaz M, Filippa G, Zanini E (2014) Vegetation influence on soil formation rate in a proglacial chronosequence (Lys Glacier, NW Italian Alps). Catena 113:122–137. https://doi.org/10.1016/j.catena.2013.10.001

Dierschke H (1994) Pflanzensoziologie: Grundlagen und Methoden. UTB, Stuttgart

Egli M, Wernli M, Burga C et al (2011) Fast but spatially scattered smectite-formation in the proglacial area Morteratsch: an evaluation using GIS. Geoderma 164:11–21. https://doi.org/10.1016/j.geoderma.2011.05.001

Eichel J, Krautblatter M, Schmidtlein S, Dikau R (2013) Biogeomorphic interactions in the Turtmann glacier forefield, Switzerland. Geomorphology 201:98–110. https://doi.org/10.1016/j.geomorph.2013.06.012

Eichel J, Corenblit D, Dikau R (2016) Conditions for feedbacks between geomorphic and vegetation dynamics on lateral moraine slopes: a biogeomorphic feedback window. Earth Surf Process Landf 41:406–419. https://doi.org/10.1002/esp.3859

Eichel J, Draebing D, Klingbeil L, Wieland M, Eling C, Schmidtlein S, Kuhlmann H, Dikau R (2017) Solifluction meets vegetation: the role of biogeomorphic feedbacks for turf-banked solifluction lobe development. Earth Surf Proc Land 42(11):1623–1635

Elkington TT (1971) Dryas Octopetala L. J Ecol 59:887–905. https://doi.org/10.2307/2258146

Ellenberg H (1996) Vegetation Mitteleuropas mit den Alpen: in ökologischer, dynamischer und historischer Sicht. UTB, Stuttgart

Erschbamer B, Mayer R (2011) Can successional species groups be discriminated based on their life history traits? A study from a glacier foreland in the Central Alps. Plant Ecol Divers 4:341–351. https://doi.org/10.1080/17550874.2012.664573

Erschbamer B, Niederfriniger Schlag R, Winkler E (2008) Colonization processes on a central Alpine glacier foreland. J Veg Sci 19:855–862. https://doi.org/10.3170/2008-8-18464

Fattet M, Fu Y, Ghestem M et al (2011) Effects of vegetation type on soil resistance to erosion: Relationship between aggregate stability and shear strength. Catena 87:60–69. https://doi.org/10.1016/j.catena.2011.05.006

Fickert T, Friend D, Grüninger F et al (2007) Did debris-covered glaciers serve as Pleistocene refugia for plants? A new hypothesis derived from observations of recent plant growth on glacier surfaces. Arct Antarct Alp Res 39:245–257. https://doi.org/10.1657/1523-0430(2007)39%5b245:DDGSAP%5d2.0.CO;2

Garibotti IA, Pissolito CI, Villalba R (2011) Spatiotemporal pattern of primary succession in relation to meso-topographic gradients on recently deglaciated terrains in the Patagonian Andes. Arct Antarct Alp Res 43:555–567. https://doi.org/10.1657/1938-4246-43.4.555

Ghestem M, Sidle RC, Stokes A (2011) The influence of plant root systems on subsurface flow: implications for slope stability. Bioscience 61:869–879. https://doi.org/10.1525/bio.2011.61.11.6

Ghestem M, Veylon G, Bernard A et al (2014) Influence of plant root system morphology and architectural traits on soil shear resistance. Plant Soil 377:43–61. https://doi.org/10.1007/s11104-012-1572-1

Graf F, Frei M, Böll A (2009) Effects of vegetation on the angle of internal friction of a moraine. For Snow Landsc Res 82:61–77

Grime JP (1977) Evidence for the existence of three primary strategies in plants and its relevance to ecological and evolutionary theory. Am Nat 111:1169–1194

Gunderson LH (2000) Ecological resilience–in theory and application. Ann Rev Ecol Syst 31:425–439

Gurnell A (2014) Plants as river system engineers. Earth Surf Process Landf 39:4–25. https://doi.org/10.1002/esp.3397

Gurnell AM, Bertoldi W, Corenblit D (2012) Changing river channels: the roles of hydrological processes, plants and pioneer fluvial landforms in humid temperate, mixed load, gravel bed rivers. Earth Sci Rev 111:129–141. https://doi.org/10.1016/j.earscirev.2011.11.005

Gurnell AM, Edwards PJ, Petts GE, Ward JV (2000) A conceptual model for alpine proglacial river channel evolution under changing climatic conditions. Catena 38:223–242. https://doi.org/10.1016/S0341-8162(99)00069-7

Harrington TJ, Mitchell DT (2002) Characterization of Dryas octopetala ectomycorrhizas from limestone karst vegetation, western Ireland. Can J Bot 80:970–982. https://doi.org/10.1139/b02-082

Harris C, Smith JS, Davies MCR, Rea B (2008) An investigation of periglacial slope stability in relation to soil properties based on physical modelling in the geotechnical centrifuge. Geomorphology 93:437–459. https://doi.org/10.1016/j.geomorph.2007.03.009

Hauenstein W (2005) Hydropower and climate change—A reciprocal relation: institutional energy issues in Switzerland. Mt Res Dev 25:321–325. https://doi.org/10.1659/0276-4741(2005)025%5b0321:HACCRR%5d2.0.CO;2

Haugland JE (2006) Short-term periglacial processes, vegetation succession, and soil development within sorted patterned ground: Jotunheimen, Norway. Arct Antarct Alp Res 38:82–89

Haugland JE, Beatty SW (2005) Vegetation establishment, succession and microsite frost disturbance on glacier forelands within patterned ground chronosequences. J Biogeogr 32:145–153

Hendry GA, Grime JP (1993) Methods in comparative plant ecology: A laboratory manual. Springer, The Netherlands

Holling CS, Gunderson LH (2002) Resilience and adaptive cycles. In: Gunderson L, Holling CS (eds) Panarchy: understanding transformations in human and natural systems. Island Press, Washington, DC, pp 63–102
Hormes A, Müller BU, Schlüchter C (2001) The Alps with little ice: evidence for eight Holocene phases of reduced glacier extent in the Central Swiss Alps. Holocene 11:255–265. https://doi.org/10.1191/095968301675275728
Jochimsen M (1962) Das Gletschervorfeld - keine Wüste. Jahrb Österr Alpenvereins 87:135–142
Jones CG (2012) Ecosystem engineers and geomorphological signatures in landscapes. Geomorphology 157–158:75–87. https://doi.org/10.1016/j.geomorph.2011.04.039
Jones CG, Lawton JH, Shachack M (1994) Organisms as ecosystem engineers. Oikos 69:373–386. https://doi.org/10.2307/3545850
Jumpponen A, Vare H, Mattson KG et al (1999) Characterization of "safe sites" for pioneers in primary succession on recently deglaciated terrain. J Ecol 87:98–105
Kaufmann R (2002) Glacier foreland colonisation: distinguishing between short-term and long-term effects of climate change. Oecologia 130:470–475. https://doi.org/10.1007/s00442-001-0815-2
Klanderud K, Totland Ø (2004) Habitat dependent nurse effects of the dwarf-shrub Dryas octopetala on alpine and arctic plant community structure. Écoscience 11:410–420. https://doi.org/10.1080/11956860.2004.11682850
Körner C (2003) Alpine plant life: functional plant ecology of high mountain ecosystems. Springer, Berlin
Kuen V, Erschbamer B (2002) Comparative study between morphology and age of Trifolium pallescens in a glacier foreland of the Central Alps. Flora Morphol Distrib Funct Ecol Plants 197:379–384. https://doi.org/10.1078/0367-2530-00054
Kutschera L, Lichtenegger E, Sobotik M (1997) Bewurzelung von Pflanzen in verschiedenen Lebensräumen. OÖ Landesmuseum, Linz
Lavorel S, Díaz S, Cornelissen J et al (2007) Plant functional types: are we getting any closer to the holy grail? In: Canadell JG, Pataki DE, Pitelka LF (eds) Terrestrial ecosystems in a changing world, pp 149–164
Lüdi W (1945) Besiedlung und Vegetationsentwicklung auf den jungen Seitenmoränen des grossen Aletschgletschers: mit einem Vergleich der Besiedlung im Vorfeld des Rhonegletschers und des Oberen Grindelwaldgletschers
Lüdi W (1958) Beobachtung über die Besiedlung von Gletschervorfeldern in den Schweizeralpen
MacArthur RH, Wilson EO (1967) The theory of island biogeography. Princeton University Press, Princeton
MacDonald G (2002) Biogeography: introduction to space, time and life. John, New York
Marcante S, Winkler E, Erschbamer B (2009) Population dynamics along a primary succession gradient: do alpine species fit into demographic succession theory? Ann Bot 103:1129–1143. https://doi.org/10.1093/aob/mcp047
Matthews JA (1992) The ecology of recently-deglaciated terrain: a geoecological approach to glacier forelands. Cambridge University Press, Cambridge
Matthews JA (1999) Disturbance regimes and ecosystem response on recently-deglaciated terrain. In: Walker LR (ed) Ecosystems of disturbed ground. Elsevier, Amsterdam, pp 17–37
Matthews JA, Vater AE (2015) Pioneer zone geo-ecological change: Observations from a chronosequence on the Storbreen glacier foreland, Jotunheimen, southern Norway. Catena 135:219–230. https://doi.org/10.1016/j.catena.2015.07.016
Matthews JA, Whittaker RJ (1987) Vegetation succession on the Storbreen glacier foreland, Jotunheimen, Norway: a review. Arct Alp Res 19:385–395. https://doi.org/10.2307/1551403
Matthews JA, Shakesby RA, Berrisford MS, McEwen LJ (1998) Periglacial patterned ground on the Styggedalsbreen glacier foreland, Jotunheimen, southern Norway: micro-topographic, paraglacial and geoecological controls. Permafr Periglac Process 9:147–166. https://doi.org/10.1002/(SICI)1099-1530(199804/06)9:2%3c147:AID-PPP278%3e3.0.CO;2-9
Mercier D, Étienne S, Sellier D, André M (2009) Paraglacial gullying of sediment-mantled slopes: a case study of Colletthøgda, Kongsfjorden area, West Spitsbergen (Svalbard). Earth Surf Process Landf 34:1772–1789. https://doi.org/10.1002/esp.1862
Milner AM, Fastie CL, Chapin FS III et al (2007) Interactions and linkages among ecosystems during landscape evolution. Bioscience 57:237–247
Minami Y, Okitsu S, Kanda H (1997) Relationship between plant community and topographic factor on the moraine at deglaciated Arctic terrain in Ny-Alesund, Svalbard. Bull Fac Agr Tamagawa Univ, 21–30
Moreau M, Laffly D, Joly D, Brossard T (2005) Analysis of plant colonization on an Arctic moraine since the end of the Little Ice Age using remotely sensed data and a Bayesian approach. Remote Sens Environ 99:244–253
Moreau M, Mercier D, Laffly D, Roussel E (2008) Impacts of recent paraglacial dynamics on plant colonization: a case study on Midtre Lovénbreen foreland, Spitsbergen (79°N). Geomorphology 95: 48–60
Moreau M, Mercier D, Laffly D (2004) Un siècle de dynamiques paraglaciaires et végétal sur les marges du Midre Lovénbreen, Spitsberg nord-occidental/a century of paraglacial and plant dynamics in the Midre Lovénbreen foreland (northwestern Spitsbergen). Géomorphol Relief Process Environ 10:157–168
Nagl F, Erschbamer B (2010) Kapitel 6. Pflanzliche Sukzessionen im Gletschervorfeld. Vegetation und Besiedlungsstrategien. In: Erschbamer B, Koch EM (eds) Glaziale und periglaziale Lebensräume im Raum Obergurgl. Innsbruck University press, Innsbruck, pp 121–142

Nicolussi K, Patzelt G (2000) Discovery of early Holocene wood and peat on the forefield of the Pasterze Glacier, Eastern Alps, Austria. Holocene 10:191–199. https://doi.org/10.1191/095968300666855842

Oliver CD, Adams AB, Zasoski RJ (1985) Disturbance patterns and forest development in a recently deglaciated valley in the northwestern Cascade Range of Washington, U.S.A. Can J For Res 15:221–232. https://doi.org/10.1139/x85-040

Paul F, Kaab A, Maisch M et al (2004) Rapid disintegration of Alpine glaciers observed with satellite data. Geophys Res Lett 31:L21402. https://doi.org/10.1029/2004GL020816

Pickett STA, White PS (1987) The ecology of natural disturbance and patch dynamics. Academic Press

Prach K, Rachlewicz G (2012) Succession of vascular plants in front of retreating glaciers in central Spitsbergen. Pol Polar Res. https://doi.org/10.2478/v10183-012-0022-3

Raab T, Krümmelbein J, Schneider A et al (2012) Initial Eecosystem processes as key factors of landscape development—a review. Phys Geogr 33:305–343. https://doi.org/10.2747/0272-3646.33.4.305

Raffl C, Mallaun M, Mayer R, Erschbamer B (2006) Vegetation succession pattern and diversity changes in a glacier valley, Central Alps, Austria. Arct Antarct Alp Res 38:421–428. https://doi.org/10.1657/1523-0430(2006)38%5b421:VSPADC%5d2.0.CO;2

Rauh W (1939) Über polsterförmigen Wuchs. Ein Beitrag zur Kenntnis der Wuchsformen der höheren Pflanzen. Nova Acta Leopold, 272–505

Raymond Pralong M, Turowski JM, Beer A et al (2011) Klimaänderung und Wasserkraft. Sektorielle Studie Wallis. Auswirkung der Klimaänderung auf die Geschiebefracht

Reisigl H, Keller R (1994) Alpenpflanzen im Lebensraum, Spektrum Akademischer Verlag

Richter M (1994) Die Pflanzensukzession im Vorfeld des Tschierva-Gletschers/Oberengadin. Geoökodynamik 15:55–88

Robbins JA, Matthews JA (2010) Regional variation in successional trajectories and rates of vegetation change on glacier forelands in South-Central Norway. Arct Antarct Alp Res 42:351–361. https://doi.org/10.1657/1938-4246-42.3.351

Robbins JA, Matthews JA (2014) Use of ecological indicator values to investigate successional change in boreal to high-alpine glacier-foreland chronosequences, southern Norway. Holocene 24:1453–1464. https://doi.org/10.1177/0959683614544067

Rydgren K, Halvorsen R, Töpper JP, Njøs JM (2014) Glacier foreland succession and the fading effect of terrain age. J Veg Sci 25:1367–1380. https://doi.org/10.1111/jvs.12184

Schröter C, Brockmann-Jerosch H, Brockmann-Jerosch MC et al (1926) Das Pflanzenleben der Alpen. Verlag von Albert Raustein, Zürich

Schweingruber FH, Münch A, Schwarz R (2007) Dendrochronologie von Kräutern und Sträuchern im Vorfeld des Morteratschgletschers. BAUHINIA Z Basl Bot Ges 20:5–17

Schwienbacher E, Navarro-Cano JA, Neuner G, Erschbamer B (2012) Correspondence of seed traits with niche position in glacier foreland succession. Plant Ecol 213:371–382. https://doi.org/10.1007/s11258-011-9981-4

Sharp RP (1958) The latest major advance of Malaspina Glacier, Alaska. Geogr Rev 48:16–26. https://doi.org/10.2307/211699

Stöcklin J, Bäumler E (1996) Seed rain, seedling establishment and clonal growth strategies on a glacier foreland. J Veg Sci 7:45–56. https://doi.org/10.2307/3236415

Tansley AG (1920) The classification of vegetation and the concept of development. J Ecol, 118–149

Thorn CE (1976) A model of stony earth circle development, Schefferville, Quebec. In: Proceedings of the Association of American Geographers, pp 19–23

Turner MG, Gardner RH, O'Neill RV (2001) Landscape ecology in theory and practice: pattern and process. Springer, New York

Vetaas OR (1997) Relationships between floristic gradients in a primary succession. J Veg Sci, 665–676

Veylon G, Ghestem M, Stokes A, Bernard A (2015) Quantification of mechanical and hydric components of soil reinforcement by plant roots. Can Geotech J 52:1839–1849. https://doi.org/10.1139/cgj-2014-0090

Viles HA (2004) Biogeomorphology. In: Goudie A (ed) Encyclopedia of Geomorphology, 1st edn. Routledge, London

Violle C, Navas ML, Vile D et al (2007) Let the concept of trait be functional! Oikos 116:882–892. https://doi.org/10.1111/j.0030-1299.2007.15559.x

Walker LR, del Moral R (2003) Primary succession and ecosystem rehabilitation. Cambridge University Press, Cambridge

Walker LR, Wardle DA, Bardgett RD, Clarkson BD (2010) The use of chronosequences in studies of ecological succession and soil development. J Ecol 98:725–736. https://doi.org/10.1111/j.1365-2745.2010.01664.x

Welker JM, Molau U, Parsons AN et al (1997) Responses of Dryas octopetala to ITEX environmental manipulations: a synthesis with circumpolar comparisons. Glob Change Biol 3:61–73. https://doi.org/10.1111/j.1365-2486.1997.gcb143.x

White PS (1979) Pattern, process, and natural disturbance in vegetation. Bot Rev 45:229–299. https://doi.org/10.1007/BF02860857

Whittaker RH (1974) Climax concepts and recognition. In: Knapp R (ed) Vegetation Dynamics. Springer, The Netherlands, pp 137–154

Wijk S (1986) Performance of Salix Herbacea in an alpine snow-bed gradient. J Ecol 74:675–684. https://doi.org/10.2307/2260390

Wilmanns O (1993) Ökologische Pflanzensoziologie. Quelle & Meyer, Heidelberg

Author Index

T. Heckmann and D. Morche (eds.), *Geomorphology of Proglacial Systems*,
Geography of the Physical Environment, https://doi.org/10.1007/978-3-319-94184-4

Subject Index

T. Heckmann and D. Morche (eds.), *Geomorphology of Proglacial Systems*, Geography of the Physical Environment, https://doi.org/10.1007/978-3-319-94184-4

H

I

J

K

L

T

MIX
Papier aus verantwortungsvollen Quellen
Paper from responsible sources
FSC® C105338

If you have any concerns about our products,
you can contact us on
ProductSafety@springernature.com

In case Publisher is established outside the EU,
the EU authorized representative is:
Springer Nature Customer Service Center GmbH
Europaplatz 3, 69115 Heidelberg, Germany

Printed by Libri Plureos GmbH
in Hamburg, Germany